FUNDAMENTALS OF BIOFUELS ENGINEERING AND TECHNOLOGY

FUNDAMENTALS OF BIOFUELS ENGINEERING AND TECHNOLOGY

VOLUME-2

Diego Rossit
Sebastian Ubal
Ioan Gergely

Kruger Brentt
Publishers
2025

Kruger Brentt Publishers UK. LTD.
Company Number 9728962
Regd. Office: 68 St Margarets Road, Edgware, Middlesex HA8 9UU

ISBN: 978-1-78715-436-0 (Vol. 2)

For information on all our publications visit our website at http://krugerbrentt.com/

PREFACE

"Fundamentals of Biofuels Engineering and Technology" offers a comprehensive exploration into the principles, processes, and technologies that underpin the production, utilization, and sustainability of biofuels. This book serves as an essential resource for students, researchers, engineers, policymakers, and industry professionals interested in understanding and advancing the field of biofuels engineering.

Biofuels represent a promising renewable energy source that can help mitigate the environmental impacts of fossil fuels and contribute to energy security and sustainability. From first-generation biofuels such as biodiesel and ethanol to advanced biofuels derived from lignocellulosic biomass and algae, biofuels offer a diverse range of options for reducing greenhouse gas emissions and dependence on finite fossil resources.

As editors of this volume, we have endeavored to compile a diverse array of perspectives, methodologies, and case studies from leading experts and practitioners in the field of biofuels engineering and technology. Through a blend of theoretical foundations, practical insights, and real-world examples, this book covers a wide range of topics, including biomass feedstocks, biofuel production processes, biorefinery concepts, biofuel characterization, and sustainability assessment.

Our overarching goal in presenting this book is to provide readers with a comprehensive understanding of the principles and technologies involved in the production and utilization of biofuels. By offering insights into the latest advancements, best practices, and emerging trends in biofuels engineering, we aim to inspire innovation, collaboration, and excellence in the development of sustainable bioenergy solutions.

We extend our sincere gratitude to all the contributors who have generously shared their expertise, experiences, and insights in the field of biofuels engineering and technology. It is our hope that this book will serve as a valuable reference and guide for students, researchers, engineers, policymakers, and industry professionals as they work to unlock the full potential of biofuels as a clean, renewable, and sustainable energy source for the future.

Diego Rossit

Sebastian Ubal

Ioan Gergely

CONTENTS

PREFACE V

VOLUME-1

1. **THE IMPACT OF THERMAL PRETREATMENT PROCESS ON BIO-FUEL CONVERSION 1**
   1. Introduction: An Overview 1
   2. Methodology 1
   3. Experimental findings and their analysis 4
   4. CONCISE SUMMARY 19
   References 19
2. **AN OVERVIEW OF THE BIOENERGIES CHALLENGE 21**
   1. Introduction: An Overview 21
   2. Plant biofuels 32
   3. Microdiesel 39
   4. Biohydrogen 40
   5. The future of transport technology 42
   6. Options for grid contributions 45
   7. Management and the ability to maintain long-term viability 47
   8. Concise Summary 51
   References 51
3. **ENHANCING THE QUALITY OF BIOGAS BY PRESSURE SWING ADSORPTION 63**
   1. Introduction: An Overview 63
   2. Carbon dioxide reduction technologies for biogas 64
   3. Principles of Pressure Swing Adsorption (PSA) for the purification of biogas 66
   4. Emerging markets and advancements in PSA technology 74
   5. Concise Summary 78
   References 78
4. **UTILIZATION OF RAPESEED STRAIGHT VEGETABLE OIL AS FUEL MANUFACTURED IN SMALL-SCALE OPERATIONS 83**
   1. Introduction: An Overview 83
   2. Production of rapeseed 83

3. Processing of rapeseed 85
4. Use of rapeseed oil as fuel 86
5. Use of rapeseed cake for animal feeding 88
6. Proposed model for crop cultivation and agricultural utilization 89
7. Environmental and economic analyses 91
8. Concise Summary 95
References 96

5. NANOTECH BIOFUELS AND FUEL ADDITIVES 99

1. Introduction: An Overview 99
2. The future of oil refining [1] and the shift of the oil industry to other energy sources [2] 100
3. Challenges related to liquid biofuels [3] 100
4. The vastness of nanotechnology 102
5. Converting algae into biofuels 103
6. The application of nanotechnology to landfill facilities[13] 104
7. The utilization of nanotechnology for the conversion of biomass into biofuels. 104
8. Nanotech liquid additives 105
9. Public apprehensions over nanotechnology encompass issues of security, health, and the environment 106
10. Concise Summary 108
References 108

6. UTILIZING BIORESOURCES FOR ADVANCED (THIRD-GENERATION) BIOFUELS 111

1. Introduction: An Overview 111
2. First-generation biofuels 111
3. Second-generation biofuels 115
4. Third-generation biofuels 120
5. Concise Summary 127
References 128

7. CORN-BASED FUEL ETHANOL COPRODUCTS: PRODUCTION AND USE 137

1. Introduction: An Overview 137
2. The study aims to achieve specific goals and objectives 140
3. Methods of production 140
4. Significance of coproducts 141
5. Evolution of coproducts 149
6. Concise Summary 154
References 154

8. BIOREFINERY PROCESSES FOR THE CONVERSION OF BIOMASS INTO LIQUID FUEL 161

1. Introduction: An Overview 161
2. Concept of Biorefinery 161
3. Production of bioethanol from lignocellulosic biomass 162

4. Production of bio-oil from lignocellulosic biomass and biomass with high moisture content 166
5. Conclusion 176
References 176

9. UTILIZATION OF DIGESTERS' WASTE FOR BIOGAS PRODUCTION 183

1. Introduction: An Overview 183
2. Significant issues 185
3. Information 186
References 209

10. PRODUCTION AND ASSESSMENT OF BIODIESEL 213

1. Introduction: An Overview 213
2. Production of biodiesel 214
3. Assessment of Biodiesel Quality 219
4. Standards for biodiesel 221
5. Concise Summary 228
References 228

11. BIOBUTANOL PRODUCTION AND UTILIZATION PERSPECTIVES 233

1. Introduction: An Overview 233
2. Theoretical framework 233
3. Biobutanol fermentation experience at ICT Prague 238
4. Biobutanol's application in road transportation 244
5. Concise Summary 249
References 250

12. ADVANCING THE PROGRESS OF ALGAL BIOFUELS THROUGH THE ESTABLISHMENT OF A GENETIC FRAMEWORK 257

1. Introduction: An Overview 257
2. Utilizing microalgal biomass for the generation of biofuels. 259
3. Advancing the creation of selectable markers for Dunaliella salina 267
4. Concise Summary 275
References 275

13. RHEOLOGICAL ANALYSIS OF BIO-OILS PRODUCED BY PILOT SCALE MICROWAVE ASSISTED PYROLYSIS 281

1. Introduction: An Overview 281
2. Materials and methods 284
3. Results and discussion 287
4. Concise Summary 294
References 294

14. SIMULTANEOUS PRODUCTION OF BIOETHANOL AND POWER 299

1. Introduction: An Overview 299
2. Energy equilibrium in the process of traditional bioethanol manufacturing 300

3. *Self-heat recuperation technology and self-heat recuperative processes* 301
4. *Integration with biomass gasification* 309
5. *Energy balance for self-heat recuperative bioethanol production* 309
6. *Concise Summary* 311
References 311
15.CONVERSION OF NON-HOMOGENEOUS BIOMASS TO ULTRACLEAN SYNGAS AND CATALYTIC CONVERSION TO ETHANOL 315
1. *Introduction: An Overview* 315
2. *Synthesis gas (syngas) production by gasification* 316
3. *Syngas conversion into methanol* 323
4. *Synthesis of Ethanol* 327
5. *Concise Summary* 330
References 331

VOLUME-2

16. INNOVATIVE APPROACHES TO BIODIESEL MANUFACTURING 335
1. *Introduction: An Overview* 335
2. *Supercritical process* 338
3. *Microwave-assisted procedure* 344
4. *Ultrasound assisted process* 349
5. *Concise Summary* 358
References 359
17. CATALYTIC HYDROTREATMENT EMPLOYED FOR OF STABILIZING PYROLYSIS OIL 369
1. *Introduction: An Overview* 369
2. *Investigation of the hydrotreatment of fast pyrolysis oil using catalysts* 371
3. *Investigations on the catalytic hydrotreatment of fast pyrolysis oils* 373
4. *Proposed paths for reactions and their potential consequences* 381
5. *Enhanced catalyst compositions for the efficient catalytic hydrotreatment of fast pyrolysis oil* 384
6. *Potential applications of enhanced oils and essential features of the product* 387
7. *Conclusions* 388
References 388
18. BIOMASS FEEDSTOCK PRE-PROCESSING-PART-1: PRE-TREATMENT 393
1. *Introduction: An Overview* 393
2. *Characterization of Lignocellulosic biomass* 394
3. *Pre-treatment of lignocellulosic biomass* 397
4. *Reduction of particle size and its impact on physical qualities* 405
5. *Concise Summary* 411
References 412

19. BIOMASS FEEDSTOCK- PRE-PROCESSING- PART-II: DENSIFICATION 421

1. Introduction: An Overview 421

2. Lignocellulosic composition and higher heating values 423

3. Lab-scale pelleting of agricultural biomass 425

4. Biomass Compression characteristics 430

5. Pilot-scale pelleting of agricultural biomass 434

6. Concise Summary 440

References 441

20. PERFORMANCES OF ENZYMATIC GLUCOSE/O_2 BIOFUEL CELLS 447

1. Introduction: An Overview 447

2. Enzymatic electrodes for glucose/O_2 biofuel cells 448

3. Abiotic catalysts for glucose/O_2 biofuel cells 455

4. Design of glucose/O_2 biofuel cells 462

5. Concise Summary 465

References 465

21. THE SIGNIFICANCE OF BIOPHYSICS IN BIOFUEL RESEARCH: A QUANTITATIVE APPROACH TO BIO-ENGINEERING 473

1. Introduction: An Ovberview 473

2. ThermoGravimetric Analysis (TGA) and Differential Scanning Calorimetry (DSC) 473

3. Cellulose crystallinity measurement 475

4. Pyro-GC/MS and GC/MS 476

5. Plant stem mechanical strength testing 478

6. Analysis of cell walls using vibrational spectroscopy and microscopy 480

7. Cell wall analysis by 2D and 3D electron microscopy 484

8. Analysis of biomass using acanning and transmission electron microscopy 487

9. Analyzing microbial populations using sophisticated electron microscopy techniques 487

10. Concise Summary 489

References 489

22. INVESTIGATION OF THE RATE OF DECOMPOSITION OF PALM OIL WASTE 499

1. Introduction: An Overview 499

2. Materials and methods 500

3. Results and discussions 502

4. Concise Summary 509

References 509

23. THE COLOMBIAN EXPERIENCE WITH BIOFUELS AND ACHIEVING ENERGY SELF-SUFFICIENCY 513

1. Introduction: An Overview 513

2. Energy replacement and consumption 514

3. Precedents 521

4. The most common raw materials 528
5. Technical considerations 531
6. Concise Summary 535
References 537
24. MICROFLUIDIC BIOFUEL CELL UTILIZING ENZYMES FOR MICROPOWER GENERATION 539
1. Introduction: An Overview 539
2. Miniature BFCs 540
3. Microfabricated devices 541
4. Concise Summary 553
References 553
25. ENERGY PATHWAYS RESULTING FROM THE BLUE TOWER PROCESS 557
1. Introduction: An Overview 557
2. Blue Tower technology 559
3. Concept of the biomass Life Cycle Assessment 565
4. Future application of bio-fuel 569
5. Concise Summary 577
References 577
26. ADVANCEMENT OF BIOETHANOL DEVELOPMENT 581
1. Introduction: An Overview 581
2. Raw materials 583
3. Production processes 586
4. Environmental issues 599
5. Conclusions 601
References 601
27. THE IMPACT OF A VARIETY OF FRIED DISHES ON CERTAIN CHARACTERISTICS OF UTILIZED PLANT OILS AS PRIMARY RESOURCES FOR THE CREATION OF AN ALTERNATIVE TO DIESEL FUEL 607
1. Introduction: An Overview 607
2. Assessment of usability of post-frying edible oils as a raw material for production of diesel fuel substitute 608
3. Result of investigation 610
4. Concise Summary 618
References 619
28. ADVANCEMENTS IN MINIATURIZED ENZYMATIC BIOFUEL CELLS 621
1. Introduction: An Overview 621
2. Miniature biofuel cells 628
3. C-MEMS 3D architecture electrodes 630
4. Surface functionalization of C-MEMS 632
5. Simulation of C-MEMS based EBFCs 634
6. Concise Summary 638
References 639

29. PROCESSING LIGNOCELLULOSIC BIOMASS USING FEEDSTOCK IMPREGNATION FOR BIOREFINING. SWIFT AND CONSECUTIVE APPLICATION OF STEAM TREATMENT 645

1. Introduction: An Overview 645

2. Biomass 647

3. Fractionation 653

4. Hydrolysis 662

5. Concise Summary 666

References 667

30. PRODUCTION OF BIOMETHANOL FROM FORAGE GRASSES, TREES AND CROP RESIDUES 673

1. Introduction: An Overview 673

2. Gasification technology and the test plants 674

3. Concise Summary 687

References 688

INDEX 691

CHAPTER-16

INNOVATIVE APPROACHES TO BIODIESEL MANUFACTURING

1. INTRODUCTION: AN OVERVIEW

In recent years, the exhaustion of fossil fuels and their detrimental impact on environmental degradation have made it imperative to utilize alternate renewable energy sources. Biodiesel is a significant alternative renewable energy source that is currently being utilized. Biodiesel is a sustainable and energy-efficient fuel that is not harmful to the environment, can be broken down naturally in water, and produces lower levels of exhaust pollutants. Additionally, it has the capacity to mitigate the greenhouse gas effect and does not contribute to global warming as a result of its lower emissions. Due to its lack of carcinogens and reduced sulphur level compared to mineral diesel, it is considered a safer alternative (Sharma & Singh, 2009; Suppalakpanya et al., 2010). Biodiesel may be efficiently utilized and securely stored as a fuel, in addition to its notable environmental advantages. Additionally, it is more cost-effective compared to fossil fuels, which have a detrimental impact on the environment. Biodiesel can be used in standard diesel engines without the need for any engine alteration or modification to the fuel system. Biodiesel is currently produced in numerous nations worldwide, including Malaysia, Germany, USA, France, Italy, Australia, Brazil, and Argentina. The European Union's biodiesel output in 2009 was documented in Table 1, as reported by the European Biodiesel Board in July 2010. Table 1 shows that European Union countries generated 9 million tons of biodiesel in 2009. Germany and France are the foremost nations in the manufacturing of biodiesel. The European Union accounts for around 65% of the global production of biodiesel. Biodiesel is the primary biofuel that is produced and sold in Europe. By 2009, over 75% of the biofuels generated in Europe were in the form of biodiesel. Figure 1 displays the global biodiesel production from 1991 to 2009. According to Figure 1, there was a significant rise in global biodiesel output after the 2000s. In 1900, Rudolph Diesel demonstrated the compatibility of diesel engines using peanut oil. Various processes, including pyrolysis, catalytic cracking, blending, and microemulsification, were employed to manufacture biodiesel from vegetable oil for use in diesel engines (Sharma & Singh, 2009; Varma & Madras, 2007). The transesterification process was ultimately devised as the optimal way for addressing the issues associated with directly using oil in diesel engines (Varma & Madras, 2007). Biodiesel is typically derived from various sources, including plant oils such as soybean oil, cottonseed oil, canola oil, sunflower oil, linseed oil, olive oil, peanut seed oil, tobacco oil, palm oil, recycled cooking oils, and animal fats. The primary economic determinant to take into account for input costs of biodiesel manufacturing is the feedstock. The feedstock accounts for 90% of the overall cost of biodiesel manufacturing. Studies to overcome this economic dilemma primarily focused on biodiesel manufacturing from cheaper raw material. Literature reports the use of agricultural wastes, high acid oils, soapstock, waste frying oil, and alg oil as raw materials for biodiesel production (Haas & Scott, 1996; Özgül & Türkay, 1993; Özgül & Türkay, 2002; Leung & Guo, 2006; Yücel et al., 2010; Özçimen & Yücel, 2010).

Table 1.Biodiesel production of EU in 2009 (EBB 2010)

Country	Production (1000 Tons)	Country	Production (1000 Tons)
Austria	310	Italy	737
Belgium	416	Latvia	44
Bulgaria	25	Lithuania	98
Cyprus	9	Luxemburg	0
Czech Republic	164	Malta	1
Denmark/Sweden	233	Netherlands	323
Estonia	24	Poland	332
Finland*	220	Portugal	250
France	1959	Romania	29
Germany	2539	Slovakia	101
Greece	77	Slovenia	9
Hungary	133	Spain	859
Ireland*	17	UK	137
TOTAL:		9.046	
(*Data include hydrodiesel production)			

The transesterification process, depicted in Figure 2 (Barnard et al., 2007), is a traditional and widely used technique for the generation of biodiesel. Homogeneous or heterogeneous catalysts can be employed in transesterification reactions. The catalysts facilitate the division of oil into glycerin and biodiesel, hence enhancing production efficiency and speed.

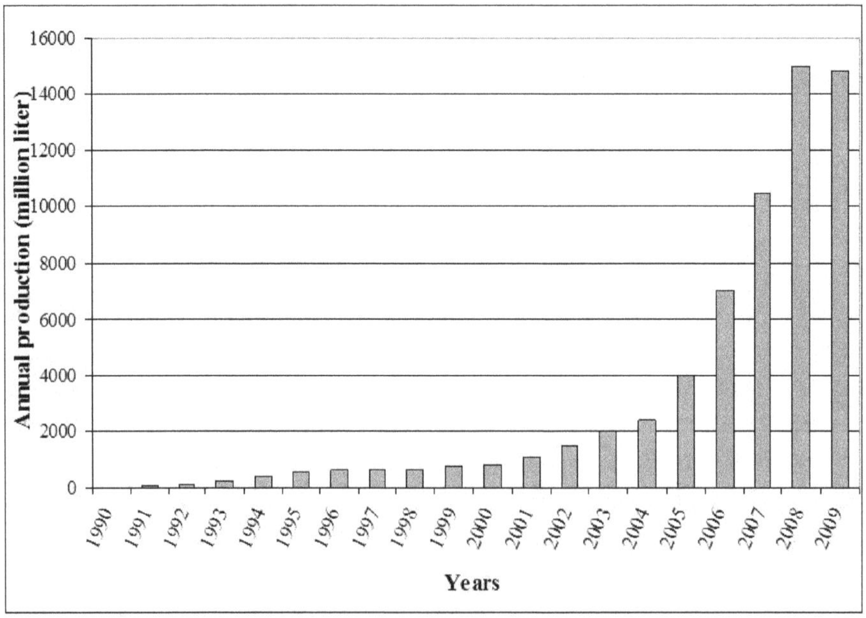

Figure 1.The world production of biodiesel between 1991 and 2009 (Licht, 2009)

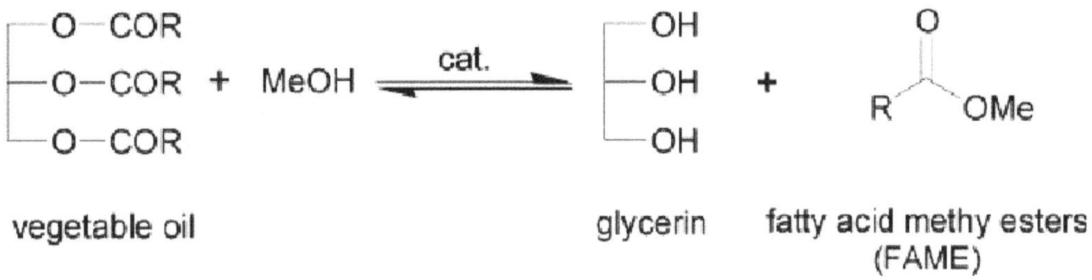

Figure 2. Biodiesel production via transesterification reaction (Barnard et al., 2007)

This process involves the synthesis of fatty acid alkyl esters by the reaction between triglycerides and an alcohol, such as ethanol or methanol, in the presence of a catalyst like alkali, acid, or enzyme. Sodium hydroxide or potassium hydroxide, when dissolved in alcohol, is commonly employed as a catalyst in transesterification reactions (Dube et al., 2007). The reaction yields fatty acid methyl esters (FAMEs), which serve as biodiesel, and glycerin (Vicente et al., 2004). Ethanol can serve as an alternative to methanol as an alcohol. When ethanol is utilized, it results in the production of fatty acid ethyl ester (FAEE) as a product (Hanh et al., 2009b). The preference for producing methyl ester instead of ethyl ester is due to the fact that methyl esters are the main product in the market, and methanol is significantly more affordable than ethanol (Zhou & Boocock, 2003). Nevertheless, the use of methanol has a significant drawback as it is derived from petroleum. While ethanol may be derived from renewable agricultural resources, this allows for complete reliance on non-petroleum-based alcohols (Saifuddin & Chua, 2004; Encinar et al. 2007). Ethanol is predominantly favored in countries that produce ethanol. Propanol and butanol have also been utilized as alcohols in the manufacturing of biodiesel. Transesterification catalyzed by alkali occurs far faster than acid-catalyzed transesterification, making it the preferred method in industrial applications (Dube et al., 2007; Freedman et al., 1984). The alkali catalysts most frequently employed include NaOH, CH_3ONa, and KOH (Vicente et al., 2004). Potassium hydroxide (KOH) and sodium hydroxide (NaOH) flakes are cost-effective, convenient for transportation and storage, and are favored by small-scale manufacturers. The catalysts of choice for large continuous-flow production operations are now commercially available as alkyl oxide solutions of sodium methoxide or potassium methoxide in methanol (Singh et al., 2006).

Sulfuric acid has been extensively studied as a catalyst for acid-catalyzed systems. However, other acids, including HCl, BF3, H3PO4, and organic sulfonic acids, have also been employed by other researchers (Lotero et al, 2005). However, in the alkali catalyzed approach, it is crucial for the glycerides and alcohol to be almost completely free of water. If there is any water present, it can cause saponification to occur (Helwani et al., 2009). Saponification leads to a decrease in catalytic effectiveness, making it challenging to separate glycerol and causing gel formation (Helwani et al., 2009). Separating the catalyst from the reaction mixture in homogeneous catalyzed processes is challenging and costly. In order to achieve this objective, a significant quantity of water is utilized to separate the catalyst from the product (Vyas et al., 2010). However, there are drawbacks to consider. Undesirable by-products, such as glycerin, might be observed during the reaction. Additionally, the reaction duration is prolonged and there is a potential for excessive energy consumption. Therefore, scientists have directed their attention on the advancement of novel techniques for producing biodiesel and enhancing the efficiency of these procedures (Sharma et al., 2008). Recently, several procedures such as supercritical process, microwave assisted method, and ultrasound assisted method have been developed. Non-conventional energy catalysts, also known as alternative energy stimulants, have been employed for an extended period to augment the rate of chemical reactions and improve the production of specific reaction products. Exploring novel approaches and integrating innovative methods and techniques pose a challenge that has the potential to yield unforeseen advancements in the field of biodiesel production techniques (Nuechter et al., 2000). This paper provides a comprehensive assessment and detailed presentation of biodiesel production using modern methods such as supercritical conditions, microwave, and ultrasonic techniques during the period from 2000 to 2011.

2. SUPERCRITICAL PROCESS

Supercritical method is one of the novel methods in biodiesel production. Biodiesel production can be easily achieved by supercritical process without catalysts. A supercritical fluid is any substance at a temperature and pressure above its critical point. It can diffuse through solids like a gas, and dissolve materials like a liquid. These fluids are environment-friendly and economic. Generally, water, carbon dioxide and alcohol are used as supercritical fluids. Supercritical fluids have different application areas. One of these applications is the biodiesel production that is firstly achieved by Saka and Kusdiana in 2001. And many studies on biodiesel production in supercritical conditions were made since 2001. All studies in the literature since 2001 were reviewed and presented in Table 2. The biodiesel production have been studied by using supercritical process from different oils such as rapeseed oil (Kusdiana& Saka, 2001; Saka et al., 2010; Saka &Kusdiana, 2002; Minami & Saka, 2006; Yoo et al., 2010), algae oil (Patil et al., 2010b), chicken fat (Marulanda et al., 2010), jatropha oil (Hawash et al., 2009; Rathore & Madras, 2007; Chen et al., 2010), soybean oil (Cao et al., 2005; He et al., 2007 ; Cheng et al., 2010; Yin et al., 2008), waste cooking oil (Patil et al., 2010a; Demirbaş, 2009), sunflower oil (Demirbaş, 2007), cottonseed oil (Demirbaş, 2008), linseed oil (Demirbaş, 2009), hazelnut kernel oil (Demirbaş, 2002), coconut oil (Bunyakiat et al, 2006), palm oil (Gui et al., 2009 ; Tan et al., 2010c; Tan et al., 2009; Song et al., 2008).

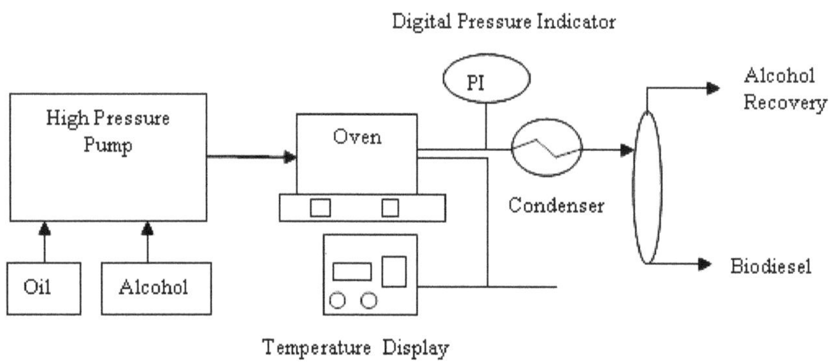

Figure 3.Biodiesel production by continuous supercritical alcohol process

In Saka's study, rapeseed oil was converted to methyl esters with supercritical methanol (molar ratio of methanol to rapeseed oil: 42 to 1) at temperature of 350°C in 240 s. The methyl ester yield of the supercritical methanol method was higher than those obtained in the conventional method with a basic catalyst. Liquid methanol is a polar solvent and has hydrogen bonding between OH oxygen and OH hydrogen to form methanol clusters, but supercritical methanol has a hydrophobic nature with a lower dielectric constant, so non-polar triglycerides can be well solvated with supercritical methanol to form a single phase oil/methanol mixture. For this reason, the oil to methyl ester conversion rate was found to increase dramatically in the supercritical state (Saka &Kusdiana, 2001; Fukuda et al., 2001).

Main factors affecting transesterification via supercritical process are the effect of temperature, pressure and effect of molar ratio between alcohol and oil sample.

Temperature is the most important factor in all parameters that affects the transesterification under supercritical condition. In the study of Kusdiana& Saka, the conversion of triglyceride to methyl esters is relatively low due to the subcritical state of methanol at temperatures of 200 and 230 °C. In these conditions, methyl esters formed are most about 70 wt% for 1 h treatment. However, a high conversion of rapeseed oil to methyl esters with the yield of 95 wt% at 350 °C for 4 min reaction time (Kusdiana& Saka, 2001).

Pressure is also very important parameter, but, reaction pressure increases with the increase of temperature. Thus the effect of pressure on the transesterification is always correlated with temperature. High pressure increases the solubility of triglyceride, thus, a contact at the molecular level between alcohol and triglyceride become closer at high pressure (Lee & Saka, 2010).

The effect of molar ratio between alcohol and oil sample is the other important parameter in supercritical condition as mentioned before. Higher molar ratio between methanol and triglyceride is favored for transesterification reaction under supercritical condition. The reason can be that contact area between methanol and triglycerides are increased at the higher molar ratios of methanol. In Kusdiana's study, the effect of the molar ratio of methanol to rapeseed oil was studied in the range between 3.5 and 42 on the yield of methyl esters formed for supercritical methanol treatments. For a molar ratio of 42 in methanol, almost complete conversion was achieved in a yield of 95% of methyl esters, whereas for the lower molar ratio of 6 or less, incomplete conversion was apparent with the lower yield of methyl esters (Kusdiana& Saka, 2001).

Advantages of supercritical process are the shorter reaction time, easier purification of products and more efficient reaction.Although higher temperature, pressure and molar ratio between methanol and triglyceride are favored for transesterification reaction under supercritical condition, energy consumption, and excess amount alcohol usage are the disadvantages for the biodiesel production in supercritical conditions (Lee & Saka, 2010).

For biodiesel production, generally supercritical methanol and supercritical ethanol is used. However, supercritical carbon dioxide can be also used for this purpose since it is cheap, non-flammable and non-toxic (Varma & Madras, 2007). In recent years, two-step transesterification processes such as both subcritical and supercritical, both enzyme and supercritical fluid conditions etc. were also developed (Saka & Isayama, 2009).

Kusdiana and Saka developed a two-step biodiesel production method "Saka–Dadan process (Kusdiana& Saka, 2004). Besides the same advantages as one-step supercritical methanol process, the two-step method is found to use milder reaction condition and shorter reaction time, which may further allow the use of common stainless steel for the reactor manufacturing and lower the energy consumption (Lee & Saka, 2010). Minami & Saka (2006), Saka et al. (2010) and Cao et al. (2005) used two-step supercritical method in their studies. Therefore, two- step method has advantages that are milder reaction conditions, high reaction rate, applicable to various feedstocks, easier separation, no catalyst needed there is no high equipment cost and high alcohol oil ratio.

Table 2.Biodiesel production studies in supercritical conditions

Raw Material	Alcohol	Alcohol/ oil molar ratio	Reaction temperature and pressure	Reaction time	Reactor type	Performance (%)	Ref.
Rapeseed oil	Supercritical methanol	42:1	350 °C,14 MPa	240 s	Batch-type vessel	35 (methyl ester yield)	Kusdiana& Saka, 2001
Wet algae	Supercritical methanol	9:1	255 °C, 1200 psi	25 min	Micro-reactor	90 (FAME yield)	Patil et al., 2010b
Rice bran oilDewaxed-degummed rice bran oil	Supercritical methanol	27:1	300 °C, 30 MPa	5 min	Stainless steel reactor	51.28 94.84 (FAME yield)	Kasim et al., 2009
Chicken fat	Supercritical methanol	6:1	400 °C, 41.1 MPa	6 min	Batch reactor	88 (FAME yield)	Marulanda et al., 2010
Jatropha oil	Supercritical methanol + propane	43:1	593 K, 8.4 MPa	4 min	Bench–scale reactor	100 (FAME yield)	Hawash et al., 2009
Soybean oil	Supercritical methanol	24:1	280 °C, 12.8 MPa	10 min	Batch-type vessel	98 (methyl ester yield)	Cao et al., 2005
Refined palm oil	Supercritical ethanol	33:1	349 °C, P”/>6.38 MPa	30 min	batch-type tubular	79.2 (biodiesel yield)	Gui et al., 2009
Rapeseed oil	Supercritical methanol	42:1	350 °C, 19 MPa	4 min	Batch-type vessel	95 (methyl ester yield)	Kusdiana& Saka, 2001
Rapeseed oil	Supercritical methanol	42:1	350 °C, 30 MPa	240 s	Batch-type vessel	95 (conversion)	Saka &Kusdiana, 2001
Rapeseed oil	Supercritical methanol	42:1	350 °C, 35 MPa	240 s	Batch-type vessel	98.5 (conversion)	Saka &Kusdiana, 2002
Rapeseed oil	Subcritical acetic acid Supercritical methanol	54:1 14:1	300 °C, 20 MPa 270 °C, 17 MPa	30 min 15 min	Batch-type vessel	92 97 (FAME yield)	Saka et al., 2010
Waste cooking oil	Supercritical methanol	10:1-50:1	300 °C, 1450 psi	10-30 min	Micro-reactor	80 (biodiesel yield)	Patil et al., 2010a
Waste cooking oil	Supercritical methanol	41:1	560 K	1800 s	Cylindrical autoclave	100 (biodiesel yield)	Demirbaş, 2009
Sunflower oil	Supercritical methanol + calcium oxide (%3 wt)	41:1	525 K, 24 Mpa	6 min	Cylindrical autoclave	100 (methyl ester yield)	Demirbaş, 2007
Cottonseed oil	Supercritical methanol Supercritical ethanol	41:1 41:1	523 K 503 K	8 min 8 min	Cylindrical autoclave	98 70 (methyl ester yield)	Demirbaş, 2008

Raw Material	Alcohol	Alcohol/ oil molar ratio	Reaction temperature and pressure	Reaction time	Reactor type	Performance (%)	Ref.
Linseed oil	Supercritical methanol Supercritical ethanol Supercritical methanol Supercritical ethanol	41:1 41:1 41:1 41:1	523 K 523 K 503 K 503 K	8 min 8 min 8 min 8 min	Cylindrical autoclave	98 89 70 65 (methyl ester yield)	Demirbaş, 2009
Hazelnut kernel oil	Supercritical methanol	41:1	350 °C	300 s	Cylindrical autoclave	95 (conversion)	Demirbaş, 2002
Jatropha oil	Supercritical methanol	40:1	350 °C, 200 bar	40 min	Small scale batch reactor	“/>90 (conversion)	Rathore & Madras, 2007
Soybean oil	Supercritical methanol	40:1	310 °C, 35 MPa	25 min	Tube reactor	96 (methyl ester yield)	He et al., 2007
Coconut oil and palm kernel oil	Supercritical methanol	42:1	350 °C, 19 MPa	400 s	Tubular reactor	95-96 (conversion)	Bunyakiat et al, 2006
Jatropha oil	Supercritical methanol	5:1	563 K, 11 MPa	15 min	Tubular reactor	100 (conversion)	Chen et al., 2010
R. sativus L. oil	Supercritical ethanol Supercritical methanol	42:1 39:1	590.5 K, 12.5 MPa 590 K, 14.1 MPa	29 min 27 min	Batch reactor	95.5 99.8 (ester yield)	Valle et al., 2010
Purified palm oil	Supercritical methanol Supercritical ethanol	40:1 33:1	372 °C, 29.7 MPa 349 °C, 26.2 MPa	16 min 29 min	Batch-type tube reactor	81.5 79.2 (biodiesel yield)	Tan et al., 2010c
Palm oil	Supercritical methanol	30:1	360 °C, 22 MPa	20 min	Batch-type tube reactor	72 (biodiesel yield)	Tan et al., 2009
Refined, bleached and deodorized palm oil	Supercritical methanol	45:1	350 °C, 40 MPa	5 min	Batch-type reactor	90 (FAME yield)	Song et al., 2008
Rapeseed oil	Subcritical water+Two-step supercritical methanol Supercritical methanol	1:1 (v/v) 1.8:1 (v/v) 1.8:1 (v/v)	270 °C, 20 MPa 320 °C, 20 MPa 380 °C, 20 MPa	60 min 10 min 15 min	Tubular reactor	90 (methyl ester yield) 80 (methyl ester yield)	Minami & Saka, 2006
Refined soybean oil	Supercritical methanol Supercritical methanol+hexane (co-solvent) Supercritical methanol+CO_2 (co-solvent) Supercritical methanol+ KOH	42:1	350 °C, 20 MPa 300 °C 300 °C 160 °C, 10 MPa	10 min 30 min 30 min 30 min	Cylindirical autoclave	95 85.5 90.6 98 (methyl ester yield)	Yin et al., 2008

Raw Material	Alcohol	Alcohol/ oil molar ratio	Reaction temperature and pressure	Reaction time	Reactor type	Performance (%)	Ref.
Waste palm cooking oil Refined palm oil	Supercritical methanol	40:1	300 °C	20 min	Batch-type tube reactor	79 80 (biodiesel yield)	Tan et al., 2010a
Free fatty acids	Supercritical methanol	1.6:1	270 °C, 10 MPa	30 min	Batch reactor	97 (FAME yield)	Alenezi et al., 2010
Rapeseed oil	Supercritical methanol +metal oxide catalysts (ZnO)	40:1 % 1 (wt) ZnO	250 °C, 105 bar	10 min	Batch- type reactor system	95.2 (FAME yield)	Yoo et al., 2010
Soybean oil	Supercritical methanol	40:1	375 °C, 15 MPa	1000 s	Vertical tubular reactor	92 (methyl ester yield)	Cheng et al., 2010

Both enzyme and supercritical fluid conditions were used in recent years (Table 3). No soap formation, no pollution, easier purification, catalyst reusable, no waste water are advantages for this mixed method. Enzymes represent an environmentally friendly alternative to chemical catalysts. Biodiesel production can further conform to environmental concerns if volatile, toxic, and flammable organic solvents are avoided and replaced enzyme with supercritical carbon dioxide (Wen et al., 2009). In recent years, it has been discovered that especially lipases can be used as catalyst for transesterification and esterification reactions. Enzyme catalyzed transesterification, using lipase as catalyst does not produce side products and involves less energy consumption (Fjerbaek et al., 2009). However, enzyme applications have also disadvantages that they are expensive and have stricted reaction conditions and some initial activity can be lost due to volume of the oil molecule (Marchetti et al., 2007).

Table 3.Enzyme usage in supercritical fluid conditions for biodiesel production

Raw Material	Alcohol+enzyme	Alcohol/oil molar ratio	Reaction temperature and pressure	Reaction time	Reactor type	Performance (%)	Ref.
Sesame oil Mustard oil	Supercritical methanol Supercritical ethanol Supercritical methanol Supercritical ethanol +Novozym 435 Candida antarctica	40:1 40:1 40:1 40:1	350 °C, 200 bar 350 °C, 200 bar 350 °C, 200 bar 350 °C, 200 bar	40 min 40 min 70 min 25 min	Batch reactors	90 100 80 100 (conversion) 70 (conversion)	Varma et al., 2010
Sunflower oil	Supercritical methanol + Novozyme 435 enzyme in supercritical CO2 Supercritical ethanol +Novozyme 435 enzyme in supercritical CO2	40:1 40:1	400 °C, 200 bar 400 °C, 200 bar	40 min 40 min	Batch reactor	96 99 (conversion)	Giridhar et al., 2004
Soybean oil Olive oil Sunflower oil Rapeseed oil Palm oil	Supercritical methanol + Candida antartica lipase enzyme in supercritical CO2	40:1 40:1	45 °C, 130 bar	6 h	Batch reactor	58 65.8 50 60 59 (conversion)	Lee et al., 2009

Table 4. Different solvents instead of methanol in supercritical processes

Raw Material	Solvent	Solvent/oil molar ratio	Reaction temperature and pressure	Reaction time	Reactor type	Performance (%)	Ref.
Rapeseed oil Oleic acid	Supercritical methyl acetate	42:1	350 °C, 20 MPa	45 min	Batch-type vessel	91	Saka & Isayama, 2009
Soybean oil Waste soybean oil Sunflower oil Jatropha curcas oil	Supercritical methyl acetate	42:1 42:1 42:1 42:1	345 °C, 20 MPa 345 °C, 20 MPa 345 °C, 20 MPa 345 °C, 20 MPa	50 min 50 min 50 min 50 min	Batch reactor	100 100100 100	Campanelli et al., 2010

Purified palm oil	Supercritical methyl acetate	30:1	399 °C	59 min	Batch-type tube reactor	97.6 (biodiesel yield)	Tan et al., 2010b
Jatropha curcas oil	Sub-critical water+ Sub-critical dimethyl carbonate	217:1 14:1	270 °C, 27 MPa 300 °C, 9 MPa	25 min 15 min	Batch-type vessel	“/> 97 (methyl ester yield)	Ilham & Saka, 2010

In supercritical processes, as solvent not only methanol but also methyl acetate and dimethyl carbonate are now good candidates. However, further researches are needed for their practical applications. Saka & Isayama (2009), Tan et al. (2010b) and Campanelli et al. (2010) studied with supercritical methyl acetate for biodiesel production (Table 4). High products recovery and no glycerol produced are advantages, however, lower reactivity than methanol is the main disadvantage for these applications of supercritical biodiesel production processes (Lee&Saka2010).

3. MICROWAVE-ASSISTED PROCEDURE

Generally, heating coils are used to heat the raw material in biodiesel production process. This treatment can be also done by microwave method. An alternative heating system “microwave irradiation” has been used in transesterification reactions in recent years. Microwaves are electromagnetic radiations which represent a nonionizing radiation that influences molecular motions such as ion migration or dipole rotations, but not altering the molecular structure (Fini & Breccia, 1999; Varma, 2001; Refaat et al., 2008). The frequencies of microwave range from 300 MHz to 30 GHz, generally frequency of 2.45 GHz is preferred in laboratory applications (Taylor et al., 2005). Microwave irradiation activates the smallest degree of variance of polar molecules and ions with the continuously changing magnetic field (Azcan&Danisman, 2007). The changing electrical field, which interacts with the molecular dipoles and charged ion, causes these molecules or ions to have a rapid rotation and heat is generated due to molecular friction (Azcan&Danisman, 2007; Saifuddin & Chua, 2004). The absorption of microwaves causes a very rapid increase of the temperature of reagents, solvents and products (Fini & Breccia, 1999).

Microwave process can be explained for the biodiesel production with transesterification reaction: the oil, methanol, and base catalyst contain both polar and ionic components. Microwaves activate the smallest degree of variance of polar molecules and ions, leading to molecular friction, and therefore the initiation of chemical reactions is possible (Nuechter et al., 2000). Because the energy interacts with the sample on a molecular level, very efficient and rapid heating can be obtained in microwave heating. Since the energy is interacting with the molecules at a very fast rate, the molecules do not have time to relax and the heat generated can be for short times and much greater than the overall recorded temperature of the bulk reaction mixture. There is instantaneous localized superheating in microwave heating and the bulk temperature may not be an accurate measure of the temperature at which the actual reaction is taking place (Barnard et al., 2007; Refaat et al., 2008).

When the reaction is carried out under microwaves, transesterification is efficiently accelerated in a short reaction time. As a result, a drastic reduction in the quantity of by-products and a short separation time are obtained (Saifuddin & Chua, 2004; Hernando et al., 2007) and high yields of highly pure products are reached within a short time (Nuechter et al., 2000). So, the cost of production also decreases and less by-products occurs by this method (Öner & Altun, 2009). Therefore, microwave heating compares very favorably over conventional methods, where heating can be relatively slow and inefficient because transferring energy into a sample depends upon convection currents and the thermal conductivity of the reaction mixture (Koopmans et al., 2006; Refaat et al., 2008). Microwave assisted transesterification process schematic diagram was presented in Figure 4.

There can be also a few drawbacks of microwave assisted biodiesel production, beside the great advantages. Microwave synthesis may not be easily scalable from laboratory small-scale synthesis to industrial production. The most significant limitation of the scale up of this technology is the penetration depth of microwave radiation into the absorbing materials, which is only a few centimeters, depending on their dielectric properties. The safety aspect is another drawback of microwave reactors in industry (Yoni & Aharon, 2008; Vyas et al., 2010).

This survey of microwave assisted transformations is abstracted from the literature published from 2000 to 2011. And studies on microwave assisted method of transesterification reaction in the literature were summarized in Table 5. The biodiesel production have been studied by using microwave assisted method from different oils such as cottonseed oil (Azcan&Danisman, 2007), safflower seed oil (Düz et al., 2011), rapeseed oil (Hernando et al., 2007; Geuens et al., 2008), soybean oil (Hernando et al., 2007; Hsiao et al., 2011; Terigar et al., 2010), corn oil (Majewski et al., 2009), macauba oil (Nogueira et al., 2010), waste frying palm oil (Lertsathapornsuk et al., 2008), micro algae oil (Patil et al., 2011), karanja oil (Venkatesh et al., 2011), jatropha oil (Shakinaz et al., 2010), yellow horn oil (Zhang et al., 2010), canola oil (Jin et al., 2011), camelina sativa oil (Patil et al., 2009), castor oil (Yuan et al., 2009), waste vegetable oils (Refaat et al., 2008), maize oil (Öztürk et al., 2010) and sunflower oil (Han et al., 2008; Kong et al., 2009).

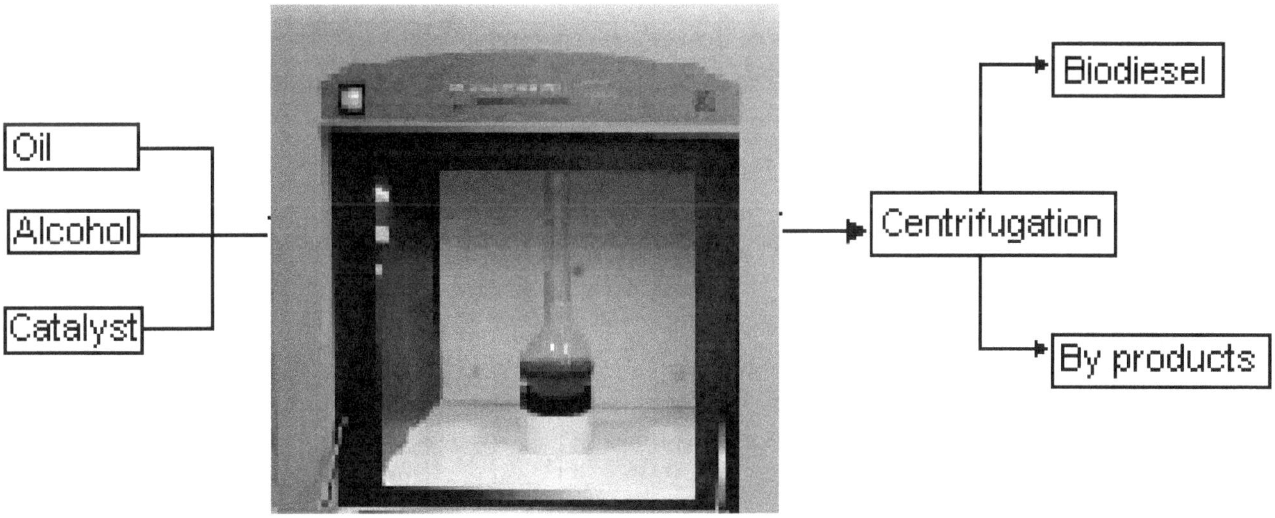

Figure 4.Microwave assisted transesterification process shematic diagram

Table 5.Microwave assisted method studies of transesterification reaction in the literature

Raw material	Catalyst	Catalyst amount (wt%)	Type of alcohol	Alcohol/ oil molar ratio	Microwawe conditions	Reaction time	Reaction tempe-rature	Performance (%)	Ref.
Cotton seed oil	KOH	1.5	Methanol	6:1	21% of 1200 W	7 min	333 K	92.4 (yield)	Azcan&Danisman, 2007
Safflower seed oil	NaOH	1	Methanol	10:1	300 W	6 min	333 K	98.4 (conversion)	Düz et al., 2011
Rapeseed oil Soybean oil	NaOH	%1.3	Methanol	18:1 1.27 ml	300 W	1 min	60 °C	97 95 (yield)	Hernando et al., 2007
Corn oil Soybean oil	Diphenylammonium salts: DPAMs (Mesylate) DPABs (Benzenesulfonate) DPATs (Tosylate) DPAMs DPABs	20 (molar) 10 (molar) 10 (molar) 109	Methanol	5 g methenol / 2 g oil	-	20 min	150°C	100 96 100 92 97 (methyl ester yield)	Majewski et al., 2009
Waste frying oil	NaOH	1	Methanol	6:1	600 W	5 min	64°C	93.36 (methyl ester content)	Yücel et al., 2010
Macauba oil	Novozyme 435 Lipozyme IM	2.5 5	Ethanol Ethanol	9:1 9:1	-	15 min 5 min	30°C 40°C	45.2 35.8 (conversion)	Nogueira et al., 2010
Waste frying palm oil	NaOH	3	Ethanol	12:1	800 W	30 s	-	97 (conversion)	Lertsathapornsuk et al., 2008
Rapeseed oil	KOH NaOH	1 1	Methanol Methanol	6:1 6:1	67 % of 1200 W	5min 3min	323 K 313 K	93.7 92.7 (yield)	Azcan&Danisman, 2008
Soybean oil	nano CaO (heterogeneous catalyst)	3	Methanol	7:1	-	60 min	338 K	96.6 (conversion)	Hsiao et al., 2011
Soybean oil Oleic acid	sulfated zirconia	5	Methanol	20:1	-	20 min	60 °C	90 (conversion)	Kim et al., 2011
Dry micro algae	KOH	2	Methanol	9:1	800 W	6 min	-	80.13 (conversion)	Patil et al., 2011
Crude karanja oil	KOH	1.33	Methanol	%33.4 (w/w)	180 W	150 s	-	89.9 (conversion)	Venkatesh et al., 2011
Jatropha oil	KOH	1.50	Methanol	7.5:1	-	2 min	65°C	97.4 (conversion)	Shakinaz et al., 2010
Crude palm oil	KOH	1.50	Ethanol	8.5:1	70 W	5 min	70°C	85 (yield) 98.1 (conversion)	Suppalakpanya et al., 2010

Raw material	Catalyst	Catalyst amount (wt%)	Type of alcohol	Alcohol/ oil molar ratio	Microwawe conditions	Reaction time	Reaction tempe-rature	Performance (%)	Ref.
Yellow horn oil	Heteropolyacid (HPA)	1	Methanol	12:1	500 W	10 min	60°C	96.22 (FAMEs)	Zhang et al., 2010
Soybean oil	NaOH	1	Methanol	6:1	900 W	1 min	303 K	97.7 (conversion)	Hsiao et al., 2011
Canola oil	$ZnO/La_2O_2CO_3$ (heterogeneous catalyst)	< 1	Methanol	1:1 (w/w)		< 5 min	<100°C	“/> 95 (yield)	Jin et al., 2011
Camelina sativa oil	Heterogeneous metal oxide catalysts (BaO, SiO)	1.5 2	Methanol	9:1	800 W	-	-	94 80 (FAME yield)	Patil et al., 2009
Castor bean oil	Al_2O_3 / 50% KOH SiO_2 / 50% H_2SO_4 SiO_2 / 30% H_2SO_4	1 1 1	Methanol Methanol Ethanol	1:6 1:6 1:6	40 W 40 W 220 W	5 min 30 min 25 min	-	95 95 95 (conversion)	Perin et al., 2008
Castor oil	H_2SO_4 / C	5	Methanol	1:12	200 W	60 min	338 K	94 (yield)	Yuan et al., 2009
Triolein	KOH NaOH	5	Methanol	1:6	25 W	1 min	323 K	98 (conversion)	Leadbeater & Stencel, 2006
Frying oil	NaOH	0.5	Ethanol	1:6	50% of 750 W	4 min	60°C	87 (conversion)	Saifuddin & Chua, 2004
Rapeseed oil	-	-	Supercritical 1-butanol	2.5:1	-	4 hour 80 bar	310°C	91 (fatty acid buthyl ester conversion)	Geuens et al., 2008
Domestic waste vegetable oil Restaurant waste vegetable oil Neat vegetable virgin sunflower oil	KOH	1	Methanol	6:1	500 W	1 h	65°C	95.79 94.51 96.15 (biodiesel yield)	Refaat et al., 2008
Safflower seed oil	NaOH	1	Methanol	10:1	300 W	16 min	60°C	98.4 (methyl ester content)	Düz et al., 2011
Soybean oil	NaOH	1	Methanol	6:1	600 W (Ultrasonic) 900 W (Microwave)	1 min 2 min	333 K	97.7 (conversion)	Hsiao et al., 2010
Maize oil	NaOH	1.5	Methanol	10:1	-	-	-	98 (conversion)	Öztürk et al., 2010

Raw material	Catalyst	Catalyst amount (wt%)	Type of alcohol	Alcohol/ oil molar ratio	Microwawe conditions	Reaction time	Reaction tempe-rature	Performance (%)	Ref.
Soybean oil Rice bran oil	NaOH	0.6	Ethanol	5:1	-	10 min 73°C	73°C	99.25 99.34 (FAME yield)	Terigar et al., 2010
Jatropha curcas	NaOH	4	Methanol	30:1	-	7 min	328 K	86.3 (conversion)	Yaakob et al., 2008
Sunflower oil	H2SO4	0.05	Methanol	10:1	400W	45 min-		96.2 (conversion)	Han et al., 2008
Sunflower oil	TiO2/SO4	0.02	Methanol	12:1	300W	-25 min		94.3 (biodiesel yield)	Kong et al., 2009

4. ULTRASOUND ASSISTED PROCESS

Ultrasonic waves are energy application of sound waves which is vibrated more than 20,000 per second. In another words, it can be defined as the sound waves beyond human hearing limit. Human hear can not hear sound waves with more high-pitched sound waves of an average of 10-12 kHz. Ultrasonic or ultrasound signals are in the order of 20 kHz- 100 kHz and above the limit of human hearing. Ultrasonic waves were used as the first for medical research and detectors in the 1930s and 1940s (Newman& Rozycki, 1998). Idea of the use of ultrasound, especially in the industry since the 1980s began to develop rapidly, and today a wide range of applications using ultrasonic waves appeared. At present, ultrasonic waves are used in areas such as Atomization: Water sprays for dust suppression and humidifiers, low velocity spray coating, spray drying nozzles. Cleaning and cleaning of engineering items, small electronic items and jeweler using aqueous based solvents. Cleaning and disinfection of medical instruments and food processing equipment. Processing: Dispersion of pigments and powders in liquid media and emulsification. Extraction: Essential oil, flavonoid, resin, Crystallization and Filtration (Cintas et al., 2010; Mason et al., 1996; Mason, 2000).

Ultrasonic irradiation has three effects according to the investigators. First one is rapid movement of fluids caused by a variation of sonic pressure. It causes solvent compression and rarefaction cycles (Mason, 1999). The second and the most important one is cavitation. If a large negative pressure gradient is applied to the liquid, the liquid will break down and cavities (cavitation bubbles) will be created. At high ultrasonic intensities, a small cavity may grow rapidly through inertial effects. So, bubbles grow and collapse violently. The formation and collapse of micro bubbles are responsible for most of the significant chemical effects (Kumar et al., 2010a). Cavitation is considered as a major factor which influences on reaction speed. Cavity collapse increases mass transfer by disrupting the interfacial boundary layers known as the liquid jet effect. The last effect of ultrasound is acoustic streaming mixing.

Ultrasound has been used to accelerate the rates of numerous chemical reactions, and the rate enhancements, mediated by cavitations, are believed to be originated from the build-up of high local pressures (up to 1000 atm) and temperatures (up to 5000 K), as well as increased catalytic surface areas and improve mass transfer (Yu et al., 2010). Low frequency ultrasonic irradiation is widely used for biodiesel production in recent years. In transesterification reaction, mixing is important factor for increasing biodiesel yield. Oil and methanol are not miscible completely in biodiesel processing. Ultrasonic mixing is an effective mixing method to achieve a better mixing and enchancing liquid–liquid mass transfer (Ji et al., 2006). Vigorous mixing increases the contact area between oil and alcohol phases with producing smaller droplets than conventional stirring (Mikkola & Salmi, 2001; Stavarache et al., 2006). Cavitation effects increase mass and heat transfer in the medium and hence increase the reaction rate and yields (Adewuyi, 2001). Ultrasonic cavitation also provides the necessary activation energy for initiating transesterification reaction.

Ultrasonic waves are produced with the power converter (transducer) which is piezoelectric material. Sound waves are converted to ultrasonic waves vibrating at high frequency with quartz crystal oscillator. If ultrasound waves are used in chemical reactions and processes it is called as sonochemistry. Industrial sonochemial reactors were designed more than 40 years ago by Sarocco and Arzono (Cintas et al., 2010). They showed that reactor geometry affected enormously the reaction kinetics. Later many rectors have been developed by researchers for different chemical reactions. For conventional biodiesel production, batch and continuous reactors have been developed in industry. Ultrasonic cleaning bath, ultrasonic probe which are usually operated at a fixed frequency are mainly used as ultrasonic apparatus. Frequency is dependent on particular type of transducer which is 20 kHz for probes and 40 kHz for bath. Figure 5 shows schematic diagram of biodiesel production via ultrasound assisted method.

Ultrasonic processing of biodiesel involves the following steps:

- Mixing vegetable oil is with the alcohol (methanol or ethanol) and catalyst,
- Heating the mixture,
- The heated mixture is being sonicated inline,
- Glycerin separation by using centrifuge. Alternative reactors have also been developed to lower energy consumption. Cintas et al., (2010) designed a flow reactor constituted by three transducers and showed that considerable energy saving could be achieved by large-scale multiple transducer sonochemical reactors operating in a continuous mode.

The factors affecting ultrasound assisted biodiesel production are: -Effect of catalyst type on ultrasound assisted biodiesel production, -Effect of alcohol type on ultrasound assisted biodiesel production, -Effect of ultrasonic power on biodiesel processing, -Frequency effect on ultrasonic assisted biodiesel production.

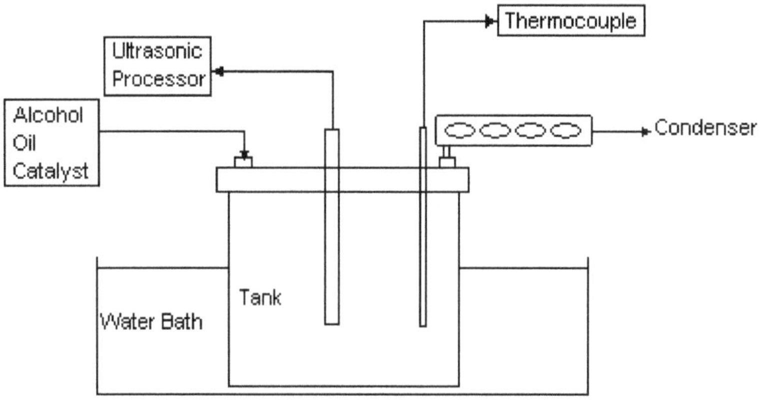

Figure 5. Scheme of biodiesel production process via ultrasound assisted method

Effect of catalyst type on ultrasound assisted biodiesel production: In ultrasonic assisted biodiesel studies homogen (alkaline, acid), heterogen and enzyme catalyst were studied with many edible and nonedible oils under ultrasonic irradiation. Transesterification reactions have been studied with KOH catalyst for corn oil (Stavarache et al., 2007a; Lee et al., 2011), grape (Stavarache et al., 2007a), canola (Stavarache et al., 2007a; Thanh et al., 2010a; Lee et al., 2011), palm (Stavarache et al., 2007a), tung (Hanh et al., 2011), beef tallow (Teixeira et al.,2009), coconut (Kumar et al., 2010), soybean (Ji et al., 2006; Mahamuni& Adewuyi, 2009;Thanh et al., 2010a; Lee et al., 2011), triolein (Hanh et al., 2008; Hanh et al., 2009b), fish oil (Armenta et al.,2007),neat vegetable oil (Stavarache et al., 2005), waste cooking oil (Thanh et al., 2010b; Hingu et al.,2010).These studies were presented in Table 6 (one step transesterification), and Table 7 (two-step esterification). Generally KOH was preferred for transesterification reactions instead of NaOH. Soybean (Ji et al., 2006), neat vegetable oil (Stavarache et al., 2005), jatropha curcas L. (Deng et al., 2010) (in the second transesterification step) and triolein (Hanh et al., 2009b) were transesterified with NaOH. KOH and NaOH were used for ultrasound assisted transesterification of neat vegetable oil. They used 0.5%, 1% and 1.5 % alkali catalyst amount, 6:1 molar ratio methanol to oil and room temperature. The researchers reported that there were no great differences in the time to complete conversion between two types of catalyst (Stavarache et al., 2005).. 98% and 96% yields were achieved with 0.5 % NaOH and KOH catalyst, respectively. They also reported that when KOH was used, high yields were obtained even for 1.5% catalyst concentration. Potassium soap is softer, more soluble in water and does not make as much foam as sodium soap. The washing of esters when using potassium hydroxide is easier and the yields of isolated product are higher. In alkali catalyzed ultrasonic transesterification for biodiesel production (Tables 6 and 7), 0.3-1.5 % alkali catalyzed amounts were used. Apart from that, Cintas et al., (2010) developed a new ultrasonic flow reactor to scale up biodiesel from soybean oil in presence of (Na or K methoxide). Na and K methoxide, are alkaline metal alkoxides (as CH3ONa for the methanolysis) are the most active catalysts because of stronger hydroxide group. In their reacton mixture of oil (1.6 L), methanol and sodium methoxide 30% in methanol (wt/wt ratio 80:19.5:0.5, respectively) was fully transesterified at about 45°C in 1 h (21.5 kHz, 600 W, flow rate 55 mL/min).

Heterogen catalysts were tried by researchers in a few studies (Ye et al., 2007; Salamatinia, 2010; Mootabadi et al., 2010;Kumar et al., 2010b). As it is known, ultrasound increase mixing of oil and alcohol with catalyst phases, as well as increase catalytic surface area. Catalyst can be broken into smaller particles by ultrasonic irradiation to create new sites of the subsequent reaction. Thus, solid catalyst is expected to last longer in the ultrasonic-assisted process (Mootabadi et al., 2010). Single component alkaline earth metal oxides (BaO, SrO, CaO) having lower solubility in alcohol catalyzed palm transesterification processes with methanol (Mootabadi et al., 2010). The catalytic activities of the three catalysts were correlated well with their basic strengths and found as the sequence of CaO<SrO<BaO. BaO catalyst achieved 95.2% of biodiesel yield within 60 min in the ultrasonic-assisted process while SrO catalyst generally demonstrated slightly lower result. CaO showed the lowest yield with 77.3%yield under optimum conditions. Although high activity of BaO as catalyst, this activity dropped severely in the BaO reusability test, especially under ultrasonic condition (compared to mechanical stirring). In another study, aluminum isopropoxide or titanium isopropoxide as heterogeneous transesterification catalysis are employed to produce nanoemulsions with large interfacial area for easy catalyst separation and enhanced reaction rate (Ye et al., 2007). These catalysts are produced by partial polymerization and metal alkoxides are connected by metal-oxygen bonds. Alkoxide parts in the polymer matrix catalyst gives the catalyst amphiphilic properties that help form and stabilize alcohol/ triglycerides nanoemulsion (Ye et al., 2007). The study showed that titanium isopropoxide also showed good catalytic activity and considerable amphiphilic properties in forming nanoemulsions. With aluminum isopropoxide or titanium isopropoxide, transparent alcohol/oil emulsions can be formed in less than four minutes and can significantly enhance the transesterification reaction rate. The micelle size was observed to be as low as 5.1 nm.

High acidity oils (Jatropha curcas L, waste frying oil) can be transesterified by two-step processes. In the first step, free fatty acids are converted to esters by direct esterification with acid catalyst. Eq. 1 shows esterification of fatty acids. In the second step, basic catalyst was used to esterify triglycerides as it was shown in Figure 2.

$$RCOOH + CH_3OH \longrightarrow RCOOCH_3 + H_2O \tag{1}$$

In production of biodiesel from Jatropha curcas L. oil (non edible oil) Deng et al., (2011) used a two-step process. The first step pretreatment (acid-esterification) of Jatropha oil was performed at 318 K an ultrasonic reactor for 1.5 h in their first study (Deng, et al., 2010). After reaction, the acid value of Jatropha oil was reduced to 0.7 mg KOH/g and 93.3% esterification rate was achieved. The second step, a base-catalyzed transesterification was performed with nano sized Mg/Al oxides under different conditions. At the optimized condition, (Table 6) 95.2% biodiesel yield was achieved, and the Jatropha oil biodiesel properties were found to be close to those of the German standard. It was reported that the catalyst could be reused for 8 times.

Although it is known that ultrasonic mixing has a significant effect on enzymatic transesterification there are a little study about using of lipases as enzyme catalyst. It has been reported that enzyme activity of Novozym 435 enhanced by ultrasound irradiation (Sinisterra, 1992; Lin & Liu, 1995). Novozym 435

(Candida antarctica lipase B immobilized on polyacrylic resin) was used in biodiesel production from soybean oil and methanol with a low frequency ultrasonic (40 kHz) waves to see enzyme activity and compare their overall effects under two different conditions—ultrasonic irradiation and vibration (Yu et al., 2010). They investigated effects of reaction conditions, such as ultrasonic power, water content, organic solvents, ratio of solvent/oil, and ratio of methanol/oil, enzyme dosage and temperature on the activity of Novozym 435. Novozym 435 activity significantly increased by ultrasonic irradiation compared with vibration and reaction rate was further increased under the condition of ultrasonic irradiation with vibration (UIV). Yu et al (2010) indicated that 96% yield of fatty acid methyl ester (FAME) could be achieved in 4 h under the optimum conditions: 50% of ultrasonic power, 50 rpm vibration, water content of 0.5%, tert-amyl alcohol/oil volume ratio of 1:1, methanol/oil molar ratio of 6:1, 6% Novozym 435 and 40 °C. Since the lipase enzyme is expensive catalyst it is important to reuse the catalyst in biodiesel industrial productions. The researchers also pointed out that Novozym 435 was not deactivated under UIV, only 4 % enzyme activity slightly decreased after five cycles.

Effect of alcohol type on ultrasound assisted biodiesel production: Methanol was mostly used in transesterification reaction under ultrasonic irradiation with oils shown in Tables 6 and 7. High conversion and yields were obtained with methanol and ethanol using. Stavarache et al., (2007a) used methanol in transesterification of commercial edible oil, corn, grapeseed, canola and palm oil. Excellent yields (99%) were obtained for all type oils in 20 minutes with 6:1 methanol to oil molar ratio at 36 °C. As it is shown in Figure 6, triglycerides are converted to di and monoglycerides to produce biodiesel to produce biodiesel and glycerin. They also examined the transesterification reaction mechanism under low frequency (40 kHz) ultrasonically driven esterification.

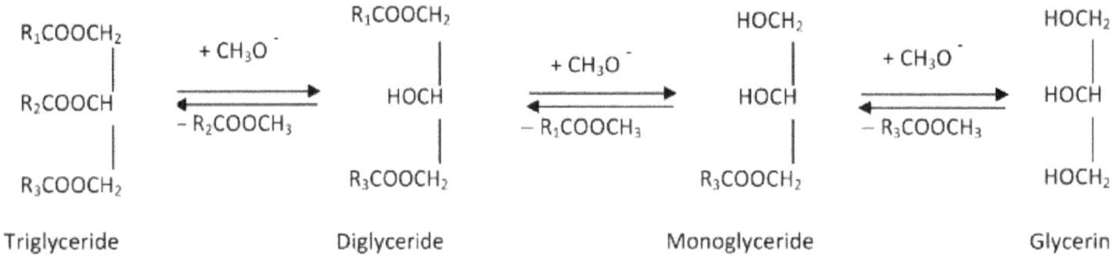

Figure 6.Alkali catalyzed transesterication steps of triglyceride with methanol

They have reported that the major part of the transesterification took place in the first 3-10 minutes of reaction if not faster and the rate- determining reaction switches from diglyceride (DG) → monoglyceride (MG) (classical mechanic agitation) to MG + ROH→Gly + ME (ultrasonically driven transesterification). In another study, the conversion of FAME greater than 99.4 % was achieved after about 15 minutes at 40 °C with ultrasonic agitation for 6:1 methanol: oil molar ratio (Calucci et al., 2005). They have also concluded that hydrolysis rate constants of DG and TG are three to five times higher than those of mechanical agitation. Ji et al., (2006) used ultrasonic transesterification process for soybean oil transesterification with methanol and reported 99% yield at 10 min reaction time with 6:1 methanol to oil molar ratio at 45°C. Oleic acid, triolein, coconut were esterified with ethanol and 90% conversion, about 99% yield and >92% yields were achieved respectively (Hanh et al., 2009a; Hanh et al., 2009b; Kumar et al., 2010a). Table 8 shows the some biodiesel yield and conversion with various monoalcohols and comparing of the alcohols.

Stravarache et al., (2005) studied effects of alcohol type on transesterification of neat vegetable oil under ultrasonic and mechanical stirring. The results of transesterfication with primary, secondary and tertiary alcohols after 60 min of reaction were presented in Table 8.

Table 6.The studies for biodiesel production from various feedstocks at different conditions under ultrasound irradiation

Raw material	Catalyst	Catalyst amount (wt %)	Alcohol type	Alcohol/oil molar ratio	Reaction temp. (°C)	Reaction time	Reactor conditions	Performance (%)	Ref.
Oleic acid	H2SO4	5	Ethanol	3:1	60	2 hour	Ultrasonic cleaner 40 kHz, 1200 W	~90 (conversion)	Hanh et al., 2009a
Commercial edible oil Corn Grape seed Canola Palm	KOH	0.5	Methanol	6:1	36 ± 2	20 min	Ultrasonic cleaner 40 kHz,1200 W	~ 99 (conversion)	Stavarache et al., 2007a
Refined soybean oil	KOH	1.5	Methanol	6:1	40	15 min	20 kHz, 14.49 W	“/>99.4 (conversion)	Colucci et al., 2005
Soybean	NaOH	1	Methanol	6:1	45	10 min	197 kHz, 150W	99 (yield)	Ji et al., 2006
Soybean	KOH	0.5	Methanol	6:1	26 - 45	30 min<	611 kHz, 139 W	“/>90 (conversion)	Mahamuni& Adewuyi, 2009
Soybean	Na or K methoxide	0.15	Methanol	6:1	45	1 h	21.5 kHz, 600 W	Fully transesterified	Cintas et al., 2010
Canola Soybean Corn	KOH	1	Methanol	6:1	55	30 min	450 W	95 (yield) 95 (yield)	Lee et al., 2011
Tung and Blended oil (20%Tung, 30%canola, 50%palm)	KOH	1	Methanol	6:1	20-30	30 min	25 kHz, 270 W	91.15 (yield) 94.03 (yield)	Hanh et al., 2011
Beef Tallow	KOH	0.5	Methanol	6:1	60	70 s	40 kHz, 1200 W	“/>92 (conversion)	Teixeira et al., 2009
Triolein	KOH	1	Methanol	6:1	25	30 min	Ultrasonic cleaner 40 kHz, 1200 W	~99 (yield)	Hanh et al., 2008

Raw material	Catalyst	Catalyst amount (wt %)	Alcohol type	Alcohol/oil molar ratio	Reaction temp. (°C)	Reaction time	Reactor conditions	Performance (%)	Ref.
Triolein	NaOH KOH	1	Methanol Ethanol	6:1	25	25 min	Ultrasonic cleaner 40 kHz, 1200 W	"/>95 (conversion)	Hanh et al., 2009b
Neat vegetable oil	NaOH KOH	0.5	Methanol	6:1	25	20 min	Ultrasonic cleaner 20 kHz 40 kHz 1200 W	98 (yield) 96 (yield	Stavarache et al., 2005
Coconut	KOH	0.75	Ethanol	6:1	-	7 min	24kHz, 200 W	"/>92 (yield)	Kumar et al., 2010a
Waste cooking oil	KOH	1	Methanol	6:1	45	40 min	20 kHz, 200 W	89 (conversion)	Hingu et al., 2010
Palm	KOH	-	Methanol	6:1	38–40	20 min	45 kHz,600 W	95 (yield)	Stavarache et al., 2007b
Palm	CaO SrO BaO	3	Methanol	9:1	65	60 min	30 kHz	77.3 (yield) 95 (yield) 95 (yield)	Mootabadi et al., 2010
Palm	BaO SrO	2.8	Methanol	9:1	65	50 min <	20 kHz, 200 W	"/>92 (yield)	Salamatinia et al., 2010
Canola	KOH	0.7	Methanol	5:1	25	50 min	20 kHz, 1000 W	"/>95 (conversion)	Thanh et al., 2010a
Soybean	Ti(Pr)4 Al(Pr)3	3	Methanol	6:1	60	2 h	-	64 (yield)	Ye et al., 2007
Soybean	Novozym 435	6	Methanol	6:1	40	4h	40 kHz, 500 W	96 (yield)	Yu et al., 2010
Jatropha oil	Na/SiO2	3	Methanol	9:1	50-70	15 min	24 kHz, 200W	98.5 (yield)	Kumar et al., 2010b
Fish oil	KOH C2H5ONa	1 0.8	Ethanol	6:1 6:1	20-60 20-60	"/>30 "/>30	25-35 kHz 25-35 kHz	"/>95 (conversion) "/>98 (conversion)	Armenta et al., 2007

Table 7.Biodiesel production with two step transterification under ultrasound irradiation

Oil	Catalyst type	Catalyst amount (wt%)	Alcohol type	Alcohol: oil ratio	Reaction temperature (°C)	Reaction time	Ultrasound conditions	Performance (%)	Ref.
Waste cooking	KOH	0.7 0.3	Methanol	2.5:1 (mol) 1.5:1	20-25	10 min 20 min	20 kHz, 1000W (For each step)	81 (yield) 99 (yield)	Thanh et al., 2010b
Jatropha curcas L.	H_2SO_4 (1. step) Mg/Al oxides (2. step)	1 (For each step)	Methanol	4:1(mol) (For each step)	40 (For each step)	1.5 h (For each step)	210W (For each step)	95.2 (total yield)	Deng et al., 2011
Jatropha curcas L.	NaOH H2SO4	1 (For each step)	Methanol	0.4 (v/v) 6:1 (mol)	60 (For each step)	1h 30 min	210W (For each step)	96.4 (total yield)	Deng et al., 2010

Table 8. The influence of alcohol on the ultrasound assisted transesterification of different oils for biodiesel production

Alcohol type	Neat vegetable oil a (Stavarache et al., 2005) Performance (%) Stirring Ultrasonic	Triolein b (Hanh et al., 2009b) Conversion (%)	Soybean oil c (Colucci et al., 2005) Conversion (%)
Methanol	80 (Yield) 98 (Yield) (60 min) (20 min)	98	99.3
Ethanol	79 (Yield) 88 (Yield) (20 min) (20 min)	~98	99.1
n- propanol	78 (Yield) 88 (Yield) (10 min) (10 min)	~93	-
Iso-propanol	No conversion Some conversion	3	29.2
n-butanol	83 (Yield) 92 (Yield) ("/>60 min) ("/>60 min)	~93	92.0
Iso- butanol	No conversion Some conversion	3	-
Tertiary- butanol	No conversion No conversion	-	

[a]Reaction conditions for neat vegetable oil: 0.5% (wt/wt) NaOH, 6:1 alcohol to oil molar ratio, 40 KHz,

[b]Reaction conditions for triolein: 25 min, 25 °C, 0.1% (wt/wt) KOH, 6:1 alcohol to triolein molar ratio, 40 KHz,

[c]Reaction conditions for soybean oil: 2h, 1.5% (wt/wt) KOH, 6:1 alcohol to oil molar ratio, 40 KHz

N- chain alcohols (methanol, ethanol, n- propanol, and n-butanol) showed the high yields between 88-98% in 10-20 min reaction time. The yields of biodiesel in ultrasound activation were higher than mechanical stirring since ultrasound produce less soap. By using ultrasound the reaction time was found much shorter than mechanical stirring. The secondary alcohols showed some conversion while transesterification reaction took place under stirring. Tertiary-butanol had no conversion with both type of procedure. Hanh et al., (2009b) produced biodiesel with triolein and various alcohols (methanol, ethanol, propanol, butanol, hexanol, octanol and decanol). The productions were performed at molar ratio 6:1 (alcohol: triolein) and 25°C in the presence of base catalysts (NaOH and KOH) under ultrasonic irradiation

(40 kHz) and mechanical stirring (1800 rot/min) conditions. The rate of ester formation depended on alcohol types; as the alcohol carbon number increased, reaction rate decreased. The secondary alcohols such as 2-propanol, 2-butanol, 2-hexanol, and 2-octanol showed 3% conversion, suggesting that the steric hindrance strongly affected the transesterification of triolein. N-propanol showed approximately 93% conversion under ultrasonic irradiation, while 75% conversion was obtained under mechanical stirring. Soybean oil was transesterifed with methanol, ethanol, n-butanol, and iso-propanol over 2 h reaction period with 1.5 % KOH as the catalyst and a 6:1 molar ratio of alcohol/oil at 60°C (Colucci et al., 2005). The similar results obtained with methanol, ethanol and n-butanol compared to other studies.

Effect of ultrasonic power on biodiesel processing: The effect of ultrasonic power on the biodiesel formation has been reported (Mahamuni& Adewuyi, 2009; Hingu et al., 2010; Lee et al.,2011). Biodiesel yield increased with increasing ultrasonic power in all the studies. Nahamuni& Adewuyi (2009) studied this effect for three different frequencies and various powers (181, 90, 181 W at 1300 kHz, 104, 139,68 W at 611 kHz, 181, 117, 81, 49 W at 581 kHz). The reactions were carried out for 60-180 minutes. The reaction rate increased with increasing ultrasound power at any given frequency and biodiesel yield was obtained above 90%. At start of the reaction, reaction rate is very low because of low interfacial area available for the reaction. As time increased the reaction rate increased. This increase is due to the amount and size of the emulsion formation varies because of ultrasonic cavitation. Ultrasonic cavitation produces finer and stable emulsion and following this higher mass transfer and hence, higher biodiesel formation. When the ultrasonic power increases acoustic amplitude increases. So, cavitation bubble will collapse each other violently resulting in high velocity and micromixing at the phase boundary between two immiscible phases. Ultrasonication can result in mean droplet sizes much lower than those generated by conventional agitation, and can be a more powerful tool in breaking methanol into small droplets (Wu et al., 2007). The emulsion droplet size of methanol/soybean oil dispersions for ultrasonic and mechanical stirring was investigated and was shown that emulsion droplet size in ultrasonic mixing 2.4 times lower than that of conventional agitation. The mean droplet sizes were 148 and 146 nm with ultrasonic energy at 50 and 70 W, respectively. However, the droplet size was about 340 nm with impeller at 1000 rpm.

Higher power levels usually gives lower conversions because of cushioning effect and hence lower cavitational activity (Ji et al., 2006; Hingu et al., 2010; Lee et al., 2011). Hingu et al. (2010) observed that while the biodiesel conversion was obtained around 66% at 150 W power 89% of conversion was obtained when the power dissipation was increased to 200 W. But further increase in power from 200 W to 250 W resulted in lower FAME conversion. FAME conversion rate also depends on the emulsification degree of reaction system (Ji et al., 2006). These authors also noted that the order of affecting factors on FAME yield was substrate molar ratio > temperature > pulse frequency > ultrasonic power.

Ultrasound pulse (few seconds on followed by second off) effects the biodiesel conversion (Hingu et al., 2010; Ji et al., 2006). Higher conversion can be obtained when higher pulse is applied to system. For example, while biodiesel conversion was obtained for the pulse 2 s ON and 2 s OFF, the conversion were 65.5% for 5 s ON and 1 s OF (Hingu et al., 2010). For a pulse duration as 1 min ON and 5 s OFF, conversion of 89.5% was obtained because of better emulsification of the methanol and oil layers. The effect of horn position on biodiesel production was investigated by same researchers. They kept the reaction parameters constant such as 6:1, methanol to waste cooking oil molar ratio, 1% catalyst concentration, 45°C temperature, 200W power ad 1 min ON and 5 s OFF pulse. Cavitation intensity depends on some parameters physicochemical properties namely viscosity, surface tension and density. Cavitation is generated due to the presence of horn in oil or methanol. According to the horn position various results can be observed. Hingu et al. (2010) applied there different positions: in the oil phase, at the interface and in methanol. While maximum conversion was achieved as 89.5% when the horn was dipped in methanol rich layer, the lowest conversion was obtained as 8.5% when the horn is dipped in the oil phase. 58.5% conversion was observed when the horn is located at the interface of two phases. Maximum ester conversion was obtained since methanol contributed cavitating conditions significantly.

Frequency effect on ultrasonic assisted biodiesel production: The effect of ultrasonic frequency was studied on the yield of transesterification reaction of vegetable oils and shortchain alcohols (Stavarache et al., 2005). NaOH or KOH were used as base catalysts. It was observed that the reaction time gets shorter (the reaction fastens) as the ultrasonicirradiation increases but the yield slightly decreases. At 40 KHz, the reaction time was shorter than 28 KHz, but the yield was obtained higher when studied at 28 kHz.. This is because of the higher formation of soap at 40 KHz and higher quantity of soap makes the purification process harder. The more soap is formed, more esters gets trapped in the soap micelles and the yield of the reaction decreases at 40 KHz as a result.

General comparison of ultrasound irradiation with conventional stirring: Ultrasonic assisted transesterification of oil presents some advantages compared to conventional stirring methods such as; reducing reaction time, increase the chemical reaction speed and decrease molar ratio and methanol, increase yield and conversion. Ultrasound irradiation reduce the reaction time compared to conventional stirring operation (Stavarache et al., 2005; Ji et al., 2006; Hanh, et al., 2008; Mootabadi, et al., 2010; Hingu et al., 2010; Lee et al., 2011). Stavarache et al. (2005) studied transesterification of vegetable oil with short-chain alcohols, in the presence of NaOH, by means of low frequency ultrasound (28 and 40 kHz). By using ultrasounds the reaction time was found much shorter (10–40 min) than for mechanical stirring. The optimal conditions for triolein methanolysiswas methanol/triolein molar ratio of 6/1, KOH concentration of 1 wt% and irradiation time of 30 min. But the optimal conditions for the conventional stirring method were found to be as were methanol/triolein molar ratio of 6/1, KOH concentration of 1 wt% and 4 h (Hanh et al.., 2008). In transesterification of waste cooking oil with methanol 89.5% conversion was obtained in 40 minutes whereas conventional stirring resulted in 57.5% conversion (Hingu et al., 2010). Palm oil was esterified with 95% yield in 60 minutes compared to 2–4 h with conventional magnetic stirring under optimal conditions. Ultrasonic irradiation method enabled to reduce the reaction time by 30 min or more comparing to conventional heating method in production of biodiesel from various vegetable oils. Also this method improved conversion rate (Hanh et al., 2007; Lee et al., 2011). In transesterification reaction, mixing is important factor for increasing biodiesel yield. Ultrasonic effect induces an effective emulsification and mass transfer compared to conventional stirring thus reaction rate increase (Hanh et al., 2009; Hingu et al., 2010). Comparison of yield and conversion of vegetable oilwith various alcohols was presented in Table 8 and also was explained in the effect of alcohol type on ultrasound assisted biodiesel production section.

Ultrasound assisted method has a similar effect as microwave assisted method that both of them reduce the separation time from 5 to 10 hours to less than 60 minutes compared to conventional transesterification method (Kumar et al., 2010). Also, during production of biodiesel via acid or base catalyst, ultrasound irradiation provides a fast and easy route (Yu et al., 2010) and the purity of glycerin increases.

The production of biodiesel from non-edible vegetable oil and waste cooking oil using ultrasonication allows under ambient operating conditions (Kumar et al., 2010a; Hingu et al., 2010). Also, biodiesel production works from vegetable oils given in Table 6 illustrates the applicability of ultrasonic irradiation under atmospheric and ambient conditions. The transesterification reaction with methanol is usually performed at 60°C with classical stirring. Roomtemperature is hardly competitive in terms of energy consumption. Room temperature is hardly competitive in terms of energy consumption. The production of biodiesel with ultrasound is effective and time and energy saving and economically functional method (Ji et al., 2006; Kumar et al., 2010a; Hanh et al., 2011). Power ultrasonic method required approximately a half of the energy that was consumed by the mechanical stirring method (Ji et al., 2006). Special mixing devices can be used to increase mass transfer. It was reported that sonochemical reactors consume only about one third the energy required for a specialty mixer for same conversion (Lifka &Ondruschka, 2004). All these results clearly indicate that ultrasonic method inexpensive, simple and efficient and would be promising to the conventional stirring method.

Table 9.The yields and reaction times of FAMEs as a result of different frequencies of ultrasonic irradiation and mechanical stirring in the presence of NaOH catalyst (1.5% wt))(Stavarache et al., 2005)

Type of alcohol		28 kHz	40 kHz	Mechanical stirring
Methanol	Reaction time (min)	10	10	10
	Yield (%)	75	68	35
Ethanol	Reaction time (min)	20	10	10
	Yield (%)	75	30	47
n-propanol	Reaction time (min)	20	10	10
	Yield (%)	75	78	79
n-butanol	Reaction time (min)	40	20	20
	Yield (%)	87	90	89

According to Table 9, the length of the alcohol chain has an impact on the reaction yield, just like the frequency of the ultrasonic irradiation affects the reaction duration. The process produces a better yield in longer alcohol chains. Increasing the length of alcohol chains enhances the solubility (miscibility) of alcohol in oil. When a faster reaction is desired, it is preferable to employ ultrasonic irradiation at a frequency of 40 kHz. However, it is important to consider that as the reaction speeds up, the yield drops due to the increased creation of soap. Ultimately, the compatibility of oil and alcohol is enhanced when subjected to ultrasonic vibrations. This phenomenon enhances the surface area, leading to greater yields of isolated methyl esters. Enhancing mass transfer improves the efficiency of soap formation reduction, leading to a more effective and simplified isolation process of methyl esters. The ultrasonic irradiation enhances the reaction rate, whereas the yield experiences a modest reduction at higher frequencies (40 kHz).

5. CONCISE SUMMARY

As a result of the increasing need for energy and concerns about the environment, research has been concentrated on renewable alternative energy sources. Biodiesel is a significant renewable energy source that serves as an alternate fuel to diesel and is widely utilized in numerous places worldwide. Biodiesel is typically synthesized through the transesterification reaction of vegetable and animal oils using a catalyst, employing both batch and continuous processes with traditional stirring. Due to economic factors, the selection of an effective transesterification technique for biodiesel production has gained significance in recent years. Within this context, the researchers have been exploring various novel methodologies, including supercritical, microwave-assisted, and ultrasound-assisted procedures, in order to circumvent inefficient processes. These methods exhibit numerous characteristics in comparison to conventional methods. Homogeneous catalysts such as sulfuric acid, sodium hydroxide, potassium hydroxide, sodium metoxide, and potassium metoxide, as well as heterogeneous catalysts like ZnO, SiO, MgO, BaO, and SrO, and enzymatic catalysts such as lipase, are readily employed in microwave and ultrasonic assisted processes. Supercritical transesterification of vegetable oils is a noncatalytic reaction that can achieve better yields compared to standard procedures. Emerging techniques for biodiesel production have numerous benefits, although they also include certain adverse consequences. The drawbacks of the supercritical process include increased energy use and excessive alcohol usage. Microwave synthesis is now limited to laboratory-scale production and is not feasible for large-scale industrial manufacturing due to the limited ability of microwave radiation to penetrate absorbing materials. Another disadvantage of microwave reactors for industrial use is their limited safety aspect. The utilization of ultrasonic technology in biodiesel production offers potential benefits for small-scale producers. However, its implementation on a larger scale may provide challenges due to the requirement of several ultrasound probes. Despite a few drawbacks, novel methods in biodiesel production offer significant advantages in the transesterification of oils. These methods effectively reduce reaction time, reaction temperature, and unwanted by-products, while increasing ester yields and making the conversion process easier compared to conventional methods. Ultimately, these methods, with their significant benefits, may be more favored over the old method.

REFERENCES

Adewuyi, Y.G. (2001). Sonochemistry: Environmental Science and Engineering Applications. Industrial & Engineering Chemistry Research, Vol. 40, pp. 4681-4715.

Alenezi, R., Leeke, G.A., Winterbottom, J.M., Santos, R.C.D.,& Khan, A.R. (2010).Esterification kinetics of free fatty acids with supercritical methanol for biodiesel production. Energy Conversion and Management, Vol. 51, pp. 1055-1059.

Armenta, R.E., Vinatoru, M., Burja, A.M., Kralovec, J.A. & Barrow, C.J. (2007).Transesterification of fish oil to producc fatty acid cthyl esters using ultrasonic energy. Journal of American Oil Chemists' Society, Vol. 84, pp. 1045-1052.

Azcan , N. &Danisman, A. (2007). Alkali catalyzed transesterification of cottonseed oil bymicrowave irradiation. Fuel, Vol. 86, pp. 2639–2644.

Azcan, N. &Danisman, A. (2008). Microwave assisted transesterification of rapeseed oil. Fuel, Vol. 87, pp. 1781-1788.

Baig, S., Faroog, A. & Rehman, F. (2010). Sonochemistry and its industrial applications. World Applied Sciences Journal, Vol. 10, No. 8, pp. 936-944, ISSN 1818-4952.

Barnard, T.M., Leadbeater, N.E., Boucher, M.B., Stencel, L.M. & Wilhite, B.A. (2007).Continuous-flow preparation of biodiesel using microwave heating. Energy & Fuels, Vol.21, No. 3, pp. 1777-1781.

Bunyakiat, K., Makmee, S., Sawangkeaw, R. & Ngamprasertsith, S. (2006). Continuous production of biodiesel via transesterification from vegetable oils in supercritical methanol. Energy & Fuels, Vol. 20, pp. 812-817.

Campanelli, P., Banchero, M. & Manna, L. (2010). Synthesis of biodiesel from edible, nonedible and waste cooking oils via supercritical methyl acetate transesterification. Fuel, Vol. 89, pp. 3675-3682.

Cao, W., Han, H. & Zhang, J. (2005). Preparation of biodiesel from soybean oil using supercritical methanol and co-solvent. Fuel, Vol. 84, pp. 347–351.

Chen C-H., Chen W-H., Chang C-M J., Lai S-M. & Tu C-H. (2010). Biodiesel production from supercritical carbon dioxide extracted Jatropha oil using subcritical hydrolysis and supercritical methylation. Journal of Supercritical Fluids, Vol. 52, pp. 228-234.

Cheng, Z., Cunwen, W., Weiguo, W., Yuanxin, W, Faquan, Y, Ru'an, C. &Junfeng, Z. (2010).Continuous production of biodiesel from soybean oil using supercritical methanolin a vertical tubular reactor: I. phase holdup and distribution of intermediate product along the axial direction. Chinese Journal of Chemical Engineering, Vol. 18, pp. 626-629.

Chung, K.H., Kim, J. & Lee, K.Y. (2009). Biodiesel production by transesterification of duck tallow with methanolonalkalicatalysts. Biomass and Bioenergy, Vol.33, pp. 155–158.

Cintas, P., Mantegna, S., Gaudino, E.C. & Cravotto, G. (2010). A new pilot flow reactor for high-intensity ultrasound irradiation. Application to the synthesis of biodiesel. Ultrasonics Sonochemistry, Vol. 17, pp. 985-989.

Colucci, J.A., Borrero, E.E. & Alape, F. (2005). Biodiesel from an alkaline transesterification reaction of soybean oil using ultrasonic mixing. Journal of American Oil Chemists' Society, Vol. 82, No. 7, pp. 525-530.

Da Cunha, M.E., Krause, L.C., Moraes, M.S.A., Faccini, C.S., Jacques , R.A., Almeida, S.R., Rodrigues, M.R. A. & Caramão, E.B. (2009). Beef tallow biodiesel produced in apilot scale. Fuel Processing Technology, Vol. 90, pp. 570–575.

Demirbaş, A. (2002). Biodiesel from vegetable oils via transesterification in supercritical methanol. Energy Conversion Management, Vol. 43, pp. 2349-4356.

Demirbaş, A. (2007). Biodiesel from sunflower oil in supercritical methanol with calcium oxide. Energy Conversion Management, Vol. 48, pp. 937-941.

Demirbaş, A. (2008). Studies on cottonseed oil biodiesel prepared in non-catalytic SCF conditions. Bioresource Technology, Vol. 99, pp. 1125–1130.

Demirbaş, A. (2009). Biodiesel from waste cooking oil via base-catalytic and supercritical methanol transesterification. Energy Conversion and Management, Vol. 50, pp. 923–927.

Deng, X., Fang, Z. & Liu, Y. (2010). Ultrasonic transesterification of Jatropha curcas L. oil to biodiesel by a two-step process. Energy Conversion and Management, Vol. 51, pp.2802-2807.

Deng, X., Fang, Z., Liu, Y. & Yu, C.L. (2011). Production of biodiesel from Jatropha oilcatalyzed by nanosized solid basic catalyst. Energy, Vol. 36, pp. 777-784.

Dube, M.A., Tremblay, A.Y. & Liu, J. (2007). Biodiesel production using a membrane reactor. Bioresource Technology, Vol. 98, pp. 639–647.

Düz, M.Z., Saydut, A. & Öztürk, G. (2011). Alkali catalyzed transesterification of safflower seed oil assisted by microwave irradiation. Fuel Processing Technology, Vol. 92, pp.308-313.

Encinar, J.M., Juan, F., Gonzalez, J.F. & Rodriguez-Reinares, A. (2007). Ethanolysis of usedfrying oils: Biodiesel preparation and characterization. Fuel Processing Technology, Vol. 88, pp. 513-522.

Fini, A. & Breccia, A. (1999). Chemistry by microwaves. Pure and Applied Chemistry, Vol. 71,No. 4, pp. 573-579.

Fjerbaek, L., Christensen, K.V. & Norddahl, B. (2009). A review of the current state of biodiesel production using enzymatic transesterification. Biotechnology and Bioengineering, Vol. 102, No. 5, pp. 1298–1315

Freedman, B., Pryde, E.H. & Mounts, T.L. (1984). Variables affecting the yields of fatty esters from transesterified vegetable oils. Journal of American Oil Chemists' Society, Vol. 61, pp. 1638–1643.

Fukuda, H., Kondo, A. & Noda, H. (2001). Biodiesel Fuel Production by Transesterification of Oils. Journal of Bioscience and Bioengineering, Vol. 92, No. 5, pp. 405-416.

Geuens, J., Kremsner, J.M., Nebel, B.A., Schober, S., Dommisse, R.A., Mittelbach, M.,Tavernier, S., Kappe, C.O. & Maes, B.U.W. (2008). Microwave-assisted catalyst-free transesterification of triglycerides with 1-butanol under supercritical conditions. Energy & Fuels, Vol. 22, No. 1, pp. 643-645.

Giridhar, M., Kolluru, C. & Kumar, R. (2004). Synthesis of biodiesel in supercritical fluids. Fuel, Vol. 83, pp. 2029-2033.

Gui, M. M., Lee, K. T. & Bhatia, S. (2009). Supercritical ethanol technology for the production of biodiesel: Process optimization studies. Journal of Supercritical Fluids, Vol. 49, pp.286–292.

Gürü, M., Artukoğlu, B.D., Keskin, A. & Koca, A. (2009). Biodiesel production from waste animal fat and improvement of its characteristics by synthesized nickel and magnesium additive. Energy Conversion and Management, Vol.50, pp. 498–502.

Gürü, M., Koca, A., Can, Ö., Cõnar, C. & ^ahin, F. (2010). Biodiesel production from waste chicken fat based sources and evaluation with Mg based additive in a diesel engine. Renewable Energy, Vol. 35, pp. 637–643.

Haas, M.J. & Scoot, K.M. (1996). Combined non enzymatic-enzymatic method for the synthesis of simple alkyl fatty acid esters from soapstock. , Journal of the American Oil Chemists' Society, Vol. 73, pp. 1393-1401.

Han, X., Chen, L., Peng, Q. (2008). Preparation of biodiesel from sunflower oil under microwave irradiation by ionic liquids H2SO4, Journal of Zhengzhou University (Engineering Science), Vol.4.

Hanh, H.D., Dong, N.T., Okitsu, K., Maeda, Y. & Nishimura, R. (2007). Effects of molar ratio, catalyst concentration and temperature on transesterification of triolein with ethanol under ultrasonic irradiation. Journal of the Japan Petroleum Institute, Vol. 50, pp. 195–199.

Hanh, H.D., Dong, N.T., Starvarache, C., Okitsu, K., Maeda, Y. & Nishimura, R. (2008). Methanolysis of triolein by low frequency ultrasonic irradiation. Energy Conversion and Management, Vol. 49, pp. 276-280.

Hanh, H.D., Dong, N.T., Okitsu, K., Nishimura, R. & Maeda, Y. (2009a). Biodiesel production by esterification of oleic acid with short-chain alcohols under ultrasonic irradiation condition. Renewable Energy, Vol. 34, pp. 780-783.

Hanh, H.D., Dong, N.T., Okitsu, K., Nishimura, R. & Maeda, Y. (2009b). Biodiesel production through transesterification of triolein with various alcohols in anultrasonic field. Renewable Energy, Vol. 34, pp. 766-768.

Hawash, S., Kamal, N., Zaher, F., Kenawi, O. & Diwani, G. (2009). Biodiesel fuel from Jatropha oil via non-catalytic supercritical methanol transesterification. Fuel, Vol.88, pp. 579-582.

He, C., Baoxiang, P., Dezheng, W. &Jinfu, W. (2007). Biodiesel production by the transesterification of cottonseed oil by solid acid catalysts. Frontiers of Chemical Engineering in China, Vol. 1, No. 1, pp. 11-15.

Helwani, Z., Othman, M.R., Aziz, N., Kim, J. & Fernando, W.J.N. (2009). Solid heterogeneous catalysts for transesterification of triglycerides with methanol: Areview. Applied Catalysis A: General, Vol. 363, pp. 1–10.

Hernando, J., Leton, P., Matia, M.P., Novella, J.L. & Alvarez-Builla, J. (2007). Biodiesel and FAME synthesis assisted by microwaves. Homogeneous batch and flow processes. Fuel, Vol. 86, No. 10–11, pp.1641–1644.

Hingu, S.M., Gogate, P.R. & Rathod, V.K. (2010). Synthesis of biodiesel from waste cooking oil using sonochemical reactors. Ultrasonics Sonochemistry, Vol. 17, pp. 827-832.

Hoda, N. (2010). Optimization of biodiesel production from cottonseed oil by transesterification using NaOH and methanol. Energy Sources, Part A: Recovery, Utilization, and Environmental Effects, Vol.32, pp. 434- 441.

Hsiao, M.C., Lin, C.C., Chang, Y.H. & Chen, L.C. (2010). Ultrasonic mixing and closed microwave irradiation-assisted transesterification of soybean oil. Fuel, Vol. 89, No: 12, pp. 3618-3622.

Hsiao, M.C., Lin, C.C., & Chang, Y.H. (2011). Microwave irradiation-assisted transesterification of soybean oil to biodiesel catalyzed by nanopowder calcium oxide. Fuel, Article in press.

Ilham, Z. & Saka, S. (2010). Two-step supercritical dimethyl carbonate method for biodiesel production from Jatropha curcas oil. Bioresource Technology, Vol. 101 pp. 2735–2740.

Issariyakul, T., Kulkarni, M.G., Meher, L.C., Dalai, A.K. & Bakhshi, N.N. (2008). Biodiesel production from mixtures of canola oil and used cooking oil. Chemical Engineering Journal, Vol. 140, pp. 77–85.

İşcan, E. (2006). Biodiesel production in supercritical conditions. MSc Thesis, Yidiz Technical University, İstanbul.

Saka, S. & Kusdiana, D. (2001). Biodiesel fuel from rapeseed oil as prepared in supercritical methanol. Fuel, Vol. 80, pp. 225-231. Ji, J., Wang, J., Li, Y., Yu, Y. & Xu, Z. (2006). Preparation of biodiesel with the help of ultrasonic and hydrodynamic cavitation. Ultrasonics, Vol. 44, pp. 411-414.

Jin, L., Zhang, Y., Dombrowski, J.P., Chen, C., Pravatas, A., Xu, L., Perkins, C. & Suib, S.L.(2011). ZnO / La2O2 CO3 layered composite: A new heterogeneous catalyst for the efficient ultra-fast microwave biofuel production. Applied Catalysis B: Environmental, Article in press.

Kaieda, M., Samukawa, T., Matsumoto, T., Ban, K., Kondo, A., Shimada, Y., Noda, H., Nomoto, F., Ohtsuka, K., Izumoto, E. & Fukuda H. (1999). Biodiesel fuel production from plant oil catalyzed by Rhizopus oryzae lipase in a water-containing system without an organic solvent. Journal of Bioscience and Bioengineering, Vol. 88, No. 6,pp. 627-631.

Kasim, N.S., Tsai T-H., Gunawan S. & Ju Y-H. (2009). Biodiesel production from rice bran oiland supercritical methanol. Bioresource Technology, Vol. 100, pp. 2399–2403.

Kaya, C., Hamamci, C., Baysal, A., Akba, O., Erdogan, S. & Saydut, A. (2009). Methyl ester of peanut (Arachis hypogea L.) seed oil as a potential feedstock for biodiesel production. Renewable Energy, Vol. 34, pp. 1257–1260.

Kim, D., Choi, J., Kim, G., Seol, S.K., Ha, Y.C., Vijayan, M., Jung, S., Kim, B.H., Lee, G.D. &Park, S.S. (2011). Microwave-accelerated energy-efficient esterification of free fattyacid with a heterogeneous catalyst. Bioresource Technology, Vol. 102, pp. 3639-3641.

Kong, J., Han, X., Chen, L. & Huo, J. (2009). Preparation of Biodiesel under Microwave Irradiation from Sunflower Oil by Solid Super Acid TiO2/SO4. Guangzhou Chemical Industry, Vol.2.

Koopmans, C., Iannelli, M., Kerep, P., Klink, M., Schmitz, S., Sinnwell, S. & Ritter, H. (2006).Microwave-assisted polymer chemistry: Heckreaction, transesterification, Baeyer–Villiger oxidation, oxazoline polymerization, acrylamides, and porous materials. Tetrahedron, Vol. 62, No. 19, pp. 4709–4714.

Köse, Ö., Tüter M. & Aksoy, H.A. (2002). Immobilized Candida antarctica lipase-catalyzed alcoholysis of cotton seed oil in a solvent-free medium. Bioresource Technology, Vol.83, No. 2, pp. 125-129.

Kumar, D., Kumar, G. & Singh, P.C.P. (2010a). Fast, easy ethanolysis of coconut oil for biodiesel production assisted by ultrasonication. Ultrasonics Sonochemistry, Vol. 17, pp. 555-559.

Kumar, D., Kumar, G. & Singh, P.C.P. (2010b). Ultrasonic-assisted transesterification of Jatropha curcus oil using solid catalyst, Na/SiO2. Ultrasonics Sonochemistry, Vol. 17, pp. 839-844.

Kusdiana, D. & Saka, S. (2001). Kinetics of transesterification in rapeseed oil to biodiesel fuelas treated in supercritical methanol. Fuel, Vol. 80, pp. 693-698.

Kusdiana, D. & Saka, S. (2004). Effects of water on biodiesel fuel production by supercritical methanol treatment. Bioresource Technology, Vol. 91, pp. 289–295.

Leadbeater, N.E. & Stencel, L.M. (2006). Fast, easy preparation of biodiesel using microwave heating. Energy & Fuels, Vol. 20, No. 5, pp. 2281-2283.

Lee, J.H., Kwon, C.H., Kang, J. W., Park, C., Tae, B. & Kim, S.W. (2009). Biodiesel production from various oils under supercritical fluid conditions by Candida antarctica lipaseB using a stepwise reaction method. Applied Biochemistry and Biotechnology, Vol. 156, pp. 454–464.

Lee, J.S. & Saka, S. (2010). Biodiesel production by heterogeneous catalysts and supercritical technologies. Bioresource Technology, Vol. 101, pp. 7191–7200.

Lee, S.B., Lee, J.D. & Hong, I.K. (2011). Ultrasonic energy effect on vegetable oil based biodiesel synthetic process. Journal of Industrial and Engineering Chemistry, Vol. 510, pp. 1-6.

Lertsathapornsuk, V., Pairintra, R., Aryusuk, K. & Krisnangkur, K. (2008). Microwave assisted in continuous biodiesel production from waste frying palm oil and its performance in a 100 kW diesel generator. Fuel Processing Technology, Vol. 89, pp.1330-1336.

Leung, D.Y.C. & Guo, Y. (2006). Transesterification of neat and used frying oil: optimization for biodiesel production. Fuel Process Technology, Vol. 87, pp. 883–90.

Licht, F.O. (2009). World-Biodiesel Production, World Ethanol and Biofuels Report, (4), 16:365.

Lifka, J. &Ondruschka, B. (2004). Influence of mass transfer on the production of biodiesel, Chemical Engineering & Technology, Vol. 27, No. 11, pp. 1156–1159.

Lin, G. & Liu, H. (1995). Ultrasound promoted lipase catalyzed reactions. Tetrahedron Letters, Vol. 36, No. 34, pp. 6067-6068.

Lotero, E., Liu, Y., Lopez, D.E., Suwannakarn, K., Bruce, D.A. & Goodwin J.G. (2005).Synthesis of Biodiesel via Acid Catalysis. Industrial & Engineering Chemistry Research, Vol. 44, pp. 5353-5363.

Madras, G., Kolluru, C. & Kumar R. (2004). Synthesis of biodiesel in supercritical fluids. Fuel, Vol. 83, pp. 2029–2033.

Mahamuni, N.N. & Adewuyi, Y.G. (2009). Optimization of the synthesis of biodiesel viaultrasound-enhanced base-catalyzed transesterification of soybean oil using a multifrequency ultrasonic reactor. Energy & Fuels. Vol. 23, pp. 2757-2766.

Majewski, M.W., Pollack, S.A. & Curtis-Palmer, V.A. (2009). Diphenylammonium salt catalysts for microwave assisted triglyceride transesterification of corn and soybean oil for biodiesel production. Tetrahedron Letters, Vol. 50, pp. 5175-5177.

Marchetti, J.M., Miguel, V.U. & Errazu, A.F. (2007). Possible methods for biodiesel production. Renewable and Sustainable Energy Reviews, Vol. 11, pp. 1300–1311.

Marulanda, V.F., Anitescu, G. &Tavlarides, L.L. (2010). Biodiesel fuels through a continuous flow process of chicken fat supercritical transesterification, Energy & Fuels, Vol. 24, No. 1, pp. 253-260.

Mason, T.J., Paniwnyk, L. & Lorimer, J.P. (1996). The uses of ultrasound in food Technology. Ultrasonics Sonochemistry, Vol. 3, pp. 253-260.

Mason, T.J. (1999). Sonochemistry, Oxford University Press, NewYork, pp. 2–30.Mason, T.J. (2000). Large scale sonochemical processing: aspiration and actuality. Ultrasonics Sonochemistry, Vol. 7, No: 4, pp. 145-149.

Manh, D.V., Chen, Y.H., Chang, C.C., Chang, M.C. & Chang, C.Y. (2011). Biodiesel production from Tung oil and blended oil via ultrasonic transesterification process. Journal of the Taiwan Institute of Chemical Engineers, Vol. 297, pp. 1876-1070.

Melero, J.A., Bautista, L., F., Morales, G., Iglesias, J. & Sánchez-Vázquez, R. (2009). Biodiesel production from crude palm oil using sulfonic acid-modified mesostructured catalysts. Chemical Engineering Journal, Article in press.

Mikkola, J.P. & Salmi, T. (2001). Three-phase catalytic hydrogenation of xylose to xylitolprolonging the catalyst activity by means of on-line ultrasonic treatment. Catalysis Today, Vol. 64, pp. 271–277.

Minami, E. & Saka S. (2006). Kinetics of hydrolysis and methyl esterification for biodiesel production in two-step supercritical methanol process. Fuel, Vol. 85, pp. 2479-2483.

Mootabadi, H., Salamatinia, B., Bhatia, S. & Abdullah, A.Z. (2010). Ultrasonic-assisted biodiesel production process from palm oil using alkaline earth metal oxides as the heterogeneous catalysts. Fuel, Vol.89, pp. 1818-1825.

Newman, P.G. & Rozycki, G.S. (1998). The history of ultrasound. Surgical Clinics of North America, Vol. 78, No: 2, pp. 179-195.

Nogueira, B.M., Carretoni, C., Cruz, R., Freitas, S., Melo, P.A., Costa-Felix, R., Pinto, J.C. & Nele, M. (2010). Microwave activation of enzymatic catalysts for biodiesel production. Journal of Molecular Catalysis B: Enzymatic, Vol. 67, pp. 117-121.

Nuechter, M., Ondruschka, B., Jungnickel, A. & Mueller, U. (2000). Organic processes initiated by non-classical energy sources. Journal of Physical Organic Chemistry, Vol.13, pp. 579–586.

Öner, C. & Altun, Ş. (2009). Biodiesel production from inedible animal tallow and anexperimental investigation of its use as alternative fuel in a direct injection dieselengine. Applied Energy, Vol. 86, pp. 2114–2120.

Özçimen, D. & Yücel, S. (2010). Biodiesel production from algae. Proceedings of 101st AOCS Annual Meeting & Expo., Phoneix-USA, May, 2010.

Özgül, S. & Türkay, S. (1993). In-situ esterification of rice bran oil with methanol and ethanol. , Journal of the American Oil Chemists' Society, Vol. 70, pp. 145-147.

Özgül-Yücel, S. & Türkay, S. (2002). Variables affecting the yields of methyl esters derived from in- situ esterification of rice bran oil. , Journal of the American Oil Chemists' Society, Vol. 79, pp. 611-614.

Öztürk, G., Kafadar, A.B., Duz, M.Z., Saydut, A. & Hamamc, C. (2010). Microwave assisted transesterification of maize (Zea mays L.) oil as a biodiesel fuel. Energy, Exploration& Exploitation, Vol. 28, No. 1, pp. 47-58, ISSN 0144-5987

Patil, P.D., Gude, V.G., Mannarswamy, A. & Deng, S. (2009). Biodiesel Production from Jatropha Curcas, Waste Cooking, and Camelina Sativa Oils, Industrial & Engineering Chemistry Research, Vol. 48, No:24, pp. 10850–10856.

Patil, P., Deng, S., Rhodes, J. I. & Lammers, P.J. (2010a). Conversion of waste cooking oil to biodiesel using ferric sulfate and supercritical methanol processes. Fuel, Vol. 89, pp.360–364.

Patil, P.D., Gude, V.G., Mannarswamy, A., Deng, S., Cooke, P., Munson-McGee S., RhodesI., Lammers, P. &Nirmalakhandan, N. (2010b). Optimization of direct conversion of wet algae to biodiesel under supercritical methanol conditions. Bioresource Technology, Vol. 102, pp. 188-122.

Patil, P.D., Gude, V.G., Mannarswamy, A., Cooke, P., Munson-McGee, S., Nirmalakhandan, N., Lammers, P. & Deng, S. (2011). Optimization of microwave-assisted transesterification of dry algal biomass using response surface methodology. Bioresource Technology, Vol. 102, pp. 1399-1405.

Perin, G., Alvaro, G., Westphal, E., Viana, L.H., Jacob, R.G., Lenardao, E.J. & D'Oca, M.G.M.(2008). Transesterification of castor oil assisted by microwave irradiation. Fuel, Vol.87, pp. 2838-2841.

Rahmanlar, İ. (2010). Alcoholysis of frying oils with microwave method. MSc. Thesis, Yildiz Technical University, Istanbul.

Rashid, U., Anwar, F. & Knothe, G. (2009). Evaluation of biodiesel obtained from cottonseed oil. Fuel Processing Technology, Vol. 90, pp. 1157–1163.

Rathore, V. & Madras, G. (2007). Synthesis of biodiesel from edible and non-edible oils in supercritical alcohols and enzymatic synthesis in supercritical carbon dioxide. Fuel, Vol. 86, pp. 2650-2659.

Refaat, A.A., El Sheltawy, S.T. & Sadek, K.U. (2008). Optimum reaction time, performance and exhaust emissions of biodiesel produced by microwave irradiation. International Journal of Enviornmental Science and Technology, Vol. 5, No. 3, pp. 315-322.

Royon, D., Daz, M., Ellenrieder, G. & Locatelli, S. (2007). Enzymatic production of biodiesel from cotton seed oil using t-butanol as a solvent. Bioresource Technology, Vol. 98, No.3, pp. 648-653.

Saifuddin, N. & Chua, K.H. (2004). Production of ethyl ester (biodiesel) from used frying oil: Optimization of transesterification process using microwave irradiation. Malaysian Journal of Chemistry, Vol. 6, No. 1, pp. 77-82.

Saka, S. & Isayama, Y. (2009). A new process for catalyst-free production of biodiesel using supercritical methyl acetate. Fuel, Vol. 88, pp. 1307–1313.

Saka, S., Isayama, Y., Ilham, Z. &Jiayu, X. (2010). New process for catalyst-free biodiesel production using subcritical acetic acid and supercritical methanol. Fuel, Vol. 89, pp. 1442-1446.

Saka, S. &Kusdiana, D. (2002). Biodiesel fuel for diesel fuel substitute as prepared by acatalyst-free supercritical methanol. Nihon Enerugi Gakkai Taikai Koen Yoshishu, pp.186-187.

Salamatinia, B., Mootabadi, H., Bhatia, S. & Abdullah, A.Z. (2010). Optimization ofultrasonic-assisted heterogeneous biodiesel production from palm oil: A response surface methodology approach. Fuel Processing Technology, Vol. 91, pp. 441-448.

Samukawa, T., Kaieda, M., Matsumoto, T., Ban, K., Kondo, A., Shimada, Y., Noda, H., &Fukuda H. (2000). Pretreatment of immobilized Candida antarctica lipase for biodiesel fuel production from plant oil. Journal of Bioscience and Bioengineering, Vol.90, No. 2, pp. 180-183.

Shakinaz, A.E.S., Refaat A.A. & Shakinaz, T.E.S. (2010). Production of biodiesel using the microwave technique. Journal of Advanced Research, Vol. 1, No. 4, pp. 309-314.

Sharma, Y.C., Singh, B. & Upadhyay, S.N. (2008). Advancements in development and characterization of biodiesel: A review. Fuel, Vol. 87, pp. 2355–2373.

Sharma, Y.C. & Singh, B. (2009). Development of biodiesel: Current scenario. Renewable and Sustainable Energy Reviews, Vol. 13, pp. 1646–1651.

Silva, C.C.C.M., Ribeiro, N.F.P., Souza, M.M.V.M. & Aranda, D.A.G. (2010). Biodiesel production from soybean oil and methanol using hydrotalcites as catalyst. Fuel Processing Technology Vol. 91, pp. 205–210.

Singh, A.B., He, C., Thompson, J. & Van Gerpen, J., (2006). Process optimization of biodiesel production using different alkaline catalysts. Applied Engineering in Agriculture, Vol.22, No.4, pp. 597-600.

Sinisterra, J.V. (1992). Application of ultrasound to biotechnology: an overview. Ultrasonics, Vol. 30, No. 3, pp. 180–185.

Song, E.S., Lim, J.W., Lee, H.S. & Lee, Y.W. (2008). Transesterification of RBD palm oil using supercritical methanol. Journal of Supercritical Fluids, Vol. 44, pp. 356–363.

Starvarache, C., Vinatoru, M., Nishimura, R. & Maeda, Y. (2005). Fatty acids methyl esters from vegetable oil by means of ultrasonic energy. Ultrasonics Sonochemistry, Vol. 12, pp. 367-372.

Stavarache, C., Vinatoru, M. & Maeda Y. (2006). Ultrasonic versus silent methylation of vegetable oils. Ultrasonics Sonochemistry, Vol. 13, No. 5, pp. 401–407.

Stavarache, C., Vinatoru, M. & Maeda, Y. (2007a). Aspects of ultrasonically assisted transesterification of various vegetable oils with methanol. Ultrasonics Sonochemistry, Vol. 14, pp. 380-386.

Stavarache, C., Vinatoru, M., Maeda, Y. & Bandow, H. (2007b). Ultrasonically driven continuous process for vegetable oil transesterification. Ultrasonics Sonochemistry, Vol. 14, pp. 413-417.

Suppalakpanya, K., Ratanawilai, S.B. &Tongurai C. (2010). Production of ethyl ester from esterified crude palm oil by microwave with dry washing by bleaching earth Applied Energy, Article in press.

Tan, K.T., Lee, K.T. & Mohamed, A.R. (2009). Production of FAME by palm oil transesterification via supercritical methanol technology. Biomass & Bioenergy, Vol.33, pp. 1096-1099.

Tan, K.T., Lee, K.T. & Mohamed, A.R. (2010a). Potential of waste palm cooking oil for catalyst-free biodiesel production. Energy, Article in press.

Tan, K.T., Lee, K.T. & Mohamed, A.R. (2010b). A glycerol-free process to produce biodiesel by supercritical methyl acetate technology: An optimization study via Response Surface Methodology. Bioresource Technology, Vol. 101, pp. 965–969.

Tan, K.T., Lee, K.T. & Mohamed, A.R. (2010c). An optimized study of methanol and ethanol in supercritical alcohol technology for biodiesel production. Journal of Supercritical Fluids, Vol. 53, pp. 82-87.

Tashtoush, G.M., Al-Widyan, M.I. & Al-Jarrah, M.M. (2004). Experimental study one valuation and optimization of conversion of waste animal fat into biodiesel. Energy Conversion and Management, Vol.45, pp. 2697–2711.

Taylor, M., Atri, B.S. & Minhas, S. (2005). Developments in microwave chemistry. Evalueserve.

Teixeira, L.S.G., Assis, J.C.R., Mendonça, D.R., Santos, I.T.V., Guimarães, P.R.B., Pontes, L.A.M. & Teixeira, J.S.R. (2009). Comparison between conventional and ultrasonic preparation of beef tallow biodiesel. Fuel Processing Technology, Vol. 90, pp. 1164–1166.

Terigar, B.G., Balasubramanian, S., Lima, M. & Boldor, D. (2010). Transesterification of soybean and rice bran oil with ethanol in a continuous-flow microwave-assisted system: yields, quality, and reaction kinetics. Energy Fuels, Vol. 24, No:12, pp. 6609–6615.

Thanh, L. T., Okitsu, K., Sadanaga, Y., Takenaka, N., Maeda, Y. & Bandow, H. (2010a).Ultrasound-assisted production of biodiesel fuel from vegetable oils in a small scale circulation process. Bioresource Technology, Vol. 101, pp. 639-645.

Thanh, L.T., Okitsu, K., Sadanaga, Y., Takenaka, N., Maeda, Y. & Bandow, H. (2010b). Atwo-step continuous ultrasound assisted production of biodiesel fuel from waste cooking oils: A practical and economical approach to produce high quality biodiesel fuel. Bioresource Technology, Vol. 101, pp. 5394-5401.

Valle, P., Velez A., Hegel P., Mabe G. &Brignole E.A. (2010). Biodiesel production using supercritical alcohols with a non-edible vegetable oil in a batch reactor. Journal of Supercritical Fluids, Vol. 54, pp. 61-70.

Varma, R. S. (2001). Solvent-free accelerated organic syntheses using microwaves. Pure and Applied Chemistry, Vol. 73, No. 1, pp. 193-198.

Varma, M. N. & Madras, G. (2007). Synthesis of biodiesel from castor oil and linseed oil in supercritical fluids. Industrial & Engineering Chemistry Research, Vol. 46, pp. 1-6.

Varma, M.N., Deshpande A. & Madras G. (2010). Synthesis of biodiesel in supercritical alcohols and supercritical carbon dioxide. Fuel, Vol. 89, pp. 1641-1646.

Veljkovic, V.B., Lakicevic, S.H., Stamenkovic, O.S., Todorovic, Z.B. & Lazic, M.L. (2006).Biodiesel production from tobacco (Nicotiana tabacum L.) seed oil with a high content of free fatty acids. Fuel, Vol. 85, pp. 2671–2675.

Venkatesh, H., Requpathi, I. &Saidutta, M.B. (2011). Optimization of two step karanja biodiesel synthesis under microwave irradiation. Fuel Processing Technology, Vol. 92, pp. 100-105.

Vicente, G., Martinez, M. & Aracil, J. (2004). Integrated biodiesel production: a comparison of different homogeneous catalysts systems. Bioresource Technology, Vol. 92, No. 3, pp. 297-305.

Vyas, A.P., Verma, J.L. & Subrahmanyam, N. (2010). A review on FAME production processes. Fuel, Vol. 89, pp. 1–9.

Wen, D., Jiang, H. & Zhang, K. (2009). Supercritical fuids technology for clean befoul production. Progress in Natural Science, Vol. 19, pp. 273–284.

Wu, P., Yang, Y., Colucci, J.A. & Grulke, E.A. (2007). Effect of ultrasonication on droplet sizein biodiesel mixtures. Journal of American Oil Chemists' Society, Vol. 84, pp. 877-884.

Yaakob, Z., Sukarmani, I.S., Kamarudini, S.K., Abdullah, S.R.S. & Mohamed, F. (2008).Production of biodiesel from jatropha curcas by microwave irradiation. 2ndWSEAS/IASME International Conference on renewable energy sources (RES'08), Corfu, Greece, October 26-28, 2008.

Ye, X., Fernando, S., Wilson, W. & Singh, A. (2007). Application of amphiphilic catalysts, ultrasonication, and nanoemulsions for biodiesel production process. Chemical Engineering & Technology, Vol. 30, No. 11, pp. 1481-1487.

Yin, J.Z., Xiao, M. & Song, J.B. (2008). Biodiesel from soybean oil in supercritical methanol with co-solvent. Energy Conversion Management, Vol. 49, pp. 908-912.

Yoni, G. & Aharon, G. (2008). Continuous ° ow, circulating microwave system and its application in nanoparticle fabrication and biodiesel synthesis. The Journal of Physical Chemistry C, Vol. 112, pp. 8802–8808.

Yoo, S.J., Lee, H-S., Veriansyah, B., Kim, J., Kim, J.D., & Lee Y.W. (2010). Synthesis of biodiesel from rapeseed oil using supercritical methanol with metal oxide catalysts. Bioresource Technology, Vol. 101, pp. 8686-8689.

Yu, D., Tian, L., Wu, H., Wang, S., Wang, Y., Ma, D. & Fang, X. (2010). Ultrasonic irradiation with vibration for biodiesel production from soybean oil by Novozym 435. Process Biochemistry, Vol. 45, pp. 519-525.

Yuan, H., Yang, B.L. & Zhu, G.L. (2009). Synthesis of biodiesel using microwave absorption catalysts. Energy & Fuels, Vol. 23, pp. 548–552.

Yücel, S., Rahmanlar, İ. & Özçimen, D. (2010). The production of methyl esters from waste frying oil with microwave method. Proceedings of Second International Symposium on Green Chemistry for Environment and Health, Mykonos, September, 2010.

Zhang, S., Zu, Y.G., Fu, Y.J., Luo, M., Zhang, D.Y. & Efferth, T. (2010). Rapid microwave assisted transesterification of yellow horn oil to biodiesel using a heteropolyacid solid catalyst. Bioresource Technology, Vol. 101, pp. 931-936

Zhang, Y., Dube, M.A., McLean, D.D. &Kates, M. (2003). Biodiesel production from waste cooking oil: 1. Process design and technological assessment. Bioresource Technology, Vol. 89, pp. 1–16.

Zhou, W. & Boocock, D.G.V. (2003). Ethyl esters from the single-phase base-catalyzed ethanolysis of vegetable oils, Journal of the American Oil Chemists' Society, Vol. 80,No.4, pp. 367-371.http://www.ebb-eu.org/EBBpressreleases/EBB, European Biodiesel Board, July 2010

CHAPTER-17

CATALYTIC HYDROTREATMENT EMPLOYED FOR OF STABILIZING PYROLYSIS OIL

1. INTRODUCTION: AN OVERVIEW

As the sole carbon-containing product that is sustainable, biomass is the exclusive alternative to crude oil derivatives formed from fossils. The research on utilizing biomass for first generation biofuels is experiencing tremendous growth, such as the production of bio-ethanol from sugar sources and starches, as well as bio-diesel from pure plant oils. Converting biomass, especially ligno-cellulosic material, into transportation fuels is a challenging process. The research is mostly centered on alternative methods such as: i) separating biomass into different components and fermenting the cellulosic and hemi-cellulosic fractions to make ethanol, and ii) converting the entire biomass into syngas through disruptive gasification, which can then be further processed into methanol or Fischer-Tropsch diesel. Traditional refinery systems, which can handle up to 100 tons per hour of crude oil, are more cost-effective. However, these systems are not suitable for processing biomass resources due to the scattered nature of biomass and the high cost of collecting. Furthermore, different types of biomass exhibit distinct variations in their structure and composition. They possess a very low energy density when compared to several fossil fuels and frequently contain substantial quantities of water and ash. To address these drawbacks, it is necessary to initially reconstruct the biomass in a decentralized manner, densify it at a rate of 2 to 10 tons per hour, and ideally disinfect it at the same time. The intermediate product is thereafter conveyed to a sizable central processing unit, where it undergoes a conversion process to yield the final product, often at a rate ranging from 50 to 200 metric tons per hour. An appealing technology that could be used for this purpose is fast pyrolysis, as discussed in a recent review by Venderbosch and Prins (2010). Pyrolysis liquids have minimal ash content and possess a volumetric energy density that is 5 to 20 times greater than that of the initial biomass. Nevertheless, the oil possesses an acidic character, is polar, and does not mix with normal crude oil. Furthermore, the oil exhibits instability due to the (re)polymerization of organic materials, resulting in a gradual increase in viscosity over time. From a historical standpoint, it is important to recognize that pyrolysis oil should not be considered as a traditional oil. Instead, it should be seen as a syrup that is abundant in carbohydrates. The validity of this assertion is supported by a solvent fractionation technique, which allows for the efficient extraction of the majority of the oil, up to 70%, using water. An illustrative instance is depicted in Figure 1 (with an analysis provided in Table 1), where pyrolysis oil is categorized into several groups of chemicals, characterized by variations in oxygen functionality and molecular size (Oasmaa et al., 2003; Oasmaa, 2003). Curiously, the sizable portion of the substance that cannot dissolve in ether contains substantial quantities of sugar-like compounds and exhibits the visual characteristics of a liquid resembling syrup. The unstable nature of the pyrolysis oil is mostly attributed to its elevated oxygen content. However, it is evident that the instability of pyrolysis oil is not caused by the oxygen itself, but rather by the characteristics and reactivity of the oxygen-containing functional chemical groups. The carbonyl compounds, particularly aldehydes and ketones, appear to be the main cause of thermal instability. Transforming them into less reactive organic groups, such as the equivalent alcohols, seems to be a feasible solution.

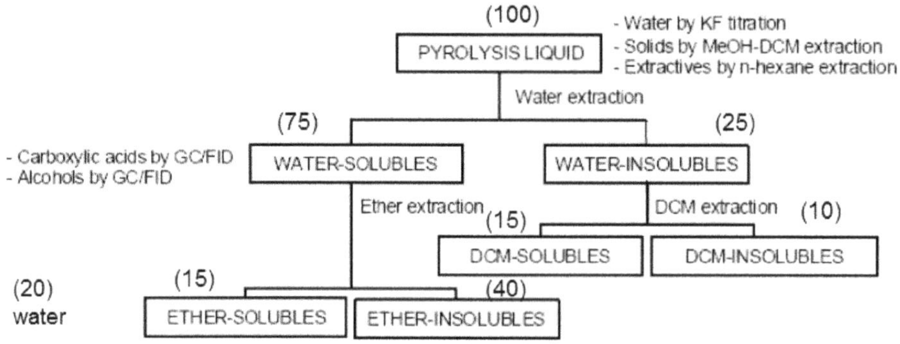

Figure 1. Fractionation scheme (Oasmaa, 2003)

Table 1.Chemical composition of reference pine oil and its fractions (Oasmaa, 2003)

	wt % (wet basis)	COMPOUND TYPES	C	H	N	O
			wt % (dry basis)			
WATER-SOLUBLES Acids, alcoholds	75-85 5-10	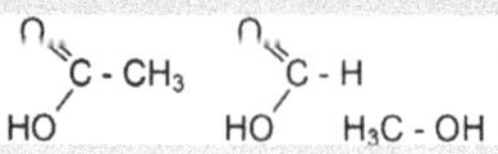	36.0	6.0	0	58.0
Ether-solubles	5-15	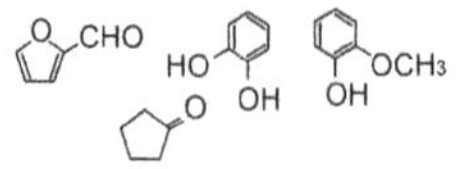	60.0	6.0	0.1	33.9
Ether-insolubles	30-40	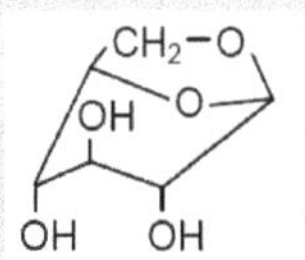	46.0	6.3	0.3	47.4
Water	20-30		0	11.1	0	88.9
WATER-INSOLUBLES	15-25		66.2	6.6	0.3	26.9
n-Hexane-solubles	2.6	$CH_3-(CH_2)_n-CO_2H$ n = 10-30 $HOCH_2-(CH_2)_n-CO_2H$ n = 10-28 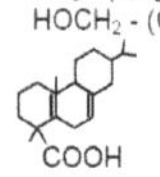	77.4	10.4	0	12.2
DCM-solubles	5-10	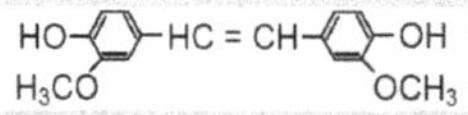	68.1	6.7	0.4	24.7
DCM-insolubles	2-10	degraded lignin	64.1	5.9	1.5	28.4

A number of catalytic approaches have been proposed to upgrade and improve the product properties of fast pyrolysis oil. A well known example is catalytic cracking of pure biomass and/or pyrolysis oil to oxygen-free products. However, this approach is accompanied by a significant amount of coke production (up to 40 wt.% of the biomass) (Horne &Williams, 1996; Vispute et al., 2010) and this issue needs to be resolved.

A number of studies have been performed with the objective to remove the bound oxygen in the form of CO and/or CO_2 by decarbonylation and decarboxylation reactions, either thermally or catalytically.

The thermal process is known as the HPTT process, a high pressure temperature treatment (De Miguel Mercader et al., 2010). However, oxygen removal beyond a level of 10% appears very difficult with this approach, even when using catalysts, and the product is highly viscous, limiting its application potential.

Catalytic hydroprocessing or hydrotreating of fast pyrolysis oil is a more promising option (Conti, 1997; Elliott, 2007; Elliott et al.; 2009; Kaiser 1997; Rep et al., 2006). In this process, the fast-pyrolysis oil is treated with hydrogen in the presence of a heterogeneous catalyst with the aim to hydro(deoxy)genate the pyrolysis oil to a product with improved product properties.

A potentially attractive outlet for the upgraded oils is its use as co-feed in existing oil refinery units. This enables partial substitution of the fossil carbon in liquid transportation fuels by renewable carbon from biomass in existing infrastructure as proposed in the European Biocoup project. As such, it is expected to lower the investment costs and to facilitate other barriers for introduction of the fast pyrolysis technology. Crude pyrolysis oil, however, cannot be used for co-feeding purposes. It is immiscible with typical petroleum feeds (such as vacuum gas oil) due to the presence of polar components, but moreover, it is highly acidic due to the presence of organic acids (up to 10 wt%) (Oasmaa, 2003) leading to corrosion issues and possibly detrimental effects on typical catalysts in the refinery units (e.g. zeolites in the FCC process) (Dimitrijevic et al., 2006). Furthermore, it has a strong tendency for coking at elevated temperatures.

An alternative product outlet for upgraded pyrolysis oils is the direct use as a green transportation fuel in internal combustion engines. This will require deep hydrotreatment of the pyrolysis oil to an upgraded oil with very low oxygen levels (preferably below 1%) in order to mimic the product properties of conventional hydrocarbon fuels. Hydrogen consumption is expected to be high for this case, and will have a negative impact on the process economics.

2. INVESTIGATION OF THE HYDROTREATMENT OF FAST PYROLYSIS OIL USING CATALYSTS

Well known catalysts for the hydrotreatment of pyrolysis oil are conventional hydrodesulfurisation catalysts such as sulfided NiMo/Al2O3, CoMo/Al2O3, and NiMo/Al2O3-SiO2 (Ferrari et al., 2002; Maity et al., 2000) to cope with the harsh reaction conditions (T > 200 C, aqueous environment with organic acids). Carbon-supported CoMo, or Mo supported on TiO2, ZrO2 and TiO2–ZrO2 mixed oxides (Satterfield&Yang, 1983; Lee & Ollis, 1984) have been tested as well. The latter catalysts all require sulphur in the feed to maintain activity. Pyrolysis oils contain only limited amounts of sulphur (Furimsky, 2000) and sulphur addition would be required during processing to maintain catalytic activity. This will lead to sulfur emissions e.g. in the flue gas, which is not preferred for an environmental point of view.

Non-sulfided, non noble metal catalysts have also been tested for the catalytic hydrotreatment of pyrolysis oil, but to a far lesser extent. The hydrodeoxygenation of phenol over a Ni/SiO2 catalyst has been reported and showed low HDO activity. Horne&Williams (1996) tested ZSM-5 zeolites as the catalyst for the deoxygenation of model compounds such as anisole. Their results, though, showed that anisole is mainly converted into phenol and methyl substituted phenols, and not to the oxygen-depleted compounds. Xu, et al. (2010) tested MoNi/γ-Al2O3 (reduced prior to reaction) for the mild hydrotreatment of pyrolysis oil (3 MPa and 200 °C).

Noble metal based catalysts have also been evaluated as substitutes for sulfided catalysts. Examples are Pd on zeolite carriers (Horne&Williams, 1996), Pd on mesoporous CeO2 and ZrO2 (Senol et al., 2005), and Rh on zirconia (Gutierrez et al., 2008). Carbon supports have also been explored (Wildschut et al., 2009b; Wildschut et al., 2010a;, Wildschut et al., 2010b; Venderbosch et al., 2010). We have recently reported exploratory catalyst studies on the upgrading of fast pyrolysis oil by catalytic hydrotreatment using a variety of heterogeneous noble metal catalysts (Ru/C, Ru/TiO2, Ru/Al2O3, Pt/C and Pd/C) and the results were compared to typical hydrotreatment catalysts (sulfided NiMo/Al2O3 and CoMoAl2O3). The reactions were carried out at temperatures in the range of 250 and 350 °C and pressures between 100 and 200 bar. The Ru/C catalyst appears superior to the classical hydrotreating catalysts with respect to oil yield (up to 60 %-wt.) and deoxygenation level (up to 90 %-wt.) (Figure 2) (Wildschut et al, 2010).

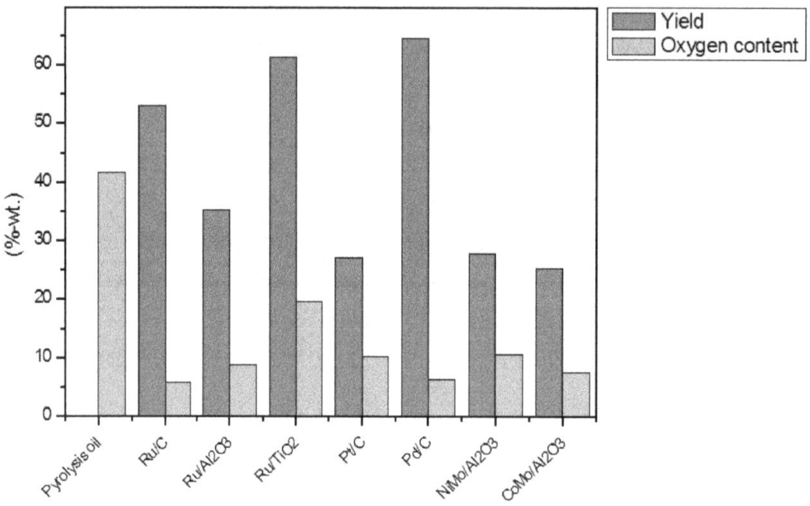

Figure 2.Catalyst screening studies for noble metal catalysts on various supports

Unfortunately, noble metal catalysts are very expensive and this limits their application potential. The catalytic hydrotreatment of fast pyrolysis oil using much cheaper bimetallic NiCu/δ-Al2O3 catalysts with various Ni/Cu ratios (0.32 – 8.1 w/w) at a fixed total metal intake of about 20 wt% was reported recently (Ardiyanti et al., 2009). Hydrotreatment reactions of fast pyrolysis oil (batch autoclave, 1 h at 150 °C followed by 3 h at 350 °C, at 200 bar total pressure) were performed and highest catalyst activity (in terms of H2 uptake per g of Ni) was observed for the catalyst with the lowest Ni loading (5.92Ni18.2Cu). Product oils with oxygen contents between 10 and 17 wt% were obtained and shown to have improved product properties when compared to the feed.

Recently, we have also shown the potential of homogeneous Ru catalysts for the catalytic upgrading of fast pyrolysis oil fractions (Mahfud et al., 2007a). Fractions were obtained by treatment of the pyrolysis oil with dichloromethane or water. The former approach leads to a dichloromethane layer enriched with the lignin fraction of pyrolysis oil. The dichloromethane fraction was hydrogenated in a biphasic system (dichloromethane/water) using a homogeneous water soluble Ru/tri-phenylphosphine-tris-sulphonate (Ru-TPPTS) catalyst. Analysis revealed that the oxygen content of was lowered considerably and that particularly the amounts of reactive aldehydes were reduced substantially.

The water-soluble fraction of pyrolysis oil, obtained by treatment of fast pyrolysis oil with an excess of water was hydrogenated in a biphasic system with a homogeneous Ru-catalyst (Ru/tri-phenylphosphine, Ru-TPP) dissolved in an apolar solvent (Mahfud et al., 2007b). Initial experiments were conducted with representative, water soluble model compounds like hydroxyacetaldehyde and acetol. The effects of process parameters (e.g. temperature, initial H2 pressure, and initial substrate concentration) were investigated and quantified for acetol using a kinetic model. The hydrogenation of the pyrolysis oil water-soluble fraction using the biphasic system at optimized conditions was performed and significant reductions in the amounts of reactive aldehydes such as hydroxyacetaldehyde and acetol were observed, demonstrating the potential of homogeneous Ru-catalysts to upgrade pyrolysis oils.

3.INVESTIGATIONSONTHECATALYTICHYDROTREATMENTOFFASTPYROLYSISOILS

3.1. Introduction

In the past, it was assumed that the catalytic hydrotreatment of pyrolysis oils shows strong resemblances with conventional hydrotreatment processes of fossil feeds (hydrodesulphurisation, hydrodenitrogenation) and typical catalysts, process conditions and reactor configuration were taken from these conventional processes (Elliott, 2007). For all, the objective is a reduction of certain elements (oxygen, sulphur, nitrogen) in the feed by the action of hydrogen and a catalyst. However, common reaction conditions for the hydroprocessing of crude oil derivatives cannot be adopted directly for pyrolysis oils. For instance, pyrolysis oil cannot be treated as such at temperatures exceeding 300 °C because of its high charring tendency.

Numerous papers on hydrotreating pyrolysis oil in packed beds and autoclaves were published in the eighties and nineties, and now very recently as well (Elliott, 2007). Specifically the older literature seems rather phenomenological in nature, and merely focuses on 'fact-finding'. Large differences in operating conditions like in pressure, temperature, residence times and hydrogen consumption are reported. In any case, without active catalyst or (high pressure) hydrogen, significant charring of pyrolysis oil occurs. Reduction of charring is possible by applying a (relatively) low temperature catalytic hydrotreatment at 175 to 250 °C, in which reactive components in the oil are 'stabilized'. Subsequently, the stable product is further processed at higher temperatures (> 300 °C) and pressures (> 150 bar). High pressures are believed to be essential to keep the water in the pyrolysis oil feed in a liquid state (and as such probably reduce the charring reactions), to promote the solubility of hydrogen in the initially polar bio-oil, and to increase the rate of the actual hydrogenation reactions.

3.2. Process studies using Ru/C

3.2.1. Batch studies

We have recently performed an in depth study in a batch set-up to determine the effects of process conditions on the catalytic hydrotreatment of fast pyrolysis oil using a Ru/C catalyst 350 °C and 200 bar pressure (Wildschut et al., 2010a). The liquid product after reaction consisted of three different phases, a slightly yellow aqueous phase and two brown oil phases, one with a density higher than water and one with a density lower than water. Furthermore, substantial amounts of solids (coke/char, about 5 wt% on pyrolysis oil intake) and gas phase organics (CO, CO2,CH4) were formed as well. The oil yield and elemental compositions of the product phases were shown to be a strong function of the reaction time. Highest oil yields (65 %-wt.) were obtained after 4 h using a 5 %-wt. intake of catalyst on fast pyrolysis oil (Figure 3). Longer reaction times lead to a reduction of the oil yield due to the formation of gas phase components (methane, ethane, propane, CO/CO2). Hydrogen uptake after 4 h reaction time was about 400 Nm3/t of pyrolysis oil (dry basis).

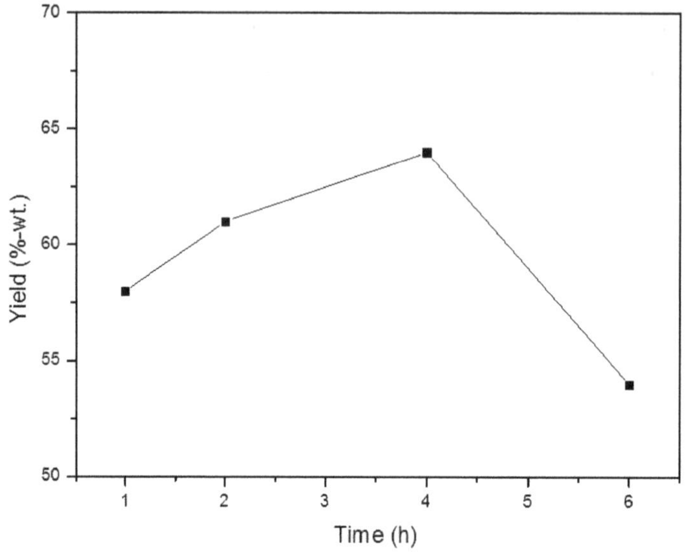

Figure 3.Total oil yield (dry basis) versus reaction time for the hydrotreatment of pyrolysis oil (350 °C and 200 bar, Ru/C).

A solvent-solvent extraction scheme based on the work of Oasmaa et al. (2003) was used to gain insight in the reactivity of various component classes (fractions) in the fast pyrolysis oil during the catalytic hydrotreatment process. The fractionation scheme (Figure 1) was applied to the original fast pyrolysis oil and product oils obtained at different reaction times (350 °C and 200 bar). The results are given in Figure 4.

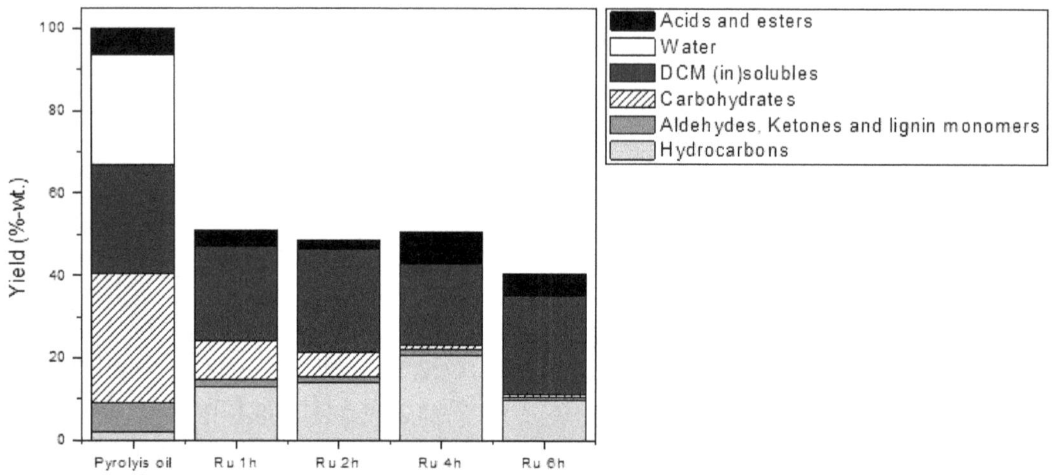

Figure 4.Composition of fast pyrolysis oil and hydrotreated product oils (Ru/C, 350 °C, 200 bar) at various reaction times using solvent-solvent extraction

It shows the amounts of the various fractions (carbohydrates, aldehydes/ketones/lignin monomers, hydrocarbons, acids and esters) as a function of the reaction time. A fast decline in the carbohydrate fraction versus time is visible. Almost complete conversion to other components within 6 h reaction time is observed, an indication of the high reactivity of this fraction.

3.2.2. Experimental studies with Ru/C in continuous set-ups

Recently, in depth catalytic hydrotreatment experiments with the Ru/C catalyst in a continuous packed bed set-up were reported (Venderbosch et al., 2010). The results of this study will be provided in detail in the following as it provides detailed insights in the effect of process conditions on product yields, product properties and the various reactions taking place on a molecular level. Some experiments were carried out in the absence of catalysts to probe thermal reactions.

The catalytic hydrotreatment reactions were carried out in a set-up consisting of 4 packed bed reactors in series. The temperature in each reactor may be varied independently, allowing experiments at different temperature profiles over the length of the reactor. Typical pressures were between 150 and 300 bar, temperatures between 150 and 400 °C and WHSV's between 2-10 kg/kg.cat.h. In the following, the thermal reactions will be discussed, followed by catalytic hydrotreatment reactions at different temperature levels.

3.2.2.1. Thermal reactions

To study the thermal, non-catalytic reaction in detail, pyrolysis oil was pumped through the reactor (without catalyst) at pressures of up to 300 bar and temperatures of maximum 350 °C for residence times in the order of tenths of second – minutes. Typically under these conditions, a single-phase pyrolysis oil is converted into a viscous organic liquid, an aqueous phase and a gas phase. The carbon content of the viscous phase is about 60 wt.% (starting with 40 wt.% in the original oil), and the oxygen content about 32 wt.%. Additional water is produced, up to 30 % compared to the water initially present in the pyrolysis oil. The water is distributed over the two layers, but most of it ends up in the aqueous phase. Energetically, 80% of the thermal energy in the pyrolysis oil is transferred to the viscous product, less than 20% and 1 % is retained by the water phase and gas phase, respectively. The gas phase in such experiments consists of CO and CO2 in a ratios varying from 1:10 to 1:3 (depending on temperature, pressure, residence time), and in yields of almost 4 wt.% of the pyrolysis feed.

Although it is unknown at a molecular level which reactions actually take place, at least two parallel pathways can be distinguished, viz. a reaction causing the formation of gas (here referred to as decarboxylation / decarbonylation, yielding CO and/or CO2), and the other causing dehydration (likely by condensation (polymerisation) reactions). Possible sources of these gases are the organic acids in the oil. For all aqueous (and organic) samples produced the pH, however, is almost similar to the pyrolysis oil feed. This indicates that either the acids are not converted or the acids are converted and simultaneously produced as well. A detailed acid analysis of the products is not available, and the precise events taking place and mechanism however remain unclear. It seems that dilution of the pyrolysis oils with 'inert' solvents suppresses the re-polymerisation. Additionally, the gas yield becomes independent of the temperature and the residence time after a certain threshold in the residence time, while the amount of water produced is increasing. This indicates that the reaction mechanism for the formation of gas is different than the polymerisation reactions.

Phase separation of the oil at these conditions may have a number of causes, e.g. an overall increase in the water content due to the formation of water by condensation reactions. It is known (but not fully explained yet) that above a certain water content pyrolysis oils phase separate into an aqueous phase and a rather nonpolar phase. Repolymerisation of some molecules / fractions in the oil is also a plausible reason, as it renders the products less soluble in water, for example caused by transformation of the polar sugar constituents behaving as bridging agents in the dissolution of hydrophilic lignin material (Diebold 2002).

3.2.2.2. Catalytic hydrotreatment reactions

The catalytic hydrotreatment reactions were carried out at three process severity levels, a mild hydrogenation at either 175 or 225 °C, a mild hydrodeoxygenation (HDO) at 225 – 275 °C and a deep hydrodeoxygenation. For the latter, samples from the mild HDO were first allowed to phase separate completely, after which the organic fraction (containing about 3 wt.% water) was treated at temperatures ranging from 350 °C in the first two reactor segments, to 400 °C in the last two.

3.2.2.3. Visual appearances of liquid phase after reaction

The catalytic hydrotreatment reaction at 175 °C resulted in a single phase oil with a visual appearance close to that of the original feed. Thus, at this temperature, phase separation does not occur. This may be related to the limited production of water at this temperature. The product has a considerable sweeter smell/odor than the original pyrolysis oil. The mild hydrogenation at 225 °C gives two liquid phases, an organic and a water rich phase. The water phase has a higher density than the aqueous phase. A similar situation was observed for experiments at higher process severities (mild HDO), see Figure 5 for details. The second stage HDO product oil has even a lower density than the aqueous phase.

The organic product yields for the various process severities are given in Figure 6. Here, the severity is expressed in terms of hydrogen consumption, and high severity is associated with high hydrogen consumption. The yield is a clear function of the temperature. A drop in the yield to about 40% is observed at about 200 oC due to the occurrence of phase separation and transfer of part of the carbon and oxygen to the aqueous phase. A further slight reduction in yield is observed at higher severities, presumably due to gasification reactions and further net transfer of components from the organic to aqueous phase.

Figure 5.Pictures of pyrolysis oil (left), mild HDO (middle) and 2nd stage HDO (right) products

Oxygen contents of the product oils are a function of the process severity, see Figure 6 for details. Phase separation between 175 and 225 °C results in a dramatic drop in the oxygen content. This is due to the loss of water and the transfer of very polar highly oxygenated components to the aqueous phase. At the highest severity, the oxygen content is about 15%, compared to about 40% for the original pyrolysis oil.

The hydrogen consumption ranges between 65 and 250 Nm3/t pyrolysis oil. Higher process severities lead to higher hydrogen uptakes (Figure 6).

A useful representation to assess the changes in the elemental composition of the product oils at various process severities is a van Krevelen diagram. Here, the ratio between O/C and H/C of the products are plotted together in a single diagram. In Figure 7, a typical plot is provided for selected literature data on pyrolysis oil hydroprocessing (Elliott, 2007; Venderbosch et al., 2010) and our results with Ru/C at different severities. Presented here are data points from e.g.:

- wood and pyrolysis oil, and for the four cases referred to in this paper (HPTT, hydroprocessing at 175 and 225 °C, Mild HDO and 2nd stage HDO);
- A selection of data points derived from literature studies (Baldauf et al. 2007; Churin et al., 1988; Conti, 1997; Diebold, 2002; Kaiser 1997; Samolada et al., 1998). Some of these data are derived from various oils from a variety of resources and processed in different reactors, different catalysts and at different conditions.

The plot also contains curves to represent the changes taking place in elemental composition during hydroprocessing, a theoretical curve for the dehydration of pyrolysis oil, and trend lines for the thermal (HPTT) route and hydroprocessing routes based upon the experimental data points.

Based on our work on the Ru/C catalysts and supported by the literature points in Figure 7, several reaction pathways can be distinguished:

- Essentially repolymerisation of the pyrolysis oil (no catalyst, no hydrogen, 'HPTT');
- Merely hydrogenation of the pyrolysis oil at mild conditions (up to 250°C, with catalyst and hydrogen, referred to as mild hydrogenation),
- Dehydration of the oil at temperatures near 250-275 °C, and
- Hydroprocessing of pyrolysis oil at temperatures up to 400 °C

Upon thermal treatment, the principal reactions are rejection of oxygen as water. Some CO_2 and CO is released as well, which shifts the trend line to slightly higher H/C ratios (but decarboxylation / decarbonylation is limited to approx. 10 wt.% of the feed). A high conversion (i.e. at high temperatures and residence times) eventually leads to a hydrogen-depleted solid material (and probably similar to conventional carbonisation processes, charcoal).

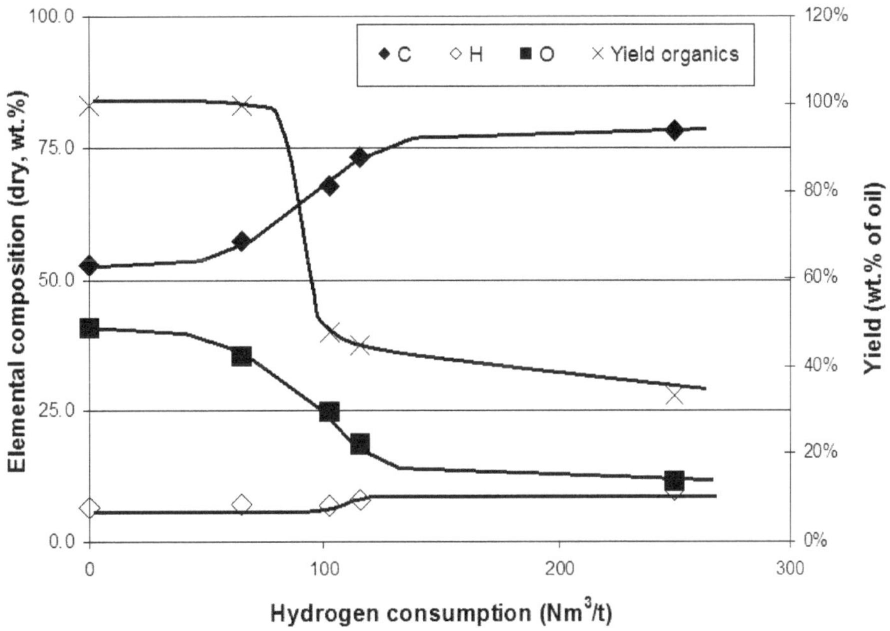

Figure 6.The elemental composition of the organic oil product (dry basis) versus the hydrogen consumption for pyrolysis oil, mild hydrogenation, mild HDO and 2nd stages HDO

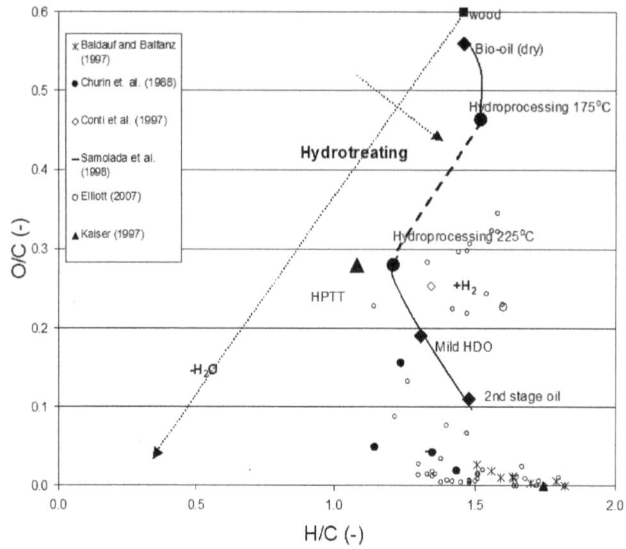

Figure 7. The Van Krevelen plot for oils derived from the thermal pathway (HPTT), mild hydrogenation, mild HDO and 2nd stages HDO, including relevant literature data points

To obtain a liquid product with a higher H/C ratio, additional hydrogen is thus required. This path is shown in Figure 7 and includes the mild hydroprocessing step, at around 175 °C (no phase separation) and 225 °C (phase separation), followed by further hydrodeoxygenation (and hydrocracking).

3.2.3. Analysis of Product oil fractionation; insights in molecular changes

The various organic products were subjected to a standardized liquid-liquid fractionation protocol (Oasmaa, 2003, Figure 1) to gain insights on the severity of the hydrotreatment process on product composition. The results are compiled in Figure 8 and show major changes in composition upon reaction. The pyrolysis oil feed mainly consist of ether solubles, ether insolubles and water. The components in these fractions originate from the cellulose and hemi-cellulose fraction in the biomass feed and particularly the ether insoluble fraction is rich in carbohydrates. The amounts of DCM solubles and insolubles, from the lignin fraction of the biomass feed, are by far lower and are about 20% in total.

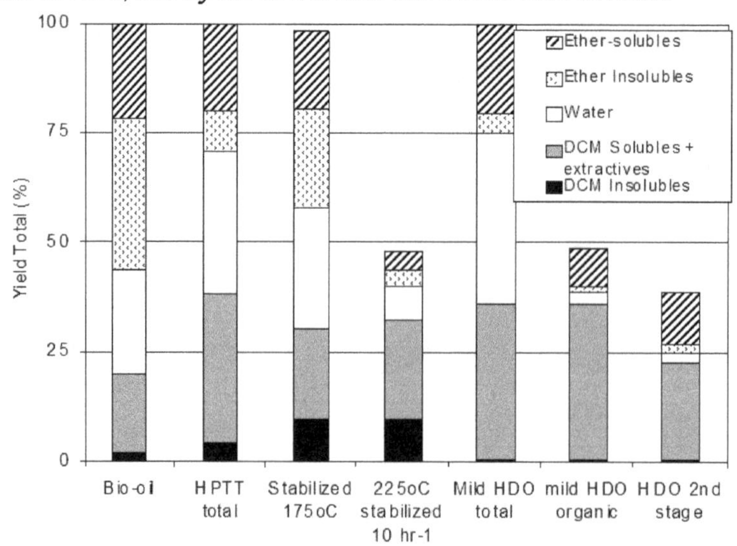

Figure 8. Comparison of the fractionation results for various process severities

3.2.3.1. Thermal reactions

Upon comparing the composition of the pyrolysis oil feed with the product obtained from the thermal route, it becomes evident that the substances that are insoluble in ether are transformed into substances that are soluble and insoluble in dichloromethane (DCM), together with the addition of water. Over time, wood oils undergo a similar transformation, resulting in the production of water insoluble substances and a decrease in the sugar content (Oasmaa&Kuoppala, 2003). At elevated temperatures and longer periods of time, particularly this portion of sugar is accountable for the process of charring, most likely due to the creation of initial DCM solubles and later DCM insolubles (known as 'char'). The formation of solid substances when an aqueous solution containing C-6 sugars (such as D-glucose and D-mannose) is heated to temperatures as high as 400 °C is a widely recognized phenomenon. Thermal degradation, which can be catalytic (mostly by acids) or non-catalytic, results in the formation of solid compounds known as humins (Girisuta et al., 2006; Watanabe et al., 2005a; Watanabe et al., 2005b). The suggested chemical pathway involves the conversion of a C-6 sugar into 5-hydroxymethyl furfural (HMF), followed by the production of levulinic acid (LA) and formic acid (FA). Both reactions are also accompanied by the creation of solids, specifically humin (Scheme 1). The creation of solids is particularly undesired and restricts the production of the two potential biobased compounds, LA and HMF. Despite extensive research endeavors, the prevention of solids/humin production during reactions in aqueous medium has not yet been achieved.

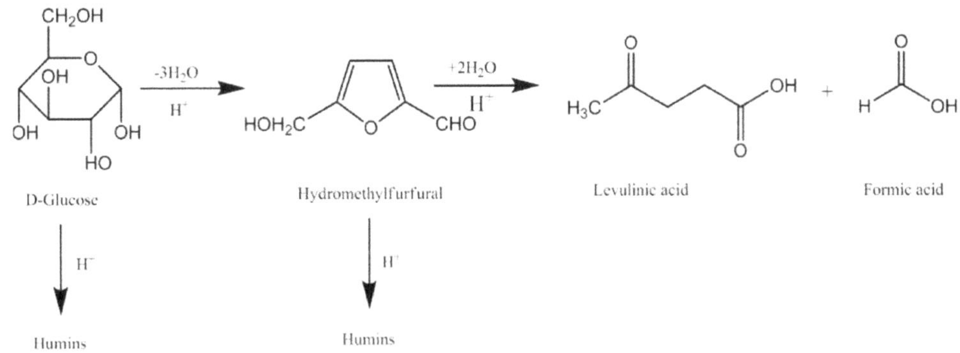

Scheme 1. Decomposition reactions of D-glucose at elevated temperatures.

Higher temperatures and the presence of acid catalysts (homogeneous and heterogeneous) increase the rate of D-glucose decomposition (Girisuta et al., 2006). Such reactions may also occur in the fast pyrolysis oil matrix. The oil is acidic in nature due to the presence of organic acids and these will catalyse the depolymerisation of oligmeric sugars to D-glucose and other C-6 sugars followed by the reaction to solids and hydroxymethylfurfural and levulinic acid/formic acid.

Knezevic et al. (2009) studied the thermal decomposition of D-glucose in hot compressed water under conditions of relevance for the catalytic hydrotreatment of pyrolysis oil (240-374 °C). It was shown that D-glucose decomposes mainly to char and some gaseous components (primarily CO2), while only a limited number of components remained in the water phase (for example formaldehyde). At these conditions, the reactions are very fast and decomposition to char takes place on the time scale of seconds to minutes.

3.2.3.2. Catalytic hydrotreatment reactions

The composition of the product from a mild hydrogenation at 175 °C (see Figure 8) differs considerably from that of the original pyrolysis oil. The amount of water increased slightly (from 25 up to about 30 wt.%), which appears insufficient for phase separation. In addition, the ether solubles (aldehydes, ketones, acids, etc) are converted, but in smaller amounts compared to HPTT. The ether insoluble (sugar fraction) is reduced considerably from 35 down to 24 wt.%, while the water insoluble fraction is increased accordingly. Simultaneously, the increase of the DCM insoluble fraction is about 8%, while the DCM soluble fraction increases with only 3 wt.%.

Similar to HPTT, we assume that the sugar fraction in the oils is (partially) converted to more water insolubles and some additional water. However, the actual components formed during mild hydrotreatment are different in nature than the HPTT oils and particularly the amount of DCM insolubles is higher.

The results of the fractionation of the product oil derived from an experiment 225 °C (mild hydrogenation) are provided in Figure 8. Phase separation occurs and as such the amount of product oil is reduced considerably. As a result, the amounts of water, ether solubles and ether insolubles in the organic phases are lowered and imply that components have been transferred to the water phase. Figure 8 also shows the result for the mild HDO reaction. Compared with the oil samples obtained at lower temperatures, the DCM insoluble fraction is now almost completely converted to DCM soluble components, evidence that some hydrocracking reaction have taken place here as well. In the 2nd stage hydroprocessing the amount of ether solubles increases, at the expense of DCM solubles and the extractives.

3.3. Characteristics of Product

In all hydrogenation experiments except those at temperatures below 200°C, the product obtained consisted of two liquid phases, viz. an aqueous phase and brown-red organic phase. For all of them, relevant (basic) characteristics were determined, viz. elemental composition (vide supra, Figure 7), water content and average molecular weight. Additionally, to get some insights in the coking tendency, the samples were analyzed using thermogravimetric analysis (TGA). Here, the residual weight of the sample, heated under N2 up to about 900°C, was taken as a measure of coking. A high residue indicates a high tendency for coking and thus a low thermal stability at elevated temperature. The residue after a TGA measurement is a strong function of the process severity, see Figure 9 for details.

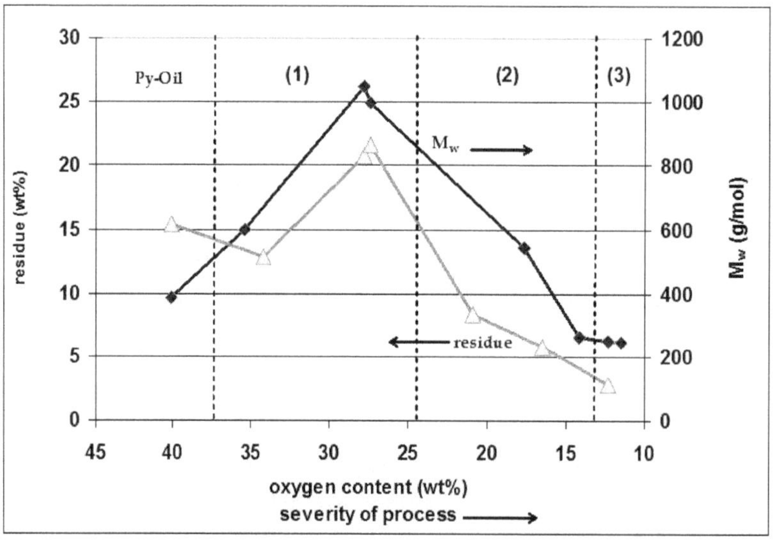

Figure 9. Mass average molecular weight and TGA residue of products from (1) stabilisation, (2) mild hydrotreating, and (3) 2-stage hydrotreating.

At low process severities, the TGA residue increases and the highest value (22%) is observed at intermediate severities. A further increase in severity leads to a strong reduction in the TGA residue. Thus, it may be concluded that intermediate severities lead to product oils with a high TGA residue and consequently have a higher tendency for coking and may be less suitable as a refinery feedstock.

The organic products were analyzed using gel permeation chromatography (GPC) to determine the average molecular weights and the results are given in Figure 9. The molecular weight of the product oils increases compared to the pyrolysis oil feed at low severity hydrotreatment reactions. Apparently, polymerisation occurs and this has also been observed when heating up pyrolysis oil to 275°C in the absence of catalysts (HPTT process) (Rep et al., 2006). A further increase in the severity (higher temperatures, shorter WHSV's) leads to a reduction of the molecular weight and a value of less than 300 is observed at the highest severities.

Of particular interest is the relation between the molecular weight of the products and the TGA residue. Products with a higher Mw also lead to higher TGA residues and this may be rationalized by assuming that the higher molecular weight fragments in the products are precursors for coke formation.

4. PROPOSED PATHS FOR REACTIONS AND THEIR POTENTIAL CONSEQUENCES

4.1. Reaction pathways

A schematic and simplified representation of relevant reactions assumed on basis of this work is presented in Figure 10. In the initial phase of the hydrotreatment process, catalytic hydrogenation and thermal, non-catalytic repolymerisation occur in a parallel mode.

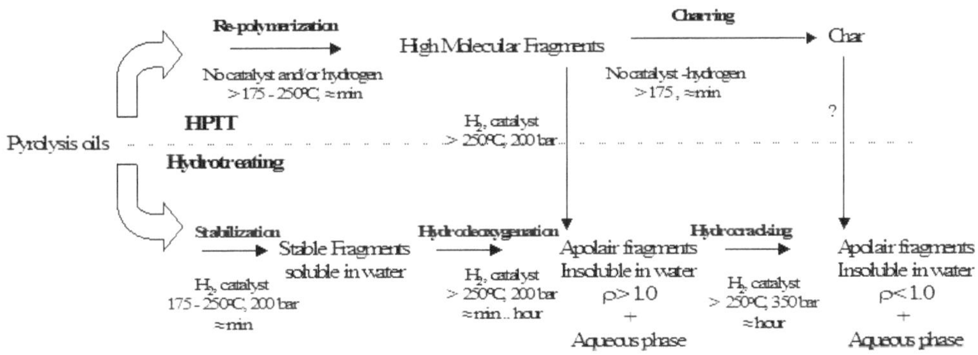

Figure 10. Proposed pathways for the catalytic hydrotreatment of pyrolysis oils

Repolymerisation results in the creation of soluble fragments with increased molecular weight, which then undergo additional condensation processes to create char. This approach is not desirable and the rate of the polymerization reactions should be minimized as much as feasible. The optimal approach entails subjecting the thermally unstable constituents in the pyrolysis oil feed to hydrogenation, resulting in the formation of stable molecules that are resistant to polymerization. Following reactions (hydrogenations and hydrocracking) occurring within a few hours result in products with lower levels of oxygen and ultimately lead to larger ratios of hydrogen to carbon (Figure 7). Figure 9 shows that the molecular weight of the organic phase increases slightly at low process severities when using Ru/C as the catalyst, indicating that the repolymerisation step cannot be avoided. Nevertheless, increased severities result in a decrease in the mean molecular weight, suggesting that soluble fragments with larger molecular weights may also undergo partial depolymerization due to the influence of hydrogen and a catalyst.

As previously mentioned, pyrolysis oil contains significant quantities of oligo- and monomeric sugars, which originate from the cellulose and hemi-cellulose components of the lignocellulosic biomass feedstock. Therefore, it is intriguing to compare the reaction pathways illustrated in Figure 10 for pyrolysis oil with those of conventional hydrogenation and heat reactions that take place for carbohydrates under different process conditions.Thermal decomposition of various monomeric sugars in aqueous media has been studies in detail and is known to lead to oligomerisation to soluble and subsequently to insoluble humins (Girisuta et al., 2006). As an example, Knezevic et al. (2009) studied the thermal decomposition of D-glucose in hot compressed water at elevated temperatures (240-374 °C), giving solids (char, humins) and some gaseous components (primarily CO2). At these conditions, the reactions are very fast and decomposition to char takes place on the time scale of seconds to minutes.

Catalytic hydrotreatment of carbohydrates using heterogeneous catalysts has been reported extensively in the literature. The main focus is on the hydrogenation of D-glucose to D-sorbitol, a well-known chemical with use in the pharmaceutical and the food industry (Kusserow et al., 2003). Catalytic hydrotreatment of D-glucose over Ni, Ru based and Pd based heterogeneous catalysts at 80 °C, 80 bar yields D-sorbitol in high yields (Crezee et al., 2003; Makkee et al., 1985). The hydrogenation reactions at these low temperature levels may be considered as the stabilisation step in fast pyrolysis oil upgrading. In the presence of hydrogen and a catalyst, D-sorbitol is not inert at elevated temperatures (above 180 °C) and may be converted to a variety of products. For instance, Huber et al. (2004) showed that D-sorbitol can be converted to n-hexane in high yield using Pd and Pt catalyst on SiO2 or Al2O3 (225-265 °C and 26-58 bar). Over Ru/SiO2, hydrogenolysis of D-sorbitol at 180-240 °C and 80-125 bar hydrogen pressure yields mainly glycerol and 1,2-propanediol (Sohounloue et al., 1982) . This implies that C-C bond cleavage occurs readily, leading to the formation of lower molecular weight products.

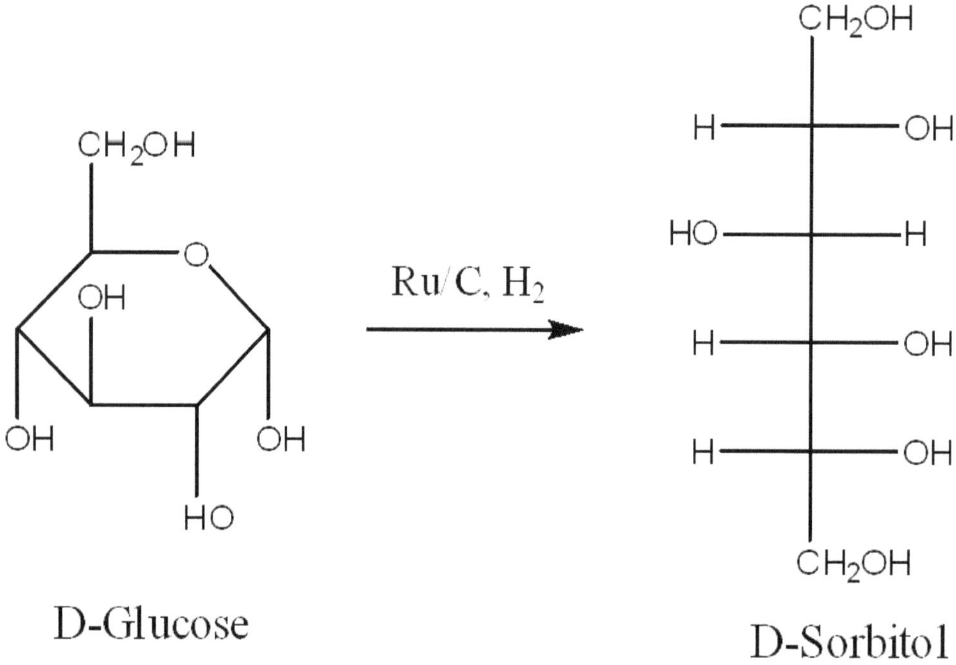

Scheme 2. Catalytic hydrogenation of D-glucose to D-sorbitol

These reactions are likely also occurring upon the catalytic hydrotreatment of fast pyrolysis and may explain the formation of more apolar lower molecular weight products at higher process severities.

Thus, it may be concluded that the typical reaction pathways for pyrolysis oils at typical low severity hydrotreatment conditions mimic those of low molecular weight sugars viz. repolymerisation reactions to solids (humins) and hydrogenation/C-C bond cleavage reactions to for instance polyols and finally to hydrocarbons. This strengthens our initial hypothesis that pyrolysis oil should be regarded as a carbohydrate rich "syrup" and not a conventional fossil derived hydrocarbon liquid.

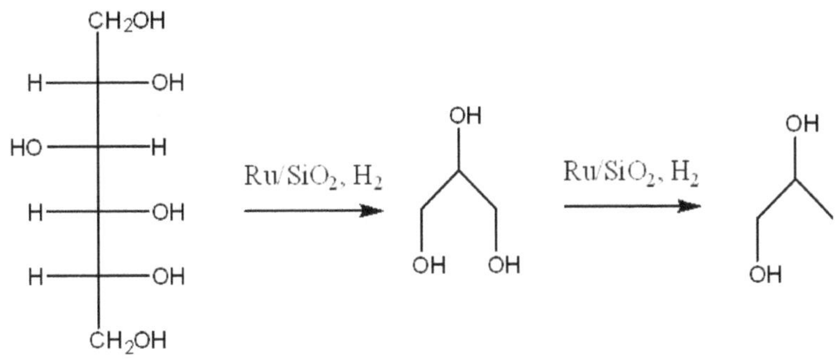

Sorbitol Glycerol 1,2-propanediol

Scheme 3.Hydrogenolysis of D-sorbitol to glycerol and 1,2-propanediol.

4.2. Process implications

The proposed reaction pathway for catalytic hydrotreatment of pyrolysis oil (Figure 10) implies that the rate of the hydrogenation route should be much higher than the rate of the repolymerisation route to obtain good quality upgraded pyrolysis oil (low molecular weight, low viscosity, low coking tendency). An obvious solution is the development of highly active hydrogenation catalysts. These studies will be reported in the next paragraph of this paper. However, a smart selection of process conditions and reactor configurations may also be considered, particularly to enhance the rate of the hydrogenation/ hydrodeoxygenation pathway compared to the repolymerisation pathway. In this respect, it is highly relevant to gain some qualitative insights in the factors that determine the rate of the individual pathways (hydrogenation versus repolymerisation).

A schematic plot is presented in Figure 11, where an envisaged reaction rate (arbitrary values, in mole reactant/min) is presented versus the actual reaction temperature. The lines drawn are taken in case (i) gas-to-liquid mass transfer determines the overall reaction rate (ii) the catalytic hydrotreatment reactions dominate, and polymerisation reactions prevail. Figure 11 is derived on basis of simplified kinetics for the glucose hydrogenation – polymerisation reactions, but a detailed outline of all assumptions made is beyond the scope of the presentation here.

For this reason the exact values on the x- and y-axes are omitted. The following relations are taken into account to derive Figure 11:

- The conversion rate due to the hydroprocessing reactions RH (mol/m3 r.s) can be simplified as a product of the intrinsic kinetic rate expression kR and the surface area available per reactor volume. Being a catalytic reaction, the influence of temperature can be rather high.
- The overall gas-to-solid mass transfer rate of hydrogen depends on reactor geometry and operating variables. In case of stirred tank reactors (including batch-wise operated autoclaves), the actual stirring rate is important, while in packed bed the catalyst particle wetness is relevant. In both, the concentration of hydrogen (thus hydrogen pressure) is important, together with catalyst particle size, and, to a limited extent, temperature.
- The rate of polymerisation, Rp, will depend largely on the temperature, and, being a reaction with order in reactant(s) > 1 (and probably up to 2 or 3), on the concentration of the reactant.

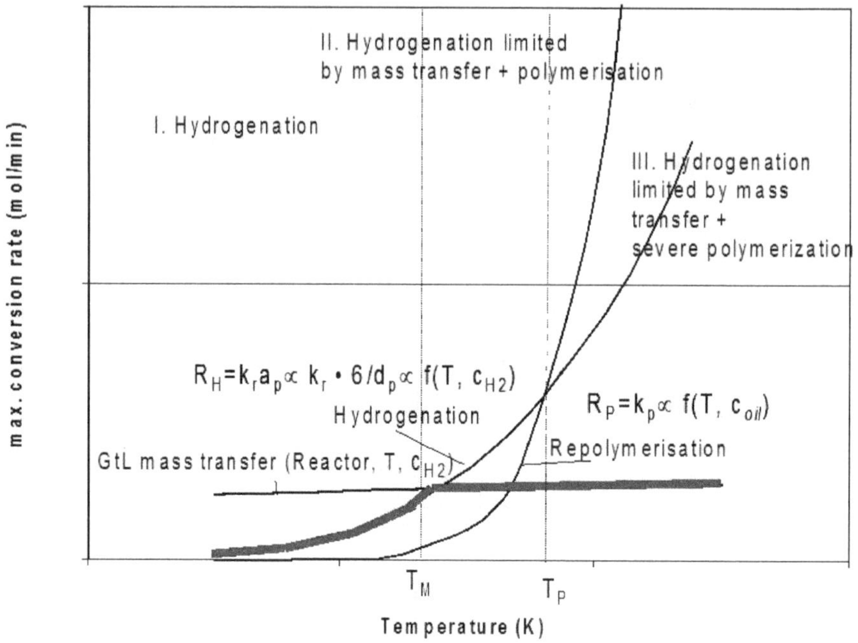

Figure 11. Schematic indication of the conversion rates in case of mass transfer, hydrogenation and polymerisation reactions as a function of temperature. The blue solid line would be the net effective conversion rate.

A number of options may be envisaged to promote the hydrogenation pathway:

- Increase the hydroprocessing reaction rate, for instance by a higher catalyst intake or by an increase in the effective hydrogen concentration in the liquid (pressure, application of a solvent with a high hydrogen solubility).
- Reducing the polymerisation reaction, a.o. by performing the initial stabilisation step at a low temperature (< 100°C) and reduction of the concentration of the reactants (a.o. by dilution).
- Increase the overall gas-to-liquid mass transfer rate in case the reaction is performed in the gas-liquid mass transfer limited regime. This may be possible by increasing the mass transfer surface area in the reactor, higher mass transfer coefficient and / or increasing the concentration difference between the gas and the liquid.

5. ENHANCED CATALYST COMPOSITIONS FOR THE EFFICIENT CATALYTIC HYDROTREATMENT OF FAST PYROLYSIS OIL

Developing highly efficient metal catalysts is crucial for minimizing repolymerization in catalytic hydrotreatment processes. The data reported in this chapter thus far are only derived from a Ru/C catalyst. Ruthenium (Ru), although a valuable precious metal, is costly. Therefore, there is a motivation to discover catalysts that are not only more efficient but also more affordable for the hydrotreatment reaction. One potential option is the utilization of more affordable bimetallic metal catalysts that are primarily composed of nickel (Ni). Nickel (Ni) exhibits significant hydrogenation activity towards various organic functional groups, especially reactive ketones and aldehydes. Consequently, it holds promise as an active metal for hydrotreatment reactions. Nevertheless, monometallic Ni catalysts, when utilized on silicon oxide, **γ**- or **δ**-alumina, or other supports, are unsuitable as hydrogenation catalysts under the typical temperature and pressures employed in this study. There are primarily two reasons: 1) Nickel (Ni) catalysts require a high temperature of reduction (usually around 700 °C) in order to be fully reduced. However, it is known that 20Ni catalysts quickly lose their effectiveness at high temperatures due to the formation of carbon deposits (known as "coking").The carbon deposition can block the nickel surface, or the pore mouths, and, eventually leading to a strong reduction in the reaction rates. These two drawbacks regarding the use of Ni were solved a.o. by using another element (metal or non-metal), also designated as a promoter. One of these proprietary catalysts was studied in detail and will be referred to in the following as catalyst D.

Figure 12 shows the liquid phase after a hydrotreatment over Cat D versus the severity of the process, showing the original oil (left) and an oil derived at the most severe conditions tested here on the right. Interestingly, the product oils obtained over cat D are much more transparent than those derived from the Ru/C catalyst.

Figure 12. Visual appearance of the liquid phase after hydrotreatment over catalyst D

A van Krevelen plot gives valuable insights in the difference in performance between catalyst D and Ru/C (Figure 13). A similar pattern for both is observed as a function of severity but the curve for Cat D is shifted to higher H/C values. Thus, at a similar oxygen content, the H/C ratio is higher for catalyst D. This is indicative for a higher hydrogenation rate for cat D and is known to be favorable regarding product properties.

Repolymerisation reactions appear to occur to a limited extent when using cat D instead of Ru/C. This is evident when comparing the average molecular weight of the final products (Figure 14a), as determined by GPC, for both Cat D and Ru/C. For Ru/C the average molecular weight shows a significant increase from 400 up to 1000 Da at low severities, but a constant value over the oxygen content interval of 400-450 Da for catalyst D is observed.

TGA residues of the product oils using cat D (Figure 14b) show carbon residues of around 5%. Surprisingly, and not expected on basis of test carried out using other catalysts, already at less severe operating conditions, a significant reduction in the TGA residues is achieved. Thus, products with a higher H/C ratio and a lower carbon residue were obtained with cat D indicating that the rate of the hydrogenation/ hydrodeoxygenation reactions over catalysts D are higher than for Ru/C.

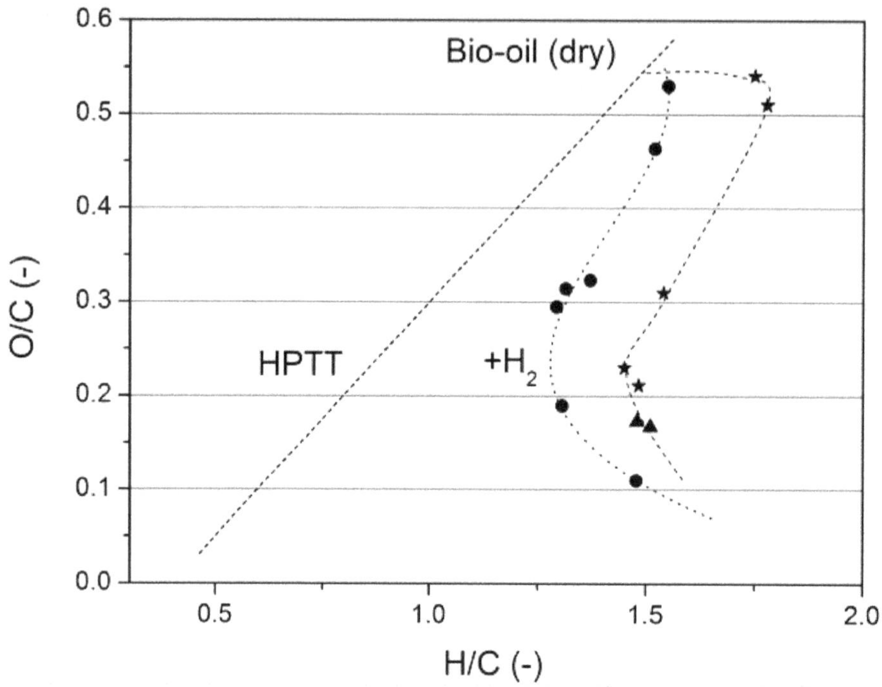

Figure 13. Van Krevelen plot for oils derived over catalyst Ru/C (circles) and over catalyst D (stripes). Lines are trendlines.

An important product property for the upgraded oils is the viscosity. In Figure 15, the viscosity profile of the product oils versus the oxygen content is compared for conventional catalysts (Ru/C and NiMo, CoMo) and catalyst D. Clearly, the viscosity in the mid range of oxygen contents is much lower for cat D. Further testing at the extreme of low oxygen content will be required to grab the full picture but it is clear that cat D gives upgraded products with a lower viscosity than for conventional catalysts. The lower viscosity is likely the result of a lower average molecular weight of the products, as shown earlier and the result of higher hydrogenation/ hydrodeoxygenation rates for Cat D compared to Ru/C.

On the basis of the product properties of the upgraded oils obtained with cat D, we can conclude that repolymerisation is not occurring to a considerable extent. As a result, product oils with a lower molecular weight and a concomitant lower viscosity, lower TGA residue is obtained. Thus, the reaction pathway for catalyst D may be simplified considerably, see Figure 16 for details.

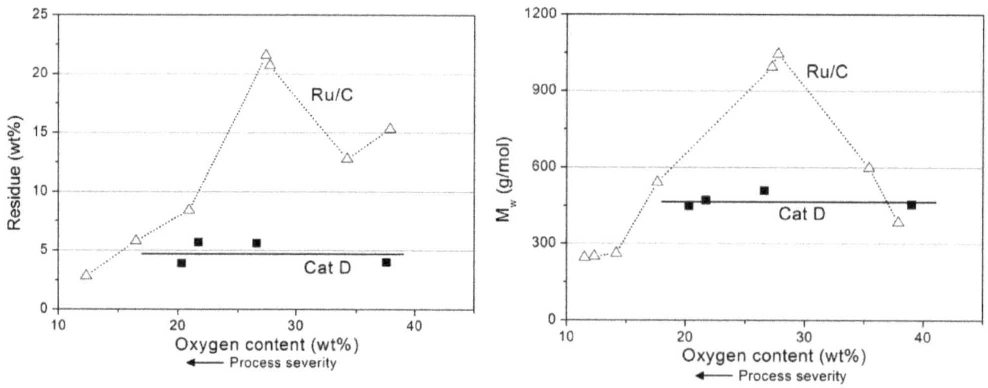

Figure 14. TGA residue (wt%) for catalyst Ru/C and the catalyst D (top); average molecular weight of final product (bottom)

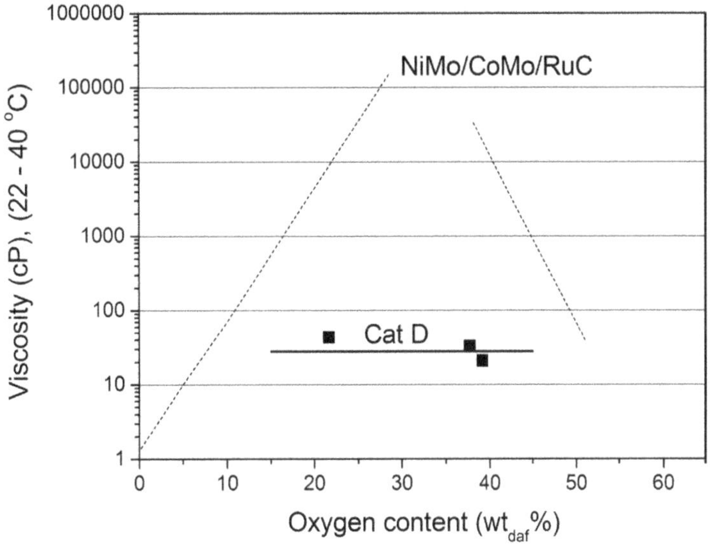

Figure 15. Viscosity of the oil versus the oxygen content for conventional catalysts and for Cat D.

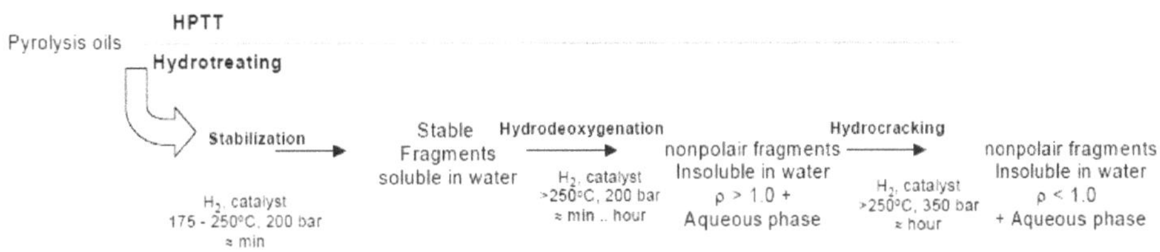

Figure 16. Proposed pathways for the mild hydrotreating of pyrolysis oils over catalyst D.

6. POTENTIAL APPLICATIONS OF ENHANCED OILS AND ESSENTIAL FEATURES OF THE PRODUCT

The primary objective of this research is to develop pyrolysis oil generated products that can be used as co-feed in crude oil refineries. It is important to examine certain product characteristics that are likely to be essential for this particular use. An essential characteristic of oil is its propensity to generate coke when heated, as assessed by the 'Conradson Carbon Residue' (CCR) or the residue obtained from thermogravimetric analysis (TGA residue) (Furimsky, 2000). Typically, pyrolysis-oils have CCR values ranging from 20 to 30% (Samolada, 1998), which restricts their direct use as a co-feed. The thermogravimetric analysis (TGA) residues for the improved oils produced using Ru/C and cat D, as described in this study, exhibit values ranging from 3 to 22 wt% (refer to Figure 14). These values can be adjusted by varying the severity of the process, such as temperature and residence/batch time. It is important to understand that hydrotreated products can have low TG residue levels even if they still have significant concentrations of bound oxygen (> 10 wt%). Therefore, it is possible to create stable oils even when they include a relatively large amount of bound oxygen. From a processing perspective, this is also beneficial since it restricts the amount of hydrogen used in the catalytic hydrotreatment process, which is a significant factor in variable costs.

A recent study has examined the co-processing of hydrotreated pyrolysis oils, which were produced using a Ru/C catalyst and contain oxygen concentrations greater than 5 wt%, in a laboratory-scale simulated FCC unit (MAT) (de Miguel Mercader et al., 2010). The hydrotreated materials were effectively dissolved and processed in Long Residue (20 wt% improved oil). The FCC gasoline yields, ranging from

44 to 46 wt.%, and the Light Cycle Oil yields, ranging from 23 to 25 wt.%, were similar to the base feed. There was no significant increase in the production of unwanted coke and dry gas. The experiments using undiluted improved oils were not as successful, resulting in much larger yields of dry gas and coke compared to co-feeding. It is evident that co-processing is essential in order to achieve favorable product yields. This study also demonstrates that contrary to early assumptions, it is probably unnecessary to target an improved oil with an oxygen level below 1 wt%.

Ongoing MAT testing is being conducted using product oils obtained from catalyst D. This testing will enable the construction of comprehensive process-product relationships for the purpose of co-feeding. The oils enhanced with cat D exhibit much greater thermal stability compared to the original pyrolysis oil, as seen by the TGA remains (Figure 14). Initial inquiries have indicated that this enables the oil to be distilled into different fractions without the excessive production of char. Ongoing comprehensive investigations are currently underway and will be disclosed in the appropriate time frame.

7. CONCLUSIONS

The upgrading of pyrolysis oil by catalytic hydrotreatment reactions using heterogeneous catalysts was studied in detail using a Ru/C catalyst. The investigations provided valuable insights in the chemical transformations occurring during catalytic hydrotreatment and include both thermal and hydrogenation/hydrodeoxygenation pathways. The repolymerisation pathway in which the oils are further condensed to soluble oligomers and eventually to char components competes with a catalytic hydrogenation reaction. In case H2 is present with a proper catalyst, these soluble oligomers may be depolymerised to stabilized components that can be further upgraded. In this respect, the pyrolysis oils show reactivity typically observed for carbohydrates, which is rationalised by considering the high amounts of oligo- and monomeric sugars in the oil. New, highly active Ni-based catalysts have been developed, which show much better performance than conventional ones and provide products with improved properties.

Experimental work is foreseen to elucidate the reaction pathways occurring during catalytic hydrotreatment in more detail and to develop efficient processes to obtain a stabilized oil with the desired product properties at the lowest manufacturing costs. These include:

- Determination of the effects of reactor configuration on the reaction rates (including mass transfer issues) and subsequent reactor selection;
- Determination of relevant physical properties (e.g hydrogen solubility);
- Optimisation of the hydroprocessing conditions and particularly the required hydrogen levels
- Determination of product-process relations. The effective hydroprocessing severity required for further co-refining must be defined;
- The source and availability for hydrogen: perhaps even syngas is applicable;
- Effects of reaction exothermicity need to be determined.

REFERENCES

Ardiyanti, A.R., Venderbosch, R.H., Heeres, H.J., (2009) Process-product studies on pyrolysis oil upgrading by hydrotreatment with Ru/C catalysts, Proceedings of the 2009 AIChE Spring National Meeting, Tampa.

Baldauf W. Balfanz U. Upgrading of fast pyrolysis liquids at VebaOel. A. G. in Biomass gasification pyrolysis state of the art future prospects Kaltschmidt and future prospects, Kaltschmidt M. and Bridgwater A.V, 392 397 1-87269-171-4

Churin E. Maggi R. Grange P. Delmon B. 1988Characterization and upgrading of a bio-oil produced by pyrolysis of biomass, in: Research in Thermochemical biomass conversion, Bridgwater A.V., Kuester J.L., 896 909Elsevier Science Publishers.

Conti L. 1997Improvement of the flash pyrolysis process and pilot plant for bio-oils upgrading, Final Report Contract AIR 1CT92-0216.

Crezee E. Hoffer B. W. Berger R. J. Makkee M. Kapteijn F. Moulijn J. A. 2003 Three-phase hydrogenation of D-glucose over a carbon supported Ruthenium catalyst-mass transfer and kinetics Appl. Catal. A-Gen., 251(1), 1 17

Diebold J. P. 2002A review of the chemical and physical mechanisms of the storage stability of fast pyrolysis biooils. In: Fast Pyrolysis of Biomass: A Handbook, 2Bridgwater, A.V., CPL Press, Newbury, UK.

Dimitrijevic R. Lutz W. Ritzmann A. 2006 Hydrothermal stability of zeolites: Determination of extra-framework species of H-Y faujasite-type steamed zeolite Journal of Physics and Chemistry of Solids, 67, 1741 1748

Elliott D.C., (2007) Historical developments in hydroprocessing bio-oils, Energy and Fuels, 21, 1792 1815

Elliott D. C. Hart T. R. Neuenschwander G. G. Rotness L. J. Zacher A. H. 2009 Catalytic hydroprocessing of Biomass fast Pyrolysis Bio-Oil to produce hydrocarbon Products Environmental Progress & Sustainable Energy, 441-449

Ferrari M. Delmon B. Grange P. 2002 Influence of the impregnation order of molybdenum and cobalt in carbon-supported catalysts for hydrodeoxygenation reactions Carbon 40, 497- 511

Furimsky E. 2000 Catalytic hydrodeoxygenation Applied Catalysis A General 199, 147- 190

GirisutaB.JanssenL.P.B.M.HeeresH.J.2006Akineticstudyonthedecompositionof5-hydroxymethylfurfural into levulinicacid Green. Chem., 8(8), 701-709

Gutierrez A. Kaila R. K. Honkela M. L. Slioor R. Krause A. O. I. 2009 Hydrodeoxygenation of guaiacol on noble metal catalystsCatal. Today, 147, 239 246

Horne P. Williams P. T. 1996 Reaction of oxygenated biomass pyrolysis compounds over a ZSM-5 catalyst, Renewable Energy 131-144

Huber G. W. Cortright R. D. Dumesic J. A. 2004 Renewable alkanes by aqueous-phase reforming of biomass-derived oxygenates. Angew. Chem., 43, 1549 1551

Kaiser M. 1997Upgrading of fast pyrolysis liquids by DMT, in: Biomass gasification & pyrolysis: state of the art and future prospects, Kaltschmidt M., and Bridgwater A.V., 399 406 1-87269-171-4

Knezevic D. van Swaaij W. P. M. Kersten S. R. A. 2009 Hydrothermal conversion of biomass: I, Glucose conversion in hot compressed water Ind. Eng. Chem. Res., 48(10], 4731 4743

Kusserow B. Schimpf S. Claus P. 2003 Hydrogenation of glucose to sorbitol over nickel and ruthenium catalysts Adv. Synth. Catal., 345, 289 299

Lee C. L. Ollis D. F. 1984 Interactions between Catalytic Hydrodeoxygenation of Benzofuran and Hydrodesulfurization of Dibenzothiophene Journal of Catalysis 332- 338

Mahfud F. H. Bussemaker S. Kooi B. ten Brink. G. Heeres H. J. 2007a The application of water soluble Ruthenium catalysts for the hydrogenation of the dichloromethane soluble fraction of fast pyrolysis oil and related model compounds in a two phase aqueous-organic system J. Mol. Cat. A. 277, 127 136

Mahfud F. H. Ghijsen F. Heeres H. J. 2007bHydrogenation of fast pyrolysis oil and model compounds in L-L biphasic systems using homogeneous ruthenium catalysts, J. Mol. Cat. A., 264, 227- 236

Maity S. K. Rana M. S. Srinivas B. N. Bej S. K. Murali Dhar. G. Prasada Rao. T. S. R. 2000Characterisation and evaluation of ZrO2 supported hydrotreating catalysts, Journal of Molecular Catalysis A: Chemical, 153, 121 127

Makkee M. Kieboom A. P. G. Van Bekkum H. 1985 Hydrogenation of D-fructose and D-glucose mixtures. Carbohydr Res., 138(2), 225 236

Mercader F. M. Groeneveld M. J. Kersten S. R. A. Way N. W. J. Schaverien J. C. Hogendoorn J. A. 2010 Production of advanced biofuels: Co-processing of upgraded pyrolysis oil in standard refinery units Applied Catalysis B: Environmental 57- 66

Miguel Mercader. F. Hogendoorn J. A. Venderbosch R. H. Kersten S. R. A. Groeneveld M. J. 2010 Pyrolysis oil upgrading by high pressure thermal treatment Fuel 2829- 2837

Oasmaa A. Kuoppala E. Solantausta Y. 2003 Fast pyrolysis of forestry residue: 2. Physicochemical composition of product liquid Energy & Fuels 433.

Oasmaa A. Kuoppala E. 2003 Fast pyrolysis of forestry residue. 3. Storage stability of liquid fuel Energy & Fuels, 17(4), 1075- 1084

Oasmaa A. 2003 Fuel Oil Quality Properties of Wood-Based Pyrolysis Liquidsthesis University of Jyväskylä, Finland.

Rep M. Venderbosch R. H. Assink D. Tromp W. Kersten S. R. A. Prins W. Van Swaaij W. P. M. 2006De-oxygenation of bio-oils, in Science in thermal and chemical biomass conversion, Bridgwater A.V and Boocock D.G.B., 1526 1535CLP Press.

Samolada M. C. Baldauf W. Vasalos I. A. 1998 Production of a bio-gasoline by upgrading biomass flash pyrolysis liquids via hydrogen processing and catalytic cracking Fuel 1667- 1675

Satterfield C. N. Yang S. H. 1983Some Effects of Sulfiding a NiMo Catalyst on its Activity for Hydrodenitrogenation of Quinoline, Journal of Catalysis, 8335- 346

Senol O. I. Viljava T. R. Krause A. O. I. 2005Hydrodeoxygenation of methyl esters on sulfided NiMo/Al2O3 and CoMo/Al2O3 catalysts, Catalysis Today, 100, 331- 335

Sohounloue D.K. Montassier C. Barbier 1982Hydrogenolysis of sorbitol. React. 22, Kinet. Catal. Lett., 391-397

Venderbosch R. H. Prins W. 2010 Fast Pyrolysis Technologies, 4(2), Biofuels Bioproducts and Biorefining 178- 208

Venderbosch R. H. Ardiyanti A. R. Wildschut J. Oasmaa A. Heeres H. J. 2010 Stabilization of biomass derived pyrolysis oils, Journal of Chemical Technology & Biotechnology, 85, pp. 674- 686

Vispute T. P. Zhang H. Y. Sanna A. Xiao R. Huber G. W. 2010 Renewable Chemical Commodity Feedstocks from Integrated Catalytic Processing of Pyrolysis Oils, Science 1222-1227

Watanabe, M., Aizawa, Y., Iida, T., Aida, T.M., Levy, C., Sue, K, Inomata, H.2005a Glucose reactions with acid and base catalysts in hot compressed water at 473 KCarbohydr. Res., 340(12),1925- 1930

Watanabe, M., Aizawa, Y., Iida, T., Levy, C., Aida, T.M., Inomata, H. (2005b). Glucose reactions within the heating period and the effect of heating rate on the reactions in hot compressed water. Carbohydr. Res., 340(12), 1931- 1939

Wildschut J. Arentz J. Rasrendra C. B. Venderbosch R. H. Heeres H. J. 2009aCatalytic Hydrotreatment of Fast Pyrolysis oil: Model studies on Reaction Pathways for the Carbohydrate Fraction, Environmental Progress & Sustainable Energy, 28(3), 450 460

Wildschut J. Mahfud F. H. Venderbosch R. H. Heeres H. J. 2009bHydrotreatment of Fast Pyrolysis Oil Using Heterogeneous Noble-Metal Catalysts, Industrial & Engineering Chemistry Research 48, 10324 10334

Wildschut J. Melián-Cabrera I. Heeres H. J. 2010aCatalyst studies on the hydrotreatment of fast pyrolysis oil, Applied Catalysis B: Environmental, 99, 298 306

Wildschut J. Iqbal M. Mahfud F. H. Melian Cabrera. I. Venderbosch R. H. Heeres H. J. 2010bInsights in the hydrotreatment of fast pyrolysis oil using a Ru on C catalyst., Energy and Environmental Science, 3, 962 970

Xu Y. Wang T. Ma L. Zhang Q. Liang W. 2010Upgrading of the liquid fuel from fast pyrolysis of biomass over MoNi/γ-Al2O3 catalysts, Applied Energy, 87, 2886- 2891

CHAPTER-18

BIOMASS FEEDSTOCK PRE-PROCESSING PART 1: PRE-TREATMENT

1. INTRODUCTION: AN OVERVIEW

1.1. Sources of biomass

The primary sources of biomass for energy production are specifically cultivated energy crops and discarded materials (Larkin et al., 2004). Energy crops, such as Miscanthus and short rotation woody crops (coppice), are primarily grown for the purpose of producing energy. These crops are often at the center of the food vs. fuels argument, which questions whether land should be allocated for fuel production instead of food production. Utilizing agricultural wastes, such as barley, canola, oat, and wheat straw, for energy production resolves the conflict between using food for fuel and enhances the worth of current crops (Chico-Santamarta et al., 2009). These leftovers are a plentiful, affordable, and easily accessible source of renewable lignocellulosic biomass (Liu et al., 2005).

1.2. Contemporary concerns about the exploitation of biomass

The primary issue with agricultural straw lies in its very low density in both its natural and baled states. The loose straw has a bulk density of around 40 kg/m3, whereas the conventional baled straw has a bulk density of about 100 kg/m3. In comparison, untreated wood residue has a bulk density of approximately 250 kg/m3 (Demirbaş, 2001; Tripathi et al., 1988). The relatively low density of straw results in higher transportation costs compared to wood and coal, as a smaller amount of straw can be transported per unit volume. Moreover, baled straw necessitates a greater storage capacity in comparison to wood chips. Pelletizing biomass increases its bulk density, leading to an increase in the net calorific content per unit volume (Bhattacharya et al., 1989). This densification also makes storage, transport, and handling of the material more convenient and cost-effective (Balatinecz, 1983; Bhattacharya et al., 1989; Kaliyan and Morey, 2006). The evaluation of fuel pellet quality often relies on its density and durability. The storage and transit capacity of pellets can be enhanced by increasing their bulk density (Adapa et al., 2007; Mani et al., 2003). In order to maintain consistent feeding of boilers and gasifiers, it is important to prevent any fluctuations in bulk density (Larsson et al., 2008). The design value for wood pellet makers, as stated by Obernberger and Thek in 2004, is a bulk density of 650 kg/m3. The limited durability of pellets leads to issues such as disruptions in pellet feeding systems, the release of dust particles, and an elevated danger of fire and explosions during the handling and storage of pellets (Temmerman et al., 2006).

Densification of straw and determining the optimal parameters involved is an art in itself. The entire process involves securing of baled straw from agricultural fields, size reduction (chopping and grinding), application of pre-treatment (chemical, physico-chemical, and biological), determining the physical and frictional properties of straw grinds, lignocellulosic characterization of straw, lab-scale and pilot-scale densification of grinds into pellets to determine the effect of various independent parameters on quality (density and durability), and energy analysis/ balance (Fig. 1). This chapter will only address the effect and need of lignocellulose characterization, pre-treatment and size reduction, and physical properties on densification of agricultural straw.

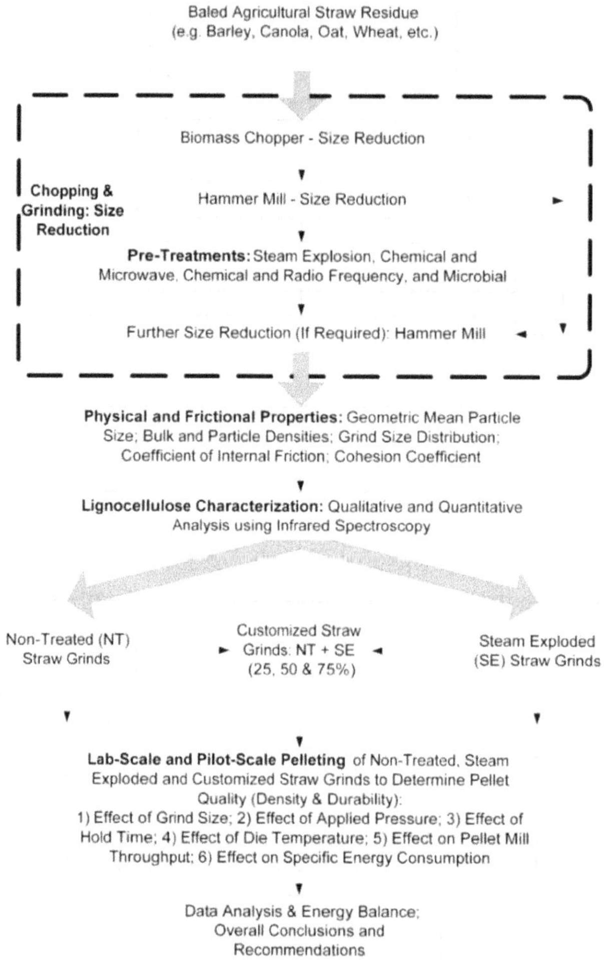

Figure 1. Processing steps involved in converting straw from field to pelletized product.

2. CHARACTERIZATION OF LIGNOCELLULOSIC BIOMASS

2.1. Structure of lignocellulosic material

Lignocellulosic material refers to plant biomass that is composed of cellulose, hemicellulose, and lignin (Fig. 2) (Lin and Tanaka, 2006). The major combustible component of non-food energy crops is cellulose, followed by lignin.

Cellulose: Cellulose is an organic polysaccharide consisting of a linear chain of several hundreds to over nine thousand **β**(1→4) linked D-glucose $(C6H10O5)n$ units (Crawford, 1981; Updegraff, 1969). Cellulose, a fibrous, tough, water-insoluble substance, is found in the cell walls of plants, particularly in the stalks, stems, trunks and all the woody portions of the plant body (Nelson and Cox, 2005). Cellulose comprises 40-60% of the dry weight of plant material (the cellulose content of cotton is 90% and that of wood is 50%) (Encyclopædia Britannica, 2008; USDE, 2006).

Zandersons et al. (2004) and Shaw (2008) reported that binding of wood material during hot pressing / densification is mainly dependent on the transition of cellulose into the amorphous state. According to Hon (1989), due to the semi-crystalline structure, hydrogen bonded cellulose cannot be dissolved easily in conventional solvents, and it cannot be melted before it burns; hence, cellulose itself is not a suitable adhesive. This can be overcome by breaking the hydrogen bonds, thus making the cellulose molecule more flexible (Hon 1989). Cellulose requires a temperature of 320°C and pressure of 25 MPa to become amorphous in water (Deguchi et al., 2006).

Hemicellulose: Hemicellulose is made of several heteropolymers (matrix polysaccharides) present in almost all plant cell walls along with cellulose (Fig. 2). While cellulose is crystalline, strong, and resistant to hydrolysis; hemicellulose has a random, amorphous structure with less strength. Hemicellulose is a polysaccharide related to cellulose and comprises 20-40% of the biomass of most plants. In contrast to cellulose, hemicellulose is derived from several sugars in addition to glucose, including especially xylose but also mannose, galactose, rhamnose and arabinose (Shambe and Kennedy, 1985). Branching in hemicellulose produces an amorphous structure that is more easily hydrolyzed than cellulose (Shaw, 2008). Also, hemicellulose can be dissolved in strong alkali solutions. Hemicellulose provides structural integrity to the cell. Some researchers believe that natural bonding may occur due to the adhesive properties of degraded hemicellulose (Bhattacharya et al., 1989).

Lignin: Lignin is a complex chemical compound most commonly derived from wood and is an integral part of the cell walls of plants (Lebo et al., 2001; Zandersons et al., 2004). The compound has several unusual properties as a biopolymer, not the least its heterogeneity in lacking a defined primary structure. Lignin fills the spaces in the cell wall between cellulose and hemicellulose (Fig. 2). It is covalently linked to hemicellulose and thereby crosslinks different plant polysaccharides, conferring mechanical strength to the cell wall and consequently to the whole plant structure (Chabannes et al., 2001).

Lignin acts as a binder for the cellulose fibres (Fig. 2). van Dam et al. (2004) have reported that lignin can be used as an intrinsic resin in binderless board production due to the fact that when lignin melts (temperatures above 140°C), it exhibits thermosetting properties. Lignin is the component that permits adhesion in the wood structure, and is a rigidifying and bulking agent (Anglès et al., 2001). Lehtikangas (2001) reported that water (8-15%) in pellets will reduce the softening temperature of lignin to 100-135°C by plasticizing the molecular chains. The adhesive properties of thermally softened lignin are thought to contribute considerably to the strength characteristics of briquettes made of lignocellulosic materials (Granada et al., 2002; Shaw, 2008).

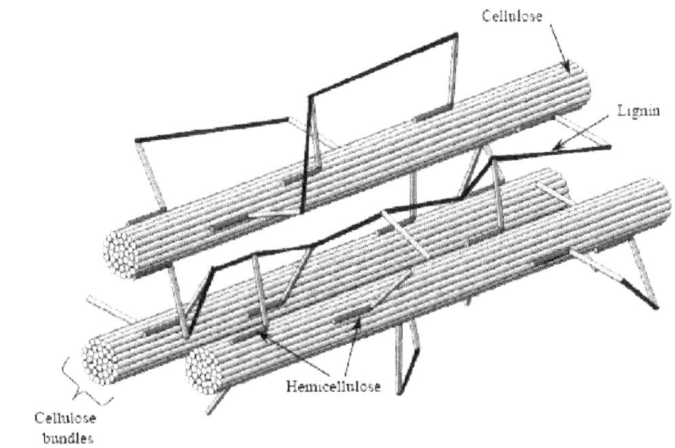

Figure 2.Location and arrangement of cellulose microfibrils in plant cell walls (Murphy and McCarthy, 2005; Shaw, 2008).

2.2. Rapid characterization of lignocellulosic materials

The molecular-level impact of several pre-processing and pre-treatment techniques (Fig. 1) on the lignocellulosic matrix remains poorly comprehended. Utilizing pre-processing methods like size reduction or increasing porosity, as well as pre-treatment techniques like steam explosion, on agricultural biomass has shown to enhance the quality of pellets by modifying the lignocellulosic components and distribution (Bagby, 1982; Focher et al., 1998). Hence, it is imperative to promptly measure the alteration in cellulose, hemicelluloses, and lignin constituents of biomass resulting from the implementation of pre-treatment techniques.

Infrared spectroscopy has the capability to generate qualitative and quantitative analytical data for materials with little or no preparation, and at a rapid pace and high efficiency (Adapa et al., 2011b and 2009; Budevska, 2002; Luypaert et al., 2003; Smola and Urleb, 2000; Tucker et al., 2000). In the past, the chemical analysis of the different parts (such as lignin) of lignocellulosic materials was done by using acid hydrolysis and then measuring the weight of the lignin (Kelley et al., 2004). These approaches have the capability to yield data of exceptional precision. Nevertheless, these techniques require a significant amount of manual effort, are time-consuming, and as a result, are costly to execute. Additionally, the rate at which samples may be processed is restricted.

2.3. Fourier transform infrared spectroscopy(FTIR)

Fourier Transform Infrared Spectroscopy (FTIR) can be used to rapidly characterize and quantify cellulose-hemicellulose-lignin composition prior to and after application of various methods of pre-processing and pre-treatment of biomass (Adapa et al., 2009). The quantitative analysis of FTIR absorption spectrometry is based on the Bouguer-Beer-Lambert law (Sherman Hsu, 1997). According to this law, the intensities of absorption bands are linearly proportional to the concentration of each component in a homogenous mixture or solution.

Figure 3.Experimental procedure followed to characterize lignocellulosic composition of agricultural straw (Adapa et al., 2011b).

Regression equations to predict the lignocellulosic content of agricultural biomass can be developed using pure cellulose, hemicelluloses and lignin as reference samples, and subsequently mixing them in different proportions to determine the change in absorption intensity at characteristic peak height (Adapa et al., 2011b). An overview of the experimental procedure to characterize the lignocellulosic composition is provided in Figure 3.

Pure cellulose has five distinct characteristic/ prominent peaks at wavenumbers of 1431, 1373, 1338, 1319 and 1203 cm-1. Similarly, hemicellulose (xylan) has prominent peaks at wavenumbers of 1606, 1461, 1251, 1213, 1166 and 1050 cm-1. The lignin spectrum has characteristic peaks at wavenumber of 1599, 1511, 1467, 1429, 1157 and 1054 cm-1. The intensity of absorption at characteristic peak heights of cellulose, hemicellulose and lignin were used to develop regression equations to predict lignocellulosic composition of any agricultural biomass (Table 1) (Adapa et al., 2011b).

Table 1.Regression equations to predict the lignocellulosic composition of agricultural biomass (Adapa et al., 2011b).

Equation	% Mean Absolute Deviation
$\%Cellulose = -135.10 + 781.35(PH\_1319) - 795.57(PH\_1431) - 135.26(PH\_1203) + 436.11(PH\_1338) - 94.24(PH\_1373)$	7.5
$\%Hemicellulose = 1638.72 - 2581.71(PH\_1251 \times PH\_1461) - 1260.90(PH\_1213) - 2518.05(PH\_1166) + 1573.69(PH\_1213 \times PH\_1251) + 118.74(PH\_1050) + 3128.51(PH\_1166 \times 1251) + 2179.65(PH\_1461) + 92.36(PH\_1606) - 2294.15(PH\_1251) - 59.29(PH\_1461 \times PH\_1606)$	2.5
$\%Lignin = 7110.87 + 388.32(PH\_1511 \times PH\_1599) - 16440.93(PH\_1467) + 447.36(PH\_1599)^2 + 19572.82(PH\_1157 \times PH\_1467) + 18374.36(PH\_1157) + 15659.98(PH\_1054 \times PH\_1429) - 4952.80(PH\_1157 \times PH\_1599) + 800.20(PH\_1511) - 3032.75(PH\_1429)^2 - 11269.16(PH\_1429) - 948.04(PH\_1511)^2 + 3444.69(PH\_1599) - 12344.90(PH\_1054) - 16689.44(PH\_1157)^2$	3.8
Note: PH – Characteristic Peak Height (Photoacoustic Units)	

3. PRE-TREATMENT OF LIGNOCELLULOSIC BIOMASS

3.1. Need for pre-treatment

Upon densification, many agricultural biomass materials, especially those from straw and stover, result in a poorly formed pellets or compacts that are more often dusty, difficult to handle and costly to manufacture. This is caused by lack of complete understanding on the natural binding characteristics of the components that make up biomass (Sokhansanj et al., 2005). The natural binding characteristics of lignocellulosic biomass can be enhanced by modifying the structure of cellulose-hemicellulose-lignin matrix by application of pre-processing and pre-treatment methods (Sokhansanj et al. 2005). It is postulated that by disrupting the lignocellulosic matrix of biomass materials via application of various chemical, physico-chemical (steam explosion, microwave, and radio frequency heating), and biological pre-treatment, the compression and compaction characteristics can be improved (Shaw 2008; Kashaninejad and Tabil, 2011). When high molecular amorphous polysaccharides are reduced to low molecular components, the polymer becomes more cohesive in the presence of moisture (Chen et al., 2004). The cellulose-hemicellulose-lignin matrix can be broken down to smaller amorphous molecules through acid or alkaline hydrolysis as well as through steam explosion (Ladisch, 1989; Vlasenko, 1997). Alkaline or acid solutions are often used for pre-treatment of biomass and the effect of pre-treatment depends on the lignin content of biomass. When biomass is treated with dilute alkaline solution, the internal surface area of the material is increased by swelling. Swelling causes a decrease in the degree of polymerization, separation of structural linkages between lignin and carbohydrates and disruption of the lignin structure (Fan et al., 1987). Increased moisture content resulting from chemical and enzymatic treatments is a problem, as the treated biomass has to be dried prior to densification. Steam explosion results in the hemicelluloses being hydrolyzed and water soluble, the cellulose is slightly depolymerized, the lignin melts and is depolymerized, which aid in binding particles together during densification. Zandersons et al. (2004) stated that activation of lignin and changes in the cellulosic structure during the steam explosion process facilitate the formation of new chemical bonds. Lam et al. (2008) reported that the quality (durability) of compacts produced from steam exploded sawdust was 20% higher than non-treated sawdust.

3.2. Physico-chemical pre-treatments

3.2.1. Steam explosion

Steam explosion is one of the most applied pre-treatment processes owing to its low use of chemicals and limited energy consumption (Harmsen et al., 2010). During steam explosion pre-treatment process, the lignocellulosic biomass is heated with high pressure saturated steam having temperatures typically in the range of 180-230°C for 2-10 minutes. Subsequently, the substrate is quickly flashed to atmospheric pressure; as a result, the water inside the substrate vaporizes and expands rapidly, disintegrating the biomass (Grous et al., 1985; Kokta and Ahmed, 1998; Zimbardi et al., 1999). This causes great reduction in the particle size of the substrate (Fig. 4). The heart of the explosion pulping process is the reactor, which allows the use of high pressure during heating and cooking. The reactor can be of either the batch (Fig. 5) (Jin and Chen, 2006) or continuous type (Fig. 6) (Kokta and Ahmed, 1998; Adapa et al., 2010a).

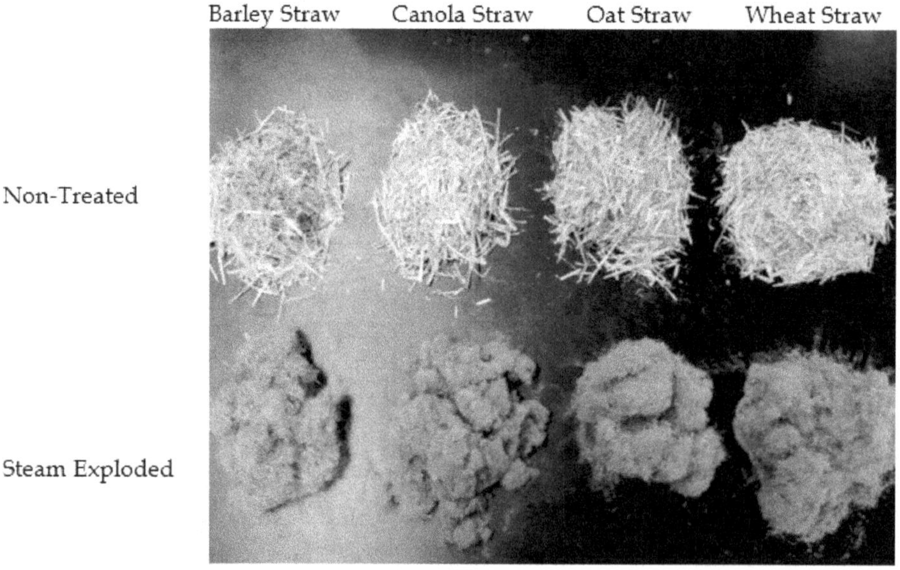

Figure 4.Photographs showing the non-treated (30 mm hammer mill screen size) and steam exploded barley, canola, oat and wheat straw grinds.

The extent of chemical and structural modifications from steam-explosion pre-treatment depends on residence time, temperature, particle size and moisture content (Sun and Cheng, 2002). However, the severity (Ro) of steam explosion is quantified as a function of retention time and reaction temperature (Equation 1) (Overend and Chornet, 1987; Viola et al. 2008).

$$\text{Ro} = t \times \exp\left(\frac{T-100}{14.75}\right) \quad (1)$$

Where T is the temperature in °C and t is the time in minutes.

According to Zimbardi et al. (1999), the simplest way to carry out steam explosion is by batch procedure, hence widely reported in literature. However, the continuous reactors are of major interest for industrial applications. They have indicated that although the products obtained at the same treatment, severity in batch and continuous reactors are macroscopically different at first sight, there is still a lack of understanding to explain these differences. Consequently, they have developed experimental relationships between the two systems useful in making the data transfer straightforward (Equation 2).

$$log(Ro)_{Batch} = 1.50 \times (log(Ro)_{Continuous} - 1) \quad (2)$$

In addition, studies have been carried out to try to improve the results of steam explosion by addition of chemicals such as acid or alkali (Cara et al., 2008; Harmsen et al., 2010; Stenberg et al., 1998; Zimbardi et al., 2007). Limitations of steam explosion include the formation of degradation products that may inhibit downstream processes (Garcia-Aparicio et al., 2006).

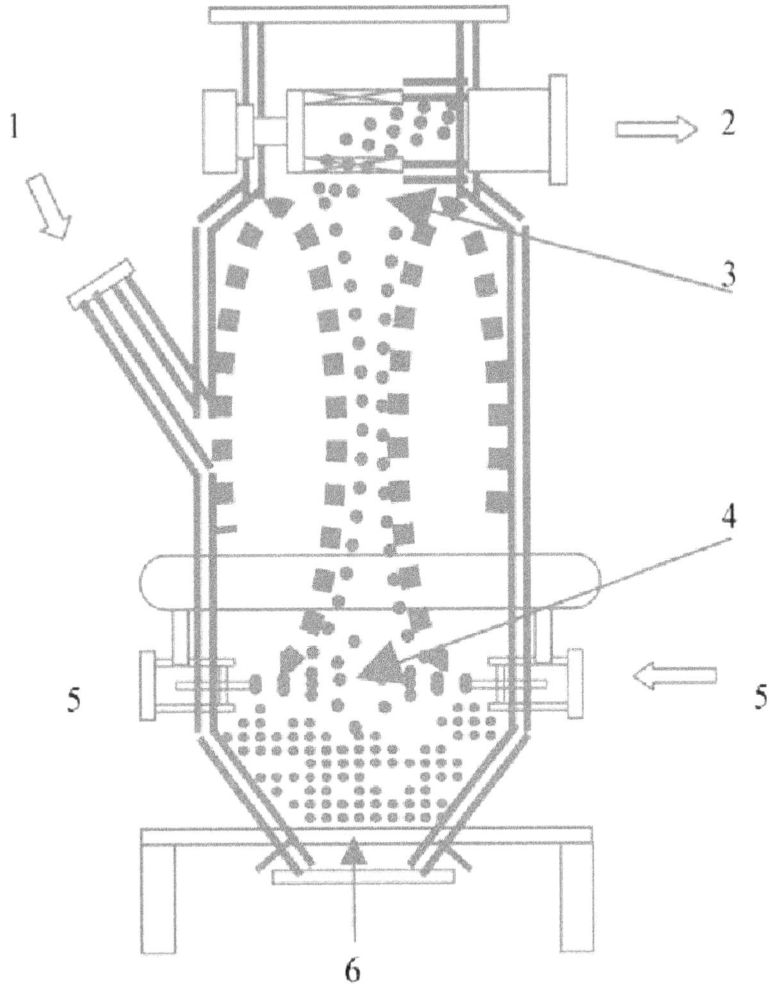

Figure 5.Schematic diagram of the FJM-200 fluidized bed opposed jet mill. 1, Infeed; 2, collection; 3, classification section; 4, grinding section; 5, compressed air; 6, discharge opening (Jin and Chen, 2006).

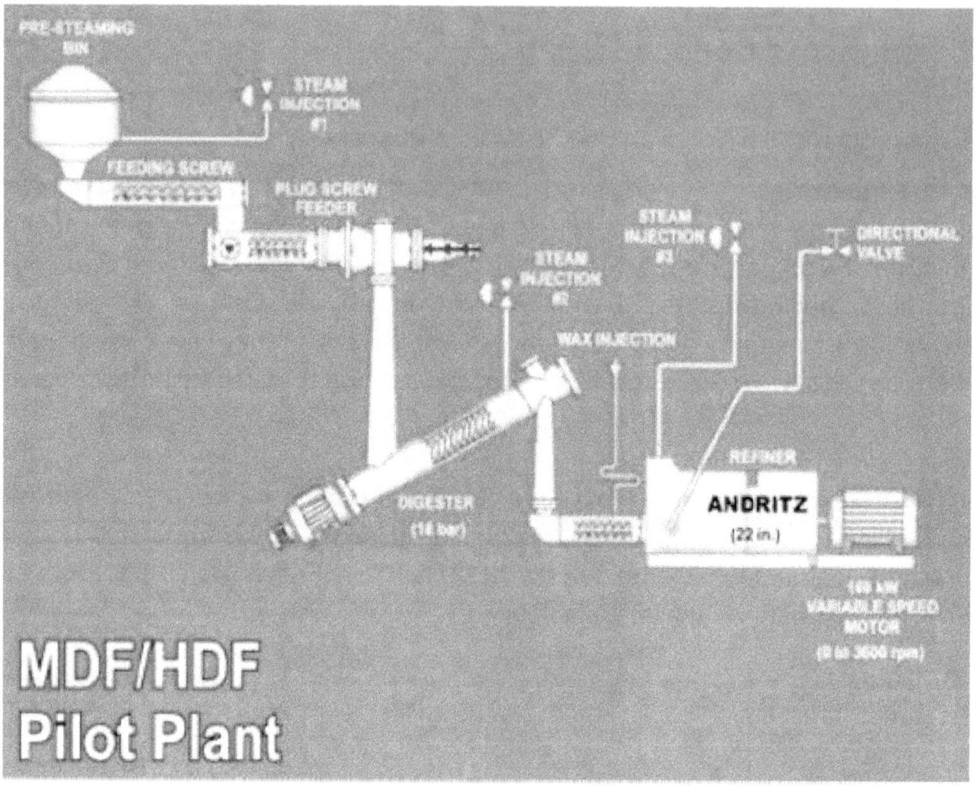

Figure 6.The Andritz (ANDRITZ AG, Graz, Austria) continuous biomass steam explosion facility for manufacturing of Medium Density and High Density Fiberboards (MDF/HDF), Forintek pilot plant at the FPInnovations, Quebec City, Quebec (Adapa et al., 2010a).

3.2.2. Microwave and radio frequency (RF) heating

Dielectric heating is an alternative method to conventional heating. Unlike conduction/convection heating, which is based on superficial heat transfer, dielectric heating is based on volumetric and rapid heat transfer (de la Hoz et al., 2005). When lignocellulosic materials are placed in an electric field for dielectric heating pre-treatment, dipole molecules such as water or other dielectric materials, rotate vigorously to orient in the field. More polar components will absorb more energy, and thus, "hot spots" will be created in non-homogeneous materials. It is hypothesized that this unique heating feature results in an "explosion" effect in the particles and improves the disruption of the lignocellulosic structures. In addition, the non thermal effects of electromagnetic field accelerate the disintegration of the crystal structures (de la Hoz et al., 2005). Dielectric heating can be categorized as microwave or radio frequency depending on the wavelength used in the heating devices (Oberndorfer et al., 2000).

Microwave is electromagnetic waves between 300 MHz (wavelength 1 m) and 300 GHz (wavelength 1 mm). This range of spectrum lies between infrared and radio frequency radiation. Microwave irradiation has been extensively used in many processes because of its high heating efficiency and easy operation. Microwave energy can penetrate into materials and heat them quickly and uniformly. Microwave irradiation is considered to create thermal and non thermal effects. It has been applied as an efficient pre-treatment technique to enhance the hydrolysis of biomass materials. Some studies have demonstrated that microwave irradiation can change the structure of lignocellulosic materials and degrade or reduce lignin content, reduce cellulose crystallinity, and increase porosity and surface area of the materials (Azuma, 1984; Zhu et al., 2006b; Kashaninejad and Tabil, 2011).

This pre-treatment involves microwave irradiation of immersed biomass in an aqueous environment. Different types of lignocellulosic materials have been pre-treated using microwave irradiation including wheat straw (Ooshima et al., 1984; Zhu et al., 2006b; 2006c; Kashaninejad and Tabil, 2011), barley straw (Kashaninejad and Tabil, 2011), rice straw (Ooshima et al., 1984; Zhu et al., 2005 ; 2006a), rice hulls (Magara and Azuma, 1989), sugarcane bagasse (Ooshima et al., 1984; Magara and Azuma, 1989; Kitchaiya et al., 2003), switchgrass (Hu and Wen, 2008; Keshwani et al., 2007) and woody materials (Azuma et al., 1984). These materials were subjected to microwave pre-treatment of 2450 MHz in the range of 250 to 1000 W. The temperature of operation ranged from 70 to 230°C, while heating time varied from 5 to 120 minutes. However, higher microwave power with short pre-treatment time and the lower microwave power with long pre-treatment time had almost the same effect on chemical composition of lignocellulosic materials (Zhu et al., 2005; 2006b; Kashaninejad and Tabil, 2011).

In order to increase the efficiency of microwave heating pre-treatment, some researchers have combined microwave treatment with alkaline treatment such as NaOH or Ca(OH)2. Some used alkaline solution during microwave heating treatment (Zhu et al., 2005; 2006a; 2006b; 2006c; Keshwani et al., 2007; Kashaninejad and Tabil, 2011) and some applied the alkaline solution before the lignocellulosic materials were subjected to microwave irradiation (Zhu et al., 2006a; Hu and Wen, 2008). Combination of microwave irradiation and alkali treatment improves the degradation of biomass by accelerating the reactions during the pre-treatment process compared with the conventional heating chemical pre-treatment process. Remarkable changes (Table 2) have been reported in the chemical composition of biomass samples after microwave-alkali pre-treatment, particularly in hemicellulose, lignin, and cellulose contents (Kashaninejad and Tabil, 2011). It has been reported that alkali treatment dissolves lignin and hemicellulose, and microwave irradiation facilitates dissolving these components in alkali solutions (Jackson, 1977; Kumar et al., 2009; Lesoing et al., 1980; Zhu et al., 2005). Biomass samples pretreated by microwave-alkali technique have lower lignin, hemicellulose, and cellulose than samples pretreated by microwave-distilled water or untreated samples. Moreover, degradation and depolymerisation of lignin to smaller phenolic components is another influence of microwave-alkali pre-treatment that could be considered as binder in densification process. The pellets made from microwave-chemical pre-treated biomass grinds have significantly higher density and tensile strength (Table 3) than the untreated or samples pre-treated by microwave alone (Kashaninejad and Tabil, 2011).

Radio frequency (RF) can penetrate more deeply into the materials compared with microwave heating because the radio frequency wavelength is up to 360 times greater than microwave (Marra et al., 2007). This unique characteristic is an advantage to treat large amount of material and it is easier to scale up the process. While radio frequency as a heating method has been widely applied in food-processing industries, there is not much report on application of radio frequency heating for lignocellulosic materials pre-treatment. Hu et al. (2008) used radio frequency heating in the NaOH pre-treatment of switchgrass to enhance its enzymatic digestibility. Because of the unique features of radio frequency heating (i.e., volumetric heat transfer, deep heat penetration of the samples, etc.), switchgrass could be treated on a large scale, at high solids content, and with a uniform temperature profile. At 20% solids content, radio frequency-assisted alkali pre-treatment (at 0.1 g of NaOH/g of biomass loading and 90°C) resulted in a higher xylose yield than the conventional heating pre-treatment. The optimal particle size and alkali loading in the radio frequency pre-treatment were determined to be 0.25-0.50 mm and 0.25 g of NaOH/g of biomass, respectively.

Table 2.Chemical composition (% dry basis) of untreated and microwave pretreated of wheat and barley straw at power 713 W.

Treatment	Wheat straw						Barley straw					
	Protein	Lignin	Ash	Starch	Cellulose	Hemicellulose	Protein	Lignin	Ash	Starch	Cellulose	Hemicellulose
Untreated 1.99b		8.33a	6.33f	1.11d	44.99b	27.96a	1.61d	11.95a	6.03d	0.79c	46.93a	27.40a
Microwave-distilled water	2.24a	8.01c	8.87e	1.48b	39.69d	22.62b	2.01a	8.85b	6.28d	1.08b	45.25b	27.21a
Microwave-NaOH (1%)	1.41e	7.82d	17.32b	1.89a	35.82e	12.32d	1.80b	6.65e	16.96b	0.60e	40.81c	8.74c
Microwave-NaOH (2%)	1.36f	7.09f	34.77a	0.27f	34.77f	4.06f	1.62d	4.52f	41.43a	0.54f	35.22d	5.46d
Microwave-Ca(OH)2 (1%)	1.85c	8.11b	12.24d	0.69e	45.66a	14.94c	1.81b	7.27d	13.21c	0.72d	41.01c	15.00b
Microwave-Ca(OH)2 (2%)	1.52d	7.55e	15.89c	1.31c	42.56c	11.10e	1.68c	8.16c	16.73b	1.19a	41.24c	8.97c

Means with the same letters designation (a, b, c, d, and e) in a column are not significantly different at P = 0.05.

Table 3.Effect of microwave-chemical pre-treatments on fracture load and tensile strength of wheat and barley straw pellets at power 713 W.

Treatment	Wheat straw		Barley straw	
	Fracture load (N)	Tensile strength (MPa)	Fracture load (N)	Tensile strength (MPa)
Untreated	19.10±5.61	0.81±0.24	16.25±5.30	0.68±0.22
Microwave- distilled water	35.00±11.93	1.48±0.46	14.25±5.31	0.61±0.21
Microwave- NaOH (1%)	85.46±22.94	3.99±0.82	57.13±12.12	2.42±0.47
Microwave- NaOH (2%)	88.00±15.86	3.69±0.66	90.75±22.42	3.59±0.98
Microwave- Ca(OH)2 (1%)	67.05±19.82	3.03±0.79	42.38±10.30	1.83±0.49
Microwave- Ca(OH)2 (2%)	78.25±25.07	3.31±1.03	67.25±19.94	2.88±0.91

3.2.3. Chemical pre-treatment

Different chemicals such as acids, alkalis, oxidizing agents and ozone have been used for chemical pre-treatment of lignocellulosic materials. Depending on the type of chemical used, pre-treatment could have different effects on structural components. Alkaline pre-treatment, ozonolysis, peroxide and wet oxidation pre-treatments were reportedly more effective in lignin removal, whereas dilute acid pre-treatment was more efficient in hemicellulose solubilization (Galbe and Zacchi, 2002; Sánchez and Cardona, 2008; Tomas-Pejo et al., 2008).

Acid Hydrolysis: Inorganic acids such as H2SO4 and HCl have been used for pre-treatment of lignocellulosic materials and have been used on a wide range of feedstocks ranging from hardwoods to grasses and agricultural residues. Acid hydrolysis can be classified as concentrated or dilute-acid hydrolysis based on the dose of acid used in the process. In the first case, the biomass is treated with high concentration of acids at ambient temperatures, which results in high conversion of lignocellulosic materials. Although concentrated acids are powerful agents for cellulose hydrolysis, they are toxic, corrosive, hazardous, and thus require reactors that are resistant to corrosion, making the pre-treatment process very expensive. In addition, the concentrated acid must be recovered after hydrolysis to make the process economically feasible (Galbe and Zacchi, 2002; Sun and Cheng, 2002).

Dilute-acid hydrolysis has been successfully developed for pre-treatment of lignocellulosic materials. Sulfuric acid at concentrations usually below 4% (wt) has been of the most interest in such studies as it is inexpensive and effective. Dilute H2SO4 pre-treatment can achieve high reaction rates and significantly improve cellulose hydrolysis (Esteghlalian et al., 1997). High temperature is favorable to attain acceptable rates of cellulose conversion. Despite low acid concentration and short reaction time, the use of high temperatures in dilute-acid hydrolysis accelerates the rate of hemicellulose sugar decomposition and increases equipment corrosion (Galbe and Zacchi, 2002; Taherzadeh and Karimi, 2007).

Alkali hydrolysis: Dilute alkali such as sodium, potassium, calcium, and ammonium hydroxides have been used for pre-treatment of lignocellulosic materials in alkali hydrolysis. The effectiveness of these agents depends on the lignin content of the materials. Temperature and pressure are lower in alkali pre-treatment compared with other pre-treatment methods (Mosier et al., 2005). Alkali pre-treatment can be conducted at ambient conditions, but process time is longer (hours or days instead of minutes or seconds). Compared with acid process, alkaline process causes less sugar degradation, and many of the caustic salts can be recovered and/or regenerated.

Sodium hydroxide has been studied more than other agents (Soto et al., 1994; Fox et al., 1989; MacDonald et al., 1983). Treatment of lignocellulosic materials using dilute NaOH results in swelling, leading to an increase in internal surface area, a decrease in the degree of polymerization, a decrease in crystallinity, separation of structural linkages between lignin and carbohydrates, and disruption of the lignin structure. However, calcium hydroxide (lime) is the least expensive hydroxide and has been shown to be an effective pre-treatment agent. The process of lime pre-treatment involves slurrying the lime with water, spraying it onto the biomass material, and storing the material in a pile for a period of hours to weeks. The particle size of the biomass is typically 10 mm or less. Elevated temperatures reduce contact time.

Oxidizing agents: In this pre-treatment, an oxidizing compound such as hydrogen peroxide (H_2O_2) or peracetic acid (CH_3CO_3H) is used to treat lignocellulosic materials and sometimes is applied in combination of an alkaline solution (e.g. NaOH) to improve effectiveness. This pre-treatment is usually carried out under mild temperature. This pre-treatment is more effective to increase crop residue digestibility compared with NaOH pre-treatment alone. Gould (1984) delignified agricultural residues using 1% H2O2 at 25°C for 18–24 h. Under this condition, more than half of the lignin and most of hemicellulose were solubilized. The pre-treatment of cane bagasse with H2O2 greatly enhanced its susceptibility to further hydrolysis. About 50% of the lignin and most of the hemicellulose were solubilized by 2% H2O2 at 30°C within 8 h, and a 95% efficiency of glucose production from cellulose was achieved in the subsequent saccharification by cellulase at 45°C for 24 h (Azzam, 1989).

Ozonolysis: In this process, ozone is used to change the structure of lignocellulosic materials and has been used for different materials such as wheat straw (Ben-Ghedalia and Miron, 1981), bagasse, green hay, peanut, pine (Neely, 1984), cotton straw (Ben-Ghedalia and Shefet, 1983) and poplar sawdust (Vidal and Molinier, 1988). Ozonolysis is carried out at room temperature and normal pressure. It can effectively remove the lignin without producing any toxic residues. In this process, hemicellulose is slightly affected, but no change in cellulose has been reported. The main restriction of this process is the large amount of ozone utilization that makes the process expensive (Sun and Cheng, 2002). Binder et al. (1980) reported 60% removal of the lignin from wheat straw using ozone pre-treatment. Enzymatic hydrolysis yield increased from 0% to 57% as the percentage of lignin decreased from 29% to 8% after ozonolysis pre-treatment of poplar sawdust (Vidal and Molinier, 1988). Garcia-Cubero et al. (2009) studied the ozonolysis pre-treatment of wheat straw in a fixed bed reactor at room conditions and concluded that enzymatic hydrolysis yield of up to 88.6% compared to 29% in non-ozonated sample.

3.3. Biological pre-treatment

Most pre-treatments require expensive instruments or equipment that require high energy requirements, depending on the process. In particular, physical and thermo-chemical processes require ample amount of energy to change the lignocellulosic structure of biomass. Biological pre-treatment using various types of rot fungi is a process that does not require high energy for lignin removal from a lignocellulosic biomass, despite extensive lignin degradation. Biological pre-treatments are safe, environmentally friendly and less energy intensive compared to other pre-treatment methods. However, the rate of hydrolysis reaction is very slow and needs a great improvement to be commercially applicable.

Biological pre-treatment comprises of using microorganisms such as brown-, white-, and soft-rot fungi for selective degradation of lignin and hemicellulose among which white-rot fungi seems to be the most effective microorganism (Fan et al., 1987). Brown rots mainly attack cellulose, while white and soft rots attack both cellulose and lignin. Lignin degradation occurs through the action of lignin-degrading enzymes such as peroxidases and laccase (Okano et al., 2005). These enzymes are regulated by carbon and nitrogen sources. The suitable fungi for biological pre-treatment should have higher affinity for lignin and degrade it faster than carbohydrate components. Hatakka et al. (1983) studied the pre-treatment of wheat straw by 19 white-rot fungi and found that 35% of the straw was converted to reducing sugars by Pleurotus ostreatus in 5 weeks. Similar conversion was obtained in the pre-treatment by Phanerochaete sordid (Ballesteros et al., 2006) and Pycnoporus cinnabarinus (Okano et al., 2005) in 4 weeks. Akin et al. (1995) also reported the delignification of bermudagrass by white-rot fungi. The biodegradation of bermudagrass stems was improved by 29-32%, after 6 weeks, using Ceriporiopsis subvermispora and by 63-77% using Cyathusstercoreus.

4. REDUCTION OF PARTICLE SIZE AND ITS IMPACT ON PHYSICAL QUALITIES

The application of pre-processing operations such as particle size reduction/ grinding is critical in order to increase the surface area of lignocellulosic biomass prior to densification (Mani et al. 2003). Particle size reduction increases the total surface area, pore size of the material and the number of contact points for inter-particle bonding in the compaction process (Drzymala, 1993). Size reduction is an important energy intensive unit operation essential for bioenergy conversion process and densification to reduce transportation costs (Bitra et al., 2009; Soucek et al., 2003). Energy consumption of grinding biomass depends on initial particle size, moisture content, material properties, feed rate of the material and machine variables (Lopo, 2002). A comprehensive summary of literature review on size reduction of lignocellulosic biomass is provided in Table 4.

Table 4.A comprehensive summary of literature review on size reduction of lignocellulosic biomass.

Type of Choppers/ Grinders	Biomass	Parameters Measured and Correlations	Observations	Reference
Hammer Mill	Non-Treated and Steam Exploded Barley, Canola, Oat and Wheat Straw	Hammer mill screen size (from 1.6 to 30.0 mm) on Specific Energy Effect of geometric mean particle size on bulk density Effect of geometric mean particle size on particle density Analysis on ground particle size distribution	Negative exponential and power correlation between geometric mean particle size with both bulk and particle density Specific energy requirement is material dependent Negative power correlation between hammer mill screen size and specific energy Shapiro-Wilk Test for normality was performed	Adapa et al., 2011a
Tub Grinder	Rectangular Wheat Straw Bales, Round Rice Straw and Corn Stover Bales	Effect of screen sizes (from 12.7 to 50.8 mm) on Specific Energy Effect of screen sizes on grinding rate Analysis on ground particle size distribution	Positive correlation between screen size and grinding rate Positive correlation between tub rotational speed and grinding rate Specific Energy is material dependent Negative power/ exponential correlation between screen size and specific energy	Arthur et al., 1982
Hammer Mill	Coastal Bermudagrass	Effect of moisture content on Specific Energy Effect of feed rate on Specific Energy	Positive correlation between moisture content and specific energy Positive correlation between feed rate and specific energy	Balk, 1964

Type of Choppers/ Grinders	Biomass	Parameters Measured and Correlations	Observations	Reference
Hammer Mill	Wheat Straw and Corn Stover	Effect of operating speeds (from 2000 to 3600 rpm) on Specific Energy Effect of hammer angles (90° and 30° hammers) on Specific Energy Analysis on ground particle size distribution	Positive correlation between operating speed and specific energy Geometric mean particle diameter decreased with an increase in hammer mill speed Specific energy increased with a decrease in hammer angle	Bitra et al., 2009
Knife and Hammer Mills	Hardwood Chips, Wheat Straw and Corn Stover	Effect of screen size (from 1.6 to 12.7 mm) on specific energy	Negative correlation between screen size and specific energy Specific Energy is material dependent	Cadoche and López, 1989
Hammer Mill	Red Winter Wheat Straw	Effect of screen size (from 1.59 to 4.76 mm) on specific energy Effect of feed rate (from 1.5 to 2.5 kg/min) on specific energy	Negative correlation was observed between screen size and specific energy Feed rate did not have significant effect on specific energy	Fang et al., 1997
Hammer Mill	Wheat Straw	Effect of screen size (3.2 and 1.6 mm) on Specific Energy Analysis on ground particle size distribution	Negative correlation between specific energy and screen size	Himmel et al., 1985
Tub Grinders	Round Bales of Corn Stover and Perennial Grasses	Effect of screen size (from 19.1 to 127.0 mm) on Specific Energy Effect of screen size on throughput Effect of screen size on bulk and particle densities	Negative correlation between screen size and bulk density Positive correlation between screen size and particle density Negative correlation between screen size and specific energy Positive correlation between screen size and throughput	Kaliyan et al., 2010
Hammer Mill	Barley Straw, Corn Stover and Switchgrass	Effect of Screen Sizes (3.2, 1.6 and 0.8 mm) on Specific Energy Effect of Moisture Content (8% and 12% wb) on Specific Energy Correlation between bulk and particle densities and geometric mean diameter Analysis on ground particle size distribution	Negative linear correlation between specific energy and hammer mill screen size at 8% mc Quadratic correlation between specific energy and hammer mill screen size at 12% mc Polynomial relations for bulk and particle densities with geometric mean particle diameters	Mani et al., 2004

Type of Choppers/ Grinders	Biomass	Parameters Measured and Correlations	Observations	Reference
Hammer Mill	Oat Straw, Rattle Grass and Miscanthus	Effect of Screen size (from 1.0 to 10.0 mm) on specific energy Effect of moisture content on specific energy	Negative power correlation between screen size and specific energy Positive correlation between moisture content and specific energy; significantly higher specific energy is required at smaller screen sizes	Soucek et al., 2003
Hammer Mill	Corn Stover	Effect of hammer thickness (6.4 and 3.2 mm) on Specific Energy Effect of hammer thickness (6.4 and 3.2 mm) on Grinding Rate Effect of hammer tip speed (54 to 86 m/s) on Specific Energy	Negative correlation between hammer thickness and Specific Energy Negative correlation between hammer thickness and Grinding Rate Positive correlation between hammer tip speed and Specific Energy	Vigneault et al., 1992
Knife Mill	Switchgrass, Corn Stover, and Wheat Straw	Effect of rotational speed (from 250 to 500 rpm) on specific energy Analysis on ground particle size distribution	Positive correlation between rotational speed and specific energy Screen size has significant effect on particle size distribution	Womac et al., 2007

4.1. Chopping

Baled agricultural biomass from the field does not have good flowing characteristics and may not flow easily into grinders such as hammer mills and disc refiners. Therefore, biomass needs to be chopped with a chopper (rotary shear shredder)/ knife mill/ tub grinder to accommodate bulk flow and uniformity of feed rate. A chopper, knife cutter, or knife mill is often used for coarse size reduction (>50 mm) of stalk, straw, and grass feed stocks (Bitra et al., 2009). Knife mills reportedly worked successfully for shredding forages under various crop and machine conditions (Cadoche and López, 1989).

Bitra et al. (2009) reported that the total specific energy (including energy to operate the knife mill) for agricultural biomass chopping increases with knife mill speed. The total specific energy for knife mill and tub grinder has been observed to have negative correlation with screen size and mass feed rate (Arthur et al., 1982; Bitra et al., 2009; Himmel et al., 1985). However, grinding rate (throughput) increases with an increase in screen size (Arthur et al., 1982).

For tub grinders, an increase in screen size results in an increase in geometric mean length of particles and throughput, but a decrease in bulk density of the particles and specific energy consumption (Kaliyan et al., 2010).

4.2. Hammer mill grinding

Typically, hammer mills are used in forage processing industry as they are relatively inexpensive, easy to operate and produces wide range of particles (Lopo, 2002). Hammer mills have achieved merit because of their ability to finely grind a greater variety of materials than any other machines (Scholten et al., 1985). The performance of a hammer mill is measured in terms of energy consumption and geometric mean diameter and particle size distribution of the ground product (Adapa et al., 2011a; Mani et al., 2004).

Screen Size: Hammer mill screen opening size was the most significant factor affecting mill performance (Fang et al., 1997) and also has significant effect on mean particle size (Pfost and Headley, 1971). The specific energy required to grind agricultural biomass significantly increases with a decrease in hammer mill screen size and shows a negative power correlation (Arthur et al., 1982; Soucek et al., 2003). Similarly, Adapa et al. (2011a) reported negative correlation between specific energy and particle size of biomass as affected by hammer mill screen sizes. However, two other studies reported a second-order polynomial relationship between the specific energy requirements for grinding biomass (Mani et al. 2004; Sitkei, 1986). Usually, the mean geometric particle size for any particular biomass decreases with a decrease in hammer mill screen size (Adapa et al., 2011a). It has been reported that wider particle size distribution is suitable for compaction (pelleting/ briquetting) process (Adapa et al., 2011a; Mani et al., 2004). During compaction, smaller (fine) particles rearrange and fill in the void space of larger (coarse) particles producing denser and durable compacts (Tabil, 1996).

Operating Speed (Peripheral Velocity): The speed has a significant effect on mean particle size (Pfost and Headley, 1971). The total specific energy of hammer mill grinding has direct correlation to an increase in hammer tip speed (Bitra et al., 2009; Vigneault et al., 1992). High speed hammer mills with smaller diameter rotors are good for fine or hard-to-grind material. However, at high tip speeds, the material moves around the mill parallel to the screen surface making the openings only partially effective. At slower speeds, the material impinges on the screen at a greater angle causing greater amounts of coarser feed to pass through (Balk, 1964).

Hammer Angles and Thickness: The direct energy input for grinding also depends on hammer angles. In general, the specific energy for grinding decreases with an increase in hammer degrees (Bitra et al., 2009). In addition, the specific energy for grinding increases with an increase in hammer thickness (Vigneault et al., 1992).

Material Moisture Content and Feed Rate: A positive correlation has been reported between moisture content and specific energy consumption for grinding of agricultural biomass (Balk, 1964; Mani et al., 2004; Soucek et al., 2003). Feeding rate also has significant effect on specific energy consumption during hammer mill grinding and has positive correlation (O'Dogherty, 1982).

Bulk and Particle Densities, and Geometric Mean Particle Size: Usually, the bulk and particle density of agricultural straw significantly increases with a decrease in hammer mill screen size (Adapa et al., 2011a). The geometric mean particle size of pre-treated straw is usually smaller than that of the non-treated straw. This could be due to the fact that application of pre-treatment disrupts/ disintegrates the lignocellulosic structure of the biomass (Sokhansanj et al., 2005) leading to lower shear strength (easier to grind the straw).

4.3. Characteristics and qualities related to the physical and frictional behavior of biomass

Bulk Density: The goal of densification is to increase the bulk density of agricultural straw to facilitate economic storage, transportation and handling of the material. In addition, densification results in an increase in the net calorific content per unit volume. The bulk density of agricultural biomass depends on the type of biomass, moisture content, grind size, and pre-treatment (Mani et al., 2006). Lower bulk densities, and concerns with uneven and low flowability of straw grinds are critical issues to sustainable production of pellets using pellet mills (Adapa et al., 2009; Larsson et al., 2008).

Typically, the bulk density of ground straw increases with a decrease in hammer mill screen size. Also, pre-treatments usually results in a decrease in bulk density since the organized lignocellulosic structure of biomass is disturbed/ disintegrated. In addition, the bulk density and geometric mean particle size of material is correlated by either power or exponential relations (Adapa et al., 2010b; Mani et al., 2004). Table 5 shows a summary of average bulk density of various agricultural biomasses ground using a hammer mill.

Particle Density: Particle size of the grinds will have direct effect on the final pellet density. Theoretically, the density of pellet can be as high as the particle density of the ground biomass. Similar to bulk density, particle density also depends on the type of biomass, moisture content, grind size, and pre-treatment (Adapa et al., 2010b). The particle density is observed to have negative correlation with hammer mill screen size. Application of pre-treatment increases the particle density since disturbance/ disintegration of lignocellulosic structure results in finer components (Adapa et al., 2010b). Table 5 shows a summary of average particle density of various agricultural biomasses ground using a hammer mill.

Geometric Mean Particle Size and Distribution: It has been reported that wider particle size distribution is suitable for compaction (pelleting/briquetting) process (Mani et al., 2004a). During compaction, smaller (fine) particles rearrange and fill in the void space of larger (coarse) particles producing denser and durable compacts (Tabil, 1996). Therefore, ideally the grinds should be normally distributed, should have near zero skewness and lower peak than expected for the normal and wider distribution of data (negative Kurtosis values). In addition, a decrease in the biomass grind size has been observed to have a positive effect on pellet mill throughput (Adapa et al., 2004).

Frictional Properties: Prior to densification, biomass grinds need to be efficiently stored, handled and transported. Physical and frictional properties of biomass have significant effect on design of new and modification of existing bins, hoppers and feeders (Fasina et al., 2006). The frictional behavior of biomass grinds in all engineering applications is described by two independent parameters: the coefficient of internal friction, and the coefficient of wall friction. The former determines the stress distribution within particles undergoing strain, and the latter describes the magnitude of the stresses between the particle and the walls of its container (Seville et al., 1997). The classic law of friction states that frictional force is directly proportional to the total force that acts normal to the shear surfaces (Chancellor, 1994; Chung and Verma, 1989; Larsson, 2010). Frictional force depends on the nature of the materials in contact but is independent of the area of contact or sliding velocity (Mohsenin, 1970). Material properties such as moisture content and particle size affect the frictional properties and densification performance of an individual feedstock (Larsson 2010; Shaw and Tabil, 2006). In addition, the determination of coefficient of friction is essential for the design of production and handling equipment and in storage structures (Adapa et al., 2010a; Puchalski and Brusewitz, 1996). A comprehensive summary of literature review on coefficient of internal friction and cohesion of agricultural biomass is provided in Table 6.

Predominantly, a linear correlation exists between normal and shear stress for agricultural straw grinds (Adapa et al., 2010a; Chevanan et al., 2008; Richter, 1954) at any specific hammer mill screen size. An increase in hammer mill screen size significantly decreases the shear stress for ground straw at any specific normal stress (Adapa et al., 2010a).

The correlation for coefficient of internal friction and cohesion with average geometric mean particle sizes for agricultural straw grinds is provided in Adapa et al. (2010a). These correlations can be used to predict the coefficient of internal friction (slope of the linear plot) and the cohesion (intercept of the linear plot) for various geometric mean particle sizes. In general, the coefficient of internal friction for ground agricultural straw decreases with an increase in average geometric mean particle diameter. The coefficient of cohesion for straw grinds increases with an increase in average geometric mean particle size (Adapa et al., 2010a).

Table 5.Average bulk and particle densities of various agricultural biomasses ground using a hammer mill

Biomass	Hammer Mill Screen Size (mm)	Moisture Content (%, wb)	Geometric Mean Particle Diameter (mm)	Bulk Density (kg/m^3)	Particle Density (kg/m^3)	Reference
Barley	6.4	8.9	0.88	96	1046	Adapa et al., 2011a
	3.2	5.3	0.46	149	1089	
	1.6	7.8	0.46	155	1149	
Canola	6.4	12.6	0.89	144	1019	Adapa et al., 2011a
	3.2	9.2	0.52	190	1192	
	1.6	8.3	0.37	203	1309	
Corn Stover	3.2	6.22	0.41	131	1170	Mani et al., 2004
	1.6	6.22	0.26	156	1330	
	0.8	6.22	0.19	158	1340	
Oat	6.4	10.9	0.94	111	873	Adapa et al., 2011a
	3.2	9.4	0.57	156	1093	
	1.6	7.7	0.40	196	1240	
Wheat	6.4	9.5	0.99	107	1078	Adapa et al., 2011a
	3.2	9.5	0.72	141	1225	
	1.6	8.6	0.45	154	1269	

Table 6.A comprehensive summary of literature review on coefficient of internal friction and cohesion of agricultural biomass

Biomass	Experimental Conditions	Observations	Reference
Corn Stover	Normal stress from 10 to 200 kPa Hammer mill screen sizes of 6.35 and 3.18 mm Moisture contents of 7, 11 and 15% (wb) Galvanized steel	Coefficient of wall friction increased from 0.18 to 0.26 with an increase in moisture from 7 to 15% No clear trend observed for the adhesion coefficient	Mani et al., 2004
Peat hull, Switchgrass and Poultry Litter	Consolidating stress from 1.5 to 12.0 kPaScreensizesfrom0.79to3.20mm Ring shear test	No effect of screen size on internal friction and cohesive properties	Fasina et al., 2006
Peat Moss, Wheat Straw, Oat Hulls and Flax Shives	Normal stress from 10 to 400 kPa Geometric mean particles sizes of 0.74 (peat moss), 0.65 (wheat straw), 0.47 (oat hulls) and 0.64 (flax shives) mm Moisture content 9-10% (wb) Mild steel surface	Coefficient of wall friction were 0.68 (peat moss), 0.45 (wheat straw), 0.39 (oat hulls), and 0.41 (flax shives) Adhesion coefficients were 0.2635 kPa (peat moss), 10.687 kPa (wheat straw), 4.719 kPa (oat hulls), and 16.203 kPa (flax shives)	Shaw and Tabil, 2006
Alfalfa, Barley Straw, Wheat Straw and Whole Green Barley	Normal stress from 200 to 735 kPa Moisture content for alfalfa, barley straw from 12.0 to 45.7%, and for wheat straw at 10.0% and whole green barley at 51.0% (wb)Chop size from 10 to 90 mm Polished steel surface	Coefficient if friction on steel surface for alfalfa and barley straw increased with moisture content and was from 0.15 to 0.26, and 0.14 and 0.27, respectively Coefficient of friction for wheat straw and whole green barley were 0.13 and 0.21, respectively No effect of chop size on coefficient of internal friction on barley straw	Afzalinia and Roberge, 2007

Biomass	Experimental Conditions	Observations	Reference
Switchgrass, Wheat straw, and Corn Stover	Normal stresses from 1.23 to 4.92 kPa Chop size of 7.81 and 13.50 mm for switchgrass, 7.09 and 10.39 mm for wheat straw, 7.80 and 14.89 mm for corn stover	No effect of chop size on friction coefficients Coefficient of internal friction varied from 0.765 to 1.586	Chevanan et al., 2008
Reed Canary Grass	Normal stresses of 0.52-7.52 kPa (low) and 23-275 MPa (high) Moisture contents from 6.7 to 17.1% (wb) for low normal stress, and 8.9 to 27.2% for high normal stress Screen sizes of 4.0 mm Ring shear test	High friction value of 0.6 was obtained at normal stress of 50 MPa and lower At low normal stresses, the coefficient of kinematic wall friction (ratio of shear stress and normal stress) was positively correlated with moisture content and negatively correlated to normal stress At high normal stresses, the coefficient of kinematic wall friction was negatively correlated to both moisture content and normal stress	Larsson, 2010
Non-Treated and Steam Exploded Barley, Canola, Oat and Wheat Straw	Normal stress from 9.8 to 39.2 kPa Moisture content of 10% (wb) Screen size from 1.6 to 6.4 mm	No effect of screen size on shear stress values Steam exploded straw had higher coefficient of internal friction than non-treated straw grinds Coefficient of friction for non-treated barley, canola, oat and wheat straw were in the range of 0.505 to 0.584, 0.661 to 0.665, 0.498 to 0.590, and 0.532 to 0.591, respectively. Coefficient of friction for steam exploded barley, canola, oat and wheat straw were in the range of 0.562 to 0.738, 0.708 to 0.841, 0.660 to 0.860, and 0.616 to 1.036, respectively	Adapa et al., 2010a

5. CONCISE SUMMARY

The present section has examined the impacts of pre-treatment (chemical, physico-chemical, and biological) and pre-processing (size reduction) methods on the compaction of agricultural straw, leading to the production of pellets with superior quality in terms of density and durability. Studies have shown that when the bulk density of biomass increases, the net calorific content per unit volume of pellets also increases. This increase in bulk density also makes it easier and more cost-effective to store, transport, and handle the biomass. Prior to treatment and processing, procedures are employed to break down the fundamental lignocellulosic structure of biomass and alter the relative content of lignin, cellulose, and hemicelluloses in the material. Furthermore, the physical and frictional properties of agricultural straw undergo changes. Pre-treatments are used to disrupt the long-chain hydrogen bond in cellulose, causing hemicelluloses to become amorphous and freeing the lignin from the lignocellulosic matrix. This leads to the production of higher quality pellets in terms of their physical properties. During this process, the large, disordered polysaccharides are broken down into smaller molecules, which then become more tightly bound together in the presence of moisture during the densification process. Reducing the size of particles enhances the overall surface area, pore size, and the number of contact points for inter-particle bonding during compaction. Research has demonstrated that Fourier Transform Infrared Spectroscopy (FTIR) is effective in quickly analyzing and measuring the composition of cellulose, hemicellulose, and lignin before and after the application of different pre-processing and pre-treatment techniques. Regression equations were formulated to forecast the lignocellulosic composition of agricultural biomass by utilizing pure cellulose, hemicelluloses, and lignin as reference materials.

REFERENCES

Adapa, P.K., Tabil L.G. & Schoenau, G.J. (2011a). Grinding Performance and Physical Properties of Non-Treated and Steam Exploded Barley, Canola, Oat and Wheat Straw. *Journal of Biomass and Bioenergy*, doi:10.1016/j.biombioe.2010.10.004, 35(2011), pp. 549-561.

Adapa, P.K., Tabil L.G., Schoenau, G.J., Canam, T., & Dumonceaux, T. (2011b). Quantitative Analysis of Lignocellulosic Components of Non-Treated and Steam Exploded Barley, Canola, Oat and Wheat Straw using Fourier Transform Infrared Spectroscopy. *The Journal of Agricultural Science and Technology*, 5(7), (In-Press).

Adapa, P.K., Tabil L.G. & Schoenau, G.J. (2010a). Physical and Frictional Properties of Ground Non-Treated and Steam Exploded Barley, Canola, Oat and Wheat Straw Grinds. *Journal of Powder Technology*, doi:10.1016/j.powtec.2010.03.038, 201(2010), pp. 230-241.

Adapa, P.K., Tabil L.G. & Schoenau, G.J. (2010b). Pelleting Characteristics of Selected Biomass with and without Steam Explosion Pre-treatment. *International Journal of Agricultural and Biological Engineering*, 3(3), pp. 62-79.

Adapa, P.K., Karunakaran, C. Tabil, L.G., & Schoenau, G.J. (2009). Potential Applications of Infrared and Raman Spectromicroscopy for Agricultural Biomass. *Agricultural Engineering International: the CIGR Ejournal*, Manuscript 1081, XI(February), pp. 1- 25.

Adapa, P.K., Schoenau, G.J., Tabil, L.G., Arinze, E.A., Singh, A.. & Dalai, A.K. (2007). Customized and value-added high quality alfalfa products - a new concept. *Agricultural Engineering International: the CIGR Ejournal*, Manuscript FP 07 003, IX(June), pp. 1-28.

Adapa, P.K., Tabil, L.G. Schoenau G.J., & Sokhansanj, S. (2004). Pelleting characteristics of fractionated sun-cured and dehydrated alfalfa grinds. *Applied Engineering in Agriculture, ASAE Transactions*, 20(6), pp. 813-820.

Afzalinia, S., & Roberge, M. (2007). Physical and mechanical properties of selected forage materials. *Canadian Biosystems Engineering*, 49, pp. 2.23-2.27.

Akin, D.E., Rigsby, L.L., Sethuraman, A., Morrison, W.H., Gamble, G.R., & Eriksson, K.E.L. (1995). Alterations in structure, chemistry, and biodegradability of grass lignocellulose treated with the white rot fungi Ceriporiopsis subvermispora and Cyathus stercoreus. *Applied Environmental Microbiology*, 61, pp. 1591–1598.

Anglès, M.N., Ferrando, F., Farriol X., & Salvadó, J. (2001). Suitability of steam exploded residual softwood for the production of binderless panels. Effect of the pretreatment severity and lignin addition. *Biomass and Bioenergy*, 21, pp. 211-224.

Arthur, J.F., Kepner, R.A., Dobie, J.B., Miller, G.E., & Parsons, P.S. (1982). Tub grinder performance with crop and forest residues. *Transactions of the ASAE*, 25, pp. 1488– 1494.

Azuma, J. I., Tanaka, F., & Koshijima, T. (1984). Enhancement of enzymatic susceptibility of lignocellulosic wastes by microwave radiation. *Journal of Fermentation Technology*, 62(4), pp. 377-384.

Azzam, M. (1989). Pre-treatment of cane bagasse with alkaline hydrogen peroxide for enzymatic hydrolysis of cellulose and ethanol fermentation. *Journal of Environmental Science and Health*, B., 24 (4), pp. 421–433.

Bagby, M. O. (1982). In International Symposium on Ethanol from Biomass. H. E. Duckworth and E. A. Thompson, eds. 561 -569. Winnipeg, MB: The Royal Society of Canada.

Balatinecz, J.J. (1983). The Potential of Densification in Biomass Utilization. In: W.A. Cote. ed. Biomass utilization. London: Plenum Press: 181–189 p.

Balk, W.A. (1964). Energy requirements for dehydrating and pelleting coastal Bermuda grass. *Transactions of the ASAE*, 4, pp. 349-1, 355.

Ben-Ghedalia, D., & Miron. J. (1981). The effect of combined chemical and enzyme treatment on the saccharification and in vitro digestion rate of wheat straw. *Biotechnology and Bioengineering*, 23, pp. 823–831.

Ben-Ghedalia, D., & Shefet, G. (1983). Chemical treatments for increasing the digestibility of cotton straw. *Journal of Agricultural Sciences*, 100, pp. 393–400.

Bhattacharya S., Sett, S., & Shrestha, R.M. (1989). State of the Art for Biomass Densification. *Energy Sources*, 11, pp. 161–182.

Binder, A., Pelloni, L., & Fiechter, A. (1980). Delignification of straw with ozone to enhance biodegradability. *Applied Microbiology and Biotechnology*, 11, pp. 1–5.

Bitra, V.S.P., Womac, A.R., Chevanan, N., Miu, P.I., Igathinathane, C., Sokhansanj, S., & Smith, D.R. (2009). Direct mechanical energy measured of hammer mill comminution of switchgarss, wheat straw, and corn stover and analysis of their particle size distributions. *Powder Technology*, 193(2009), pp. 32-45.

Budevska, B.O. 2002. Handbook of Vibrational Spectroscopy, Vol. 5, edited by J. M. Chalmers and P. R. Griffiths, New York: John Wiley and Sons: 3720–3732 p.

Cadoche, L. & López, G.D. (1989). Assessment of size reduction as a preliminary step in the production of ethanol for lignocellulosic wastes. *Biological Wastes*, 30, pp. 153–157.

Cara, C., Ruiz, E., Ballesteros, M., Manzanares, P., Jose Negro, M., & Castro, E. (2008). Production of fuel ethanol from steam-explosion pretreated olive tree pruning. *Fuel*, 87(6), pp. 692-700.

Chabannes, M., Ruel, K., Yoshinaga, A,. Chabbert, B., Jauneau, A., Joseleau, J.P., & Boudet, A.M. (2001). In situ analysis of lignins in transgenic tobacco reveals a differential impact of individual transformations on the spatial patterns of lignin deposition at the cellular and subcellular levels. *Plant Journal*, 28, pp. 271-282, doi:10.1046/j.1365- 313X.2001.01159.x

Chancellor, W.L. (1994). Friction between soil and equipment materials. ASAE Paper No. 941034. St. Joseph, MI.: American Society of Agricultural and Biological Engineers.

Chen, W., Lickfield, G.C., & Yang, C.Q. (2004). Molecular modeling of cellulose in amorphous state. Part I: model building and plastic deformation study. *Polymer*, 45, pp. 1063-1071.

Chevanan, N., Womac, A.R. Bitra, V.S. & Yoder, D.C. (2008). Flowability parameters for chopped switchgrass, wheat straw and corn stover. ASABE Annual International Meeting, Rhode Island, Paper No. 084162. St. Joseph, MI: American Society of Agricultural and Biological Engineers, June 29 – July 2, 12 pages.

Chico-Santamarta, L., Humphroes, A., White, D., Chaney, K., & Godwin, R.J. (2009). Effect of Pre- and Post-Pelletization Storage of Canola (Oilseed Rape) Straw on the Quality and Properties of Pellets. ASABE Annaul International Meeting, Grand Sierra Resort and Casino, Reno, Nevada, Paper No. 096105, June 21 – 24, 17 pages.

Chung, J.H., & Verma, L.R. (1989). Determination of friction coefficients of beans and peanuts. *Transactions of the ASAE* 32(2), pp. 745-750.

Crawford, R. L. (1981). Lignin biodegradation and transformation. New York: John Wiley and Sons. ISBN 0-471-05743-6.

de la Hoz, A., Diaz-Ortiz, A. & Moreno, A. (2005). Microwaves in organic synthesis: Thermal and non-thermal microwave effects. *Chemical Society Reviews*, 34, pp. 164–178.

Deguchi, S., Tsujii, K., & Horikoshi, K. (2006). Cooking Cellulose in Hot and Compressed Water. *Chemical Communications*, pp. 3293 – 3295, doi:10.1039/b605812d

Demirbaş, A. (2001). Biomass Resource Facilities and Biomass Conversion Processing for Fuels and Chemicals. *Energy Conversion and Management*, 41, pp. 1357-1378.

Drzymala, Z. (1993). Industrial briquetting – fundamentals and methods. Studies in mechanical engineering, Vol. 13 Warszawa: PWN-Polish Scientific Publishers.

Encyclopædia Britannica. (2008). Cellulose. Retrieved May 23, 2008, from Encyclopædia Britannica Online.

Esteghlalian, A., Hashimoto, A.G., Fenske, J.J., & Penner, M.H. (1997). Modeling and optimization of the dilute-sulfuric-acid pre-treatment of corn stover, poplar and switchgrass. *Bioresource Technology*, 59, pp. 129–136.

Fan, L. T., Gharpuray, M. M., & Lee, Y. H. (1987). In Cellulose Hydrolysis. *Biotechnology Monographs*, 57, pp. 149-187. New York, NY: Springer-Verlag.

Fang, Q., Bolloni, I., Haque, E., & Spillman, C.K. (1997). Comparison of energy efficiency between a roller mill and a hammer mill. *Applied Engineering in Agriculture*, 13(35), pp. 631-635.

Fasina, O.O, Bransby D., & Wood, C.W. (2006). Flow properties of biomass feedstock. ASABE Annual International Meeting, Portland, Oregon, Paper No. 066149, July 9- 12, 12 pages.

Focher, B., Marzetti, A., Marsano, E., Conio, G., Tealdi, A., Cosani, A., & Terbojevich, M. (1998). Regenerated and Graft Copolymer Fibres from Steam-Exploded Wheat Straw: Characterization and Properties. *Journal of Applied Polymer* Science, 67, pp. 961-974.

Fox, D.J., Gray, P.P., Dunn, N.W., & Warwick, L.M. (1989). Comparison of alkali and steam (acid) pre-treatments of lignocellulosic materials to increase enzymic susceptibility: Evaluation under optimized pre-treatment conditions. *Journal of Chemical Technology & Biotechnology*, 44, pp. 135–146.

Galbe, M., & Zacchi, G. (2002). A review of the production of ethanol from softwood. *Applied Microbiology and Biotechnology* 59, pp. 618–628.

Garcia-Cubero, M.T., Gonzalez-Benito, G. Indacoechea, I. Coca, M. & Bolado, S. (2009). Effect of ozonolysis pre-treatment on enzymatic digestibility of wheat and rye straw. *Bioresource Technology*, 100, pp. 1608–1613.

Garcia-Aparicio, M. P., Ballesteros, I., Gonzalez, A., Oliva, J.M., Ballesteros, M., & Negro, M.J. (2006). Effect of inhibitors released during steam-explosion pre-treatment of barley straw on enzymatic hydrolysis. *Applied Biochemistry and Biotechnology*, 129(1- 3), pp. 278-288.

Gelbrich, J., Mai, C. & Militz, H. (2009). Evaluation of Bacterial Wood Degradation by Fourier-Transform-Infrared (FTIR) Measurements, The study has been done within the EU-project BACPOLES (EVK4-CT-2001-00043) at the University of Gottingen (2009), 6 pages.

Gould, J.M. (1984). Alkaline peroxide delignification of agricultural residues to enhance enzymatic saccharification. *Biotechnology and Bioengineering*, 26, pp. 46–52.

Granada, E., López González, L.M., Míguez J.L., & Moran, J. (2002). Fuel lignocellulosic briquettes, die design and products study. *Renewable Energy*, 27, pp. 561-573.

Grous, W.R., Converse, A.O., & Grethlein, H.E. (1985). Effect of Stem Explosion Pretreatment on Pore Size and Enzymatic Hydrolysis or Poplar. *Enzyme Microbial Technology*, 1986(8), pp. 274-280.

Harmsen, P.F.H., Huijgen, W.J.J., Bermudez Lopez, L.M., & Bakker, R.R.C. (2010). Literature Review of Physical and Chemical Pre-treatment Processes for Lignocellulosic Biomass. Energy Research Centre of the Netherlands. Report accessed on January 2010. Web Address: http://www.ecn.nl/docs/library/report/2010/e10013.pdf

Hatakka, A.I. (1983). Pre-treatment of wheat straw by white-rot fungi for enzymic saccharification of cellulose. *Applied Microbiology and Biotechnology*, 18, pp. 350– 357. Himmel, M., Tucker, M., Baker, J., Rivard, C., Oh, K., & Grohmann, K. (1985). Comminution of biomass: Hammer and knife mills. *Biotechnology and Bioengineering Symposium*, 15, pp. 39-58.

Hon, D.N.-S. (1989). Cellulosic adhesives. In Adhesives from Renewable Resources, Eds. R.W. Hemmingway and A.H. Conner, 289-304. Washington, DC: American Chemical Society.

Hu Z., & Wen, Z. (2008). Enhancing enzymatic digestibility of switchgrass by microwave assisted alkali pre-treatment. *Biochemical Engineering Journal*, 38, pp. 369-378

Hu, Z., Wang, Y. & Wen, Z. (2008). Alkali (NaOH) Pre-treatment of switchgrass by radio frequency-based dielectric heating. *Applied Biochemistry and Biotechnology*, 148, pp. 71–81.

Igathinathane, C., Womac, A.R., Sokhansanj, S., & Narayan, S. (2008). Knife grid size reduction to pre-process packed beds of high- and low-moisture switchgrass. *Bioresource Technology*, 99, pp. 2254–2264.

Jackson, M. G. (1977). The alkali treatment of straws. *Animal Feed Science and Technology*, 2, pp. 105-130.

Jin, S. & Chen, H. (2006). Superfine Grinding of Steam-Exploded Rice Straw and its Enzymatic Hydrolysis. *Biochemical Engineering Journal*, 30(2006), pp. 225-230.

Kaliyan, N., Schmidt, D.R., & Morey, R.V. (2010). Commercial Scale Grinding of Corn Stover and Perennial Grasses. ASABE Annual International Meeting, David L. Lawrence Convention Centre, Pittsburgh, Pennsylvania, June 20-23, Paper No. 1009062, 2950 Niles Road, St. Joseph, MI 49085-9659 USA.

Kaliyan, N., & Morey, R.V. (2006). Factors affecting strength and durability of densified products. ASABE Annual International Meeting, American Society of Agricultural and Biological Engineers, Portland, Oregon July 9-12, Paper Number 066077, 2950 Niles Road, St. Joseph, MI 49085-9659 USA.

Kashaninejad, M., & Tabil. L.G. (2011). Effect of microwave-chemical pre-treatment on compression characteristics of biomass grinds. *Biosystems Engineering*, 108, pp. 36- 45.

Kelley, S.S., Rowell, R.M., Davis, M., Jurich, C.K., & Ibach, R. (2004). Rapid Analysis of the Chemical Composition of Agricultural Fibres Using Near Infrared Spectroscopy and Pyrolysis Molecular Beam Mass Spectrometry. *Biomass and Bioenergy*, 27(2004), pp. 77-88.

Keshwani, D.R., Cheng, J.J., Burns, J.C., Li, L., & Chiang, V. (2007). Microwave pre-treatment of switchgrass to enhance enzymatic hydrolysis. ASABE Annual Meeting Paper No. 077127, Minneapolis, MN, USA.

Kitchaiya, P., Intanakul, P., & Krairiksh, M. (2003). Enhancement of enzymatic hydrolysis of lignocellulosic wastes by microwave pre-treatment under atmospheric pressure. *Journal of Wood Chemistry and Technology*, 23(2), pp. 217- 225.

Kokta, B.V., & Ahmed, A. (1998). Chapter 6: Steam Explosion Pulping. Environmentally Friendly Technologies for the Pulp and Paper Industry, edited by Raymond A. Young and Masood Akhtar, ISBN 0-471-15770-8, John Wiley and Sons, Inc.

Kumar, P., Barrett, D.M., Delwiche, M.J., & Stroeve, P. (2009). Methods for pre-treatment of lignocellulosic biomass for efficient hydrolysis and biofuel production. *Industrial and Engineering Chemistry Research*, 48, pp. 3713–3729.

Ladisch, M. R. (1989). Hydrolysis. In: Biomass Handbook. O. Kitani and C. W. Hall, eds. 434-451. New York: Gordon and Breach.

Lam, P.S., Sokhansanj, S., Bi, X., Lim, C.J., Mani, S., & Melin, S. (2008). Energetic feasibility of biomass densification with steam explosion pretreatment. CSBE Annual General Meeting; July 14-16, Poster Number CSBE08305, Vancouver, BC.

Larkin, S., Ramage, J., & Scurlock, J. (2004). Bioenergy. In: G. Boyle. ed. Renewable energy. Power for a sustainable future: 106-146 p.

Larsson, S.H. (2010). Kinematic Wall Friction Properties of Reed Canary Grass Powder at High and Low Normal Stresses. *Powder Technology*, 198, pp. 108-113.

Larsson, S.H., Thyrel, M., Geladi, P., & Lestander, T.A. (2008). High Quality Biofuel Pellet Production from Pre-Compacted Low Density Raw Materials. *Bioresource Technology*, 99(2008), pp. 7176-7182.

Lebo, S. E. Jr., Gargulak, J.D., & McNally, T.J. (2001). "Lignin". Kirk-Othmer Encyclopedia of Chemical Technology. John Wiley & Sons, Inc. DOI:10.1002/0471238961.12090714120914.a01.pub2. Retrieved on 2007-10-14.

Lehtikangas, P. (2001). Quality properties of palletized sawdust, logging residues and bark. *Biomass and Bioenergy*, 20, pp. 351-360.

Lesoing, G., Klopfenstein, T., Rush, I., & Ward, J. (1980). Chemical treatment of wheat straw. *Journal of Animal Science*, 51, pp. 263-269.

Lin, Y., & Tanaka, S. (2006). Ethanol fermentation from biomass resources: current state and prospects. *Applied Microbiology and Biotechnology*, 69.6, pp. 627-642.

Liu, R., Yu, H., and Huang, Y. (2005). Structure and Morphology of Cellulose in Wheat Straw. *Cellulose*, 12, pp. 25-34.

Lopo, P. (2002). The right grinding solution for you: roll, horizontal or vertical. *Feed Management*, 53(3), pp. 203-6.

Luypaert, J., Zhang, M.H., & Massart, D.L. (2003). Feasibility Study for the Use of Near Infrared Spectroscopy in the Qualitative and Quantitative Analysis of Green Tea. *Analytca Chimica Acta*, 478(2003), pp. 303-312.

MacDonald, D.G., Bakhshi, N.N., Mathews, J.F., Roychowdhury, A., Bajpai, P., & Moo-Young, M. (1983). Alkali treatment of corn stover to improve sugar production by enzymatic hydrolysis. *Biotechnology and Bioengineering*, 25, pp. 2067–2076.

Magara K., & Azuma, J. (1989). Microwave-irradiation of lignocellulosic materials X: Conversion of microwave-irradiated agricultural wastes into ethanol. *Wood Research*, 76, pp. 1–9

Mani, S., Tabil, L.G., & Sokhansanj, S. (2006). Effects of Compressive Force, Particle Size, and Moisture Content on Mechanical Properties of Biomass Pellets from Grasses. *Biomass and Bioenergy*, 30(2006), pp. 648-654.

Mani, S., Tabil, L.G., & Sokhansanj, S. (2004). Grinding performance and physical properties of wheat and barley straws, corn stover and switchgrass. *Biomass and Bioenergy*, 27(2004), pp. 339-52.

Mani, S., Tabil, L.G., & Sokhansanj, S. (2003). An overview of compaction of biomass grinds. Powder Handling & Process, 15(3), pp. 160-168.

Marra, F., Lyng, J., Romano, V., & McKenna, B. (2007). Radio-frequency heating of foodstuff: Solution and validation of a mathematical model. *Journal of Food Engineering*, 79, pp. 998–1006.

McMullen, J., Fasina, O.O., Wood, C.W., & Feng, Y. (2005). Storage and handling characteristics of pellets from poultry litter. *Applied Engineering in Agriculture*, 21(4), pp. 645-651.

Mohsenin, N.N. (1970). Physical Properties of Plant and Animal Materials. New York, N.Y.: Gordon and Breach Science Publishers.

Mosier, N., Wyman, C., Dale, B., Elander, R., Lee, Y.Y., Holtzapple, M., & Ladisch, M. (2005). Features of promising technologies for pre-treatment of lignocellulosic biomass. *Bioresource Technology*, 96, pp. 673–686.

Murphy, J.D and K. McCarthy. (2005). Ethanol production from energy crops and wastes for use as a transport fuel in Ireland. *Applied Energy*, 82, pp. 148-166.

Neely, W. C. (1984). Factors affecting the pre-treatment of biomass with gaseous ozone. *Biotechnology and Bioengineering*, 26, pp. 59–65.

Nelson, D.L., & Cox, M.M. (2005). Lehninger Principles of Biochemistry. New York, NY: W.H. Freeman and Company.

Obernberger, I., & Thek, G. (2004). Physical characterization and chemical composition of densified biomass fuels with regard to their combustion behavior. *Biomass and Bioenergy*, 27, pp. 653-669.

Oberndorfer, C., Pawelzik, E., & Lucke, W. (2000). Prospects for the application of dielectric heating processes in the pre-treatment of oilseeds. *European Journal of Lipid Science and Technology*, 102, pp. 487–493.

O'Dogherty, M.J. (1982). A review of research on forage chopping. *Journal of Agricultural Engineering Research*, 27(4), pp. 267-289.

Okano, K., Kitagawa, M., Sasaki, Y., & Watanabe, T. (2005). Conversion of Japanese red cedar (Cryptomeria japonica) into a feed for ruminants by white-rot basidiomycetes. *Animal Feed Science and Technology*, 120, pp. 235–243

Ooshima, H., Aso, K. Harano, Y., & Yamamoto, T. (1984). Microwave treatment of cellulosic materials for their enzymatic hydrolysis. *Biotechnology Letters*, 6(5), pp. 289-294.

Overend, R.P., & Chornet, E. (1987). Fractionation of lignocellulosics by steam-aqueous pretreatment. *Philosophical Transactions of the Royal Society*, London A 321, pp. 523–536.

Pfost, H.B., & Headley, V.E. (1971). Use of logarithmic normal distribution to describe hammer mill performance. *Transactions of the ASABE*, 14(3), pp. 531-5.

Puchalski, C., & Brusewitz, G.H. (1996). Coefficient of friction of watermelon. *Transactions of the ASABE*, 39(2), pp. 589-594.

Sánchez, Ó.J., & Cardona, C.A. (2008). Trends in biotechnological production of fuel ethanol from different feedstocks. *Bioresource Technology*, 99, pp. 5270–5295.

Schell, D.J., & Harwood, C. (1994). Milling of Lignocellulosic Biomass: Results of Pilot-Scale Testing. *Applied Biochemistry and Biotechnology*, 45(46), pp. 159-168.

Scholten, R.L., & McEllhiney, R.R. (1985). The effects of prebreaking in hammer mill particle size reduction. American Society of Agricultural Engineers; Paper no. 85-3542, St. Joseph MI.

Seville, J.P.K., Tuzun, U., & Clift, R. (1997). Processing of Particulate Solids. London, U.K.: Chapman and Hall.

Shambe, T., & Kennedy, J.F. (1985). Acid and enzymatic hydrolysis of chaotropically pretreated millet stalk, acha and rice straws and conversion of the products to ethanol. *Enzyme and Microbial Technology*, 7, pp. 115-120.

Shaw, M. (2008). Feedstock and Process Variables Influencing Biomass Densification. M.Sc. Thesis, Department of Agricultural and Bioresource Engineering, University of Saskatchewan. Website: http://library2.usask.ca/theses/available/etd-03092008- 115636/unrestricted/Thesis.MarkShaw.FinalR3.pdf

Shaw, M.D., & Tabil, L.G. (2006). Mechanical properties of selected biomass grinds. ASABE Annual International Meeting, Oregon Convention Center, Portland, Oregon, Paper Number 066175, July 9-12: 12 pages.

Sherman Hsu, C.P. (1997). Chapter 15: Infrared Spectroscopy, in: F. Settle, Editor, Handbook of Instrumental Techniques for Analytical Chemistry, Prentice-Hall, NJ 270. Sitkei, G. (1986). Mechanics of agricultural materials. Amsterdam: Elsevier.

Smola, N., & Urleb, U. (2000). Qualitative and Quantitative Analysis of Oxytetracycline by Near-Infrared Spectroscopy. *Analytica Chimica Acta*, 410(2000), pp. 203-210.

Sokhansanj, S., Mani, S., Bi, X., Zaini, P., & Tabil, L.G. (2005). Binderless pelletization of biomass. ASAE Annual International Meeting, Tampa Convention Centre, Tampa, Florida; July 17-20, Paper Number 056061, 2950 Niles Road, St. Joseph, MI 49085- 9659 USA.

Soto, M.L., Dominguez, H., Nunez, M.J., & Lema, J.M. (1994). Enzymatic saccharification of alkali-treated sunflower hulls. *Bioresource Technology*, 49, pp. 53–59.

Soucek, J., Hanzlikova, I., & Hulta, P. (2003). A fine disintegration of plants suitable for composite biofuels production. *Research in Agricultural Engineering*, 49(1), pp. 7-11.

Stenberg, K., Tengborg, C., Galbe, M., & Zacchi, G. (1998). Optimisation of steam pretreatment of SO2-impregnated mixed softwoods for ethanol production. *Journal of Chemical Technology and Biotechnology*, 71(4), pp. 299-308.

Sun, F., & Chen, H. (2007). Evaluation of enzymatic hydrolysis of wheat straw pretreated by atmos-pheric glycerol autocatalysis. *Journal of chemical technology and biotechnology*, 82, pp. 1039-1044.

Sun, Y., & Cheng, J. (2002). Hydrolysis of lignocellulosic materials for ethanol production: a review. *Bioresource Technology* 83, pp. 1–11.

Tabil, L.G. (1996). Pelleting and binding characteristics of alfalfa. Unpublished Ph.D. thesis, Department of Agricultural and Bioresource Engineering, Saskatoon, SK Canada: University of Saskatchewan.

Taherzadeh, M.J., & Karimi, K. (2007). Acid-based hydrolysis processes for ethanol from lignocellulosic materials: a review. *Bioresources*, 2, pp. 472–499.

Temmerman, M., Rabier, F., Jensen, P.D., Hartmann, H., & Bohm, T. (2006). Comparative Study of Durability Test Methods for Pellets and Briquettes. *Biomass and Bioenergy*, 30(11), pp. 964.

Tomas-Pejo, E., Oliva, J.M., & Ballesteros, M. (2008). Realistic approach for full-scale bioethanol production from lignocellulose: a review. *Journal of Scientific & Industrial Research*, 67, pp. 874–884.

Tripathi, K., Iyer, P.V.R., and Kandpal, T.C. (1988). A Techno-Economic Evaluation of Biomass Briquetting in India. *Biomass and Bioenergy*, 14, pp. 479–488.

Tucker, M.P., Mitri, R.K., Eddy, F.P., Nguyen, Q.A., Gedvilas, L.M., & Webb, J.D. (2000). Fourier Transform Infrared Quantification of Sugars in Pretreated Biomass Liquors. *Applied Biochemistry and Biotechnology*, 84-86(2000), pp. 39-50.

United States Department of Energy (USDE). (2006). Energy Efficiency and Renewable Energy – Biomass Program. Website Accessed on January 21, 2011. Web Address: http://www1.eere.energy.gov/biomass

Updegraff, D.M. (1969). Semimicro Determination of Cellulose in Biological Materials. *Analytical Biochemistry*, 32, pp. 420 – 424.

van Dam, J.E.G., van den Oever, M.J.A., Teunissen, W., Keijsers, E.R.P., & Peralta, A.G. (2004). Process for production of high density/high performance binderless boards from whole coconut husk. Part 1: lignin as intrinsic thermosetting binder resin. *Industrial Crops and Products*, 19, pp. 207-216.

Vazquez, G., Gonzalez, J., Freire, S., & Antorrena, G. (2002). Characterization of Eucalyptus Globules and Pinus Pinaster Acetosolv Pulps Prebleached with O2 by FTIR and DRIFT Spectroscopy. *Holz als Roh- und Werkstoff*, 60(2002), pp. 25-30.

Vlasenko, E.Y., Ding, H., Labavitch J.M., & Shoemaker, S.P. (1997). Enzymatic Hydrolysis of Pretreated Rice Straw. *Bioresource Technology*, 59, pp. 109-119.

Vidal, P.F., & Molinier, J. (1988). Ozonolysis of lignin-improvement of in vitro digestibility of poplar sawdust. *Biomass*, 16, pp. 1–17.

Vigneault, C., Rothwell, T.M., & Bourgeois, G. (1992). Hammer mill grinding rate and energy requirements for thin and conventional hammers. *Canadian Biosystems Engineering*, 34, pp. 203-6.

Viola, E., Zimbardi, F., Cardinale, M., Cardinale, G., Braccio, G., & Gambacorta, E. (2008). Processing Cereal Straws by Steam Explosion in a Pilot Plant to Enhance Digestibility in Ruminants. *Bioresource Technology*, 99(2008), pp. 681-689.

Womac, A.R., Igathinathane, C., Bitra, P., Miu, P., Yang, T., Sokhansanj, S., & Narayan, S. (2007). Biomass Pre-Processing Size Reduction with Instrumented Mills. ASABE Annual International Meeting, Minneapolis Convention Centre, Minneapolis, June 17-20, Paper No. 076046, 2950 Niles Road, St. Joseph, MI 49085-9659 USA.

Zandersons, J., Gravitis, J., Zhurinsh, A., Kokorevics, A., Kallavus, U., & Suzuki, C.K. (2004). Carbon materials obtained from self-binding sugar cane bagasse and deciduous wood residues plastics. *Biomass and Bioenergy*, 26, pp. 345-360.

Zimbardi, F., Viola, E., Nanna, F., Larocca, E., Cardinale, M., & Barisano, D. (2007). Acid impregnation and steam explosion of corn stover in batch processes. *Industrial Crops and Products*, 26(2), pp. 195-206.

Zimbardi, F., Viggiano, D., Nanna, F., Demichele, M., Cuna, D., & Cardinale, G. (1999). Steam Explosion of Straw in Batch and Continuous Systems. *Applied Biochemistry and Biotechnology*, 77-79, pp. 117-125.

Zhu, S., Wu, Y., Yu, Z., Zhang, X., Li, H., & Gao, M. (2006a). The effect of microwave irradiation on enzymatic hydrolysis of rice straw. *Bioresource Technology*, 97, pp. 1964–1968.

Zhu, S., Wu, Y., Yu, Z., Chen, Q., Wu, G., Yu, F., Wang, C., & Jin, S. (2006b). Microwaveassisted alkali pre-treatment of wheat straw and its enzymatic hydrolysis. Biosystems Engineering, 94, pp. 437–442.

Zhu, S., Wu, Y., Yu, Z., Zhang, X., Wang, C., Yu, F., & Jin, S. (2006c). Production of ethanol from microwave-assisted alkali pretreated wheat straw. *Process Biochemistry*, 41, pp. 869 – 873.

Zhu, S., Wu, Y., Yu, Z., Liao, J., & Zhang, Y. (2005). Pre-treatment by microwave/alkali of rice straw and its enzymic hydrolysis. *Process Biochemistry*, 40, pp. 3082-3086.

CHAPTER-19

BIOMASS FEEDSTOCK PRE-PROCESSING: PART-II-DENSIFICATION

1. INTRODUCTION: AN OVERVIEW

1.1. The need for densification

Agricultural biomass residues has the capacity to be utilized for the sustainable generation of bio-fuels and to counterbalance the release of greenhouse gases (Campbell et al., 2002; Sokhansanj et al., 2006). Straw derived from crop production and agricultural residues found in the waste streams of commercial crop processing plants have less intrinsic worth and have historically been a challenge for proper disposal. These leftovers are a plentiful, affordable, and easily accessible source of renewable lignocellulosic biomass (Liu et al., 2005). Novel approaches must be devised to process biomass, rendering it a suitable feedstock for the generation of bio-fuels. Furthermore, the economic utilization of agricultural crop leftover is hindered by several obstacles, including the inconsistent quality of the residue, the expenses associated with its collection, and challenges related to transportation and storage (Bowyer and Stockmann, 2001; Sokhansanj et al., 2006). Efficient processing and handling of biomass is necessary to decrease operational costs in the industry and fulfill the raw material needs for biofuel generation. Because of its elevated moisture content, uneven shape and size, and low bulk density, biomass poses significant challenges in terms of handling, transportation, storage, and utilization in its natural state (Sokhansanj et al., 2005). Compacting biomass into durable compacts is an efficient solution to these issues and can decrease material waste. Densification is a process that can raise the bulk density of biomass from an initial range of 40-200 kg/m3 to a final compact density of 600-1200 kg/m3. This information is supported by studies conducted by Adapa et al. (2007), Holley (1983), Mani et al. (2003), McMullen et al. (2005), and Obernberger and Thek (2004). The densification process involves applying pressures between 400-800 MPa to biomass, resulting in compressed and stable densities that are 7-10 times higher than conventional bales (Demirbas and Sahin, 1998). Due to their consistent shape and size, densified products can be conveniently managed using standard equipment for handling and storage. They can also be readily utilized in various biomass-based conversions, such as direct-combustion or co-firing with coal, gasification, pyrolysis, and biochemical processes (Kaliyan and Morey, 2006a). When agricultural biomass materials, particularly straw and stover, are compressed, they often create poorly shaped pellets or compacts that are dusty, hard to handle, and expensive to produce. The lack of comprehensive understanding regarding the inherent binding properties of the various constituents of biomass is the root cause of this issue (Sokhansanj et al., 2005).

1.2. Fuel pellet quality parameters

The quality of fuel pellet is usually assessed based on its density and durability. High density of pellet represents higher energy per unit volume of material, while durability is the resistance of pellets to withstand various shear and impact forces applied during handling and transportation. High bulk density increases storage and transport capacity of pellets. Since feeding of boilers and gasifiers generally is volume-dependent, variations in bulk density should be avoided (Larsson et al., 2008). A bulk density of 650 kg/m3 is stated as design value for wood pellet producers (Obernberger and Thek, 2004). Low durability of pellets results in problems like disturbance within pellet feeding systems, dust emissions, and an increased risk of fire and explosions during pellet handling and storage (Temmerman et al., 2006). Other quality factors of biomass for thermo-chemical conversion include (FAO, 2011; Rajvanshi, 1986):

- **Energy content:** The choice of a biomass for energy conversion will in part be decided by its heating value. The method of measurement of the biomass energy content will influence the estimate of efficiency of a given gasifier. The only realistic way of presenting fuel heating values for gasification purposes is to give lower heating values (excluding the heat of condensation of the water produced) on an ash inclusive basis and with specific reference to the actual moisture content of the fuel.
- **Moisture content:** High moisture contents reduce the thermal efficiency since heat is used to drive off the water and consequently this energy is not available for the reduction reactions and for converting thermal energy into chemical bound energy in the gas. Therefore, high moisture contents result in low gas heating values during thermo chemical processes.
- **Volatile matter:** The amount of volatiles in the feedstock determines the necessity of special measures (either in design of the gasifier or in the layout of the gas cleanup train) in order to remove tars from the product gas in engine applications.
- **Ash content and slagging characteristics:** The mineral content in the biomass that remains in oxidation form after complete combustion is usually called ash. The ash content of a fuel and the ash composition have a major impact on trouble free operation of a gasifier or a burner. Slagging or clinker formation in the reactor, caused by melting and agglomeration of ashes, at the best will greatly add to the amount of labour required to operate the gasifier. If no special measures are taken, slagging can lead to excessive tar formation and/or complete blocking of the reactor.
- **Reactivity:** The reactivity is an important factor determining the rate of reduction of carbon dioxide to carbon monoxide in a gasifier. Reactivity depends in the first instance on the type of fuel. For example, it has been observed that fuels such as wood, charcoal and peat are far more reactive than coal.
- **Size and size distribution:** Low bulk density feedstock may cause flow problems in the gasifier or burner as well as an inadmissible pressure drop over the reduction zone and a high proportion of dust in the gas. Large pressure drops will lead to reduction of the gas load, resulting in low temperatures and tar production. Excessively large sizes of particles or pieces give rise to reduction in reactivity of the fuel, resulting in start-up problems and poor gas quality, and to transport problems through the equipment. A large range in size distribution of the feedstock will generally aggravate the above phenomena. Too large particle sizes can cause gas channelling problems. Fluidized bed gasifiers are normally able to handle fuels with particle diameters varying between 0.1 and 20 mm (FAO, 2011).

- **Bulk density:** Fuels with high bulk density are advantageous because they represent a high energy-for-volume value. Consequently, these fuels need less bunker space for a given refuelling time. Low bulk density fuels sometimes give rise to insufficient flow under gravity, resulting in low gas heating values and ultimately in burning of the char in the reduction zone. Inadequate bulk densities can be improved by briquetting or pelletizing.

All of the abovementioned biomass properties could be altered by subjecting raw biomass to various processing methods and forming composites. Before choosing a gasifier, it is important to ensure that the individual biomass meets the requirements of the gasifier or that it can be treated to meet these requirements.

1.3. Effect of moisture content on pellet quality

The moisture in biomass both acts as a facilitator of natural binding agents and a lubricant (Kaliyan and Morey, 2006a). Many studies have indicated that the production of high quality pellets is possible only if the moisture content of the feed is between 8 and 12% (wb). Moisture contents above or below this range would lead to lower quality pellets (Hill and Pulkinen, 1988; Kashaninejad et al., 2011; Li and Liu, 2000; Obernberger and Thek, 2004; Shaw and Tabil, 2007). In general, an increase in moisture content from 10 to 44% could result in up to 30-40% decrease in pellet densities of biomass (Chancellor, 1962; Grover and Mishra, 1996; Gustafson and Kjelgaard, 1963; Kaliyan and Morey, 2006a; Mani et al., 2002 and 2006b; Smith et al., 1977). However, the percentage decrease in density depends on the type of biomass. Therefore, a moisture content of 10% (w.b.) is considered as optimal moisture content to obtain high density and durability pellets.

1.4. Effect of grind size on pellet quality

In general, finer grinds produces higher quality pellets since they can readily absorb moisture than large particles, and therefore, undergo a higher degree of conditioning. In addition, finer grinds have higher surface area of contact to form bonds/solid bridges during the compaction processes. Also, large particles are fissure points that cause cracks and fractures in compacts (MacBain, 1966). A reduction in hammer mill screen size from 3.2 to 0.6 mm can result in an increase in pellet densities from 5 to 16% (Kaliyan and Morey, 2006b; Kashaninejad et al., 2011; Mani et al., 2002 and 2004a). However, no significant trend in change in density were observed at geometric mean particles size of 0.6 mm and lower (Kaliyan and Morey 2006b; Mani et al., 2002). The change in pellet density depends on the type of biomass.

This chapter will address various factors that directly or indirectly effect densification of agricultural biomass residue into high quality pellets. The compression and compaction characteristics of ground biomass will be dealt in detail that will provide a comprehensive understanding of the behaviour of biomass as influenced by various factors. The compression studies will explore the affect of independent variables such as biomass, treatment, grind size, and moisture content on pellet density and durability, while compaction studies will study the effect of various machine variables on the pellet quality. In addition, overall specific energy requirements will be established and techno-economic models will be explained.

2. LIGNOCELLULOSIC COMPOSITION AND HIGHER HEATING VALUES

The experimental lignocellulosic composition of agricultural straw can be determined using the modified NREL LAP method for "Determination of Structural Carbohydrates and Lignin in Biomass" (Table 1) (Adapa et al., 2011; Sluiter et al., 2008). This procedure uses a two-step acid hydrolysis to fractionate the biomass into forms that are more easily quantified. During this process, the lignin fractionates into acid insoluble material and acid soluble material, while the polymeric carbohydrates are hydrolyzed into the monomeric forms, which are soluble in the hydrolysis liquid and subsequently can be measured using HPLC. The Percentage cellulose in the samples can be measured using the percentage glucan content, while the percentage hemicelluloses can be measured by adding the percentage mannose, galactose, xylose and arabinose content in the biomass samples.

Table 1 shows the lignocellulosic composition and higher heating values of non-treated and steam exploded barley, canola, oat and wheat straw samples. In general, the cellulose, hemicelluloses and lignin content of steam exploded straw was higher than non-treated straw. This may be due to other components (soluble lignin, loosely-bound sugars) being washed away during steam explosion, thereby leaving the proportion of insoluble lignin, cellulose and hemicellulose in the resulting dried sample higher than for the non-treated samples (i.e. higher percent of dry mass).

Table 1.Lignocellulosic composition and higher heating values of non-treated and steam exploded agricultural straw (Adapa et al., 2011)

Properties of Biomass	Barley Straw		Canola Straw		Oat Straw		Wheat Straw	
	NT	SE	NT	SE	NT	SE	NT	SE
Composition (% of dry matter)								
Celluloseb	22.7 ± 0.9a	25.3 ± 1.8	22.4 ± 0.8	27.5 ± 1.1	25.4 ± 1.0	27.4 ± 2.4	27.1 ± 1.0	29.9 ± 1.4
Hemicellulosec	21.2 ± 0.5	21.0 ± 1.4	16.9 ± 0.5	20.2 ± 0.7	21.7 ± 0.9	18.8 ± 1.2	21.1 ± 0.5	19.7 ± 0.9
Galactose	0.9 ± 0.0	0.7 ± 0.0	1.0 ± 0.0	0.9 ± 0.1	0.8 ± 0.0	0.7 ± 0.0	0.8 ± 0.0	0.9 ± 0.1
Mannose	1.6 ± 0.2	1.5 ± 0.0	2.3 ± 0.1	1.9 ± 0.4	1.4 ± 0.0	1.7 ± 0.1	1.6 ± 0.1	2.8 ± 0.2
Xylose	14.4 ± 0.3	15.3 ± 1.0	11.5 ± 0.5	14.3 ± 0.2	15.1 ± 0.8	13.3 ± 1.0	14.9 ± 0.4	13.5 ± 0.4
Arabinose	4.4 ± 0.2	3.5 ± 0.5	2.0 ± 0.1	3.2 ± 0.0	4.4 ± 0.2	3.1 ± 0.2	3.9 ± 0.1	2.6 ± 0.2
Total Lignind	21.0 ± 0.6	21.6 ± 0.6	19.6 ± 0.6	22.3 ± 0.2	19.5 ± 0.6	23.7 ± 0.2	22.5 ± 0.7	24.2 ± 0.3
Soluble Lignin	1.6 ± 0.1	1.4 ± 0.1	1.6 ± 0.1	1.2 ± 0.1	1.5 ± 0.1	1.3 ± 0.1	1.4 ± 0.0	1.0 ± 0.1
Insoluble Lignin	19.4 ± 0.6	20.2 ± 0.6	18.0 ± 0.6	21.1 ± 0.1	17.9 ± 0.7	22.4 ± 0.1	21.0 ± 0.7	23.3 ± 0.4
Higher Heating Values (MJ/kg of dry matter)								
HHV (MJ/kg)	16.4±0.3‡†	17.4±0.1	16.7±0.3	18.3±0.0	16.4±0.1	17.8±0.0	17.0±0.2	17.8±0.0
DM – Dry Matter; NT – Non-Treated; SE – Steam Exploded; a Average and standard deviation of 3 replicates at 95% confidence interval; b %Cellulose = %glucan; c %Hemicellulose = %(mannose + galactose + xylose + arabinose); d %Total Lignin = %(soluble lignin + insoluble lignin); HHV – Higher Heating Values (measured using Parr 1281 Bomb Calorimeter); ‡3 replicates; † 95% confidence interval								

The calorific (heating) value of biomass feedstocks are indicative of the energy they possess as potential fuels. The gross calorific value (higher heating value, HHV) and the net calorific value (lower heating value, LHV) at constant pressure measures the enthalpy change of combustion with and without water condensed, respectively (Demirbaş, 2007). A bomb calorimeter can be used to determine the HHV of the non-treated and steam exploded straw in MJ/kg. In addition, the ASTM Standard D5865-03 (ASTM, 2003) test method for gross calorific value of coal and coke, can be used as a guideline for heating value testing (Table 1).

Cellulose, hemicelluloses and lignin are major components of a plant biomass. Therefore, a change in composition could potentially lead to change in HHV of the biomass (Adapa et al., 2010a). The Net combined percentage change of cellulose, hemicelluloses and lignin in steam exploded barley, canola, oat and wheat straw is 5%, 19%, 5% and 4% higher than non-treated straw, respectively. As a result, the average HHV of steam exploded barley, canola, oat and wheat straw was 6%, 10%, 9% and 5% higher than non-treated straw, respectively (Table 1).

3. LAB-SCALE PELLETING OF AGRICULTURAL BIOMASS

3.1. Compression test

A compression apparatus having a close fit plunger die assembly can be used to make a single compact in one stroke of the plunger from ground straw samples (Adapa et al., 2006 and 2010a; Mani et al., 2004). The compression test should be performed to study the effect of independent variables such as biomass, treatment, grind size, and moisture content on pellet density and durability. In order to simulate frictional heating during commercial pelleting operation, the compression die should be maintained at pre-heat temperatures of 75 to 100oC (Adapa et al., 2006; Kaliyan and Morey, 2009; Mani et al., 2006). Different levels of pre-set compressive forces can be applied using the Instron testing machine. Typical pre-set loads in the range of 31.0 to 150.0 MPa are applied to make pellets. Figure 1 represents the photographs of pellets made from barley, canola, oat and wheat straw grinds from hammer mill screen sizes of 3.2, 1.6 and 0.8 mm (Adapa et al., 2010a).

3.2. Single-pellet density

The density of pellet is calculated from the mass and volume (measuring the length and diameter) of compacts. In general, the density of pellets from agricultural straw significantly increases with an increase in applied pressure at any specific hammer mill screen size. An increase in pressure results in plastic deformation of ground particles and consequently leads to pellets that have densities closer to their respective particle densities (Adapa et al., 2010a; Kaliyan and Morey, 2009; Mani et al., 2004). The Application of pre-treatment has been observed to significantly increase the pellet density since pre-treated straw has lower geometric particle diameters and significantly higher particle densities (Adapa et al., 2010a; Kashaninejad and Tabil, 2011). Usually, it has been reported that an increase in moisture content from 10% and up results in a significant decrease in pellet quality (Hill and Pulkinen, 1988; Li and Liu, 2000; Obernberger and Thek, 2004; Shaw and Tabil, 2007). In general, a decrease in hammer mill screen size results in an increase in pellet density (Adapa et al., 2010a; Kaliyan and Morey, 2009; Kashaninejad et al., 2011; Mani et al., 2004). A comprehensive literature on various single-pellet compression test data is provided in Table 2.

Adapa et al. (2010a) reported that the type of agricultural biomass did not have any significant effect on pellet density, while steam explosion pre-treatment, applied pressure and screen size had significant effects. In addition, correlation for pellet density with applied pressure and hammer mill screen size having highest R2 values were developed (Table 3). Similarly, Kaliyan and Morey (2009) indicated that the pellet density of corn stover or switchgrass briquettes was significantly affected by pressure, particle size, moisture content and preheating temperature. Kashaninejad and Tabil (2011) also indicated that the pellets made from microwave-chemical pretreated biomass grinds had a significantly higher density and tensile strength than the untreated or samples pretreated by microwave-distilled water.

The densities of pellets should also be measured after a storage period of one week to one month to ascertain its dimensional stability, and associated handling and storage costs (Adapa et al., 2010b; Kaliyan and Morey, 2009). Adapa et al. (2010b) reported that a reduction in pellet density is usually expected due to relaxation of grinds in the pellet after release of pressure. They have observed that the relaxation was higher for larger hammer mill screen sizes and lower applied pressures. In some cases, the average reduction in density was negative giving the impression that pellet density actually increased during storage period. However, these negative values are primarily due to higher standard deviations in pellet density measurements. Therefore, from a practical manufacturing point of view, these values should be considered as a zero percent change in pellet density (Adapa et al., 2010b).

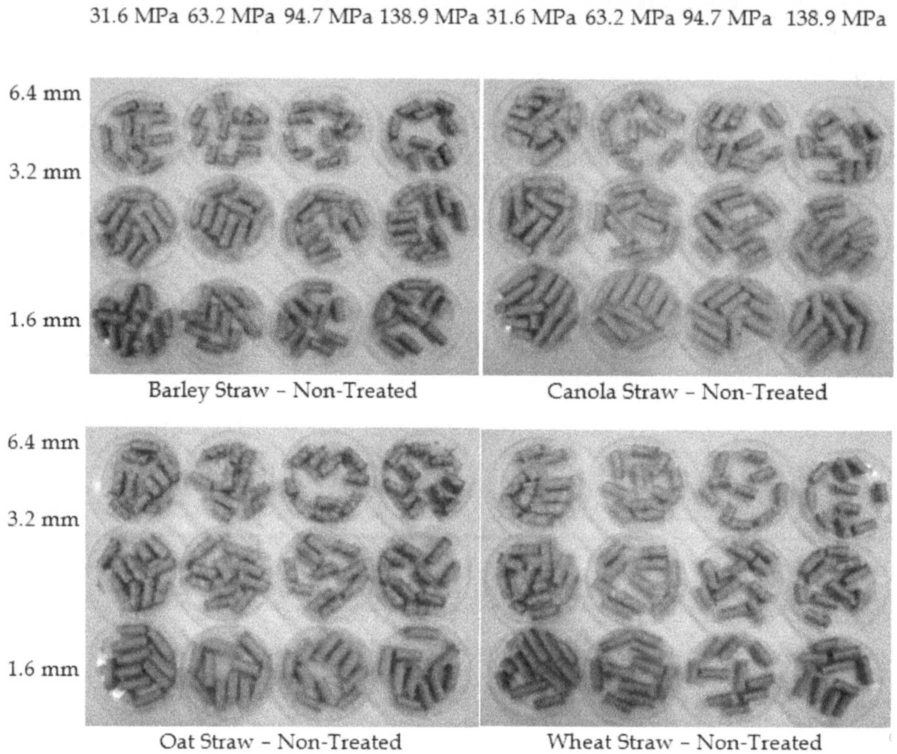

Figure 1.Photograph of pellets made from barley, canola, oat and wheat straw grind from hammer mill screen sizes of 3.2, 1.6, and 0.8 mm.

Table 2.Comprehensive literature review on single-pellet compression tests for agricultural biomass as feedstock for biofuel

Biomass	Independent Variables	Maximum Pellet Density	Reference
Barley, Canola, Oat and Wheat Straw	Hammer Mill Screen Size: 3.2, 1.6 and 0.8 mm Applied Pressure: 31.6, 63.2, 94.7 and 138.9 MPa Moisture Content: 10% (w.b.) Treatment: Non-Treated (NT) and Steam Exploded (SE)	Barley-NT 1003 kg/m3 Barley-SE 1169 kg/m3 Canola-NT 1035 kg/m3 Canola-SE 1165 kg/m3 Oat-NT 1024 kg/m3 Oat-SE 1165 kg/m3 Wheat-NT 1009 kg/m3 Wheat-SE 1180 kg/m3	Adapa et al., 2010a
Corn Stover and Switchgrass	Hammer Mill Screen Size: 3.0 and 4.6 mm Applied Pressure: 100 and 150 MPa Pre-Heat Temperature: 25, 75 and 100oC Moisture Content: 10 and 15% (w.b.)	Corn Stover 1197 kg/m3 Switchgrass 1098 kg/m3	Kaliyan and Morey, 2009
Barley and wheat straw	Hammer Mill Screen Size: 1.6 mm Applied Pressure: 126.3 MPa Moisture Content: 12% (w.b.) Treatment: Non-Treated (NT), Microwave Pretreated (MT) and Microwave-Chemical Pretreated (MCT)	Barley-NT 995 kg/m3 Barley-MT 984 kg/m3 Barley-MCT 1440 kg/m3 Wheat -NT 950 kg/m3 Wheat -MT 1032 kg/m3 Wheat -MCT 1431 kg/m3	Kashaninejad and Tabil, 2011

Biomass	Independent Variables	Maximum Pellet Density	Reference
Barley and Wheat Straw, Corn Stover, and Switchgrass	Hammer Mill Screen Size: 3.2, 1.6 and 0.8 mm Applied Pressure: 31.6, 63.2, 94.7, 126.3 and 138.9 MPa Moisture Content: 6.22 to 8.30% (w.b.)	Barley Straw 1245 kg/m3 Wheat Straw 1344 kg/m3 Corn Stover 1399 kg/m3 Switchgrass 1173 kg/m3	Mani et al., 2004
Poplar Wood and Wheat Straw	Hammer Mill Screen Size: 3.2 and 0.8 mm Applied Pressure: 31.6, 63.2, 94.7 and 126.3 MPa Moisture Content: 9 and 15% (w.b.) Pre-Heat Die Temperature: 70 and 100oC Treatment: Non-Treated (NT) and Steam Exploded (SE)	Poplar-NT 1100 kg/m3 Poplar-SE 1341 kg/m3 Wheat-NT 1005 kg/m3 Wheat-SE 1324 kg/m3	Shaw, 2008

Table 3.Correlation for pellet density (ρ, kg/m3) with applied pressure (P, MPa) and hammer mill screen size (S, mm) for non-treated and steam exploded straw grinds.

Independent Variables and Interactions	Estimated Coefficients	R2 value	Coefficient of Variation	Root Mean Square Error
pBarleyNT =587.19+6.29(P)-0.025(P)2				
Intercept	587.19	0.91	2.99	27.59
P	6.29			
P*P	-0.025			
pBarleySE=727.38+7.41(P)-0.03(P)2-0.08(P×S)+2.84(S)2				
Intercept	727.38	0.89	3.30	34.89
P	7.41			
P*P	-0.03			
P*S	-0.08			
S*S	2.84			
pCanolaNT=545.50+7.41(P)-0.03(P)2-0.11(P×S)				
Intercept	545.50	0.89	3.49	32.62
P	7.41			
P*P	-0.03			
P*S	-0.11			
pCanolaSE=738.77+7.54(P)-0.03(P)2+0.11(P×S)-28.84(S)				
Intercept	738.77	0.92	3.20	22.55
P	7.54			
P*P	-0.03			
P*S	0.11			
S	-28.84			
pOatNT=666.24+5.79(P)-0.02(P)2-0.12(P×S)+3.54(S)2				

Independent Variables and Interactions	Estimated Coefficients	R2 value	Coefficient of Variation	Root Mean Square Error
Intercept	666.24	0.87	3.13	29.00
P	5.79			
P*P	-0.02			
P*S	-0.12			
S*S	3.54			
pOatSE=818.41+6.56(P)-0.03(P)2+0.12(P×S)-32.48(S)				
Intercept	818.41	0.92	2.66	28.37
P	6.56			
P*P	-0.03			
P*S	0.12			
S	-32.48			
pWheatNT=700.76+5.99(P)-0.02(P)2-36.06(S)+3.48(S)2				
Intercept	700.76	0.90	3.01	27.91
P	5.99			
P*P	-0.02			
S	-36.03			
S*S	3.48			
pWheatSE=717.60+6.77(P)-0.03(P)2				
Intercept	717.60	0.91	2.95	31.55
P	6.77			
P*P	-0.03			
Note: ρ – Density, kg/m3; NT – Non-Treated; SE – Steam Exploded; P – Pressure, MPa; S – Hammer Mill Screen Size, mm				

3.3. Durability

Durability represents the measure of shear and impact forces that a pellet could withstand during handling, storing and transportation process. The durability of pellets is usually measured following the ASABE Standard S269 (ASABE, 2007), which require about 50-100 g of pellets/ compacts. However, due to the limited number of pellets obtained during single-pellet compression test, it is not feasible to use this method. Instead, the durability of pellets can be measured by following the drop test method (Al-Widyan and Al-Jalil, 2001; Khankari et al., 1989; Sah et al., 1980; Shrivastava et al., 1989), where a single pellet is dropped from a 1.85 m height on a metal plate. The larger intact portion of the mass retained is expressed as the percentage of the initial weight.

Adapa et al. (2010) reported that the type of agricultural biomass, steam explosion pretreatment, applied pressure and screen size all had significant effect on pellet durability. Statistically, no significant correlation (R2 values) was obtained for change in durability with applied pressure and hammer mill screen sizes. In general, pellet durability increases with an increase in applied pressure and grind size, and application of pre-treatment. Similarly, Kaliyan and Morey (2009) indicated that the durability of corn stover or switchgrass briquettes was significantly affected by pressure, moisture content and preheating temperature, while particle size did not have any significant effect. Kashaninejad et al. (2011) also reported the mean durability of pellets made of giant wild rye and mixed forage increased from 63.08

to 89.26% and from 61.47 to 89.21%, respectively when the hammer mill screen size increased from 0.8 to 3.2 mm. This could be primarily due to mechanical interlocking of relatively long fibers at higher grind sizes. They also indicated that at any specific compressive load, the pellet durability of biomass grinds with 12% moisture content was significantly higher than samples with 9 and 15% and demonstrates the moisture contents above or below 12% would lead to lower quality pellets.

3.4. Specific energy for compaction and extrusion of pellet

During the compression and extrusion processes of individual biomass compacts, the force-displacement data is recorded and can be used to calculate the specific compression and extrusion energies following the methodology reported by Adapa et al. (2006) and Mani et al. (2006). The area under the force-displacement curve can be integrated using the trapezoid rule (Cheney and Kincaid, 1980); when combined with the pellet mass, the specific energy values in MJ/t can be calculated.

During single-pellet compression and extrusion, the pellets are prepared by densifying material against a base plate (representing the specific energy required to overcome friction within the straw grinds) as opposed to commercial operation where compacts are formed due to back-pressure effect in the die. Therefore, the specific energy required to extrude the compact should be included, which will closely emulate the specific energy required to overcome the friction between the ground compressed biomass and the die. Mani et al. (2006) have indicated that the extrusion (frictional) energy required to overcome the skin friction was roughly half of the total energy (12-30 MJ/t) for corn stover. Mewes (1959) showed that roughly 40% of the total applied energy was used to compress the materials (straw and hay) and the remaining 60% was used to overcome friction. Faborode and O'Callaghan (1987) studied the energy requirement for compression of fibrous agricultural materials. They reported that chopped barley straw at 8.3% (wb) moisture content consumed 28-31 MJ/t of energy, while un-chopped material consumed 18-27 MJ/t. Shaw (2008) reported that between 95 and 99% of the total specific energy was required to compress the grinds, whereas between 1 and 5% of the total specific energy was required to extrude the compact in single pellet tests. Shaw (2008) also reported that the mean values of specific compression energy ranged from 7.2 (pretreated wheat straw using steam explosion) to 39.1 MJ/t (wheat straw). Kashaninejad and Tabil (2011) indicated that microwave-distilled water and microwave-NaOH pre-treatments significantly increased the specific energy required for compression of wheat straw grinds so that it increased from 16.60 MJ/t to as high as 29.04 and 27.84 MJ/t after pre-treatment by microwave-distilled water and microwave-NaOH, respectively. They also reported less specific energy was required to compress wheat straw pre-treated by combination of microwave and Ca(OH)2. More specific energy was required to eject the pre-treated wheat straw grinds than the untreated wheat straw grinds and it increased from 3.20 MJ/t to 23.08 after pre-treatment by microwave-NaOH. Data analysis showed that the total energy required for compression and ejection of wheat straw grinds pre-treated by microwave-distilled water or microwave-alkaline was higher than untreated samples.

Adapa et al. (2010b) reported that the type of agricultural biomass, steam explosion pretreatment, applied pressure and screen size all had significant effect on specific energy required to form a pellet. In addition, they have developed correlations for specific energy with applied pressure and hammer mill screen size having highest R2 values for barley, canola, oat and wheat straw (Table 4). In general, the total and compression specific energy for compaction of non-treated and steam exploded barley, canola, oat and wheat straw at any particular hammer mill screen size significantly increased with an increase in applied pressure and significantly decreased with a decrease in hammer mill screen size.

Adapa et al. (2010b) also reported that the specific energy values obtained from the single-pellet compression tests should be used to compare the densification variables. However, these values may not have practical applications since the energy consumed by commercial densification machines / pilot-scale pellet mills may be higher.

4. BIOMASS COMPRESSION CHARACTERISTICS

4.1. Compression mechanism

The compression characteristics of ground agricultural biomass vary under various applied pressures. It is important to understand the fundamental mechanism of the biomass compression process, which is required in the design of energy efficient compaction equipment to mitigate the cost of production and enhance the quality of the product (Mani et al., 2004). To a great extent, the strength of manufactured pellets depends on the physical forces that bond the particles together (Tabil and Sokhansanj, 1996). These physical forces come in three different forms during pelleting operations: a) thermal; b) mechanical; and c) atomic forces (Adapa et al., 2002).

Pellets are formed by subjecting the biomass grinds to high pressures, wherein the particles are forced to agglomerate. It is generally accepted that the compression process is categorized in several distinct stages and difficult to let one simple monovariate equation to cover the entire densification region (Sonnergaard, 2001). Compression of grinds is usually achieved in three stages (Holman, 1991). In the first stage, particles rearrange themselves under low pressure to form close packing. The particles retain most of their original properties, although energy is dissipated due to inter-particle and particle-to-wall friction. During the second stage, elastic and plastic deformation of particles occurs, allowing them to flow into smaller void spaces, thus increasing inter-particle surface contact area and as a result, bonding forces like van der Waal forces become effective (Rumpf, 1962; Sastry and Fuerstenau, 1973; Pietsch, 1997). Brittle particles may fracture under stress, leading to mechanical interlocking (Gray, 1968). Finally, under high pressure the second stage of compression continues until the particle density of grinds has been reached. During this phase, the particles may reach their melting point and form very strong solid bridges upon cooling (Ghebre-Sellassie, 1989). Figure 2 shows the deformation mechanisms of ground particles under compression (Comoglu, 2007; Denny, 2002).

Table 4.Correlation for pellet density (ρ, kg/m3) with applied pressure (P, MPa) and hammer mill screen size (S, mm) for non-treated and stem exploded straw grinds.

Independent Variables and Interactions	Estimated Coefficients	R2 value	Coefficient of Variation	Root Mean Square Error
ρBarleyNT =587.19+6.29(P)–0.025(P)2				
Intercept	587.19	0.91	2.99	27.59
P	6.29			
P*P	-0.025			
ρBarleySE=727.38+7.41(P)–0.03(P)2–0.08(P×S)+2.84(S)2				
Intercept	727.38	0.89	3.30	34.89
P	7.41			
P*P	-0.03			
P*S	-0.08			
S*S	2.84			
ρCanolaNT=545.50+7.41(P)–0.03(P)2–0.11(P×S)				
Intercept	545.50	0.89	3.49	32.62

Independent Variables and Interactions	Estimated Coefficients	R2 value	Coefficient of Variation	Root Mean Square Error
P	7.41			
P*P	-0.03			
P*S	-0.11			
pCanolaSE=738.77+7.54(P)-0.03(P)2+0.11(P×S)-28.84(S)				
Intercept	738.77	0.92	3.20	22.55
P	7.54			
P*P	-0.03			
P*S	0.11			
S	-28.84			
pOatNT=666.24+5.79(P)-0.02(P)2-0.12(P×S)+3.54(S)2				
Intercept	666.24	0.87	3.13	29.00
P	5.79			
P*P	-0.02			
P*S	-0.12			
S*S	3.54			
pOatSE=818.41+6.56(P)-0.03(P)2+0.12(P×S)-32.48(S)				
Intercept	818.41	0.92	2.66	28.37
P	6.56			
P*P	-0.03			
P*S	0.12			
S	-32.48			
pWheatNT=700.76+5.99(P)-0.02(P)2-36.06(S)+3.48(S)2				
Intercept	700.76	0.90	3.01	27.91
P	5.99			
P*P	-0.02			
S	-36.03			
S*S	3.48			
pWheatSE=717.60+6.77(P)-0.03(P)2				
Intercept	717.60	0.91	2.95	31.55
P	6.77			
P*P	-0.03			
Note: ρ – Density, kg/m3; NT – Non-Treated; SE – Steam Exploded; P – Pressure, MPa; S – Hammer Mill Screen Size, mm				

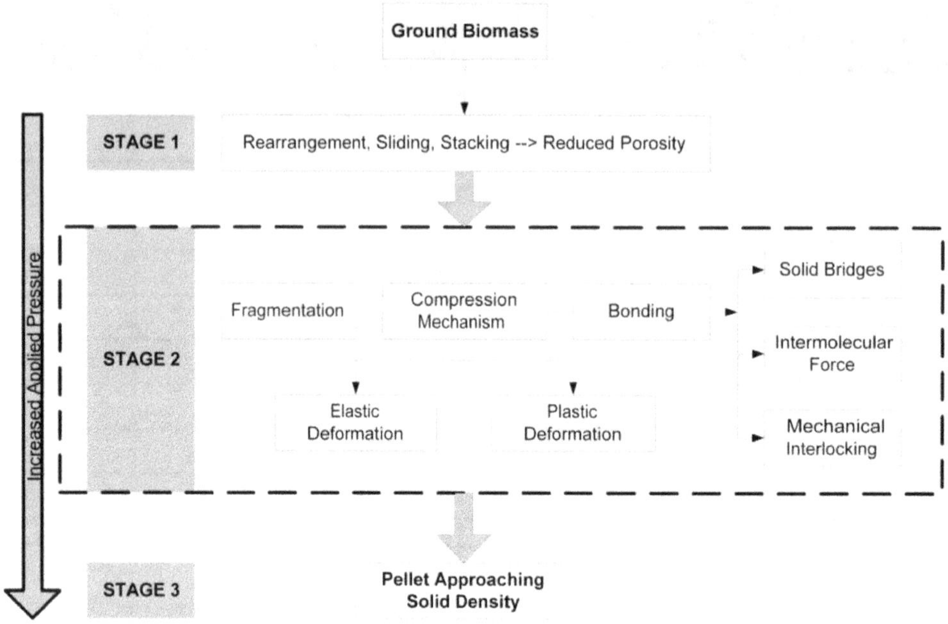

Figure 2.The deformation mechanisms of ground particles under compression (Comoglu, 2007; Denny, 2002)

4.2. Compression characteristics models

The technique of densifying or compacting agricultural biomass into pellets is crucial for the generation of biofuels. Ground biomass particles exhibit distinct behaviors when subjected to varying levels of applied pressure (Adapa et al., 2002 and 2009a). Hence, it is crucial to examine the alteration in compact density and volume at varying levels of pressure. The primary objective of fitting experimental data to an equation is often to provide linear graphs, facilitating comparisons across various data sets (Comoglu, 2007). Adapa et al. (2002 and 2009a), Denny (2002), and Mani et al. (2003) have extensively explored and reviewed the majority of compression models used for biomass materials. According to Adapa et al. (2009a), the models proposed by Kawakita and Ludde (1971), Cooper and Eaton (1962), and Jones (1960) were shown to have the most accurate compression and deformation properties for agricultural biomass.

4.2.1. Jones model

Jones (1960) expressed the density-pressure data of compacted powder in the form of equation 1.

$$ln(\rho) = mln(P) + b \qquad (1)$$

where, ρ is bulk density of compact powder mixture, kg/m^3, P is applied compressive pressure, MPa; *m and b are model constants.*

The constants b and m are determined from the intercept and slope, respectively, of the extrapolated linear region of the plot of ln(ρ) vs ln(P). The constant m has been shown to be equal to the reciprocal of the mean yield pressure required to induce plastic deformation (York and Pilpel, 1973). A large m value (low yield pressure) indicates the onset of plastic deformation at relatively low pressure, thus, the material is more compressible.

4.2.2. Cooper-Eaton model

Cooper and Eaton (1962) studied the compaction behavior of four ceramic powders. In each case it, was assumed that compression is attained by two nearly independent probabilistic processes, namely, the filling of voids having equal size as particles and filling of voids smaller than particles. Based on these assumptions, the following equation (2) was given:

$$\frac{V_O - V}{V_O - V_S} = a_1 e^{-\frac{k_1}{P}} + a_2 e^{-\frac{k_2}{P}} \tag{2}$$

where, V O = volume of compact at zero pressure, m^3; V = volume of compact at pressure P, m^3; V S = void free solid material volume, m^3; a^1 , a^2 , k^1, and k^2 = Cooper-Eaton model constants.

The difficulty in practical use of equation (2) is the assignment of some physical significance to the constant parameters. In addition, another drawback of this model is its applicability to only one-component system (Comoglu, 2007).

4.2.3. Kawakita-Ludde model

Kawakita and Ludde (1971) performed compression experiments and proposed an equation for compaction of powders based on observed relationship between pressure and volume (Equation 3).

$$\frac{P}{C} = \frac{1}{ab} + \frac{P}{a} \tag{3}$$

Where,

$$C = \frac{V_O - V}{V_O}$$

C = degree of volume reduction or engineering strain; *a and b = Kawakita-Ludde model constants related to characteristic of the powder.*

The linear relationship between P/C and P allows the constants to be evaluated graphically. This compression equation holds true for soft and low bulk density powders (Denny, 2002; Kawakita and Ludde, 1971), but particular attention must be paid on the measurement of the initial volume of the powder. Any deviations from this expression are sometimes due to fluctuations in the measured value of V0. The constant a is equal to the values of C = C∞ at infinitely large pressure P.

$$C_\infty = \frac{V_O - V_\infty}{V_O}$$

Where, V∞= net volume of the powder, m3.

It has been reported that the constant a is equal to the initial porosity of the sample, while constant 1/b is related to the failure stress in the case of piston compression (Mani et al., 2004).

4.3. Compressibility of various biomass samples

The constant m in the Jones (1960) model can provided valuable information about the onset of plastic deformation of the ground agricultural biomass. It has been observed that ground particles obtained from larger hammer mill screen sizes has higher compressibility. In addition, application of pre-treatment also improves the compressibility of the agricultural biomass (Adapa et al., 2010).

The dimensionless coefficients, a 1 and a 2 in Cooper and Eaton (1962) model represent the densification of powdered material by particle rearrangement and deformation, respectively. If the sum of coefficients (a1 + a2) is less than unity, it is an indication that other process must become operative before complete compaction is achieved. For agricultural biomass grinds, the a1 values were higher than a 2 values, hence the material was primarily densified through the process of particle rearrangement. Occasionally, the sum of coefficients (a1 + a2) for agricultural biomass was observed to be above unity. The phenomenon of having sum of coefficient more than unity was also observed by Adapa et al. (2002 and 2009a), and Shivanand and Sprockel (1992), which implies that the densification could not be fully attributed to the two mechanisms of compression as assumed by the Cooper and Eaton (1962) model (Adapa et al., 2010a).

In the Kawakita and Ludde (1971) model, constant a represents the initial porosity of the sample. It has been reported that the porosity and hammer mill screen sizes (corresponding geometric mean particle diameter) are positively correlated. In addition, porosity increases with application of pre-treatment since organized lignocellulosic structure of biomass disintegrates during this process. The parameter 1/b in the Kawakita-Ludde model indicates the yield strength or failure stress of the compact. In general, the yield strength has negative correlation with hammer mill screen sizes. Also, application of pre-treatment lowers the yield strength of ground agricultural biomass. Statistically, the Kawakita and Ludde (1971) model has been observed to provide accurate representation of the compression and deformation characteristics of agricultural biomass (Adapa et al., 2010a).

5. PILOT-SCALE PELLETING OF AGRICULTURAL BIOMASS

Pilot-scale densification of biomass is required to demonstrate the feasibility of production of pellets by application of various variables studied during single-pellet experiments. A pilot-scale pellet mill such as CPM CL-5 pellet mill (Figure 3) (California Pellet Mill Co., Crawfordsville, IN) can be used for processing of agricultural straw grinds into pellets. The pellet mill usually consists of a corrugated roller and ring die assembly, which compacts and extrudes the biomass grinds from the inside of a ring-shaped die by pressure applied by rolls where either the die or the roll suspension is rotating. Rolls are mounted close to the die surface, but still leaving room for a compacted feed layer to enter the roll gap. Friction between feed layer and rolls makes the rolls rotate (Larsson et al., 2008). In addition to variables indicated in the single-pellet testing, the quality of pellets also depends on machine variables such as the ring die size (radius), length (thickness, l), ring hole diameter (d), l/d ratio, and the rotational speed of the pellet mill (Adapa et al., 2002; Hill and Pulkinen, 1988; Tabil and Sokhansanj, 1996). A monitoring study of commercial pellets was done by Hill and Pulkinen (1988), on variables such as die geometry, conditioning temperatures, natural moisture of the grind, forage quality, bulk density of the grinds, and the use of binding agents. Similarly, Larsson et al. (2008) studied the effect of raw material moisture content, steam addition, raw material bulk density, and die temperature on production of high quality pellets. Also, Serrano et al. (2011), determined the effect of grind size, moisture content and customization of barley straw by adding pine dust to the mixture (blended pellets).

The feed rate of ground biomass to the pellet mill can be controlled using a vibratory feeder (Figure 3). The feed rate should be optimized according to the pellet mill capacity, which will directly affect the throughput. The pilot-scale pelleting test should be performed for a predefined period and the manufactured pellets should be collected and weighed to determine the pellet mill throughput (kg/h). In addition, the pellet mill energy consumption (kWh) should be recorded in real time using a data logger connected to a computer and should be used to calculate the specific energy (MJ/t) required to manufacture pellets from ground agricultural biomass.

Raw materials causing uneven pellet production have low bulk density compared to other milled biofuel pellet raw materials. Low raw material bulk density will put higher demands on the die feeding system of the pelletizer with greater volume throughput for maintained production level. Larsson et al. (2008) investigated the pre-compaction of reed canary grass as an alternative to avoid low and intermittent production of biofuel pellets. They have observed that the process of pre-compaction can increase the bulk density of raw material from 150 kg/m3 to 270 kg/m3, which resulted in the continuous production of pellets at a moisture content of 13.8% (w.b.). Pressurized steam conditioners are used in the feed pellet industry to decrease raw material porosity and to improve pellet hardness/ durability (Thomas et al., 1997). Adapa et al. (2010b) were unable to produce any pellets due to the low bulk density of both non-treated and pre-treated agricultural straw grinds at 10% moisture content (w.b.). Therefore, they have added moisture and oil to increase bulk density of grinds to a level of 17.5% (w.b.) and 10% (by weight), respectively, which resulted in production of pellets. Similar observation was made by Serrano et al. (2011) where they have to increase the grind moisture content in the range of 19-23% (w.b.) to produce pellets in a pellet mill. However, addition of pine sawdust to barley straw resulted in high quality pellets at a lower moisture content of 12% (w.b.).

Testing and, if required improving the durability of pellets is important for the industry to evaluate pellet quality and minimize losses during handling and transportation. The concept is not to add any external binders to enhance pellet quality, but rather activate the natural binders in the agricultural biomass by application of various variables, pre-processing techniques and pre-treatments. Biomass pellets can be customized based on proximate analysis data to make them suitable for direct combustion and thermo-chemical conversion applications. Customization can be achieved by forming composites of different straws to control important variables such as energy and ash content of pellets. Similarly, addition of biomass having good binding characteristics to straw with less cohesive characteristics may enhance particle bonding resulting in durable pellets.

Adapa et al. (2010b) reported pellet mill tests on both non-treated and steam exploded agricultural biomass at different hammer mill screen sizes. They have successfully produced pellets from ground non-treated barley, canola, oat and wheat straw at hammer mill screen sizes of 0.8 and 1.6 mm having moisture content of 17.5% (wb) and flax seed oil of 10% by weight. The non-treated ground straw at 3.2 and 6.4 mm screen size did not produce pellets. Similar pelleting process was followed for ground steam exploded straw. Due to very low bulk density and poor flowability, the steam exploded grinds did not produce pellets at any of the hammer mill screen sizes used in the investigation. However, the customized barley, canola, oat and wheat straw having 25% steam exploded material by weight at 0.8 mm screen size successfully produced pellets. Addition of higher percentage of steam exploded straw and customization at screen sizes of 1.6, 3.2, and 6.4 mm did not produce pellets, which could be due to the fact that adding steam exploded (having very low bulk density) to non-treated straw (having relatively higher bulk density) decreased the overall bulk density and flowability of the grinds, thus hindering the production of pellets in the pilot scale mill. The pilot scale pellet mill in this test is constrained with a small motor (3.7 kW (5 hp)) running it, whereas in a commercial pellet mill, the motors are much bigger and more tolerant to changes in feed bulk density. Shaw et al. (2007) reported similar trends where the quality of wheat straw pellets increased with an increase in moisture content to 15.9% (wb).

Figure 4 shows the photograph of pellets manufactured from barley, canola, oat and wheat straw from non-treated grinds at 0.8 and 1.6 mm screen sizes, and customized straw grinds at 0.8 mm having 25% steam exploded straw by weight (Adapa et al., 2010b).

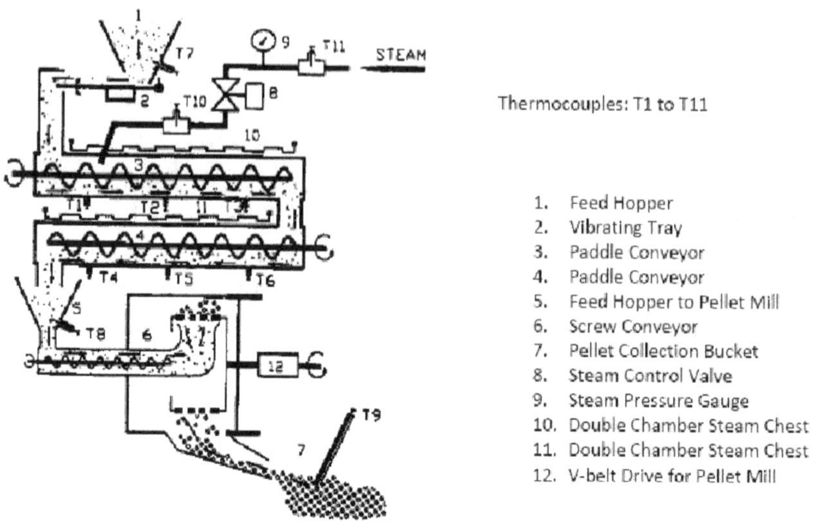

Figure 3.Schematic diagram of CPM CL-5 pellet mill

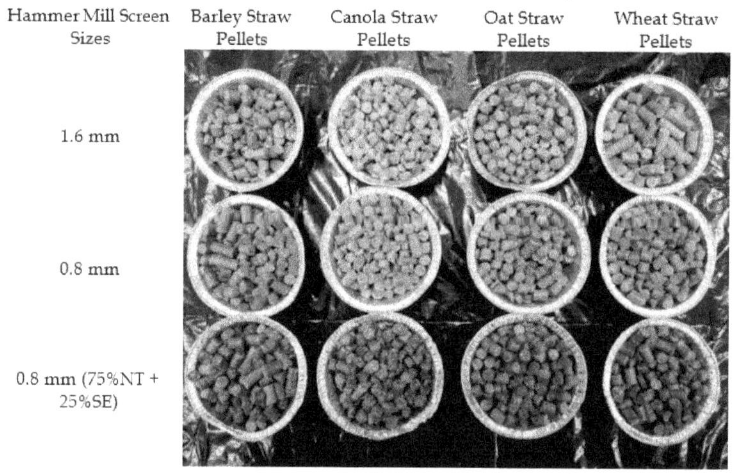

Figure 4.Photograph of pellets manufactured using a pilot scale pellet mill for non-treated (NT) straw at 1.6 and 0.8 mm hammer mill screen size, and customized grinds at 0.8 mm screen size having 25% steam exploded (SE) straw.

5.1. Pellet bulk density

To calculate the density of each pellet in kg/m3, the mass, length, and diameter of each pellet should be taken into account. The bulk density of produced pellets can be determined by measuring the mass of pellets filled in a cylindrical container with a known volume. In their study, Larsson et al. (2008) found that the moisture content of the raw material had the greatest impact on the bulk density of the pellets. They observed a negative association between these two factors. In a similar vein, two more studies have found that the density of wheat straw, big bluestem grass, maize stover, sorghum stalk, and switchgrass reduced when the moisture content increased (Colley et al., 2006; Theerarattananoon, et al., 2011). Larsson et al. (2008) observed no association between pellet bulk density and die temperature, which contradicts the findings of Hill and Pulkinen (1988) and Tabil and Sokhansanj (1996). The study conducted by Serrano et al. (2011) found no significant impact of hammer mill screen size (4 mm and 7 mm) on pellet density. Nevertheless, the density of the pellets dropped as the moisture content increased. Adapa et al. (2010b) provided data on the density of pellets made from untreated straw samples, which were measured at 1.6 and 0.8 mm. They also included data on pellets made from a tailored sample containing 25% steam exploded straw, which were measured at a screen size of 0.8 mm (Table 5). Typically, the density of pellets rose as the screen size decreased from 1.6 to 0.8 mm. Nevertheless, there were no notable disparities in density measurements between untreated samples at 0.8 mm and tailored samples, with the exception of canola and oat straw. This may be attributed to significant variations in the density values of individual pellets. The pellet density values closely approached the densities of individual biomass particles at the corresponding grind sizes (Adapa et al., 2010b). The bulk density of pellets made from barley, canola, oat, and wheat straw varied significantly depending on the grind size and customisation, except for wheat straw pellets with a grind size of 0.8 mm in both non-treated and customized samples (Table 5). Generally, the average bulk densities of the tailored straw samples were greater (except for barley straw), which is in line with the increase in particle densities. The bulk densities of the generated pellets exceeded the minimal design value of 650 kg/m3 recommended by Obernberger and Thek (2004) for wood pellet producers.

5.2. Pellet durability

The durability of pellets can be measured following the ASABE Standard S269 (2007). The method states that 100 g of pellet sample should be weighed and placed in a dust-tight enclosure/ chamber, and tumbled for 10 min at 50 r/min. A 5.70 mm sieve should used to determine the fines produced by the pellets during the tumbling process. The mass of pellets left on the sieve, as percentage of the total mass of pellet sample used during the test, was considered as the durability of the pellets.

Hill and Pulkinen (1988) reported an increase in pellet durability by about 30 to 35% with an increase in pellet temperature from 60 to 104ºC. A die length-to-diameter ratio (l/d) of 8 to 10 was also reported to be ideal for making high quality pellets. Similarly, Tabil and Sokhansanj (1996) conducted a study for improving the physical quality of alfalfa pellets by controlling and optimizing the manufacturing process. The process conditions investigated were steam conditioning temperatures, die geometry (length to diameter or l/d ratio), hammer mill screen sizes used in grinding dry chops, and die speed. They reported that a higher conditioning temperature (95ºC) resulted in improved durability of processed pellets. The durability of samples was generally better using the smaller die (higher l/d ratio). The hammer mill screen size did not show any affect on pellet durability. Finally, they reported that high durable pellets are obtained at low die speed (250 rpm). Theerarattananoon et al. (2011) also observed that an increase in length / thickness of die resulted in significant increase in durability of biomass pellets.

Larsson et al. (2008) and Serrano et al. (2011) reported that the most influential factor affecting pellet durability was raw material moisture content and showing a positive correlation. The maximum durability was obtained at moisture content of 14.9% without steam addition, and at 13.7% (w.b.) with steam addition of 2.6% (Larsson et al., 2008). Serrano et al. (2011) found that the highest mechanical durability reached for barley straw pellets was 95.5% at moisture content of 19-23% (w.b.), while no pellets were formed below the 19% moisture content. In addition, they observed that the durability of barley straw pellets increased with addition of pine sawdust at 2, 7 and 12% bu weight. However, the percentage increase in pine sawdust did not have significant effect on durability. Colley et al. (2006) observed highest durability of 95.9% for pellets from switchgrass at a moisture content of 8.6% (w.b.). The moisture content in the range of 9-14%, 9-11% and 14-16% (d.b.) did not have any significant effect on maximum durability of 96.8%, 96.8% and 89.5% for pellets from wheat straw and corn stover, big bluestem grass, and sorghum stalk, respectively (Theerarattananoon, et al., 2011); however, further increasing the moisture content reduced pellet durability for respective agricultural biomass. Also, the durability was observed to be positively correlated to die temperature (Larsson et al., 2008).

Adapa et al. (2010b) reported that the durability of pellets obtained from non-treated straw samples at 1.6 and 0.8 mm, and customized sample having 25% steam exploded straw at 0.8 mm screen size were significantly different (Table 5). In general, higher durability values were observed for non-treated straw samples at 0.8 mm hammer mill screen size. The durability of pellets significantly increased with a decrease in grind size for non-treated samples from 1.6 to 0.8 mm. However, addition of steam exploded straw to non-treated straw at 0.8 mm screen size resulted in a decrease in durability, except for wheat straw. This could be due to the fact that steam exploded material has lower soluble lignin content and higher cellulose and hemicelluloses content compared to non-treated straw (Table 1). This observation is in contrast to Lam et al. (2008), who reported that the quality (durability) of pellets produced from steam exploded sawdust was 20% higher than non-treated sawdust. Though, it is important to note that high durability values (>80%) were obtained for all pilot scale pelleting tests.

5.3. Analysis of specific energy and energy balance in the process of pelleting

The durability of pellets was negatively correlated to pellet mill throughput and was positively correlated to specific energy consumption (Table 5). The specific energy values obtained from pilot scale pellet mill are 10-25 times higher than reported by Mani et al. (2006b) and Adapa et al. (2010a and 2009b) for agricultural straw, using a single pellet Instron testing machine. The higher pellet mill specific energy numbers could be due to higher friction values and practical pelleting conditions, which are closer to industrial operations.

An overall specific energy analysis is desired in order to understand the net amount of energy available for the production of biofuels after postharvest processing and densification of agricultural straw. The specific energy analysis was performed for pilot-scale pelleting of non-treated and customized (75% non-treated + 25% steam exploded) barley, canola, oat and wheat straw at 1.6 and 0.8 mm hammer mill screen sizes (Table 6). The specific energy for grinding of straw at 0.8 mm was calculated using regression equations reported in Adapa et al. (2011b). The specific energy for chopping and grinding of biomass, production of pellets using pellet mill and higher heating values for straw were obtained from experimental data (Adapa et al., 2011b and 2010b). In addition, the specific energy required for operating the chopper, hammer mill and pellet mill were 337, 759 and 429 W, respectively. On average, the operation of biomass chopper required five times more energy than chopping of biomass. On the other hand, the grinding of biomass required on an average three times more energy than operation of hammer mill. Interestingly, almost the same amount of energy was required to operate the pellet mill and production of pellets. Total specific energy required to form pellets increased with a decrease in hammer mill screen size from 1.6 to 0.8 mm, however, the total specific energy for the process decreased for customized straw compared to non-treated straw at 0.8 mm screen size (Table 6). It has been determined that the net specific energy available for production of biofuel is a significant portion of original agricultural biomass energy (92-94%) for all agricultural biomass (Table 6).

5.4. Analysis of the financial and environmental impact of biomass densification

Sultana et al. (2010) performed a techno-economic analysis and developed a model for a plant that can produce agricultural straw (barley, oat and wheat) pellets for 30 years. They have included the cost of obtaining the straw, transporting straw to the pellet plant, and producing pellets. Costs incurred by the plant for the production of pellets included capital cost, energy cost, labor cost, and consumable cost. The biomass procurement area was determined to estimate the transportation cost. The scale factors for all the equipment related to pellet production were determined based on the data of previous studies (Sultana et al., 2010). To develop the model, minimum, average and maximum yields of wheat, barley and oats straw in Western Canada were considered. They have determined that the cost of pellets does not change much for capacities over 70,000 tonnes per year (cost of production per tonne is $170.89). Therefore, the optimum size is the same for both average and maximum yield cases. In addition, it was observed that the total cost of pellet production is most sensitive to field cost followed by the transportation cost.

Life cycle assessment (LCA) study was performed on wheat straw production system and densification system in the Canadian Prairies using the LCA modelling software tool SimaPro 7.2 to determine the environmental burdens of manufacturing the wheat straw bale and wheat straw pellet (Li et al., 2011). The factors taken into consideration were greenhouse gas emission, acidification, eutrophication, ozone layer depletion, abiotic depletion, human toxicity, and photochemical oxidation. Li et al. (2011) reported that the production of biomass pellet has higher global warming effect than biomass bale, especially in CO2 andCH4 emissions from fossil fuel consumption, which is very high in densification system due to machinery usage. It was also reported that the production of wheat straw pellet has higher environmental impact on acidification, eutrophication, human toxicity and other categories than biomass bale. The dominant factors determining most environmental impacts in agricultural system are fertilizer use and production, while machinery use, manufacturing and energy consumption are main contributors to greenhouse gas emission and other environmental burdens in the densification system (Li et al., 2011).

Table 5.Pellet density, durability, throughput and specific energy data for non-treated and steam exploded barley canola, oat and wheat straw at 17.5% moisture content (wb) and 10% flaxseed oil content

Agricultural Biomass	Hammer Mill Screen Size (mm)	Pellet Density (kg/m3)	Pellet Bulk Density (kg/m3)	Durability (%)	Throughput (kg/h)	Specific Energy (MJ/t)
Barley Straw	1.6 (100% NT)	1158±109*†£ aD	665±01‡ aD	91±00‡ aD	4.88	293
	0.8 (100% NT)	1174±46 aD	700±07 bD	93±01 bD	4.21	353
	0.8 (75% NT + 25% SE)	1184±63 aD	714±02 cD	87±01 cD	3.46	301
Canola Straw	1.6 (100% NT)	1023±85 aE	629±01 aE	90±01 aD	3.86	385
	0.8 (100% NT)	1204±43 bDE	720±04 bE	95±00 bE	3.63	440
	0.8 (75% NT + 25% SE)	1144±50 cD	641±01 cE	82±00 cE	5.51	265
Oat Straw	1.6 (100% NT)	1140±63 abD	631±03 aE	89±01 aE	4.48	340
	0.8 (100% NT)	1188±78 aDE	649±02 bF	93±00 bD	3.81	344
	0.8 (75% NT + 25% SE)	1071±101 bE	676±06 cF	89±01 aF	4.03	335
Wheat Straw	1.6 (100% NT)	1163±57 aD	673±02 aF	94±01 aF	5.44	381
	0.8 (100% NT)	1278±136 bE	721±04 bE	95±01 bE	3.81	297
	0.8 (75% NT + 25% SE)	1213±88 abD	722±04 bG	95±00 cG	4.08	342

NOTE – Non-treated Straw Samples;SE – Steam Exploded Straw Samples; *10 replicates; ‡3 replicates; † 95% confidence interval; £ Student-Neuman-Keuls test at 5% level of significance for same sample biomass at various hammer mill screen sizes (a, b and c); at same hammer mill screen size for different sample biomass (D, E, F and G)

Table 6.Overall specific energy analysis to show net energy available for production of biofuels after postharvest processing and densification of agricultural straw.

Treatment	Hammer Mill Screen Size (mm)	Specific Energy (MJ/t) Chopping Biomass	Grinding Biomass	Pilot-Scale Pelleting	Total£	HHV (MJ/t)	Net Energy γ (MJ/t)
Barley							
NT*	1.6	11.3	90.4	293	924	16400	15476
NT	0.8	11.3	206.6	353	1100	16400	15300
75% NT + 25% SE*	0.8	11.3	189.3	301	1030	16650	15620
Canola							
NT	1.6	7.1	128.5	385	987	16700	15713
NT	0.8	7.1	363.3	440	1277	16700	15423
75% NT + 25% SE	0.8	7.1	341.6	265	1080	17100	16020
Oat							
NT	1.6	9.9	149.5	340	1029	16400	15371
NT	0.8	9.9	253.6	344	1137	16400	15263
75% NT + 25% SE	0.8	9.9	245.2	335	1120	16750	15630
Wheat							
NT	1.6	8.2	153.3	381	1048	17000	15952
NT	0.8	8.2	382.7	297	1194	17000	15806
75% NT + 25% SE	0.8	8.2	332.1	342	1188	17200	16012

*NOTE – Non-Treated; SE – Steam Exploded

£ Total Specific Energy = Specific Energy (Chopping Biomass + Operating Chopper + Grinding Biomass + Operating Hammer Mill + Pilot-Scale Pelleting)

γNet Energy = HHV – Total

6. CONCISE SUMMARY

Compacting biomass into durable forms is an efficient method to fulfill the raw material needs for biofuel production. The compressibility of agricultural biomass on the ground varies depending on the amount of pressure applied. Understanding the fundamental mechanism of biomass compression is crucial for designing energy-efficient compaction equipment that reduces production costs and improves product quality. The strength of manufactured compacts mostly relies on the physical forces that bind the particles together. The compaction procedures generate physical forces in three distinct forms: a) thermal forces, b) mechanical forces, and c) atomic forces. In order to produce and personalize superior products that can endure different stresses during transportation and handling, it is crucial to anticipate the desired and interdependent quality characteristics (such as density and durability) in relation to various independent factors (such as pre-treatment, grind size, applied pressure, hold time, die temperature, and moisture content). Furthermore, it is crucial to define precise energy needs for the manufacturing of biomass pellets. This information will help determine the economic feasibility of the densification process. An increase in applied pressure and a decrease in hammer mill screen size have been found to cause a considerable rise in the density of biomass pellets. Furthermore, the implementation of pre-treatment has been found to greatly enhance the density of the pellets. This is due to the fact that pre-treated straw has smaller particle sizes and considerably greater particle densities. According to statistical analysis, the presence of agricultural biomass did not have a noticeable impact on pellet density. However, factors such as steam explosion pre-treatment, applied pressure, moisture content, pre-heat temperature, and screen size did have a significant influence. An inverse relationship has been noted between the bulk density of the pellets and their moisture content, but a direct relationship is detected between the bulk density and the temperature of the pellet mill die. Typically, the average bulk density of pellets made from tailored straw samples is higher due to an increase in particle densities. Factors such as agricultural biomass, steam explosion pre-treatment, applied pressure, moisture content, pre-heat temperature, and screen size all have a substantial influence on pellet durability. Typically, the durability of pellets improves as the applied pressure and grind size increase, and when pre-treatment is applied. Higher pellet mill die temperature, steam conditioning temperature, and die thickness led to improved pellet durability. There was no discernible pattern in the durability of straw when it was customized by combining untreated and steam-exploded straw grinds. The specific energy needed to create a pellet has been greatly influenced by factors such as the type of agricultural biomass, steam explosion pre-treatment, applied pressure, and screen size. The overall and compression-specific energy required for compacting untreated and steam-exploded barley, canola, oat, and wheat straw significantly increased as the applied pressure increased, and significantly decreased as the hammer mill screen size decreased. The durability of pellets exhibited a negative correlation with pellet mill throughput and a positive correlation with specific energy consumption. A comprehensive energy balance analysis revealed that a substantial proportion (92-94%) of the initial energy derived from agricultural biomass can be utilized for the generation of biofuels.

REFERENCES

Adapa, P.K., Tabil L.G., Schoenau, G.J., Canam, T., & Dumonceaux, T. (2011a). Quantitative Analysis of Lignocellulosic Components of Non-Treated and Steam Exploded Barley, Canola, Oat and Wheat Straw using Fourier Transform Infrared Spectroscopy. *The Journal of Agricultural Science and Technology*, 5(7), (In-Press).

Adapa, P.K., Tabil, L.G., & Schoenau, G.J. (2011b). Grinding Performance and Physical Properties of Non-Treated and Steam Exploded Barley, Canola, Oat and Wheat Straw. *Journal of Biomass and Bioenergy, doi:10.1016/j.biombioe.2010.10.004*, 35(2011), pp. 549-561.

Adapa, P.K., Tabil, L.G., & Schoenau, G.J. (2010a). Compression Characteristics of Non-Treated and Steam Exploded Barley, Canola, Oat and Wheat Straw Grinds. *ASABE Journal of Applied Engineering in Agriculture*, 26(4), pp. 617-632.

Adapa, P.K., Tabil, L.G., & Schoenau, G.J. (2010b). Pelleting Characteristics of Selected Biomass with and without Steam Explosion Pretreatment. *International Journal of Agricultural and Biological Engineering*, 3(3), pp. 62-79.

Adapa, P.K., Tabil, L.G., & Schoenau, G.J. (2009a). Compression Characteristics of Selected Ground Agricultural Biomass. *Agricultural Engineering International: the CIGR Ejournal*, Manuscript 1347, XI(June), pp. 1-19.

Adapa, P.K., Tabil, L.G., & Schoenau, G.J. (2009b). Compaction Characteristics of Barley, Canola, Oat and Wheat Straw. *Biosystems Engineering, doi:10.1016/j.biosystemseng.2009.06.022*, 104(2009), pp. 335-344.

Adapa, P.K., Schoenau, G.J., Tabil, L.G., Arinze, E.A., Singh, A., & Dalai, A.K. (2007). Customized and value-added high quality alfalfa products - a new concept. *Agricultural Engineering International: the CIGR Ejournal*, Manuscript FP 07 003, IX(June), pp. 1-28.

Adapa, P.K, Singh, A., Schoenau, G.J., & Tabil, L.G. (2006). Pelleting Characteristics of Fractionated Alfalfa Grinds - Hardness Models. *International Journal of Powder Handling and Processing*, 18(5), pp. 294-299.

Adapa, P.K., Tabil, L.G., Schoenau, G.J., Crerar, B., & Sokhansanj, S. (2002). Compression Characteristics of Fractionated Alfalfa Grinds. *Powder Handling and Processing*, 14(4), pp. 252-259.

Al-Widyan, M.I., & Al-Jalil, H.F. (2001). Stress-density relationship and energy requirement of compressed only cake. *Applied Engineering in Agriculture*, 17(6), pp. 749-753.

ASABE Standards 269.4. (2007). Cubes, Pellets and Crumbles – Definitions and methods for determining density, durability and moisture content. In: ASABE Standards. St. Joseph, MI.: American Society of Agricultural and Biological Engineers.

ASTM. ASTM D5865-03. (2003). Standard test method for gross calorific value of coal and coke. In: Annual book of ASTM standards, Vol. 05.06, 517-527. West Conshohocken, PA.: American Society for Testing and Materials.

Bowyer, J.L., & Stockmann, V.E. (2001). Agricultural residues: An exciting bio-based raw material for the global panel industry. *Forest Products Journal*, 51(1), pp. 10-21.

Campbell, C.A., Zentner, R.P., Gameda, S., Blomert, B. & Wall, D.D. (2002). Production of annual crops on the canadian prairies: trends during 1976 – 1998. *Canadian Journal of Soil Science*, 82, pp. 45-57.

Chancellor, W.J. (1962). Formation of hay wafers with impact loads. *Agricultural Engineering*, 43(3), pp. 136-138, 149.

Cheney, W., & Kincaid, D. (1980). Numerical mathematics and computing. Monterey, CA: Brooks/Cole Publishing Company.

Colley, Z., Fasina, O.O., Bransby, D., & Lee, Y.Y. (2006). Moisture effect on the physical characteristics of switchgrass pellets. Transactions of ASABE, 49, pp. 1845–1851.

Comoglu, T. (2007). An Overview of Compaction Equations. Journal of Faculty of Pharmacy, Ankara 36(2), pp. 123-133.

Cooper, A.R., & Eaton, L.E. (1962). Compaction Behavior of Several Ceramic Powders. Journal of the American Ceramic Society, 45(3), pp. 97-101.

Demirbaş, A. (2007). Effect of moisture and hydrogen content on the heating value of fuels. Energy Sources Part A, 29, pp. 649-655.

Demirbaş, A. & Sahin, A. (1998). Evaluation of biomass residue 1. Briquetting waste paper and wheat straw mixtures. Fuel Processing Technology, 55, 175–183.

Denny, P.J. (2002). Compaction Equations: A Comparison of the Heckel and Kawakita Equations. Powder Technology, 127, pp. 162-172.

Faborode, M.O., & O'Callaghan, J.R. (1987). Optimizing the compression/ briquetting of fibrous agricultural materials. Journal of Agricultural Engineering Research, 38, pp. 245–262.

FAO, (2011). Gasification Fuels. Food and Agriculture Organization of the United Nations. Website accessed on March 15, 2011. Web Address: http://www.fao.org/docrep/t0512e/t0512e0b.htm

Ghebre-Sellassie, I. (1989). Mechanism of Pellet Formation and Growth. Pharmaceutical Pelletization Technology, ed. I. Ghebre-Sellassie, 123-143. New York, NY: Marcel Dekker Inc.

Gray, W.A. (1968). Compaction after Deposition. In The Packing of Solid Particles, 89-107. New York, NY: Marcel Dekker Inc.

Grover, P.D. & Mishra, S.K. (1996). Biomass Briquetting Technology and Practices. Food and Agriculture Organization (FAO), UN, Document No. 46.

Gustafson, A.S. & Kjelgaard, W.L. (1963). Hay pellet geometry and stability. *Agricultural Engineering*, 44(8), pp. 442–445.

Hill, B., & Pulkinen, D.A. (1988). A study of the factors affecting pellet durability and pelleting efficiency in the production of dehydrated alfalfa pellets. Saskatchewan, Canada: Saskatchewan Dehydrators Association.

Holley, C.A. (1983). The densification of biomass by roll briquetting. Proceedings of the Institute for Briquetting and Agglomeration (IBA), 18, pp. 95-102.

Holman, L.E. (1991). The Compaction Behaviour of Particulate Materials: An Elucidation based on Percolation Theory. *Powder Technology*, 66, pp. 265.

Jones, W.D. (1960). Fundamental Principles of Powder Metallurgy; London UK; Edward Arnold Publishers Ltd.: 242-370.

Kaliyan, N. & Morey, R.V. (2006a). Factors affecting strength and durability of densified products. ASABE Annual International Meeting, American Society of Agricultural and Biological Engineers, Portland, Oregon July 9-12, Paper Number 066077, 2950 Niles Road, St. Joseph, MI 49085-9659 USA.

Kaliyan, N. & Morey, R.V. (2006b). Densification characteristics of corn stover and switchgrass. ASABE Annual International Meeting, American Society of Agricultural and Biological Engineers, Portland, Oregon July 9-12, Paper Number 066174, 2950 Niles Road, St. Joseph, MI 49085-9659 USA.

Kaliyan, N. & Morey, R.V. (2009). Densification Characteristics of Corn Stover and Switchgrass. *Transactions of the ASABE*, 52(3), pp. 907-920.

Kashaninejad, M., & Tabil. L.G. (2011). Effect of microwave-chemical pre-treatment on compression characteristics of biomass grinds. *Biosystems Engineering*, 108, pp. 36- 45.

Kashaninejad, M., Tabil, L.G. & Tabil, X. (2011). Compression characteristics of giant wild rye and mixed forage grinds, *Renewable Energy*, Under review.

Kawakita, K., & Ludde, K.-H. (1971). Some Considerations on Powder Compression Equations. *Powder Technology*, 4, pp. 61-68.

Khankari, K.K., Shrivastava, M. & Morey, R.V. (1989). Densification characteristics of rice hulls. ASAE Paper No. 89-6093, St. Joseph, Mich., ASABE.

Lam, P.S., Sokhansanj, S., Bi, X., Lim, C.J., Mani, S., & Melin, S. (2008). Energetic feasibility of biomass densification with steam explosion pretreatment. CSBE Annual General Meeting; July 14-16, Poster Number CSBE08305, Vancouver, BC.

Larsson, S.H., Thyrel, M., Geladi, P., & Lestander, T.A. (2008). High quality biofuel pellet production from pre-compacted low density raw materials. *Bioresource Technology*, 99 (2008), pp. 7176-7182.

Li, Y., & Liu, H. (2000). High-pressure densification of wood residues to form an upgraded fuel. *Biomass and Bioenergy*, 19, pp. 177-186.

Liu, R., Yu, H., & Huang, Y. (2005). Structure and morphology of cellulose in wheat straw. *Cellulose*, 12, pp. 25-34.

MacBain, R. (1966). Pelleting animal feed. Chicago, IL: American Feed Manufacturing Association.

Mani, S., Tabil, L.G. & Sokhansanj, S. (2006). Effects of compressive force, particle size and moisture content on mechanical properties of biomass pellets from grasses. *Biomass and Bioenergy*, 97(2006b), pp. 1420-1426.

Mani, S., Tabil, L.G. & Sokhansanj, S. (2006b). Specific energy requirement for compacting corn stover. *Bioresource Technology*, 97, pp. 1420-1426.

Mani, S., Tabil, L.G. & Sokhansanj, S. (2004). Evaluation of compaction equations applied to four biomass species. *Canadian Biosystems Engineering*, 46(3), pp. 3.55-3.61.

Mani, S., Tabil, L.G. & Sokhansanj, S. (2003). An overview of compaction of biomass grinds. *Powder Handling & Process*, 15(3), pp. 160-168.

Mani, S., Tabil, L.G. & Sokhansanj, S. (2002). Compaction behavior of some biomass grinds. AIC Paper No. 02-305. Saskatoon, Saskatchewan: AIC 2002 Meeting, CSAE/SCGR Program.

McMullen, J., Fasina, O.O., Wood, C.W., & Feng, Y. (2005). Storage and handling characteristics of pellets from poultry litter. *Applied Engineering in Agriculture*, 21(4), pp. 645-651.

Mewes, E. (1959). Verdichtungsgesetzmassigkeiten nach presstopfversuchen (Compression relationships as a result of experiments in pressure chambers). *Landtechnische Forschung*, 9(3), pp. 68–76.

Obernberger, I., & Thek, G. (2004). Physical characterization and chemical composition of densified biomass fuels with regard to their combustion behavior. *Biomass and Bioenergy*, 27, pp. 653-669.

Pietsch, W. (1997). Size Enlargement by Agglomeration. In Handbook of Powder Science and Technology, 2nd edition, eds. M.E. Fayed and L. Otten, 202-377. Florence, KY: International Thomson Publishing.

Rajvanshi, A. (1986). Published as Chapter 4 in the book Alternative Energy in Agriculture, Vol. II, Ed. D. Yogi Goswami, CRC Press, Boca Raton, FL. Rumpf, H. (1962). The Strength of Granules and Agglomerates. In Agglomeration, ed. W.A. Knepper, 379-419. New York, NY: John Wiley and Sons.

Sah, P., Singh, B., & Agrawal, U. (1980). Compaction behavior of straw. *Journal of Agricultural Engineering-India*, 18(1), pp. 89-96.

Sastry, K.V.S., & Fuerstenau, D.W. (1973). Mechanisms of Agglomerate Growth in Green Pelletization. *Powder Technology*, 7, pp. 97-105.

Serrano, C., Monedero, E., Laupuerta, M., & Portero, H. (2011). Effect of Moisture Content, Particle Size and Pine Addition on Quality Parameters of Barley Straw Pellets. *Fuel Processing Technology*, 92(2011), pp. 699-706.

Shaw, M.D., & Tabil, L.G. (2007). Compression and relaxation characteristics of selected biomass grinds. ASAE Annual International Meeting, Minneapolis, MN, June 17-20 Paper Number 076183, 2950 Niles Road, St. Joseph, MI 49085-9659 USA.

Shaw, M.D., Tabil, L.G., & Opoku, A. (2007). Pelleting characteristics and related properties of poplar wood and wheat straw grinds. A report prepared for SunOpta Bioprocess Inc., Brampton, ON, Canada, 16 p.

Shaw, M.D. (2008). Feedstock and Process Variables Influencing Biomass Densification. M.Sc. dissertation. Saskatoon, SK, Canada: University of Saskatchewan.

Shivanand, P. & Sprockel, O.L. (1992). Compaction Behavior of Cellulose Polymers. *Powder Technology*, 69, pp. 177-184.

Shrivastava, M., Shrivastava, P., & Khankari, K.K. (1989). Densification characteristics of rice husk under cold and hot compression. In Agricultural Engineering: Proceedings of the 11th International

Congress on Agricultural Engineering, Dublin, 4-8 September, 2441-2443. V.A. Dodd and P. M. Grace, eds. Rotterdam: A.A. Balkema Pub.

Sluiter, A., Hames, B., Ruiz, R., Scarlata, C., Sluiter, J., Templeton, D., & Crocker, D. (2008). Determination of Structural Carbohydrates and Lignin in Biomass, A Technical Report: NREL/TP-510-42618, National Renewable Energy Laboratory, USA, 2008. Available online at: http://www.nrel.gov/biomass/pdfs/42618.pdf

Smith, E., Probert, S., Stokes, R., & Hansford, R. (1977). The briquetting of wheat straw. *Journal of Agricultural Engineering Research*, 22, pp. 105–111.

Sokhansanj, S., Mani, S., Stumborg, M., Samson, R., & Fenton, J. (2006). Production and distribution of cereal straw on the canadian pariries. *Canadian Biosystems Engineering*, 48, pp. 3.39-3.46.

Sokhansanj, S., Mani, S., Bi, X., Zaini, P., & Tabil, L.G. (2005). Binderless pelletization of biomass. ASAE Annual International Meeting, Tampa Convention Centre, Tampa, Florida July 17-20, Paper Number 056061, 2950 Niles Road, St. Joseph, MI 49085- 9659 USA.

Sonnergaard, J.M. (2001). Investigation of a New Mathematical Model for Compression of Pharmaceutical Powders. *European Journal of Pharmaceutical Sciences*, 14, pp. 149-157.

Sultana, A., Kumar, A., & Harfield, D. (2010). Development of Agri-Pellet Production Cost and Optimum Size. *Bioresource Technology*, 101(2010), pp. 5609-5621.

Tabil, L.G., & Sokhansanj, S. (1996). Compression and Compaction Behavior of Alfalfa Grinds Part 1 – Compression Behaviour. *Journal of Powder Handling and Processing*, 8(1), pp. 17-23.

Temmerman, M., Rabier, F., Jensen, P.D., Hartmann, H., & Bohm, T. (2006). Comparative study of durability test methods for pellets and briquettes. *Biomass and Bioenergy*, 30 (11), pp. 964.

Theerarattananoon, K., Xu, F., Wilson, J., Ballard, R., McKinney, L., Staggenborg, S., Vadlani, P., Pei, Z.J., and Wang, D. (2011). Physical Properties of Pellets Made from Sorghum Stalk, Corn Stover, Hwaet Straw and Big Bluestem. *Industrial Crops and Products*, 33(2011), pp. 325-332.

Thomas, M., van Zuilichem, D.J., & van der Poel, A.F.B. (1997). Physical quality of pelleted animal feed. 2. Contribution of processes and its conditions. *Animal Feed Science and Technology*, 64 (2–4), pp. 173–192.

Xue, L., Mupondwa, E., Panigrahi, S., Tabil, L., & Adapa, P. (2011). Life Cycle Assessment of Biomass Densification System in Canadian Prairies. Personal Communication.

York, P., & Pilpel, N. (1973). The Tensile Strength and Compression Behavior of Lactose, Four Fatty Acids and their Mixture in relation to Tableting. *Journal of Pharmacy and Pharmacology*, 25, pp. 1-11.

CHAPTER-20

PERFORMANCES OF ENZYMATIC GLUCOSE/O_2 BIOFUEL CELLS

1. INTRODUCTION: AN OVERVIEW

Currently, there are many research publications discussing the creation of reliable devices that can convert chemical energy into electrical energy. These devices are used to power implanted devices and microelectronic equipment. Some notable studies on this topic include those by Bullen et al. (2006), Davis & Higson (2007), and Minteer et al. (2007). This topic is highly intriguing as it has the potential to provide valuable knowledge for the treatment of several illnesses, including diabetes, deafness, and heart disease (Heller, 2004; Katz & Willner, 2003). In addition, it can also serve as a cost-effective method for supplying energy to microelectromechanical systems (Calabrese Barton et al., 2004) or for treating wastewater (Fishilevich et al., 2009). If at least one of the catalysts employed in these devices is derived from biological resources, such as enzymes or microorganisms, they are referred to as "biofuel cells". Like any fuel cell, a biofuel cell is composed of two electrodes, an anode and a cathode, which may or may not be separated. The study issue on biofuel cells is extensive and can be categorized into three subsections. One category consists of microbial fuel cells, which are bio-electrochemical systems that generate an electric current by imitating bacterial interactions observed in nature. The second one pertains to enzymatic biofuel cells that employ enzymes as catalysts. Enzymes in this device prevent the separation of each compartment of the cell, which allows for the system to be made smaller (Heller, 2004). Hybrid biofuel cells involve the merging of an enzyme catalyst with an abiotic catalyst. These applications allow for the manipulation of the cell's working conditions, such as pH value and reactant concentration, in order to enhance the power density. This chapter will exclusively concentrate on the second and third types of biofuel cells. For the latter devices, the optimal fuel is glucose, which is found in all organic tissues, and the oxidant is dioxygen. Currently, there is a requirement to enhance both the lifespan (often a few days) (Calabrese Barton et al., 2004) and the power density (typically less than 1 mW cm-2) (Neto et al., 2010) of these devices. This article discusses the factors that impact the stability and power density of biofuel cells, as well as the recommended remedies in terms of electrode assembly, catalysts, and cell design. Furthermore, given the primary method for enhancing the longevity and effectiveness of enzymatic electrodes is to enhance the bond between the enzyme and the electrode surface, we will specifically examine the many immobilization approaches that have been documented in current literature. Furthermore, recent evidence has clearly shown that abiotic catalysts have significantly enhanced the stability of the device (Kerzenmacher et al., 2008) and have exhibited rapid substrate conversion kinetics (Choi et al., 2009). Therefore, we will specifically focus on examining novel abiotic nanocatalysts and their application in hybrid biofuel cells.

2. ENZYMATIC ELECTRODES FOR GLUCOSE/O_2 BIOFUEL CELLS

Except the lack of stability of enzyme molecules due to their proteic nature, one of the major problems encountered with enzymatic electrodes concerns electron transfer between the enzyme and the electrode surface. In the next part we will describe the different electron transfer mechanisms occurring between an enzyme and the electrode as well as the immobilization techniques of the protein.

2.1. Electron transfer occurring between an enzyme and an electrode

Enzymes are proteins which have high molecular weights. The active sites of these molecules are located in the organic matrix at a depth of several angstroms from the surface. It is thus easy to understand that kinetically fast electron transfer between enzymes and electrodes surface is difficult to obtain because of great insulation of the active centers (Armstrong et al., 1985). Different strategies have been used by the past to make efficient electrical connections between the enzyme and the electrode surface. Corresponding electron transfer mechanisms can be arranged in two different classes: mediated electron transfer (MET) and direct electron transfer (DET).

The major interest in directly transferring electrons between enzymes and electrodes is to reduce the electrode overpotential which is of particular importance for biofuel cells applications. DET is possible as soon as the distance between the active center of the enzyme and the electrode surface is in the order of a tunneling one (Degani & Heller, 1987). Different evidences for DET between enzymes classically used in glucose/O2 biofuel cells and electrodes have already been given. Actually, laccase (Gupta et al., 2004), bilirubin oxidase (Shleev et al., 2005) and glucose oxidase (Wang et al., 2009) are capable of exhibiting non-negligible catalytic current densities without the presence of a redox mediator.

In the case of MET, a redox molecule acts as a substrate and is able to transfer electrons between the electrode surface and the active center of the enzymatic molecule. Let's notice that current densities obtained with MET are generally higher that what can be delivered in the case of DET. However, to get efficient MET, the redox mediator must possess some properties which can be deduced from Marcus theory as it was already mentioned by Rusling et al. (Rusling et al., 2008). This theory is used to describe outer sphere electron transfer between an electron donor (D) and an electron acceptor (A) as depicted in Fig. 1.

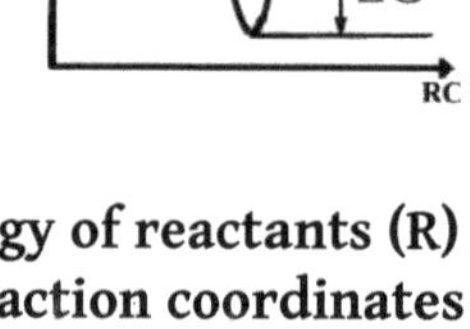

Figure 1.Curves presenting potential energy of reactants (R) and products (P) (A δ --D δ +) as a function of reaction coordinates (RC).

The rate (k) of electron transfer can be described as follows (Eq.1.) by an Arrhenius type law.

$$k = AKe^{\left(-\frac{(\Delta G^{\circ}+\lambda)^2}{4RT\lambda}\right)} \tag{1}$$

where A is the collision frequency, K is the electronic transmission factor, ΔG0 is the Gibbs free energy, R is the gas constant and T the temperature. λ is the reorganization energy (energetic cost associated to the reorganization of both solvent and molecules and necessary to proceed in electronic transfer between the donor and the acceptor). From this relation it can be deduced that to have an efficient electron transfer between enzyme and mediator, it is essential that the redox mediator used presents a highly reversible redox system to minimize λ value. It is also fundamental to minimize the ΔG 0 value. Thus it is very important that formal potentials of mediator and enzyme are close. Moreover, since active centers of enzyme are greatly insulated in high molecular weight molecules it is necessary to use small mediator molecules to reduce the distance of electron transfer and to guarantee a high k value.

2.2. Immobilization of enzymes on electrode surfaces

It is of great interest to develop new non-damaging immobilization techniques of enzyme for the development of stable biofuel cells. In fact, it is very difficult to propose a technique that does not affect the stability of biomolecules. Enzymes are proteins which possess tridimensional structures in which active centers are insulated. To keep the stability of the molecule and to preserve its catalytic efficiency it is necessary not to modify this tridimensional structure and particularly not to affect the environment of the active center. Different combining techniques can be used to immobilize enzymes onto the surface of solid electrodes:

- Immobilization into a polymer network
- Adsorption on an electrode material
- Covalent grafting to an electrode
- Immobilization within a membrane

The first technique consists in immobilizing enzyme in an electropolymerized thin film. It is a very simple technique since it only needs to dip the electrode into a solution containing both monomers and biomolecules. Then the growth of polymer film can be realized by different ways: chronoamperometry (Brunel et al., 2007), chronopotentiometry or cyclic voltammetry (Fei et al., 2007). Different monomers such as pyrrole (Habrioux et al., 2008), aniline (Timur et al., 2004) or phenol (Bartlett et al., 1992) can be electropolymerized. This kind of films can be either conductive or not. The main advantage in using tridimensional conductive films lies in their ability to transfer electrons. Moreover, to increase the number of enzymatic molecules immobilized close to the electrode surface, a first adsorption step of enzymatic molecules can be performed (Merle et al., 2009). Other non-electropolymerized films can be used for enzyme holding. Currently, both chitosan and Nafion® are commonly used (Habrioux et al., 2010; Klotzbach et al., 2008). These two polymers possess surfactant properties interesting to immobilize enzymes in micellar structures (Moore et al., 2004). Moreover, the hydrophobic/hydrophilic property of the polymers can be tuned by modifying the chemical structure of these molecules (Klotzbach et al., 2008; Thomas et al., 2003). It is also possible to simply use retention properties of the Nafion® film for buffering its sulfonic groups (Habrioux et al., 2010). The main problem associated with the use of these polymers lies in the non-control of the film thickness. One of the most promising immobilization techniques has been proposed by Heller's group. This approach consists in immobilizing enzymes in an osmium-based redox polymer (Mao et al., 2003) which is able to swell in contact with water. It acts both as an immobilizing network and an electrochemical mediator. The whole structure of the film leads to very fast electron transfer between the active centers of enzymes and the electrode surface. Another smart technique consists in covalent grafting of enzyme to the electrode surface. Thus Merle et al. (Merle et al., 2008) realized the grafting of amino groups on a carbon electrode before coupling these functions with amino-groups of enzymes using glutaraldehyde. This seems to confer a remarkable stability to the resulting electrode. Another well-known approach has been proposed by Willner et al. (Willner et al., 1996) that consisted in the reconstitution of the enzyme after the grafting of its active center on a gold electrode.

2.3. Enzymatic oxidation of glucose

The development of efficient enzymatic electrodes to oxidize glucose is of relevance for the development of implantable glucose/O2 biofuel cells. Nowadays, enzymes classically used to perform efficient oxidation of glucose to gluconic acid are either glucose oxidase (GOD) or glucose dehydrogenase (GDH). In the next part the properties of these two enzymes will be explained in details.

2.3.1. Glucose oxidation catalyzed by glucose dehydrogenase

Contrary to GOD, GDH is non-sensitive towards oxygen (Zhang et al., 2007). This is an attractive property for its use in glucose oxygen biofuel cells. However, GDH is an NAD-dependant enzyme. It is well-known that the oxidation of glucose catalyzed by GDH is rather limited by oxidation kinetics of NADH into NAD+. Even if the use of modified electrodes allows to reduce the overpotential associated to the oxidation of NADH into NAD+ (Delecouls-Servat et al., 2001), the stability of the electrodes remains poor. Another solution based on the use of a pyrroloquinoline quinone (PQQ) cofactor proposes to suppress the NAD dependence. Nevertheless, the PQQ cofactor has a limited stability (Wang, 2007).

2.3.2. Glucose oxidation catalyzed by glucose oxidase

2.3.2.1 Properties of the enzyme

GOD is by far the most used anode catalyst in glucose/O2 biofuel cells. Its molecular weight (155 kDa) and molecular size (60 Å × 52 Å × 77 Å) are high (Alvarez-Icaza et al., 1995). This constitutes a limitation for current densities obtained with a solid electrode since the footprint of the enzyme is great. This enzyme possesses two identical linked FAD (Flavine Adenine Dinucleotide) subunits which are responsible for β-D-glucose oxidation (Zhu et al., 2006) to gluconic acid (two electrons reaction product). The redox potential of FAD-FADH2 cofactor is ca. -0.36 V vs. Ag/AgCl/KCl(sat.) at a pH value of 7.2 (Stankovich et al., 1978), which is of particular interest for biofuel cells applications since it allows low-potential glucose oxidation. In this reaction, dioxygen is the natural electrons acceptor. Therefore, during the oxidation process, dioxygen is reduced towards hydrogen peroxide. The formation of H2O2 leads to inhibition of the enzyme because it modifies the amino groups in the vicinity of the active center (Kleppe, 1966). The pH value and the temperature have also an effect on GOD performances. Temperatures higher than 40 °C lead to a drastic decrease of activity (Kenausis et al., 1997). The pH value which optimizes GOD activity greatly depends on the electron acceptor. This value is equal to 5.5 and 7.5 when oxygen (Kenausis et al., 1997) methylene blue (Wilson & Turner, 1992) are used, respectively.

2.3.2.2 Performances of GOD electrodes towards β-D-glucose oxidation

In the case of MET, the use of suitable electrochemical mediators is of importance to increase the rate of electron transfer between the enzyme and the electrode surface since it allows to raise current densities. The second interest lies in the possibility to inhibit the formation of peroxide. Actually, it is just necessary to use a mediator which is able to realize faster electron transfer with GOD than oxygen can do. One of the most efficient systems has been developed by Heller's group (Mano et al., 2005; Mao et al., 2003). It consists of a tridimensional matrix of an osmium based redox polymer containing GOD. The formal potential of the polymer is -195 mV vs. Ag/AgCl at pH 7.2. The covalent chain composed of thirteen atoms long allows the increase of the electron diffusion coefficient (Mao et al., 2003) by increasing the collision probability between reduced and oxidized forms of the osmium centers. The reticulation with PEGDGE (polyethyleneglycoldiglycydilether) allows the formation of a redox hydrogel capable of swelling in contact with water. It is probable that the matrix structure is responsible for a weak deformation of the protein structure. Such electrodes are able to deliver a catalytic current at potentials as low as -360 mV vs. Ag/AgCl in a physiologic medium containing 15 mM glucose (Mano et al., 2004).

2.4. Enzymatic reduction of oxygen to water

Generally, enzymes used to catalyze the reduction of oxygen into water are either laccase or bilirubin oxidase (BOD). The main property of these enzymes is their ability to directly reduce oxygen to water at potentials higher than what can be observed with platinum based electrodes (Soukharev et al., 2004). These two enzymes are classified in "multicopper oxidases" class and contain four Cu^{2+}/Cu^{+} active centers which are commonly categorized in three types: T1, T2 and T3. T1 site is responsible for the oxidation of the electron donor. The trinuclear center composed both of T2 center and two equivalent T3 centers is the place where oxygen reduction occurs (Palmer et al., 2001). The associated mechanism is proposed in Fig. 2.

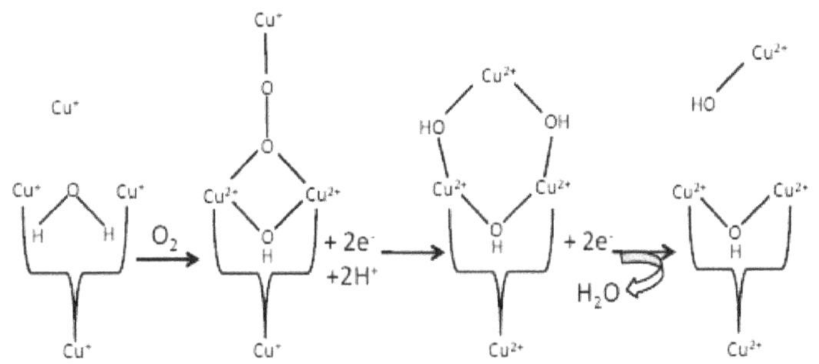

Figure 2.Oxygen reduction catalyzed by "multicopper oxidases"

In the next part the different properties and performances of both laccase and BOD electrodes will be discussed.

2.4.1. Reduction of oxygen catalyzed by laccase

Laccase is able to oxidize phenolic compounds and to simultaneously reduce oxygen into water. The microorganism from which it is extracted greatly determines the redox potential of the T1 site which can vary from 430 mV vs. NHE up to 780 mV vs. NHE (Palmore & Kim, 1999). Laccase from Trametes versicolor is the most attractive one since redox potential of its T1 site is ca. 780 mV vs. NHE (Shleev et al., 2005). Nowadays, the best performances with laccase electrodes are obtained with osmium based polymers as redox mediators (Mano et al., 2006). Actually these electrodes are able to deliver a current density of 860 µA cm^{-2} at only -70 mV vs. O_2/H_2O at pH 5. In the same conditions, the identical current density is obtained at -400 mV vs. O_2/H_2O with a platinum wire as catalyst. Nevertheless, performances of laccase (from Pleurotus Ostreatus) electrodes drop drastically in the presence of chloride ions (Barton et al., 2002) what constitutes both a major problem and a great challenge for its use in implantable glucose/O_2 biofuel cells.

2.4.2. Reduction of oxygen catalyzed by bilirubin oxidase

BOD is naturally capable of catalyzing the oxidation of bilirubin into biliverdin and to simultaneously reduce dioxygen (Shimizu et al., 1999). BOD is very similar to laccase. Performances of BOD electrodes are greatly related to the amino-acids sequence around T1 site of the enzyme (Li et al., 2004). It is clearly reported that the most efficient BOD enzyme comes from Myrothecium verrucaria. Redox potential of its

T1 site is included between 650 and 750 mV vs. NHE, and the enzyme is thermally stable up to 60 °C (Mano et al., 2002b). It is thus possible to use it at physiological temperature without denaturing the protein. To build efficient BOD electrodes intended in working at physiological pH value, it is judicious to use positively charged mediator molecules since the isoelectric point of BOD is close to pH = 4. Actually, during oxygen reduction reaction, the use of an osmium based redox polymer has lead to performances such as 880 µA cm-2 at 0.3 V vs. Ag/AgCl (physiological conditions) at a scan rate of 1 mV s-1 (Mano et al., 2002a). Additionally, the redox osmium based hydrogel conferred a very favorable environment to stabilize BOD since 95% of the initial activity of a BOD electrode can be preserved after three weeks storage (Mano et al., 2002a). This remarkable stability probably results in auspicious electrostatic interactions between the swelling matrix and the enzyme. Performances of BOD electrodes are furthermore unaffected in the presence of chloride ions. In fact BOD remains active for chloride concentrations lower than 1 M (Mano et al., 2002a). This property is of major interest for the development of implantable microscale glucose/O2 biofuel cells using BOD as cathode catalysts. The major encountered problem with BOD electrodes is the relative lack of stability of the enzyme in physiological serum. Cupric centers of BOD are indeed capable of binding with one urea oxidation product, oxidation reaction catalyzed by the enzyme (Kang et al., 2004). This phenomenon can nevertheless be limited by spreading a Nafion® film on the catalyst (Kang et al., 2004). It is moreover reported that chemically modified Nafion® is capable of constituting a favorable environment to stabilize BOD (Topcagic& Minteer, 2006). Consequently, it seems of interest to immobilize BOD in Nafion® films. A promising technique for the development of efficient BOD electrodes has already been reported in literature (Habrioux et al., 2010). It consists in firstly adsorbing BOD/ABTS2- (2,2-azinobis-3-ethylbenzothiazoline-5-sulfonic acid) complex on a carbon powder, Vulcan XC 72 R in order to increase both enzyme loading, the stability of the protein and the quality of the percolating network in the whole thickness of the polymer film. Actually, to realize the electrochemical reaction, a triple contact point (between the catalytic system, the electrolyte and the electronic conductor) is required. Once the catalytic system is adsorbed, a buffered Nafion® solution is added. The whole system is then immobilized onto a solid carbon electrode (Fig. 3).

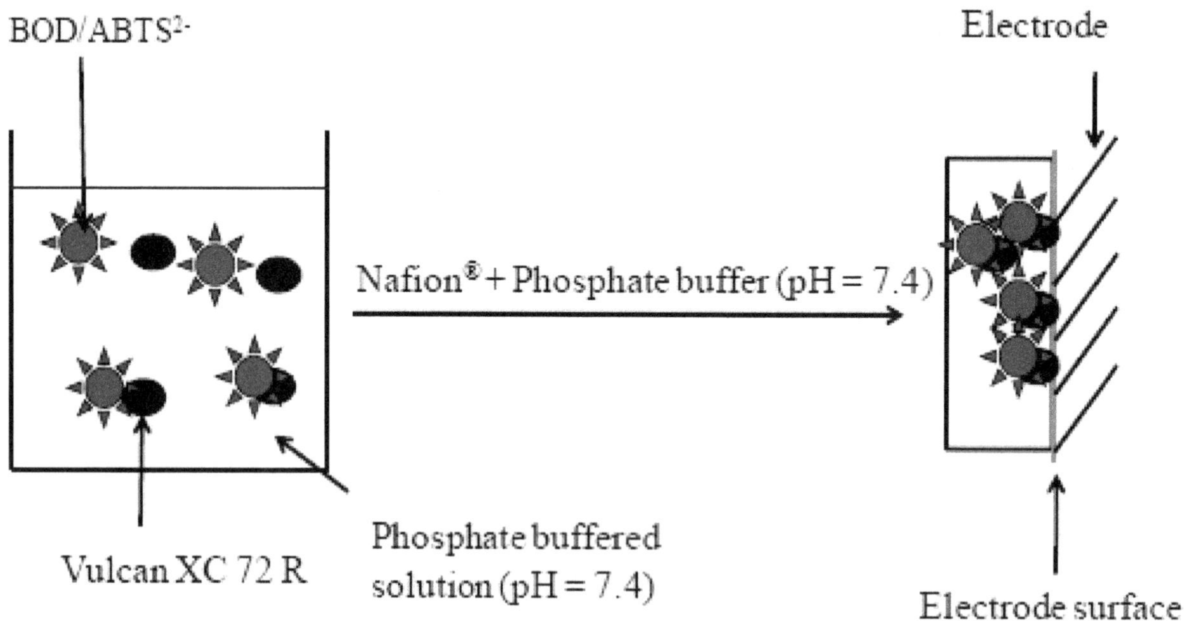

Figure 3.Method used for the preparation of BOD cathodes according to the process described in Ref. (Habrioux et al., 2010)

Previous studies have shown the interest lying in the use of ABTS2- as redox mediator in combination with multicopper oxidases. One of them was carried out by Karnicka et al. who have shown that wiring laccase to glassy carbon through a ABTS2-/carbon nanotube system was a very efficient pathway to reduce molecular oxygen into water (Karnicka et al., 2008). The combination of ABTS2- with BOD is also known to exhibit a high electrochemical activity towards oxygen reduction reaction (Tsujimura et al., 2001). These observations are confirmed by electrochemical studies performed on electrodes previously described (Fig. 3). Results are shown in Fig. 4.

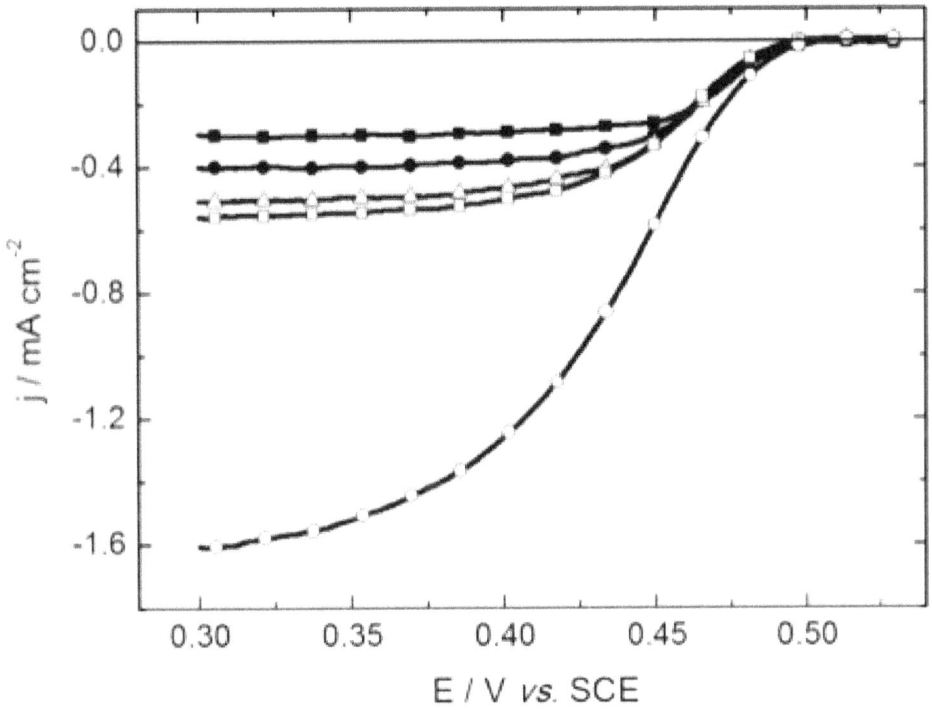

Figure 4.Oxygen reduction reaction catalyzed by BOD/ABTS2-/Nafion® electrode in a phosphate buffered solution (pH = 7.4, 0.2 M) at 25 °C. Curves registered at different rotation rates (Ω), in an air-saturated electrolyte at Ω = 100 rpm (■); Ω = 200 rpm (•); Ω = 400 rpm (Δ); Ω = 600 rpm (□) and in an oxygen saturated electrolyte at Ω = 600 rpm (○). Scan rate 3 mV s-1.

Curves of Fig.4 clearly show the interest of such electrodes that exhibit a catalytic current from potentials as high as -50 mV vs. O2/H2O (0.536 V vs. SCE). Furthermore the half-wave potential is only 100 mV lower than the reversible redox potential of O2/H2O. This value is in good agreement with that reported by Tsujimura et al. (0.49 V vs. Ag/AgCl/KCl(sat.) at pH = 7.0) (Tsujimura et al., 2001). Let's notice that the half-wave potential value is very close to the redox potential of T1 site of BOD (0.46 V vs. SCE). This has already been explained by the fact that the reaction between ABTS2- and BOD is an uphill one (Tsujimura et al., 2001). Fig. 4 also shows that electrochemical performances of BOD/ABTS2-/Nafion® clearly depend on the amount of oxygen dissolved in the electrolyte. The limiting current is a plateau and increases from 0.56 mA cm-2 in an air saturated electrolyte to 1.61 mA cm-2 in an oxygen saturated (at a rotation rate of 600 rpm). Dependence of limiting current with oxygen concentration in the electrolyte is presented in Fig.5. In this figure, current obtained at 0.2 V vs. SCE is plotted versus oxygen saturation.

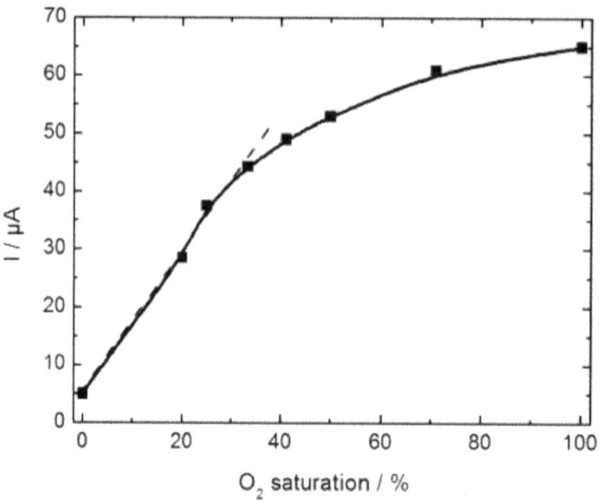

Figure 5.Electrochemical activity of BOD/ABTS2-/Nafion® electrode: dependence of the current value at 0.2 V vs. SCE with oxygen concentration

The current linearly increases with the oxygen concentration from low values to around 35%. This linearity suggests that the reaction is of a first order with oxygen concentration thereby, the Koutecky-Levich plots can be considered. Assuming that the rate determining step is an enzymatic intramolecular electron transfer step, it is possible to express the current density of a BOD/ABTS2-/Nafion® electrode working in an air saturated solution as follows (Schmidt et al., 1999):

$$\frac{1}{|j|} = \frac{1}{|j_0|\frac{\theta}{\theta_c}e^{\left(\frac{\alpha nF}{RT}\right)|\eta|}} + \frac{1}{|j_L^{ads}|} + \frac{1}{|j_L^{film}|} + \frac{1}{|j_L^{diff}|} \quad (2)$$

In Eq.2, JLdiff diff represents the diffusion limiting current density expressed by Levich equation:

$$j_L^{diff} = 0.2nFD^{2/3}\nu^{-1/6}C_0\sqrt{\Omega} \quad (3)$$

In Eq.3, n is the number of electrons exchanged, D the diffusion coefficient, C0 is the oxygen concentration, Ω is the rotation rate, F the Faraday constant and ν is the kinematic viscosity. Then, j Lfilm corresponds to the limitation due to oxygen diffusion in the catalytic film and jLads is the limiting current density due to oxygen adsorption on the catalytic site. Since these two last contributions to the total current density do not depend on Ω, it is impossible to separate them. They will be described according to Eq.4.

$$\frac{1}{j_L} = \frac{1}{j_L^{ads}} + \frac{1}{j_L^{film}} \quad (4)$$

In Eq.2, η is the overpotential (η = E-Eeq), j 0 the exchange current density, α the transfer coefficient, R = 8.31 J mol-1 K-1, F=96500 C mol-1 and T the temperature. Θ and Θ c are the covering rates of the active sites of the enzyme at E and Eeq, respectively. We will assume that Θ ≈ Θ c for all potential values. From Eq.2, when Ω→∞, the limit of 1/j can be expressed as follows:

$$\frac{1}{|j_k|} = \frac{1}{|j_L|} + \frac{1}{|j_0|e^{\left(\frac{\alpha nF}{RT}\right)|\eta|}} \quad (5)$$

In Eq.5, when η→∞, 1/jk→1/jL. It is thus possible to determine j L value by extrapolating and reporting the 1/j k values as a function of the potential value E. Transforming Eq.5 (Grolleau et al., 2008), it becomes as follows:

$$\eta = E - E_{eq} = -b\left[\ln\frac{j_L}{j_0} + \ln\frac{j_k}{j_L - j_k}\right] \tag{6}$$

where b = RT/αnF is the Tafel slope. The plot of the η values vs. ln(j K /(j L -j K)) (Fig. 6) permits the calculation of b and j 0 values.

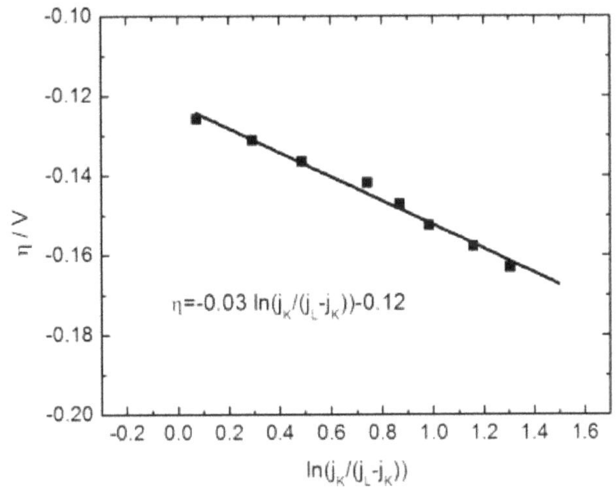

Figure 6.Curve obtained from Koutecky-Levich treatment on oxygen reduction reaction catalyzed by BOD/ABTS2-/Nafion® system.

Under these experimental conditions, calculated values for both Tafel slope and exchange current density are respectively of 69 mV/decade and 25 μA cm-2. The high value obtained for j 0 confirms the ability of BOD/ABTS2-/Nafion® system to activate molecular oxygen in a physiological type medium. Moreover, it also certifies that the oxygen reduction reaction starts at very high potentials. The reference catalyst classically used to reduce molecular oxygen is platinum. It can be noticed that under similar conditions, the exchange current density is only of 5 μA cm-2 when we used platinum nanoparticles as catalyst. This clearly shows the great interest lying in these electrodes to reduce oxygen in glucose/O2 biofuel cells. Nowadays, the major problem encountered with these electrodes is the lack of stability of the redox mediator (ABTS2-) (Tsujimura et al., 2001).

3. ABIOTIC CATALYSTS FOR GLUCOSE/O2 BIOFUEL CELLS

This section provides a comprehensive overview of non-enzymatic catalysts that are already being utilized or have the potential to be used in glucose/O2 biofuel cell systems. The primary challenge in using abiotic catalysts for such applications stems from their limited specificity. As a result, it is challenging to use them in implantable microscale devices. However, they frequently exhibit rapid substrate conversion kinetics and possess significantly greater stability compared to enzymes. Therefore, they can be utilized as catalysts in biocompatible devices designed to provide sustained high power densities over an extended period of time.

3.1. Non-enzymatic oxidation of glucose

3.1.1. Different offered possibilities

An effective strategy involves utilizing metallophthalocyanines for the purpose of glucose oxidation. Cobalt phthalocyanines, in particular, display intriguing characteristics (Zagal et al., 2010). Moreover, the reactivity of these electrodes can be altered by making minor modifications to the complicated structure, which is advantageous for the advancement of electrode technology. These catalysts have the potential to be utilized for glucose electrooxidation in glucose/O_2 biofuel cells, while their development is still pending. Another method involves utilizing metallic nanoparticles as catalysts. The study of glucose oxidation on metallic surfaces has been extensively researched. Several investigations have focused on understanding the catalytic influence of platinum on the process of glucose oxidation (Kokoh et al., 1992a; Kokoh et al., 1992b; Sun et al., 2001). The experiments yielded the conclusion that the primary oxidation product is gluconic acid (Kokoh et al., 1992b; Rao & Drake, 1969). The oxidation process entails the removal of hydrogen atoms from the anomeric carbon of the glucose molecule (Ernst et al., 1979). The primary advantage of including platinum into the catalyst composition is its capacity to oxidize glucose at extremely low potentials (below 0.3 V vs. RHE). Nevertheless, it is widely recognized that platinum surfaces are highly susceptible to being poisoned by chemisorbed intermediates (Bae et al., 1990; Bae et al., 1991). In order to enhance the electrochemical activity of platinum, researchers have utilized various heavy atoms (Tl, Pb, Bi, and W) as adatoms to change the platinum surfaces (Park et al., 2006). Additional research has investigated the process of glucose oxidation on platinum alloys, with the inclusion of other second metals such as Rh, Pd, Au, Pb (Sun et al., 2001), Bi, Ru, and Sn (Becerik&Kadirgan, 2001). The most effective catalysts seem to be Pt-Pb or Pt-Bi, as stated by Becerik and Kadirgan in 2001. Nevertheless, these catalysts exhibit sensitivity to the dissolution of the second metal, rendering them unsuitable for utilization in fuel cell systems. Furthermore, the majority of the aforementioned compounds are hazardous. Gold is the only material that has the potential to be environmentally friendly, despite the little understanding of the oxidative stress nanoparticles may wreak on living cells. In addition, the synthesis of alloyed materials can greatly enhance the catalytic activity of pure metals through a synergistic effect. This phenomenon has been clearly demonstrated in platinum-gold nanoalloys (Möller & Pistorius, 2004).

3.1.2. Oxidation of glucose on gold-platinum nanoparticles

The oxidation of glucose on gold-platinum nanoparticles has been investigated in numerous studies (Habrioux et al., 2007; Sun et al., 2001). Jin and Chen (Jin & Chen, 2007) examined glucose oxidation catalyzed by Pt-Au prepared by a co-reduction of metallic salts. An oxidation peak of glucose was visible at much lower potentials than on gold electrode. Moreover, they showed that both metals favored the dehydrogenation of the glucose molecule. They concluded that the presence of gold prevents platinum from chemisorbed poisonous species. The efficiency of such catalysts towards glucose oxidation is thus not to be any more demonstrated, and greatly depends on the synthesis method used to elaborate the catalytic material.

3.1.2.1 Synthesis of gold-platinum nanoparticles

Various gold-platinum nanoparticles synthesis methods have been already studied: Polyol (Senthil Kumar & Phani, 2009), sol-gel (Devarajan et al., 2005), water-in-oil microemulsion (Habrioux et al., 2007), electrodeposition (El Roustom et al., 2007) and Bönnemann (Atwan et al., 2006). Among all these methods, the water-in-oil microemulsion technique produces particles that exhibit high catalytic activity towards glucose electrooxidation (Habrioux et al., 2007). It consists in mixing two microemulsions, one containing the reducing agent in the aqueous phase and the other containing one or several metallic precursors in the aqueous phase. Collisions of water nanodroplets permit to obtain metallic nanoparticles which can be then cleaned and dispersed onto a carbon support. The choice of the different components of the microemulsions is not unique and influences the physical properties of the obtained nanoparticles. Actually, both surfactant molecules and oil-phase chemical nature have an effect on interfacial tension of the surfactant film that determines water solubility in micelles (Paul & Mitra, 2005). This greatly affects intermicellar exchanges. Moreover, the chemical nature of the reducing agent controls the rate of the nucleation step and subsequently the kinetic of particles formation. In the system described herein, n-heptan is used as oil phase, non-ionic polyethyleneglycol-dodecylether as emulsifier molecule and sodium borohydride as reducing agent. The synthesized particles have been dispersed onto Vulcan XC 72 R and then washed several times with acetone, ethanol and water, respectively to remove surfactant from their surface (Habrioux et al., 2009b). The removal of surfactant molecules from all the catalytic sites without modifying structural properties of the catalyst is currently a great challenge (Brimaud et al., 2007). Since electrocatalysis is a surface phenomenon depending on the chemical nature of the surface of the catalyst, on its crystalline structure and on the number of active sites, it is useful to precisely know the physico-chemical properties of the used nanoparticles to understand their electrochemical performances.

3.1.2.2 The electrochemical behavior of gold-platinum nanoparticles in relation to the electrooxidation of glucose

This part aims at showing the importance to realize a correlation between the structural properties of the catalysts and their electrocatalytic activities towards glucose oxidation. The use of nanocatalysts indeed involves a deep structural characterization of the nanoparticles to fully understand the whole of the catalytic process. Therefore, in order to show the presence and the proportion of gold and platinum at the surface of the catalysts, electrochemical investigations have been carried out (Burke et al., 2003). It is indeed possible to quantify surface compositions of the catalysts by using cyclic voltammetry and by calculating the amount of charge associated with both reduction of platinum and gold oxides (Woods, 1971). The charge calculated for pure metals was 493 μC cm-2 and 543 μC cm-2 for Au and Pt, respectively, for an upper potential value of 250 mV vs. MSE (Habrioux et al., 2007) in a NaOH (0.1 M) solution. The atomic ratio between gold and platinum can be thus determined according to Eq. 7 and Eq. 8 assuming that for all bimetallic compositions, the oxidation takes place only on the first atomic monolayer.

$$\%Au = \frac{S_{Au}}{S_{Au} + S_{Pt}} \times 100 \quad (7)$$

and

$$\%Pt = \frac{S_{Pt}}{S_{Au} + S_{Pt}} \times 100 \quad (8)$$

Both voltammograms used and results of the quantification are shown in Fig. 7. Mean diameter of the different nanoparticles weighted to their volume (obtained from transmission electron microscopy measurements) as well as their mean coherent domain size weighted to the volume of the particles (obtained from X-ray diffraction measurements) are also presented in Fig. 7.

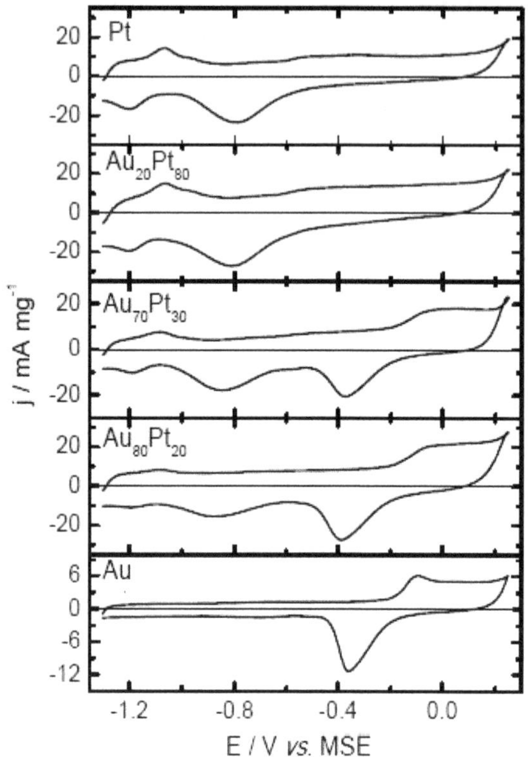

% At. Pt	%At. Au	D_V (nm)	L_V (nm)
100	0	4.7	3.5
100	0	4.6	4.0
44	56	5.2	4.3
29	71	5.3	5.0
0	100	9.4	14.1

Figure 7.Voltammograms (after 19 cycles) of gold-platinum nanoparticles recorded at 25 °C in alkaline media (0.1 M NaOH). Scan rate = 20 mV s-1. The surface composition of the used catalyst is given on the right of the corresponding voltammogram.

In Fig. 7 it is noticed that for all compositions, desorption of oxygen species occurs in two peaks. The reduction of the gold surface takes place at -0.38 V vs. MSE whereas the potential for which platinum surface is reduced depends on the amount of gold in the alloy. Indeed, for pure platinum nanoparticles this potential is ca. -0.8 V vs. MSE (reduction of platinum oxides). The potential at which oxygen species desorption occurs, shifts to lower potentials when the atomic ratio of gold increases in the composition of alloys. The deformation of this peak increases with the amount of gold probably because of the formation of more complex platinum oxides. The quantification realized on the different bimetallic compositions, clearly shows a platinum enrichment of nanoparticles surfaces. Desorption of gold oxides is indeed invisible for low gold containing samples (i.e. with gold content lower than 40%). These nanoparticles exhibit a typical core-shell structure composed of a gold core and a platinum shell (Habrioux et al., 2009b), while high gold content samples (i.e. with gold content higher than 80%) possess a surface composition that is close to the nominal one. This results in a purely kinetic effect. Actually, reduction of gold precursor is considerably faster than reduction of platinum cation. Consequently, there is firstly formation of a gold seed on which platinum reduction occurs. So, the natural tendency of these systems is to form core-shell particles. Furthermore, let's notice that both mean diameter of nanoparticles weighted to their volume and their mean coherent domain size weighted to their volume increase with gold content but ever stay in the nanometer range. That is only the result of differences in reduction kinetics of the particles since the ratio water to surfactant remains constant whatever the synthesized sample. To correlate surface composition with efficiency to oxidize glucose for all gold-platinum catalysts compositions, voltammograms were first recorded in alkaline medium. Results are shown in Fig. 8.

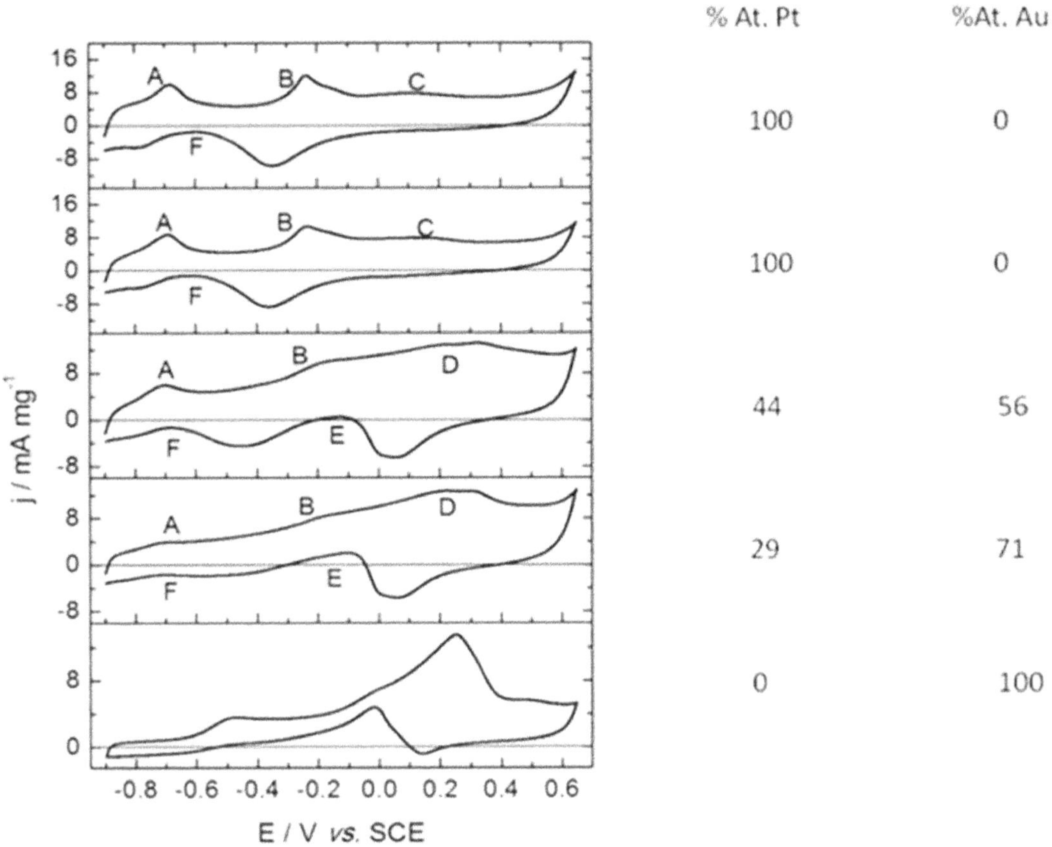

Figure 8.Voltammograms (after 19 cycles) of gold-platinum nanoparticles recorded at 3 °C in alkaline medium (0.1 M NaOH) in the presence of 10 mM glucose. Scan rate = 20 mV s-1. Surface composition of the used catalyst is given on the right of the corresponding voltammogram.

In Fig. 8, different oxidation peaks appear during the oxidation process on gold-platinum nanocatalysts. When platinum content decreases in the bimetallic surface composition, intensity of peak A, located at ca. -0.7 V vs. SCE, diminishes. For pure gold catalyst, this peak is furthermore invisible. It is thus related to the oxidation phenomenon on platinum. It has already been attributed to dehydrogenation of anomeric carbon of glucose molecule (Ernst et al., 1979). Peaks B and C correspond to the direct oxidation of glucose molecule (Habrioux et al., 2007) and are located both in gold and platinum oxides region. In the case of catalysts with nominal compositions such as Au70Pt30 or Au80Pt20, the different oxidation peaks located between -0.3 V vs. SCE and 0.4 V vs. SCE are not well-defined. For these catalysts, the presence of platinum at their surface allows a low potential oxidation of glucose molecule, which starts earlier than on pure gold. Moreover, on these catalysts, after the dehydrogenation step, current densities raise rapidly. Furthermore, in the potential region where formation of both gold hydroxides and platinum oxides occurs, current densities are very high (i.e. 12 mA mg-1 at 0.2 V vs. SCE). This is the result of a synergistic effect between the two oxidized metals at the bimetallic catalyst surface (Habrioux et al., 2007). Such effect between gold and platinum has already been observed for CO oxidation (Mott et al., 2007). On these catalysts, during the negative going scan, two oxidation peaks, E and F, are visible.

During the reduction of both oxidized gold and platinum clusters, oxygenated species are desorbed from the surface and stay at its vicinity. Subsequently, there is desorption of adsorbed lactone from the electrode surface what implies the formation of both peak E and peak F (Beden et al., 1996). Fig. 9 shows the reactions involving in the oxidation of glucose on the catalyst surface.

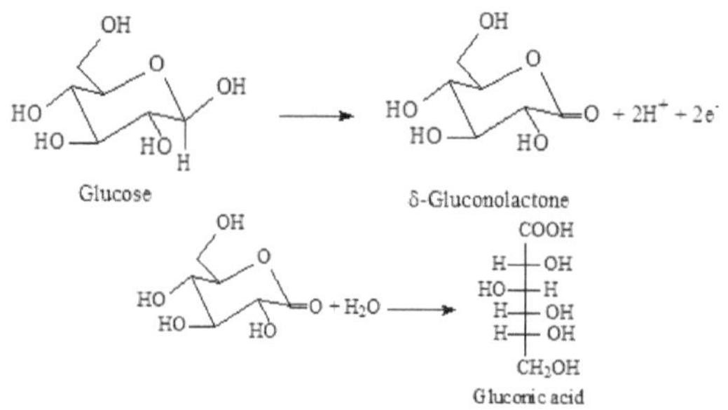

Figure 9.Oxidation of glucose on gold-platinum catalysts

The remarkable electrocatalytic activity of both Au80Pt20 and Au70Pt30 nanocatalysts towards glucose electrooxidation is probably the result of a suitable surface composition combined with a convenient crystallographic structure. An X-ray diffraction study (Fig. 10) based on Warren's treatment of defective metals and previously described (Vogel et al., 1998; Vogel et al., 1983) combined with high resolution transmission electron microscopy (HRTEM) measurements allowed to exhibit the peculiar structure of high gold content catalysts (Habrioux et al., 2009b).

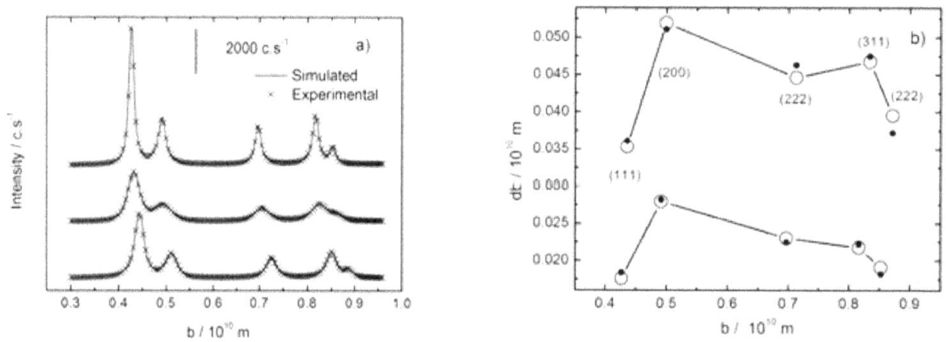

Figure 10.a) Experimental and simulated diffractograms obtained with Au, Au70Pt30 and Pt nanoparticles (from top to bottom), b) Experimental (•) and simulated (○) Williamson-Hall diagrams obtained with Au30Pt70 and Au nanoparticles (from top to bottom).

Each experimental diffractogram has been fitted with five Pearson VII functions what gives two important parameters: the accurate peak position b ($b = 2\sin\theta/\lambda$) and the integral line width db. The value of db is plotted versus b in Fig.10b. As a result of best fits, it can be assumed that line profiles of diffractograms are lorentzian. This implies that all contributions to the integral line width can be added linearly and can be expressed as follows:

$$db = db_{size} + db_{stacking\ fault} + db_{strain} \tag{9}$$

With

$$db_{size} = \frac{1}{L_v} \tag{10}$$

$$db_{stacking\ fault} = \frac{\alpha V_{hkl}}{a} \tag{11}$$

and

$$db_{strain} = \frac{2\sigma b}{E_{hkl}} \tag{12}$$

where Lv is the mean coherent domain size weighted to the volume of the particles, α the stacking fault probability, V hkl a parameter depending on the miller indexes, σ the mean internal stress and E hkl the young modulus. The fit of Williamson-Hall diagrams with the expression given by Eq.7 leads to the determination of L v , α and σ for each catalyst. It has been concluded that for catalysts with nominal compositions of Au70Pt30 and Au80Pt20, both σ and α values were high (Habrioux et al., 2009b). For Au80Pt20, these values were indeed of 510 N.mm-2 and 8.2%, respectively for σ and α. In the case of Au70Pt30, these values were of 490 N.mm-2 and 7.4%. HRTEM observations have confirmed the results of the fit since the observed particles present numerous twins and stacking faults, as shown in Fig. 11.

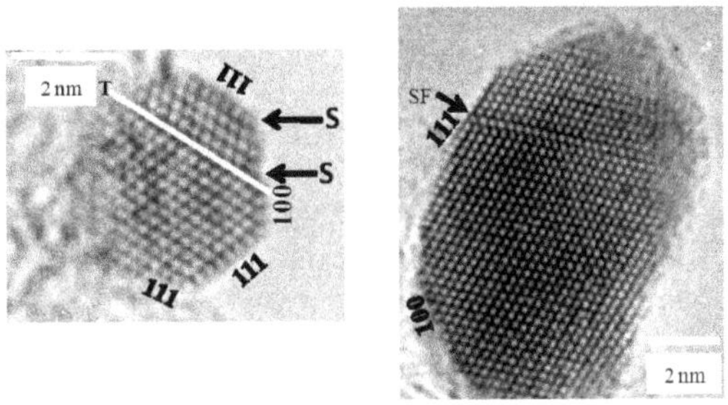

Figure 11. HRTEM observations of Au70Pt30 nanoparticle (left image) and Au nanoparticle (right image).

As a result of the high internal mean strain existing in these particles, there is an important strain energy which leads to the formation of twins and stacking faults. Consequently the equilibrium shape of the particles is modified and the interaction between the different surface atoms is changed. Accordingly, the catalytic behaviour of these particles is greatly affected. This can also explain the remarkable activity of these particles towards glucose oxidation both in alkaline medium as shown in Fig. 8, and in physiological type medium, as shown in Fig. 12.

Let's notice that at low potential values, current densities obtained with Au70Pt30 and Pt catalysts are similar. Competitive adsorption between phosphate species and glucose molecules can be involved to explain this phenomenon. Actually, de Mele et al. (de Mele et al., 1982) showed that phosphate species are capable of creating oxygen-metal bonds with platinum surfaces and thus inhibiting glucose oxidation. This engenders the low current density observed at low potentials on pure platinum. On Au70Pt30 catalyst, it is possible that modification of 5d band center of platinum due to the presence of gold allows discriminating the adsorption of phosphate species. Furthermore, the oxidation of glucose on high gold content catalysts starts at a very low potential value (i.e. -0.5 V vs. SCE), which can easily be compared with values observed for catalysts such as Pt-Bi, Pt-Sn (Becerik&Kadirgan, 2001) or Pt-Pd (Becerik et al., 1999).

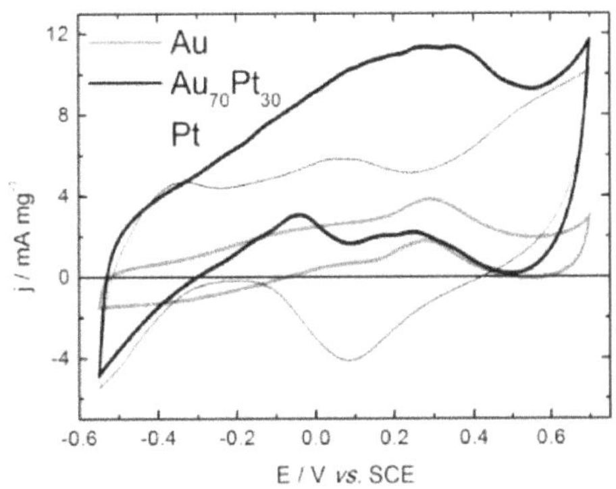

Figure 12.Voltammograms (after 19 cycles) of gold-platinum nanoparticles recorded at 37 °C in a phosphate buffered solution (0.1 M pH 7.4) in the presence of 10 mM glucose. Scan rate = 20 mV s-1.

3.2. Catalytic oxygen reduction process using inorganic catalysts

It is difficult to tailor non-enzymatic catalyst, capable of exhibiting electrochemical performances similar to those shown by laccase or BOD in physiological type media. The major problem with enzymes lies in the natural lack of stability of the proteins. One of the possibilities to tailor new efficient and stable cathode catalysts for glucose/O2 biofuel cells is to artificially reproduce active centers of enzymes and to stabilize their environment by mimicking the structure of enzymatic proteins and by removing all organic parts responsible for instability of enzymes. The possibility of designing this kind of catalyst has already been discussed (Ma & Balbuena, 2007).

4. DESIGN OF GLUCOSE/O2 BIOFUEL CELLS

The global reaction associated to the glucose/O2 biofuel cell can be described according to Eq. 13 :

$$C_6H_{12}O_6 + \frac{1}{2}O_2 \rightarrow C_6H_{10}O_6 + H_2O \tag{13}$$

Gibbs free energy associated to this reaction is Δr G0 = -251 kJ mol-1. This implies that the theoretical cell voltage is E0 = 1.3 V (Kerzenmacher et al., 2008). Furthermore, when the cell delivers a current j, the cell voltage E(j) can be expressed as follows:

$$E(j) = E_{eq} - \eta_a - \eta_c - Rj \tag{14}$$

where ηa is the anodic overvoltage, η c the cathodic one, R the cell resistance and E eq the equilibrium cell voltage. In Eq.14, it clearly appears that both values of η a, η c and R must be very low in order to increase the cell performances.

Since the development of the first biofuel cell realized by Yahiro et al. (Yahiro et al., 1964) that consisted in a two-compartment anionic membrane cell in which two platinum foils were used as conducting supports, numerous progress have been realized in designing devices. Nowadays, four main designs are developed. The first one has been developed by Heller's group. It simply consists in using two carbon fibers of 7 μm diameter as electrode materials. On these fibers, enzymes are immobilized in a redox osmium based hydrogel capable of immobilizing enzymes. These two electrodes are directly dipped into the electrolyte. In a physiological medium containing 15 mM glucose, the device was primarily able to deliver a power density of 431 μW cm-2 at a cell voltage of 0.52 V (Mano et al., 2002c). The device exhibited a high stability, since after one week of continuous working, it was still capable of delivering 227 μW cm-2. Based on this study, and by replacing carbon fibers by newly engineered porous microwires comprised of assembled and oriented carbon nanotubes, Mano's group (Gao et al., 2010) recently made the most efficient glucose/O2 biofuel cell ever designed. It indeed achieved a remarkably high power density of 740 μW cm-2 at a cell voltage of 0.57 V. The success of the experiment probably lies in the increase of the mass transfer of substrates. Other promising but presently less performing designs of glucose/O2 biofuel cells have been developed in the recent past years. The first one consists in using a microfluidic channel to build a glucose/O2 biofuel cell. The laminar flow obtained in the channel at low Reynold's number prevents the electrodes from depolarization phenomena and/or from degradation. The mixing of the reactants indeed occurs only on a very small distance in the middle of the channel. The development of such glucose/O2 biofuel cells seems of great interest for various applications. It is very simple to use abiotic and non-specific materials as catalysts. Moreover, it offers the possibility of working with two different pH values for the catholyte and the anolyte what can be interesting to improve electrochemical performances of each electrode (Zebda et al., 2009a). Nowadays, these devices are capable of delivering 110 μW cm-2 for a cell voltage of 0.3 V (Zebda et al., 2009b) by using GOD and laccase as catalysts. Glucose/

O2 biofuel cells realized with classical fuel cell stacks have also been carried out (Habrioux et al., 2010). Both the used system and the obtained performances are described in Fig. 13.

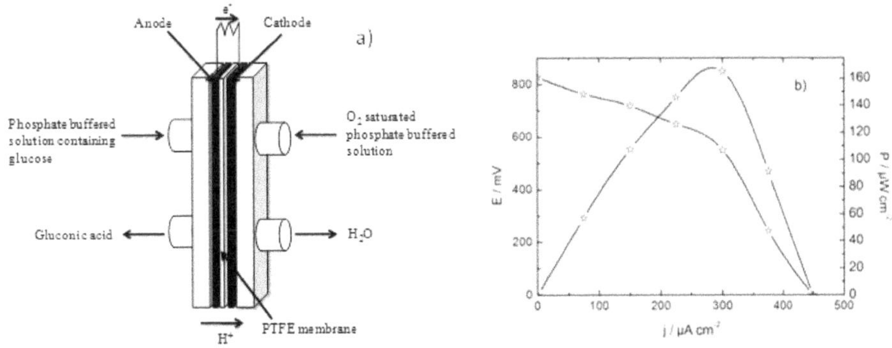

Figure 13.a) Description of the glucose/O2 biofuel cell design, b) Characteristic E vs. j of glucose/O2 cell performed at 20 °C: anode (Au70Pt30/Vulcan XC 72R, metal loading 40%); cathode (BOD/ABTS/ Vulcan XC 72 R system). Test realized in the presence of a phosphate buffered solution (0.2 M; pH 7.4) containing 0.3 M glucose. The cathodic compartment contains an oxygen saturated phosphate buffered solution (pH 7.4; 0.2 M).

Fig. 13 shows that the maximum power density obtained is 170 µW cm−2 for a cell voltage of 600 mV. However, let's notice that performances of the biofuel cell rapidly decrease for current densities higher than 300 µA cm-2. This is clearly due to a very low ionic exchange rate between the two compartments of the cell since this value is too weak to correspond to mass transfer limitation of glucose molecule. The last design of glucose/O2 biofuel cell developed in the last past years is the concentric device (Habrioux et al., 2008; Habrioux et al., 2009a). It is based on concentric carbon tubes as electrodes and operates at physiological pH. An oxygen saturated solution circulates inside the internal tube composed of porous carbon, which is capable of providing oxygen diffusion. The whole system is immersed in a phosphate buffered solution (pH 7.4, 0.1 M) containing various glucose concentrations. Oxygen consumption occurs at the cathode such that no oxygen diffuses towards the anode. This allows to use in this device both abiotic and enzymatic materials as anode and cathode catalysts, respectively. BOD/ABTS/Vulcan XC 72 R system is immobilized on the internal surface of the inner tube whereas Au-Pt nanocatalysts are immobilized on the internal surface of the outer tube. The surfaces of the cathode and anode were 3.14 and 4.4 cm2, respectively. The system is fully described in Fig. 14.

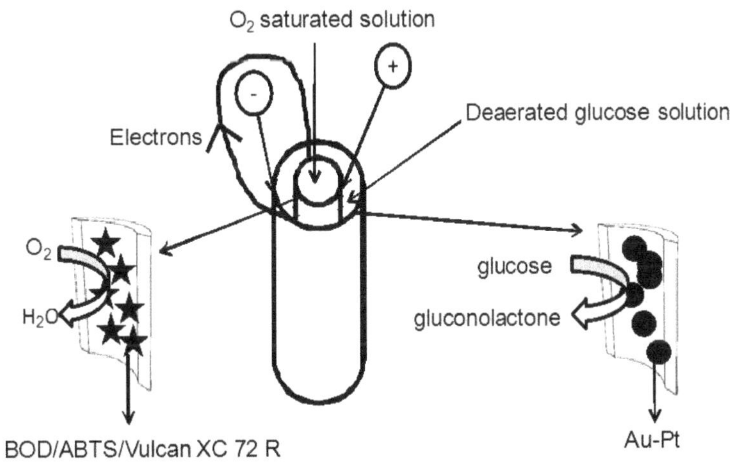

Figure 14.Schematic view of the glucose/O_2 biofuel cell system

Different fuel cell tests realized by using various nominal compositions of Au-Pt nanomaterials have been realized. The best performances are obtained with Au70Pt30 as anodic catalyst. Actually, the maximum power density achieved is approximately of 90 μW cm-2 for a cell voltage of 0.45 V. Results are shown in Fig. 15.

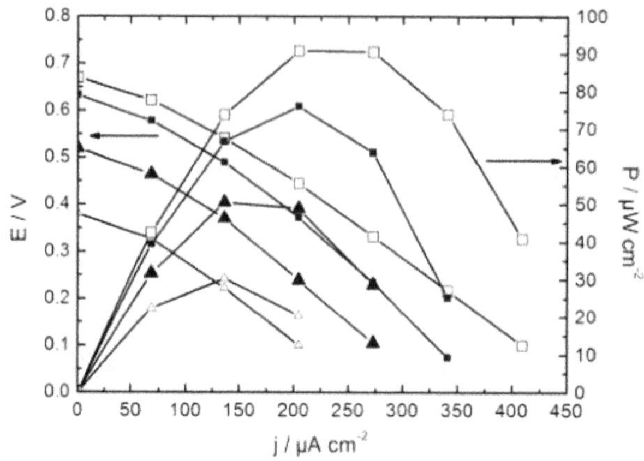

Figure 15.Fuel cell performances obtained with Au (▲), Au80Pt20 (■), Au70Pt30 (□) and Pt (Δ) nanoparticles as anode catalysts. These performances were obtained in a phosphate buffered solution (0.2 M, pH 7.4) containing 10 mM glucose at 37 °C. A saturated oxygen solution circulated in the inner tube of the device.

When Au80Pt20 is used as anode catalyst, the open circuit voltage is lower (i.e. 0.64 V). This is clearly explained by the surface composition of the catalyst which only contains 29 at.% of platinum. In the case of pure platinum, the open circuit voltage is very low due to strong competitions between phosphate species and glucose for adsorption. Such competition also occurs on other Au-Pt catalysts but the presence of gold allows a weaker interaction between phosphate species and the metallic surface. Consequently, higher glucose concentrations were used so as to improve biofuel cell performances. The obtained results are given in Fig. 16.

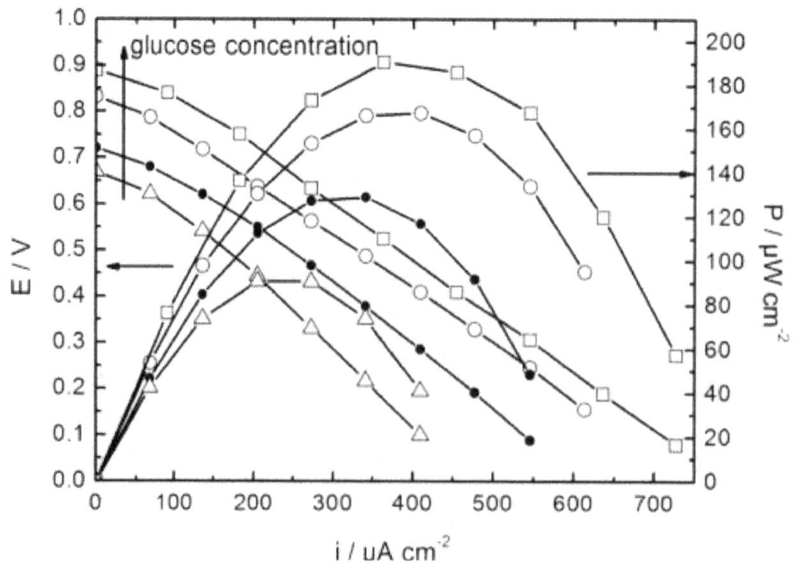

Figure 16.Fuel cell performances obtained with 10 mM glucose (Δ), 100 mM glucose (•), 300 mM glucose (○) and 700 mM glucose (□), with Au70Pt30 nanoparticles as anode catalyst. Performances obtained in a phosphate buffered solution (0.2 M, pH 7.4) at 37 °C. A saturated O2 solution circulated in the inner tube.

The data show a strong increase in cell voltage with glucose concentration. The raise observed in cell voltage between 0.1 M and 0.3 M can be attributed to the slow adsorption of phosphate species due to the presence of a higher glucose concentration. The maximum power density was also increased from 90 μW cm-2 (for a glucose concentration of 10 mM) up to 190 μW cm-2 (for a glucose concentration of 0.7 M). Nevertheless, in all cases, the fuel cell performances are greatly limited by resistance of the cell.

5. CONCISE SUMMARY

This chapter demonstrates the significance of both the assembly of electrodes and the overall design of the cell on the power output of the glucose/O2 biofuel cell. Furthermore, it appears that selecting well-defined nanocatalyst materials can enhance both the efficiency and durability of the cells, ultimately contributing to the inorganic properties of these materials. The utilization of an abiotic anode catalyst and an enzyme for a four electrons reduction is a method that can certainly create new possibilities for the applications of biofuel cells. This hybrid biofuel cell integrates the enhanced process of fuel electrooxidation, as established in conventional fuel cells, with the full conversion of dioxygen into H2O without generating H2O2. Furthermore, employing a concentric membrane-free configuration combined with a suitable immobilization of the catalysts can eliminate the need for an expensive cell separator. However, further advancements are still required to produce an optimal cell architecture.

REFERENCES

Alvarez-Icaza, M.; Kalisz, H.M.; Hecht, H.J.; Aumann, K.D.; Schomburg, D. & Schmid, R.D. (1995). The design of enzyme sensors based on the enzyme structure. *Biosensors and Bioelectronics,* Vol. 10, No. 8, pp. 735-742, ISSN 0956-5663.

Armstrong, F.A.; Hill, H.A.O. & Walton, N.J. (1985). Reactions of electron-transfer proteins at electrodes. *Quartely Reviews of Biophysics,* Vol. 18, No. 03, pp. 261-322, ISSN 0033- 5835.

Atwan, M.H.; Macdonald, C.L.B.; Northwood, D.O. & Gyenge, E.L. (2006). Colloidal Au and Au-alloy catalysts for direct borohydride fuel cells: Electrocatalysis and fuel cell performance. *Journal of Power Sources,* Vol. 158, No. 1, pp. 36-44, ISSN 0378-7753.

Bae, I.T.; Xing, X.; Liu, C.C. & Yeager, E. (1990). In situ Fourier transform infrared reflection absorption spectroscopic studies of glucose oxidation on platinum in acid. *Journal of Electroanalytical Chemistry and Interfacial Electrochemistry,* Vol. 284, No. 2, pp. 335-349, ISSN 0022-0728.

Bae, I.T.; Yeager, E.; Xing, X. & Liu, C.C. (1991). In situ infrared studies of glucose oxidation on platinum in an alkaline medium. *Journal of Electroanalytical Chemistry and Interfacial Electrochemistry,* Vol. 309, No. 1-2, pp. 131-145, ISSN 0022-0728.

Bartlett, P.N.; Tebbutt, P. & Tyrrell, C.H. (1992). Electrochemical immobilization of enzymes. 3. Immobilization of glucose oxidase in thin films of electrochemically polymerized phenols. *Analytical Chemistry,* Vol. 64, No. 2, pp. 138-142, ISSN 0003-2700.

Barton, S.C.; Pickard, M.; Vazquez-Duhalt, R. & Heller, A. (2002). Electroreduction of O2 to water at 0.6 V (SHE) at pH 7 on the "wired" Pleurotus ostreatus laccase cathode. *Biosensors and Bioelectronics,* Vol. 17, No. 11-12, pp. 1071-1074, ISSN 0956-5663.

Becerik, I. & Kadirgan, F. (2001). Glucose sensitivity of platinum-based alloys incorporated in polypyrrole films at neutral media. *Synthetic Metals,* Vol. 124, No. 2-3, pp. 379-384, ISSN 0379-6779.

Becerik, I.; Süzer, S. & Kadirgan, F. (1999). Platinum-palladium loaded polypyrrole film electrodes for the electrooxidation of d-glucose in neutral media. *Journal of Electroanalytical Chemistry,* Vol. 476, No. 2, pp. 171-176, ISSN 1572-6657.

Beden, B.; Largeaud, F.; Kokoh, K.B. & Lamy, C. (1996). Fourier transform infrared reflectance spectroscopic investigation of the electrocatalytic oxidation of -glucose: Identification of reactive intermediates and reaction products. *Electrochimica Acta,* Vol. 41, No. 5, pp. 701-709, ISSN 0013-4686.

Brimaud, S.; Coutanceau, C.; Garnier, E.; Léger, J.M.; Gérard, F.; Pronier, S. & Leoni, M. (2007). Influence of surfactant removal by chemical or thermal methods on structure and electroactivity of Pt/C catalysts prepared by water-in-oil microemulsion. *Journal of Electroanalytical Chemistry,* Vol. 602, No. 2, pp. 226-236, ISSN 1572-6657.

Brunel, L.; Denele, J.; Servat, K.; Kokoh, K.B.; Jolivalt, C.; Innocent, C.; Cretin, M.; Rolland, M. & Tingry, S. (2007). Oxygen transport through laccase biocathodes for a membrane-less glucose/O2 biofuel cell. *Electrochemistry Communications,* Vol. 9, No. 2, pp. 331-336, ISSN 1388-2481.

Bullen, R.A.; Arnot, T.C.; Lakeman, J.B. & Walsh, F.C. (2006). Biofuel cells and their development. *Biosensors and Bioelectronics,* Vol. 21, No. 11, pp. 2015-2045, ISSN 0956- 5663.

Burke, L.D.; Moran, J. & Nugent, P. (2003). Cyclic voltammetry responses of metastable gold electrodes in aqueous media. *Journal of Solid State Electrochemistry,* Vol. 7, No. 9, pp. 529-538, ISSN 1432-8488.

Calabrese Barton, S.; Gallaway, J. & Atanassov, P. (2004). Enzymatic Biofuel Cells for Implantable and Microscale Devices. *Chemical Reviews,* Vol. 104, No. 10, pp. 4867- 4886, ISSN 0009-2665.

Choi, Y.; Wang, G.; Nayfeh, M.H. & Yau, S.-T. (2009). A hybrid biofuel cell based on electrooxidation of glucose using ultra-small silicon nanoparticles. *Biosensors and Bioelectronics,* Vol. 24, No. 10, pp. 3103-3107, ISSN 0956-5663.

Davis, F. & Higson, S.P.J. (2007). Biofuel cells--Recent advances and applications. *Biosensors and Bioelectronics,* Vol. 22, No. 7, pp. 1224-1235, ISSN 0956-5663.

de Mele, M.F.L.; Videla, H.A. & Arvia, A.J. (1982). Potentiodynamic Study of Glucose Electro-Oxidation at Bright Platinum Electrodes. *Journal of The Electrochemical Society,* Vol. 129, No. 10, pp. 2207-2213, ISSN 1945-7111.

Degani, Y. & Heller, A. (1987). Direct electrical communication between chemically modified enzymes and metal electrodes. I. Electron transfer from glucose oxidase to metal electrodes via electron relays, bound covalently to the enzyme. *The Journal of Physical Chemistry,* Vol. 91, No. 6, pp. 1285-1289, ISSN 0022-3654.

Delecouls-Servat, K.; Bergel, A. & Basseguy, R. (2001). Surface-modified electrodes for NADH oxidation in oxidoreductase-catalyzed synthesis. *Journal of applied electrochemistry,* Vol. 10, No. 31, pp. 1095-1101, ISSN 1572-8838.

Devarajan, S.; Bera, P. & Sampath, S. (2005). Bimetallic nanoparticles: A single step synthesis, stabilization, and characterization of Au-Ag, Au-Pd, and Au-Pt in sol-gel derived silicates. *Journal of Colloid and Interface Science,* Vol. 290, No. 1, pp. 117-129, ISSN 0021-9797.

El Roustom, B.; Siné, G.; Fóti, G. & Comninellis, C. (2007). A novel method for the preparation of bi-metallic (Pt–Au) nanoparticles on boron doped diamond (BDD) substrate: application to the oxygen reduction reaction. *Journal of applied electrochemistry,* Vol. 37, No. 11, pp. 1227-1236, ISSN 0021-891X.

Ernst, S.; Heitbaum, J. & Hamann, C.H. (1979). The electrooxidation of glucose in phosphate buffer solutions : Part I. Reactivity and kinetics below 350 mV/RHE. *Journal of Electroanalytical Chemistry,* Vol. 100, No. 1-2, pp. 173-183, ISSN 0368-1874.

Fei, J.; Song, H.-K. & Palmore, G.T.R. (2007). A biopolymer composite that catalyzes the reduction of oxygen to water. *Chemistry of Materials,* Vol. 19, No. 7, pp. 1565-1570, ISSN 0897-4756.

Fishilevich, S.; Amir, L.; Fridman, Y.; Aharoni, A. & Alfonta, L. (2009). Surface display of redox enzymes in microbial fuel cells. *Journal of the American Chemical Society,* Vol. 131, No. 34, pp. 12052-12053, ISSN 0002-7863.

Gao, F.; Viry, L.; Maugey, M.; Poulin, P. & Mano, N. (2010). Engineering hybrid nanotube wires for high-power biofuel cells. *Nature Communications,* Vol. 1, No. 1, pp. 1-7, ISSN 2041-1723.

Grolleau, C.; Coutanceau, C.; Pierre, F. & Léger, J.M. (2008). Effect of potential cycling on structure and activity of Pt nanoparticles dispersed on different carbon supports. *Electrochimica Acta,* Vol. 53, No. 24, pp. 7157-7165, ISSN 0013-4686.

Gupta, G.; Rajendran, V. & Atanassov, P. (2004). Bioelectrocatalysis of oxygen reduction reaction by laccase on gold electrodes. *Electroanalysis,* Vol. 16, No. 13-14, pp. 1182- 1185, ISSN 1521-4109.

Habrioux, A.; Merle, G.; Servat, K.; Kokoh, K.B.; Innocent, C.; Cretin, M. & Tingry, S. (2008). Concentric glucose/O2 biofuel cell. *Journal of Electroanalytical Chemistry,* Vol. 622, No. 1, pp. 97-102, ISSN 1572-6657.

Habrioux, A.; Napporn, T.; Servat, K.; Tingry, S. & Kokoh, K.B. (2010). Electrochemical characterization of adsorbed bilirubin oxidase on Vulcan XC 72R for the biocathode preparation in a glucose/O2 biofuel cell. *Electrochimica Acta,* Vol. 55, No. 26, pp.7701-7705, ISSN 0013-4686.

Habrioux, A.; Servat, K.; Tingry, S. & Kokoh, K.B. (2009a). Enhancement of the performances of a single concentric glucose/O2 biofuel cell by combination of bilirubin oxidase/Nafion cathode and Au-Pt anode. *Electrochemistry Communications,* Vol. 11, No. 1, pp. 111-113, ISSN 1388-2481.

Habrioux, A.; Sibert, E.; Servat, K.; Vogel, W.; Kokoh, K.B. & Alonso-Vante, N. (2007). Activity of Platinum-Gold Alloys for Glucose Electrooxidation in Biofuel Cells. *The Journal of Physical Chemistry B,* Vol. 111, No. 34, pp. 10329-10333, ISSN 1520-6106.

Habrioux, A.; Vogel, W.; Guinel, M.; Guetaz, L.; Servat, K.; Kokoh, K.B. & Alonso-Vante, N. (2009b). Structural and electrochemical studies of Au-Pt nanoalloys. *Physical Chemistry Chemical Physics,* Vol. 11, No. 18, pp. 3573-3579, ISSN 1463-9076.

Heller, A. (2004). Miniature biofuel cells. *Physical Chemistry Chemical Physics,* Vol. 6, No. 2, pp. 209-216, ISSN 1463-9076.

Jin, C. & Chen, Z. (2007). Electrocatalytic oxidation of glucose on gold-platinum nanocomposite electrodes and platinum-modified gold electrodes. *Synthetic Metals,* Vol. 157, No. 13-15, pp. 592-596, ISSN 0379-6779.

Kang, C.; Shin, H.; Zhang, Y. & Heller, A. (2004). Deactivation of bilirubin oxidase by a product of the reaction of urate and O2. *Bioelectrochemistry,* Vol. 65, No. 1, pp. 83-88, ISSN 1567-5394.

Karnicka, K.; Miecznikowski, K.; Kowalewska, B.; Skunik, M.; Opallo, M.; Rogalski, J.; Schuhmann, W. & Kulesza, P.J. (2008). ABTS-Modified Multiwalled Carbon Nanotubes as an Effective Mediating System for Bioelectrocatalytic Reduction of Oxygen. *Analytical Chemistry,* Vol. 80, No. 19, pp. 7643-7648, ISSN 0003-2700.

Katz, E. & Willner, I. (2003). A Biofuel Cell with Electrochemically Switchable and Tunable Power Output. *Journal of the American Chemical Society,* Vol. 125, No. 22, pp. 6803-6813, ISSN 0002-7863.

Kenausis, G.; Chen, Q. & Heller, A. (1997). Electrochemical glucose and lactate sensors based on "wired" thermostable soybean peroxidase operating continuously and stably at 37 °C. *Analytical Chemistry*, Vol. 69, No. 6, pp. 1054-1060, ISSN 0003-2700.

Kerzenmacher, S.; Ducrée, J.; Zengerle, R. & von Stetten, F. (2008). An abiotically catalysed glucose fuel cell for powering medical implants: Reconstructed manufacturing protocol and analysis of performance. Journal of Power Sources, Vol. 182, No. 1, pp. 66-75, ISSN 0378-7753.

Kleppe, K. (1966). The effect of hydrogen peroxide on glucose oxidase from Aspergillus Niger. *Biochemistry*, Vol. 5, No. 1, pp. 139-143, ISSN 0006-2960.

Klotzbach, T.L.; Watt, M.; Ansari, Y. & Minteer, S.D. (2008). Improving the microenvironment for enzyme immobilization at electrodes by hydrophobically modifying chitosan and Nafion® polymers. *Journal of Membrane Science*, Vol. 311, No. 1-2, pp. 81-88, ISSN 0376-7388.

Kokoh, K.B.; Léger, J.M.; Beden, B.; Huser, H. & Lamy, C. (1992a). "On line" chromatographic analysis of the products resulting from the electrocatalytic oxidation of d-glucose on pure and adatoms modified Pt and Au electrodes-Part II. Alkaline medium. *Electrochimica Acta*, Vol. 37, No. 11, pp. 1909-1918, ISSN 0013- 4686.

Kokoh, K.B.; Léger, J.M.; Beden, B. & Lamy, C. (1992b). "On line" chromatographic analysis of the products resulting from the electrocatalytic oxidation of d-glucose on Pt, Au and adatoms modified Pt electrodes-Part I. Acid and neutral media. *Electrochimica Acta*, Vol. 37, No. 8, pp. 1333-1342, ISSN 0013-4686.

Li, H.; Webb, S.P.; Ivanic, J. & Jensen, J.H. (2004). Determinants of the Relative Reduction Potentials of Type-1 Copper Sites in Proteins. *Journal of the American Chemical Society*, Vol. 126, No. 25, pp. 8010-8019, ISSN 0002-7863.

Ma, Y. & Balbuena, P.B. (2007). Designing oxygen reduction catalysts: Insights from metalloenzymes. *Chemical Physics Letters*, Vol. 440, No. 1-3, pp. 130-133, ISSN 0009-2614.

Mano, N.; Kim, H.-H. & Heller, A. (2002a). On the Relationship between the Characteristics of Bilirubin Oxidases and O2 Cathodes Based on Their "Wiring". *The Journal of Physical Chemistry B*, Vol. 106, No. 34, pp. 8842-8848, ISSN 1520-6106.

Mano, N.; Kim, H.-H.; Zhang, Y. & Heller, A. (2002b). An Oxygen Cathode Operating in a Physiological Solution. *Journal of the American Chemical Society*, Vol. 124, No. 22, pp.6480-6486, ISSN 0002-7863.

Mano, N.; Mao, F. & Heller, A. (2002c). A Miniature Biofuel Cell Operating in A Physiological Buffer. *Journal of the American Chemical Society*, Vol. 124, No. 44, pp.12962-12963, ISSN 0002-7863.

Mano, N.; Mao, F. & Heller, A. (2004). Electro-oxidation of glucose at an increased current density at a reducing potential. *Chemical Communications*, Vol. No. 18, pp. 2116-2117, ISSN 1359-7345.

Mano, N.; Mao, F. & Heller, A. (2005). On the parameters affecting the characteristics of the "wired" glucose oxidase anode. *Journal of Electroanalytical Chemistry*, Vol. 574, No. 2,pp. 347-357, ISSN 1572-6657.

Mano, N.; Soukharev, V. & Heller, A. (2006). A laccase-wiring redox hydrogel for efficient catalysis of O2 electroreduction. *The Journal of Physical Chemistry B*, Vol. 110, No. 23, pp. 11180-11187, ISSN 1520-6106.

Mao, F.; Mano, N. & Heller, A. (2003). Long Tethers Binding Redox Centers to Polymer Backbones Enhance Electron Transport in Enzyme "Wiring" Hydrogels. *Journal of the American Chemical Society*, Vol. 125, No. 16, pp. 4951-4957, ISSN 0002-7863.

Merle, G.; Brunel, L.; Tingry, S.; Cretin, M.; Rolland, M.; Servat, K.; Jolivalt, C.; Innocent, C. & Seta, P. (2008). Electrode biomaterials based on immobilized laccase. Application for enzymatic reduction of dioxygen. *Materials Science and Engineering: C*, Vol. 28, No. 5-6, pp. 932-938, ISSN 0928-4931.

Merle, G.; Habrioux, A.; Servat, K.; Rolland, M.; Innocent, C.; Kokoh, K.B. & Tingry, S. (2009). Long-term activity of covalent grafted biocatalysts during intermittent use of a glucose/O2 biofuel cell. *Electrochimica Acta*, Vol. 54, No. 11, pp. 2998-3003, ISSN 0013-4686.

Minteer, S.D.; Liaw, B.Y. & Cooney, M.J. (2007). Enzyme-based biofuel cells. *Current Opinion in Biotechnology*, Vol. 18, No. 3, pp. 228-234, ISSN 0958-1669.

Möller, H. & Pistorius, P.C. (2004). The electrochemistry of gold-platinum alloys. *Journal of Electroanalytical Chemistry*, Vol. 570, No. 2, pp. 243-255, ISSN 1572-6657.

Moore, C.M.; Akers, N.L.; Hill, A.D.; Johnson, Z.C. & Minteer, S.D. (2004). Improving the environment for immobilized dehydrogenase enzymes by modifying Nafion with tetraalkylammonium bromides. *Biomacromolecules*, Vol. 5, No. 4, pp. 1241-1247, ISSN 1525-7797.

Mott, D.; Luo, J.; Njoki, P.N.; Lin, Y.; Wang, L. & Zhong, C.-J. (2007). Synergistic activity of gold-platinum alloy nanoparticle catalysts. *Catalysis Today*, Vol. 122, No. 3-4, pp. 378-385, ISSN 0920-5861.

Neto, S.; Forti, J. & De Andrade, A. (2010). An Overview of Enzymatic Biofuel Cells. *Electrocatalysis*, Vol. 1, No. 1, pp. 87-94, ISSN 1868-2529.

Palmer, A.E.; Lee, S.K. & Solomon, E.I. (2001). Decay of the Peroxide Intermediate in Laccase: Reductive Cleavage of the O-O Bond. *Journal of the American Chemical Society*, Vol. 123, No. 27, pp. 6591-6599, ISSN 0002-7863.

Palmore, G.T.R. & Kim, H.-H. (1999). Electro-enzymatic reduction of dioxygen to water in the cathode compartment of a biofuel cell. *Journal of Electroanalytical Chemistry*, Vol. 464, No. 1, pp. 110-117, ISSN 1572-6657.

Park, S.; Boo, H. & Chung, T.D. (2006). Electrochemical non-enzymatic glucose sensors. *Analytica Chimica Acta*, Vol. 556, No. 1, pp. 46-57, ISSN 0003-2670.

Paul, B.K. & Mitra, R.K. (2005). Water solubilization capacity of mixed reverse micelles: Effect of surfactant component, the nature of the oil, and electrolyte concentration. *Journal of Colloid and Interface Science*, Vol. 288, No. 1, pp. 261-279, ISSN 0021-9797.

Rao, M.L.B. & Drake, R.F. (1969). Studies of Electrooxidation of Dextrose in Neutral Media. *Journal of The Electrochemical Society*, Vol. 116, No. 3, pp. 334-337, ISSN 1945-7111.

Rusling, J.F.; Wang, B. & Yun, S.E., (2008). Electrochemistry of redox enzymes, In: *Bioelectrochemistry : Fundamentals, Experimental Techniques and Applications*, P.N. Bartlett, (Ed.), 39-85, Wiley, ISBN 978-0-470-84364-2, New-York, U. States.

Schmidt, T.J.; Gasteiger, H.A. & Behm, R.J. (1999). Rotating Disk Electrode Measurements on the CO Tolerance of a High-Surface Area Pt/Vulcan Carbon Fuel Cell Catalyst. *Journal of The Electrochemical Society*, Vol. 146, No. 4, pp. 1296-1304, ISSN 1945-7111.

Senthil Kumar, S. & Phani, K.L.N. (2009). Exploration of unalloyed bimetallic Au-Pt/C nanoparticles for oxygen reduction reaction. *Journal of Power Sources*, Vol. 187, No. 1, pp. 19-24, ISSN 0378-7753.

Shimizu, A.; Kwon, J.-H.; Sasaki, T.; Satoh, T.; Sakurai, N.; Sakurai, T.; Yamaguchi, S. & Samejima, T. (1999). Myrothecium verrucaria Bilirubin Oxidase and Its Mutants for Potential Copper Ligands. *Biochemistry*, Vol. 38, No. 10, pp. 3034-3042, ISSN 0006- 2960.

Shleev, S.; Tkac, J.; Christenson, A.; Ruzgas, T.; Yaropolov, A.I.; Whittaker, J.W. & Gorton, L. (2005). Direct electron transfer between copper-containing proteins and electrodes. *Biosensors and Bioelectronics,* Vol. 20, No. 12, pp. 2517-2554, ISSN 0956-5663.

Soukharev, V.; Mano, N. & Heller, A. (2004). A four-electron O2-electroreduction biocatalyst superior to platinum and a biofuel cell operating at 0.88 V. *Journal of the American Chemical Society,* Vol. 126, No. 27, pp. 8368-8369, ISSN 0002-7863.

Stankovich, M.T.; Schopfer, L.M. & Massey, V. (1978). Determination of glucose oxidase oxidation-reduction potentials and the oxygen reactivity of fully reduced and semiquinoid forms. *Journal of Biological Chemistry,* Vol. 253, No. 14, pp. 4971-4979, ISSN 0021-9258.

Sun, Y.; Buck, H. & Mallouk, T.E. (2001). Combinatorial Discovery of Alloy Electrocatalysts for Amperometric Glucose Sensors. *Analytical Chemistry,* Vol. 73, No. 7, pp. 1599-1604, ISSN 0003-2700.

Thomas, T.J.; Ponnusamy, K.E.; Chang, N.M.; Galmore, K. & Minteer, S.D. (2003). Effects of annealing on mixture-cast membranes of Nafion® and quaternary ammonium bromide salts. *Journal of Membrane Science,* Vol. 213, No. 1-2, pp. 55-66, ISSN 0376-7388.

Timur, S.; Pazarlioglu, N.; Pilloton, R. & Telefoncu, A. (2004). Thick film sensors based on laccases from different sources immobilized in polyaniline matrix. *Sensors and Actuators B: Chemical,* Vol. 97, No. 1, pp. 132-136, ISSN 0925-4005.

Topcagic, S. & Minteer, S.D. (2006). Development of a membraneless ethanol/oxygen biofuel cell. *Electrochimica Acta,* Vol. 51, No. 11, pp. 2168-2172, ISSN 0013-4686. Tsujimura, S.; Tatsumi, H.; Ogawa, J.; Shimizu, S.; Kano, K. & Ikeda, T. (2001).

Bioelectrocatalytic reduction of dioxygen to water at neutral pH using bilirubin oxidase as an enzyme and 2,2'-azinobis (3-ethylbenzothiazolin-6-sulfonate) as an electron transfer mediator. *Journal of Electroanalytical Chemistry,* Vol. 496, No. 1-2, pp. 69-75, ISSN 1572-6657.

Vogel, W.; Bradley, J.; Vollmer, O. & Abraham, I. (1998). Transition from five-fold symmetric to twinned FCC gold particles by thermally induced growth. *The Journal of Physical Chemistry B,* Vol. 102, No. 52, pp. 10853-10859, ISSN 1520-6106.

Vogel, W.; Tesche, B. & Schulze, W. (1983). X-ray structure and optical properties of silver films prepared by condensation of pre-formed Ag-clusters. *Chemical Physics,* Vol. 74, No. 1, pp. 137-142, ISSN 0301-0104.

Wang, J. (2007). Electrochemical Glucose Biosensors. *Chemical Reviews,* Vol. 108, No. 2, pp. 814-825, ISSN 0009-2665.

Wang, Z.; Liu, S.; Wu, P. & Cai, C. (2009). Detection of glucose based on direct electron transfer reaction of glucose oxidase immobilized on highly ordered polyaniline nanotubes. *Analytical Chemistry,* Vol. 81, No. 4, pp. 1638-1645, ISSN 0003-2700.

Willner, I.; Heleg-Shabtai, V.; Blonder, R.; Katz, E.; Tao, G.; Backmann, A.F. & Heller, A. (1996). Electrical wiring of glucose oxidase by reconstitution of FAD-modified monolayers assembled onto Au-electrodes. *Journal of the American Chemical Society,* Vol. 118, No. 42, pp. 10321-10322, ISSN 0002-7863.

Wilson, R. & Turner, A.P.F. (1992). Glucose oxidase: an ideal enzyme. *Biosensors and Bioelectronics,* Vol. 7, No. 3, pp. 165-185, ISSN 0956-5663.

Woods, R. (1971). The surface composition of platinum-gold alloys. *Electrochimica Acta,* Vol. 16, No. 5, pp. 655-658, ISSN 0013-4686.

Yahiro, A.T.; Lee, S.M. & Kimble, D.O. (1964). Bioelectrochemistry: I. Enzyme utilizing biofuel cell studies. *Biochimica et Biophysica Acta (BBA) - Specialized Section on Biophysical Subjects,* Vol. 88, No. 2, pp. 375-383, ISSN 0926-6577.

Zagal, J.H.; Griveau, S.; Silva, J.F.; Nyokong, T. & Bedioui, F. (2010). Metallophthalocyaninebased molecular materials as catalysts for electrochemical reactions. *Coordination Chemistry Reviews,* Vol. 254, No. 23-24, pp. 2755-2791, ISSN 0010-8545.

Zebda, A.; Renaud, L.; Cretin, M.; Innocent, C.; Pichot, F.; Ferrigno, R. & Tingry, S. (2009a). Electrochemical performance of a glucose/oxygen microfluidic biofuel cell. *Journal of Power Sources,* Vol. 193, No. 2, pp. 602-606, ISSN 0378-7753.

Zebda, A.; Renaud, L.; Cretin, M.; Pichot, F.; Innocent, C.; Ferrigno, R. & Tingry, S. (2009b). A microfluidic glucose biofuel cell to generate micropower from enzymes at ambient temperature. *Electrochemistry Communications,* Vol. 11, No. 3, pp. 592-595, ISSN 1388-2481.

Zhang, M.; Mullens, C. & Gorski, W. (2007). Coimmobilization of dehydrogenases and their cofactors in electrochemical biosensors. *Analytical Chemistry,* Vol. 79, No. 6, pp. 2446-2450, ISSN 0003-2700.

Zhu, Z.; Momeu, C.; Zakhartsev, M. & Schwaneberg, U. (2006). Making glucose oxidase fit for biofuel cell applications by directed protein evolution. *Biosensors and Bioelectronics,* Vol. 21, No. 11, pp. 2046-2051, ISSN 0956-5663.

CHAPTER -21

THE SIGNIFICANCE OF BIOPHYSICS IN BIOFUEL RESEARCH: A QUANTITATIVE APPROACH TO BIO-ENGINEERING

1. INTRODUCTION: AN OVBERVIEW

The dwindling accessibility and rising demand for fossil energy sources, along with concerns about irreversible climate change, have triggered a search for alternative energy sources, such as carbon-neutral transportation fuels. An example of an alternative to fossil fuels is biofuel derived from currently untapped plant material. Due to its inability to serve as a food source for humans, lignocellulosic biomass is an excellent candidate for biofuel generation, unlike corn-starch. At present, the process of creating biofuels from this unused biomass is not financially feasible and hindered by the expensive conversion of biomass into sugars, as well as the limited range of microbes currently available to produce various essential transportation fuels beyond basic ethanol fermentation. Currently, there is a significant amount of research being conducted to address the obstacles that hinder the development of a financially sustainable lignocellulose biofuel business. Progress needs to be achieved in the field of feedstocks engineering, optimizing deconstruction processes such as chemical, enzymatic, or microbial pretreatment and saccharification methods. Additionally, advancements are necessary in the area of fuels synthesis, which involve a range of biophysical approaches. This text will delve into some of these approaches in greater detail.

2. THERMOGRAVIMETRIC ANALYSIS (TGA) AND DIFFERENTIAL SCANNING CALORIMETRY (DSC)

Thermogravimetric analysis is a method of thermal analysis that quantifies the alteration in mass of a material when it is exposed to varying temperatures. The temperature of both the sample and a blank are systematically raised or lowered, while the resulting change in weight is monitored as a function of the sample's temperature. Due to the extreme sensitivity of these tests to temperature, they are carried out in chambers that have a high level of insulation. In TGA experiments, it is necessary to employ sample holders that have high conductivity in order to prevent any delay between the temperature measured outside and inside the sample holder. Minimal sample preparation is necessary for this approach, which is applicable to solids, liquids, or a combination of both. A minute quantity of the sample (less than 10 mg) is carefully weighed and its change in mass is compared to the change in mass of an empty sample container. The sample holders are available in two variations: with or without a hole in the lid. This design feature determines whether gases can pass through the holder or not. The TGA studies can be carried out in either an inert atmosphere (using Nitrogen or Argon) or under oxidative conditions (using Oxygen or Air). The change in weight of the sample as the temperature rises may suggest the material's loss caused by evaporation, breakdown, or oxidation. The weight loss curve produced by the TGA instrument accurately reflects the changes in mass as a result of temperature variations. TGA curves are typically analyzed using a derivative of the weight loss curves for better readability. Differential thermogravimetric (dTG) curves are valuable for determining the temperature at which vaporization or breakdown occurs for both individual substances and combinations of substances. The sequence and activation energy of the breakdown reactions can be determined by analyzing TGA curves obtained at various temperature ramp rates. TGA curves can also be utilized to determine the overall moisture content of the samples by measuring the weight loss in the temperature range of 80°C to 120°C, which corresponds to the evaporation of water. TGA/DSC procedures are especially valuable for mixes containing diverse elements. We can construct a weight loss zone for each constituent based on the temperature at which it decomposes or vaporizes. Subsequently, we may utilize the weight % loss in each region to determine the alterations in composition of the individual constituents in the original sample. Differential scanning calorimetry (DSC) is a thermoanalytical technique commonly employed alongside TGA. The method quantifies the amount of energy needed to raise or lower the temperature of the sample compared to a reference. Similar to TGA, this method likewise necessitates minimal sample preparation and can be used to solids, liquids, or a combination of the two. DSC employs comparable sample holders and equivalent sample weights to those used in TGA. The experimental settings, such as the pace at which temperature increases and the type of gas present, can be adjusted depending on the specific analysis being conducted. A DSC cure is characterized by a wealth of data, providing essential information such as the enthalpy of melting, boiling, breakdown, and oxidation. By analyzing both TGA (Thermogravimetric Analysis) and DSC (Differential Scanning Calorimetry) curves, one can ascertain the specific temperature at which the system undergoes boiling, decomposition, or oxidation. Boiling or endothermic decomposition can cause a loss of mass due to absorption of heat. An exothermic mass shift often occurs as a result of exothermic breakdown or oxidation of the substrate. Therefore, DSC data is also utilized to determine the caloric value of the breakdown and oxidation reactions. Differential scanning calorimetry (DSC) data is utilized to determine the amount of energy needed for the dehydration of a sample in the region where moisture loss occurs. This data can be utilized to determine if the moisture is physically absorbed into the substrate or chemically bonded with the substrate. DSC curves can be employed to determine the glass transition temperatures and, consequently, the degree of polymerization (DP) of polymer substrates (Couchman 1981). Amorphous to crystalline transitions can also be identified by analyzing the temperature at which weight loss occurs and the enthalpy of breakdown. According to Kim, Eom, and Wada (2010), amorphous substrates degrade at a lower temperature than crystalline substrates made of the same material. Amorphous materials experience the process of crystallization prior to melting. The percentage of crystalline material in the sample can be determined by comparing the energy needed for crystallization to the energy needed for melting. TGA/DSC techniques are highly advantageous for biomass characterisation because to the intricate composition of biomass, which consists of a diverse array of polymers such as cellulose, hemicelluloses, and lignin. Serapiglia et al. (2008), Kaloustian et al. (2003), and Jaffe, Collins, and Mencze (2006) have demonstrated that the dTG (differential thermogravimetry) curves obtained from biomass may be categorized into three distinct zones of weight loss, with hemicelluloses being the first to degrade, followed by cellulose and lignin. The technique of thermogravimetric analysis (TGA) is highly sensitive and capable of distinguishing between mutants of comparable types of feedstocks. Serapiglia et al. (2008) employed high throughput thermogravimetric analysis (TGA) to identify distinctions among different mutants of shrub willow. TGA was employed to determine the overall lignin content of different feedstocks (Ghetti et al., 1996). The TGA/DSC curves were employed to distinguish and comprehend the impacts of dilute acid steam explosion on Eucalyptus, as described by Emmel et al. in 2003. It was reported that lignin underwent fragmentation and recondensation during the procedure. Shao et al. (2009) reported a reduction in the softening temperature of bamboo lignin that underwent steam explosion, leading to the formation of a polymer with a decreased molecular weight. Differential scanning calorimetry (DSC) was employed to distinguish between hardwoods and softwoods by analyzing the glass transition temperature of the recovered lignin from the respective woods (Kubo and Kadla 2005). The study conducted by Reh et al. (1987) investigated the process of lignin degradation by white-rot fungus using DSC. The data demonstrates that the separation and sharpening of peaks in each region occur as the links between lignin and carbohydrates are disrupted. The variations in lignin carbohydrate linkages in biomass can also be identified through the use of Differential Scanning Calorimetry (DSC) (Reh et al. 1987; Tsujiyama and Miyamori 2000). In addition to these qualitative measurements, TGA (Thermogravimetric Analysis) and DSC (Differential Scanning Calorimetry) can be employed to determine the enthalpy, activation energy (Flynn and Wall 1966; Paul et al. 2010), reaction order, and the percentage of cellulose crystallinity in the samples.

3. CELLULOSE CRYSTALLINITY MEASUREMENT

Measuring the percent crystallinity of the biomass samples is particularly important as it affects rate and yield of enzymatic saccharification (Hall et al. 2010). It is challenging to measure the percent crystallinity of the biomass using DSC, as cellulose decomposes even before it undergoes melting (Paul et al. 2010; Kaloustian et al. 2003; Soares et al. 1995). As enthalpy of crystallization and enthalpy of melting are required to find the percent crystallinity of the sample, it is not possible to measure percent crystallinity of the biomass based on this technique. However, the enthalpy of dehydration of cellulose and biomass samples has been correlated to percent cellulose crystallinity of the samples based on the hydrogen bonds formed by the crystalline and amorphous cellulose. A detailed method for measuring cellulose crystallinity (CC) by DSC has been reported by Bertran and Dale (1986). They have reported that DSC provides a better way of measuring CC than other traditional approaches like XRD, especially for substance with very low crystallinity index. The amount of moisture absorbed by cellulose containing substances is dependent on the amount of crystalline and amorphous cellulose present in the substance. Amorphous cellulose absorbs a high amount of moisture while pure crystalline cellulose has no absorption capacity. Hence the endothermic dehydration peak appearing in DSC thermograms can be used to estimate the percent crystallinity of the cellulose present in the substances. Using completely amorphous cellulose measured under the same experimental conditions, the percent CC of the sample can be estimated from the following equation where ΔH0 is the heat of dehydration for a completely amorphous cellulose sample and ΔHs is the heat required to dehydrate the sample.

$$\%CC = \frac{\Delta H_0 - \Delta H_s}{\Delta H_0} x\,100$$

In spite of the many advantages of using TGA and DSC analysis they cannot be used to find bond specific information about the substrate. Knowledge of the chemical composition of the substrate is required ahead of time. Analysis through this technique always requires one to correlate the original substrate to the pure components of the substrate. The changes to DP of a polymer can only be visualized but cannot be quantitatively correlated to changes in the molecular weight of the polymer. Also if the substrates decompose very close to each other, we cannot differentiate between the increase of one of the substrate or decrease in DP of the polymer degrading at a higher temperature. As these techniques affect the samples physically and chemically the sample used in this technique cannot be recovered or reused.

4. PYRO-GC/MS AND GC/MS

Gas Chromatography/ Mass Spectrometry (GC/MS) is a very popular technique used to separate and identify each organic compound from mixtures. It provides both qualitative and quantitative information about the sample. Gas Chromatography (GC) separates the molecules based on their molecular weight and their volatility. Gas chromatography uses a conventional oven with a column of very small internal diameter. The temperature of the oven is changed and based on the volatility of the compounds they travel through the column at different rates. Hence they separate and reach the mass spectrometer at different times. The time a compound takes to reach the mass spectrometer from the time it is injected into the column is called retention time. The retention time of a compound depends on its molecular weight and structure (thus its volatility). The retention time of the compounds is dependent on the initial temperature and temperature ramp rate of the GC. The compounds injected into the GC reach the MS at different times based on their volatility and complex mixtures are thus separated, but this separation is not enough to understand their composition. Two compounds of different molecular weight and structure may reach the MS at the same time if they have similar volatility. A mass spectrometer is used to identify these compounds eluting from the GC column. The mass spectrometer charges the compounds, accelerates them under high magnetic fields and then breaks them into ions. Based on their mass and charge, the ions hit the detector at various positions and are thus detected. Unlike it is the case for TGA and DSC mass spectrometry analysis does not require a a thorough understanding of the mixture composition of the mixture is not necessary before analysis. However confirming the identity of the predicted compound, after the analysis of the unknown sample, with a known standard is necessary. The mass spectrometer is usually run in scan mode where it detects all the ions that hit the detector. However, scans of samples containing known compounds can be made very sensitive by forcing the mass spectrometer to detect the ions specific to the compounds of interest (selective ion mode). Based on the fragmented ions and the retention time, the chemical structure of the compound can be determined.

The mass spectrometer is very sensitive and detects even a small concentration of impurities. Hence the sample needs to be extremely pure and free of any oxygen or moisture. This implies the need for tedious sample preparation and purification step before the introduction of the sample into the instrument. It is also important to maintain the instrument itself and the column used in the GC. There is good chance of introduction of moisture and column degradation products into the sample without proper care. The quadruple of the MS itself needs to be cleaned and tested for its sensitivity frequently.

In liquid GC/MS the samples are usually extracted into volatile organic solvents and are then injected into the GC. The compounds thus detected are biased towards non-polar organic molecules of low molecular weight. However, it is also possible to detect polar organic molecules like organic acids after derivatization thus rendering them non polar. For quantitative analysis one injects a know quantity of a compound called internal standard (IS). An internal standard is a compound, which is similar in all respects to the compounds of interest but can be separated from the original compounds in the column. Deuterated counterparts of the original compounds are usually used as internal standards for the GC/MS analysis. Internal standards are injected into the original solid or liquid prior to extraction. The compounds of interest and the original solids are then extracted into a volatile organic solvent through soxhlet extraction or liquid-liquid extraction. Specific ASTM (American Society for Testing Materials) guidelines exist for the extraction procedures for quantitative analysis of various volatile organic pollutants. A standard mixture of known quantities of the internal standard and the compound of interest is made in similar solvent. This standard mixture is then analyzed in GC/MS using the same protocol used for the sample and the standard.

Differences in efficiency of the extraction of both the compounds and the internal standard can lead to erroneous results such as incorrect concentration predictions. Only samples which contain extractible substances of lower molecular weight can be analyzed using this technique.

These shortcomings can be overcome by using pyrolysis GC-MS. In this technique, the sample is pyrolyzed at high temperatures for very short amount of time and the products thus produced are directly injected into GC/MS. This technique is usually used to find the structures of heteropolymers (polystyrene and PVC blends, lignin, etc.). The primary advantage of this technique is that it requires a small sample volume and no sample preparation, but due to measurement errors of such very small volumes pyrolysis GC-MS is difficult to quantify and inaccurate. Hence the ratio of breakdown products is often a better measurement estimate to describe the original sample.

Both GC/MS and pyrolysis GC/MS are used extensively in biofuel research. Lignin as a heteropolymer present in the biomass is often analyzed for its chemical composition using pyrolysis GC/MS (Galletti et al. 1997; Mills et al. 2009). Lignin typically contains three major constituents called p-hydroxyphenyl lignin (H-lignin), guaiacyl lignin (G-lignin) and syringyl-lignin (S-lignin). The structure and strength of lignin is dependent on the ratio of these three kinds of lignin (Boudet et al. 1998). Lignin rich in G-lignin is supposed to be highly condensed when compared to lignin rich in S- lignin (Chiang and Funaoka 1990). Ralph and Hatfield (1991) were one of the first ones to analyze the composition of lignin using pyrolysis GC/MS. They show that the pyrolysis products of lignin can be a variety of products derived from H, G and S-lignin. They suggest using a ratio of the total amount of H to the total amount of G to the total amount of S products to find the H:G:S ratio. The S: G ratio of a feedstock has been shown to affect the efficiency of both the pretreatment technology and the enzymatic saccharification. Pyro-GC/MS has been used to find the changes in S:G ratio due to different pretreatments (Samuel et al. 2010; Huyen et al. 2010; Ibarra et al. 2007; Jung et al. 2010; Papatheofanous et al. 1995). The S:G ratio decreases as a result of kraft pulping (Ibarra et al. 2007) and thus hardwoods (S:G ratio >1) have always been a choice for the paper and pulp industry. Ammonia treated Miscanthus showed a decrease in the S:G ratio as result of pretreatment (Huyen et al. 2010). Samuel et al. (2010) have shown that the S:G ratio of switchgrass decreased due to dilute acid pretreatment of biomass, showing it is easier to break S-lignin when compared to G-lignin. Based on this hypothesis genes have been identified which increase the amount S-lignin present in the plant biomass (Marita et al. 1999; Ralph 2007). The mechanism of how lignin is formed and how each gene effects the formation of these individual components of lignin has been found by both NMR and pyro-GC/MS. Pyro-GC/MS is used to differentiate between various plant mutants with different S:G ratios. The changes to S:G ratio as function of harvesting time has also been identified using pyro-GC/MS (Huyen et al. 2010). Feedstocks with similar lignin content have been shown to have different saccharification yields based on their S:G ratio (Davison et al. 2006). Feedstocks with higher amount S-lignin were easier to pretreat but had lower xylose yields when compared to feedstocks with higher G-lignin (Davison et al. 2006). The differences in S:G ratio has also been correlated to in-rumen digestibility of plant biomass (Guo et al. 2001; Baucher et al. 1999; Vailhé et al. 1998). As the S:G ratio decreased the in-rumen digestibility increased, showing that feedstocks such as grass are ideal for animal feed. The lignin degradation products produced during the pretreatment process are also dependent on the S:G ratio of the original lignin polymer (Chiang and Funaoka 1990). S:G ratio, as measured from pyro-GC/MS, offers a platform to correlate its effects on different biofuel production steps and thus helps us design better feedstocks.

Lignin and some of its monomers have been shown to inhibit both cellulases during enzymatic saccharification (Sewalt et al. 1997; Ximenes et al. 2010; Ximenes et al. 2011) and microbes during fuel production (Mills et al. 2009). These lignin degradation products produced during the pretreatment may have an inhibitory role on the downstream processes. These products can be identified and quantified using GC/MS (Pecina et al. 1986). Identifying these lignin degradation products is also important for the biofuel industry as it gives us an idea about the byproducts produced during the pretreatment process itself (Ehara et al. 2005). As pyro-GC/Ms can give a ratio of the amount of G-lignin and S-lignin present in the lignin, derivatization followed by the reductive cleavage (DFRC) method has been developed by Ralph et al. to measure the quantitatively the various constituents of lignin in the biomass (Lu and Ralph 1997; Lu and Ralph 1998; Lu and Ralph 1999; Peng et al.1998; Ralph and Lu 1998). This method uses GC/MS to quantify various derivatized and reduced products. Genetic pathway engineering, also known as synthetic biology, is gaining popularity as it aims to produce microbes which may in turn act as biofuel factories to produce the desired kind of fuel from the glucose units. Qualitative analysis of the composition of the fuel allows the identification of the right modification of the microbes to produce fuel. Quantification of the products is also necessary to understand the most efficient mechanism and the toxicity limit of the fuel until which the microbes can sustain. GC/MS is also used to qualitatively and quantitatively measure the fatty acids and fuels produced from the microbes ((Akiyama et al. 2008; Atsumi et al. 2008; Dai et al. 2007; Lu et al. 2008; Tang et al. 2007).

5. PLANT STEM MECHANICAL STRENGTH TESTING

As a successful feedstock will have the ability to grow tall and hence provide high yield per area of land used, it is key to measure the mechanical stability for any plant derived by breeding, mutagenesis or targeted cell wall engineering. Mechanical strength is a very important property for materials and defines the use of materials as support systems. A tensile testing instrument measures the mechanical strength, the breaking force required and the strain that the material underwent during the tensile test. A simple tensile testing instrument consists of a motor to apply the tensile force and a force sensor to find the tensile force applied by the motor. The approximate stress and strain during the tensile testing is calculated based on the total length and cross sectional area of the sample before the tensile testing. Tensile stress is calculated as a ratio of force to the cross-sectional area. In ductile materials, as the cross sectional area changes due to deformation under tensile stress the measurement of tensile stress using the above method gives only an approximate value for stress. As brittle materials undergo little or no deformation during tensile testing, the above method provides a better approximation for the stress in case of brittle materials. The total strain that a material underwent during tensile testing can be calculated as the ratio of the total change in length of the material to the initial length of the material. The relative brittle or ductile nature of the samples can be inferred from the total strain they undergo during tensile testing. A brittle material undergoes little deformation and shows a lower amount of strain when compared to ductile materials, which undergo a higher deformation and hence show higher amount of strain. The breaking strength of material, defined as the mechanical stress at the breaking point, is also an important property. The maximum mechanical strength and the breaking strength of a material need not be the same. During tensile testing a material elongates and reaches a maximum stress value, this is the maximum force the material can endure. The material may not break at this point, but on further increasing the strain, the material becomes weaker and breaks at a force lower than the maximum strength. This difference in behavior is also related to brittle and ductile nature of the materials. In ductile materials, like metals, the crystals lattices expand and rearrange when stress is applied, but brittle materials like ceramics are very closely packed and have no room for rearrangement when stress is applied. For the same reason, in case of brittle materials the maximum tensile strength and the breaking strength are usually the same. Tensile testing of the materials thus provides in depth knowledge about maximum mechanical strength, total strain and the nature of the material. Standard

ASTM procedures have been developed for measuring the mechanical strength and the above parameters for a variety of materials (Czichos et al. 2006).

Physical pretreatment or grinding is a major part of the pretreatment that would be applied to the biomass during the production of biofuel in order to provide higher surface area, which can increase the efficiency of enzymatic saccharification (Li et al. 2010) and improve the efficiency of the chemical pretreatment by providing smaller particles. Materials of lower mechanical strength are easier to handle and degrade compared to materials of higher mechanical strength. Thus the energy required during the grinding process can be co-related the mechanical strength of the biomass and this property plays an important role for biofuel production. Studies have been conducted which find the mechanical strength of the plant stems and total mechanical energy required for grinding the plant stems to be correlated. Decreasing the mechanical strength of the plant stems would in turn decrease the total energy required during the grinding process. Yu et al. (2006) studied the tensile stress and ultimate shear stress profile of switchgrass as a function of the plant maturity, the time after harvest and the moisture content. The tensile stress was reported to increase with an increase in the elapsed time after harvest and a decrease in the moisture content. Based on this they recommend tensile dominant size reduction to be carried out just after the harvest process of biomass rich in moisture. Igathinathane et al. (2008) showed the dependence of ultimate shear stress on the total moisture content of the switchgrass.

The measurement of the mechanical strength of plant mutants has been studied to find the effect of genetic manipulations on the overall mechanical strength of the plants. An in-house instrument was developed by Prat and Paresys (1989) for measuring different mechanical strength properties of plant stems. A decreased mechanical strength may be desirable for both reduced grinding energy and easier chemical pretreatment of the feedstocks (Wilkinson and Santillana 1978; Lin et al. 1985; Grethlein and Converse 1991; Ibrahim and Pearce 1983).

Lignin present in the plant cell walls provides the plant with support. It is reported to be the most resistant to pretreatment processes (Vinzant et al. 1997; Vailhé et al. 1998; Chen and Dixon 2007). Lignin also inhibits enzymatic saccharification (Chen and Dixon 2007; Jackson et al. 2008) and microbial growth during the fuel production process (Mills et al. 2009). Hence there is a major effort by plant biologists to understand and modify lignin in plant cell wall to ease the conversion of cellulose to fuel (Himmel et al. 2007; Chen and Dixon 2007; Lee et al. 2009). In this effort, mutants are developed with modified/ lowered lignin content (Davison et al. 2006). These modifications may have a direct effect on the plant cell wall and hence on the systems that provide mechanical strength to the plant. However, the decreased mechanical strength can have an adverse effect on the growth of the plant and its ability to sustain under natural conditions. A mutagenized plant may be too weak to sustain a strong gust of wind or even a heavy downpour. This initiates the need to reach an optimum level for mechanical strength to ease the pretreatment process and not lose the ability to sustain natural calamities. Mechanical strength can also be used to find the condition under which a plant would or would not sustain under natural environmental conditions (Coops and Van der Velde 1996). An Arabidopsis mutant grown in the presence of dichlorobenzonitrile, which inhibited the production of cellulose, was also reported to have lower strength than the wild type (Ryden et al. 2003). Ethyl methanesulfonate-mutagenized Arabidopsis populations showed the absence of interfascicular fibers in stems, decreased L-Fucose content and reduced mechanical strength only in the inflorescence part of the stem (Zhong et al. 1997; Reiter et al. 1993). Zhong et al. (2004) showed that the fragile fiber3 (fra3) gene of Arabidopsis thaliana, which encodes type II inositol polyphosphate 5-phosphatases (5PTases), resulted in a decrease in secondary cell wall thickness and hence a reduction in stem strength. An expression of the Heynh root hair defective3 (RHD3) gene in an Arabidopsis fragile fiber4 (fra4) mutant also exhibited reduction in cell wall thickness and alteration in cell wall composition (Hu et al. 2003). This also resulted in lowered mechanical strength of the plant stems. Burk et al. (2001) isolated a group of weaker Arabidopsis plant mutants based on their mechanical strength and found a katanin-like protein, which regulated cell wall biosynthesis and cell elongation. Arabidopsis thaliana fragile fiber8 (fra8), which was defective in xylan synthesis, was also found to have lower mechanical strength than the wild type. Arabidopsis thaliana stems down regulated in secondary cell wall–associate NAC domain protein1 (SND1) and NAC secondary wall thickening promoting factor1 (NST1) showed loss of secondary cell wall formation (Zhong et al. 2007), and as a result had lower amount of cellulose, hemicelluloses and lignin. They also showed a considerable decrease in mechanical strength. Fragile fiber (fra1) mutation of Arabidopsis exhibited a dramatic reduction in mechanical strength while no defects were found in cell wall thickness or composition or in the organization of fiber cells (Zhong et al. 2002). This lowered mechanical strength was attributed to differences in cellulose microfibril orientation in the walls of fra1. Irregular xylem8 (irx8), irx9, and irx14 mutations of Arabidopsis thaliana resulting in defects in glucuronoxylan biosynthesis (Keppler and Showalter 2010; Pena et al. 2007; Lee et al. 2010). These defects lead to decreased mechanical strength of the plant stems. An overexpression of these genes rescued the plants of only irx8 and irx9 (but not irx14) and the mechanical strength of the resulting plants was comparable to the wild type plants (Lee et al. 2010; Pena et al. 2007). A classic rice mutant brittle clum1 was shown to have reduced mechanical strength and the mutants were characterized to display increased lignin content, a reduced cell wall thickness and decreased cellulose content (Li et al. 2003). The consequence of crowding on the structural behavior and response from marsh plants was studied by Harley and Bertness (1996). They reported that due to crowding the total amount of plant biomass (above the ground) decreased and resulted in thinner and weaker plants.

The effect of various gene expressions and the environmental conditions on the overall plant strength can be easily determined by mechanical strength analysis. It is also a very sensitive technique, which responds to the changes in the orientation of fibers, in plants with no changes in cell wall composition and thickness. The data from these tests would be very useful for the development of feedstock in biofuel research, which are environmentally sustainable, with desired cell wall composition and lowered grinding energies. However, this is a physical characterization technique and can be only be used to understand the effects of various parameters (like genetic modifications or natural effects) on the plant strength. It cannot be used to understand the reason for the observed increase or decrease in the mechanical strength. We are currently using this approach to swiftly characterize the mechanical properties of plants derived from a random-mutagenesis screen.

6. ANALYSIS OF CELL WALLS USING VIBRATIONAL SPECTROSCOPY AND MICROSCOPY

Vibrational spectroscopy and microscopy techniques have been broadly used to analyze chemical and biological materials. The two main vibrational spectroscopic techniques, infrared spectroscopy and Raman spectroscopy, detect vibrations including both simple bond vibrations and group vibrations in molecules and thus identify these molecules by their spectral fingerprints originated from various vibrational modes. Infrared absorption typically involves photon absorption with the molecule excited to a higher vibrational energy level when the photon energy matches the energy difference between the two vibrational energy levels. This process depends on changes of dipole moments, and hence asymmetric vibrations cause the most intense infrared absorption. On the other hand, Raman scattering is the inelastic scattering of a photon that interacts with molecular vibrations, resulting in an energy shift of the exciting photon. This process depends on changes in polarizability of the electron cloud around the vibrating bonds or groups. Usually, symmetric vibrations cause the largest polarizability changes and thus render the greatest scattering. Therefore, these two techniques often provide complementary information about the molecules (Smith and Dent, 2005). All major cell wall biopolymers are both IR and Raman active. Nowadays both techniques are widely used in plant cell wall research (Dokken et al., 2005; Gierlinger and Schwanninger, 2007).

Unlike most of the current techniques for cell wall compositional analysis, such as wet chemistry assays, chromatography methods, mass spectrometry etc., which are destructive and involve breakdown of the cell wall components or extensive chemical treatment of the plant cell walls, IR and Raman spectroscopy can characterize cell wall components in their native form with minimal requirement for sample preparation. Moreover, a reliable measurement of a single sample can usually be completed from seconds to a few minutes. With the aid of a high throughput platform, such as 96-well plates, rapid screening of a large number of samples can be realized to meet the requirements of bioenergy feedstock development and biomass conversion process optimization for efficient biofuel production.

To acquire cellular level understanding of plant cell walls, various microscopic techniques have been employed, such as bright/dark field microscopy (D'Haeze et al., 2007), polarized light microscopy (Baskin et al., 2004), transmission electron microscopy (Fromm et al., 2003), scanning electron microscopy (Persson et al., 2007b), etc. However, to localize molecules of interest, histochemical and cytochemical staining and labeling methods have to be applied (Cavalier et al., 2008; Grünwald et al., 2002; Persson et al., 2007a). Although autofluorescence of lignin can be utilized to visualize distribution of lignin in the cell wall by fluorescence microscopy (Cavalier et al., 2008; De Micco and Aronne, 2007; Singh et al., 2009), chemical information of cellulose cannot be obtained without additional techniques and quantitative analysis is difficult. On the other hand, using IR and Raman microspectroscopy, chemical maps of specific cell wall components can be generated based on their spectroscopic fingerprints without any disruption of plant tissues and necessities of staining or any other extensive treatment of the cell walls. Localized and tissue/cell specific chemical information can be revealed, which enables acquisition of chemical information on the ultrastructure of plant cell walls. Thus, IR and Raman imaging can detect important compositional changes in cell walls by mutations not necessarily reflected in their average contents and spatial chemical changes during processing, which is very difficult to achieve by other chemical analysis and microscopic methods. Between these two vibrational spectroscopic techniques, IR spectroscopy has been established for decades as a useful tool for plant cell wall research. Integrated with attenuated total reflectance (ATR) technique, synchrotron sources, focal plane array (FPA) infrared detector, and chemometric analysis, IR technique has become increasingly powerful. However, this technique is limited by low sensitivity due to non-background-free detection, low spatial resolution associated with the long infrared wavelengths, and water absorption of the infrared light (Evans and Xie, 2008). The Raman technique, on the other hand, does not have these limitations. Recent advancement in laser technology, optics and detectors has led to a rapid growth in the applications of the Raman technique. Although auto-fluorescence from lignin may sometimes interfere with Raman measurement, the use of near infrared or UV excitation or a fluorescence quencher and effective baseline correction afterwards can alleviate this problem generally. In this section, the applications of IR and Raman techniques for cell wall analysis will be discussed.

Applications of IR microspectroscopy in plant cell wall research: The most prevalent type of IR spectrometer is a Fourier Transform Infrared Spectrometer (FTIR) by recording the raw data as an interferogram and then using Fourier transform to turn this raw data into a spectrum. Attenuated total reflection (ATR) objectives containing an internal reflection element usually made of ZnSe, Ge, or diamond are often used with IR spectroscopy to study plant materials due to faster sampling, improved reproducibility and less impedance by water absorption. IR spectra are most commonly obtained from 4000 to 400 cm-1, the mid-infrared region, where the peaks in the spectra can be associated with fundamental vibrations and the peak intensities are proportional to concentrations. Chemical compositions of various lignocellulosic biomass have been studied by mid-IR spectroscopy, such as wood samples, grasses and herbaceous plants (Dokken et al., 2005). Peak assignments for major cell wall components of representative biomass were summarized by Adapa et al. (2009). Compositional changes by biochemical and chemical treatments of cell walls were also investigated by FTIR. For example, behaviors of cell wall components of oak wood and barley straw treated with cellulase, acidic sodium chlorite, acid and base were studied (Stewart et al., 1995). Dilute acid pretreatment and ionic liquid pretreatment were compared using switchgrass as a model bioenergy feedstock (Li et al., 2010). Moreover, with the aid of chemometric techniques, FTIR can be used as a rapid method for cell wall mutant screening (Chen et al., 1998; Mouille et al., 2003). In addition, using polarizers with FTIR, the orientation of particular functional groups or cell wall components can be determined (McCann et al., 1993; Wilson et al., 2000).

The chemical imaging capability has made IR microscopy a powerful tool to reveal spatial distribution of cell wall components by collecting spectra at each spatial position in the defined area for imaging. FTIR microscopy was also demonstrated by Gierlinger et al. (2008a) to monitor in situ the enzymatic degradation of poplar wood and fast and selective degradation of the gelatinous layer in tension wood was observed. This method could be used for enzyme screening and working condition optimization for enzymes. The source intensity from conventional IR thermal sources can only provide a spatial resolution of tens of micrometers, which is limited by both signal-to-noise (S/N) ratio and diffraction, thus restricting plant analysis to the tissue level. Coupling a synchrotron IR source with a small effective source size to IR microscopy can overcome this difficulty due to the high source brightness that allows smaller regions to be probed with acceptable S/N, and thus only diffraction controls spatial resolution in this case (Carr, 1999). Using the synchrotron radiation-based FTIR microspectroscopy (SR-FTIR), imaging structures of plant tissue at cellular level with high S/N at ultraspatial resolutions (3-10 μm) was achieved (Dokken and Davis, 2007; Yu et al., 2003). The main drawbacks of SR-FTIR are the expense and limited access to synchrotrons and slow imaging acquisition by the point-by-point process. A more recent technique known as focal plane array (FPA) based FTIR imaging becomes available and promises many advantages. The FPA-FTIR imaging technique is laboratory based and employs two-dimensional detector arrays to collect spectra at marked positions simultaneously. Thus, FPA-FTIR imaging can acquire the chemical map in a fraction of time required by SR-FTIR imaging. Heraud et al. (2007) compared results produced by FPA- FTIR imaging and SR-FTIR imaging using Eucalyptus botryoides leaves as a model sample. They found that the two methods produced similar infrared images allowing differentiation of all tissue types in the leaves. While SR-FTIR imaging provided superior S/N ratio and better spatial resolution, it only took approximate 2 min for FPA-FTIR to map a 350 μm2 of tissue area, which took approximate 8 h for the SR-FTIR imaging to complete.

In addition to mid-IR spectroscopy, near infrared (NIR) spectroscopy has also been used as a useful tool for compositional analysis of lignocellulosic materials. Absorption spectra in the NIR region, i.e. 14000 – 4000 cm-1, are derived from the overtone or harmonic vibrations. Krongtaew et al. (2010a) has summarized the NIR peak assignments for major cell wall components. Nevertheless, multivariate methods are often implemented for NIR data analysis and both qualitative and quantitative information can be derived. For example, a partial least-squares regression (PLSR) model was developed based on the NIR spectra to assess lignin composition (p-hydroxyphenyl to guaiacyl ratio) in maritime pine wood (Alves et al., 2006). Key properties influencing the enzymatic hydrolysis yield and rates, such as lignin content, hemicellulose content, and cellulose crystallinity, influenced by different pretreatment methods were resolved in FT-NIR spectra and successfully evaluated by principal component analysis (Krongtaew et al., 2010a). Total residual lignin content, enzymatically released reducing sugars, total solids, volatile solids, and biogas yield can be assessed quantitatively by FT-NIR spectroscopy combined with partial least-squares regression models (Krongtaew et al., 2010b).

Applications of Raman microspectroscopy in plant cell wall research: Classical dispersive Raman spectrometers are usually composed of laser with wavelength in the visible range for excitation, a dispersive spectrometer and a charge coupled device detector (CCD) for detection. This system is often coupled to a confocal microscope equipped with objectives with high numerical apertures to achieve high spatial resolution (Gierlinger and Schwanninger, 2007; Smith and Dent, 2005). For example, a Raman spectrometer with a 514.5 nm laser was used to study the concentration of lignocellulosics in the cell corner middle lamella of both birch and spruce (Tirumalai et al., 1996). However, for plant cell wall research, the strong autofluorescence from lignocellulosic materials may mask the Raman spectra. Therefore, near infrared Fourier Transform Raman spectrometers (NIR-FT Raman) with laser radiation at 1064 nm coupled with interferometers is often utilized for cell wall analysis, because fluorescence is much less in this region. FT-Raman was used to characterize cell wall components of milled black spruce wood (Agarwal and Ralph, 1997) and in various anatomical parts of flax (Himmelsbach and Akin, 1998). Alternatively, the excitation laser can be shifted to the UV region (below 300 nm) where fluorescence is nearly absence. The utilization of UV excitation also leads to the resonance enhancement of aromatic structures and thus is very sensitive for lignin analysis. Nuopponen et al. (2004) has used UV resonance Raman (UVRR) spectroscopy to analyze the extractable compounds and solid wood samples of Scots pine. Saariaho et al. (2003) has characterized Raman peaks for p-hydroxyphenyl, guaiacyl and syringyl structures of lignin using UVRR. Raman peak assignments for major cell wall components were summarized by Agarwal and Ralph (1997) and Adapa et al. (2009). In addition to compositional analysis, a Raman spectroscopy-based method was also developed to obtain mechanical properties of plant cell walls (Ryden et al., 2003), which may serve as an indicator for the ease of cell wall deconstruction during pretreatment or enzymatic saccharification.

Like IR microspectroscopy, one of the major advantages of Raman technique exists in its chemical mapping capabilities. With a much higher lateral spatial resolution than IR (~ 1 μm), ultrastructure of plant cell walls with the corresponding compositional information can be revealed. Using confocal Raman imaging, the distribution of lignin and cellulose in black spruce wood was investigated (Agarwal, 2006) and changes of molecular composition in secondary plant cell wall tissues of poplar wood were illustrated (Gierlinger and Schwanninger, 2006). Raman imaging of Arabidopsis thaliana, one of the most important model plants, was recently demonstrated (Schmidt et al., 2009). Principal component analysis (PCA) and partial least square (PLS) modeling can be incorporated in image analysis to provide a more detailed comparison of cell wall compositions at different mapped regions (Gierlinger et al., 2008b). Raman imaging technique was also used to compare lignification in wild type and lignin-reduced 4-coumarate-CoA ligase (4CL) transgenic Populus trichocarpa stem wood (Schmidt et al., 2009). Raman imaging was further implemented to provide a more complete picture of the effects of alkaline treatment on Miscanthus x giganteus, a potential energy crop and a model lignocellulosic material. Longitudinal and transversal-section images of the parenchyma cells were generated, which revealed that lignin is removed preferentially from the inner surface of the cell wall and that cellulose is largely undisturbed (Chu et al., 2010).

Very recently our laboratory has developed a Raman imaging method to provide complete tissue/cell type specific compositional information for the first time (Sun et al., 2011). The method was demonstrated on stem sections of corn stover ranging from the epidermis to the pith area by both one-dimensional and two-dimensional chemical mapping. Lignin and cellulose abundance was determined in various cell types in the following order: sclerenchyma cells and tracheids (~5 times) > epidermal cells (~3 times) > bundle sheath cells > parenchyma cells. Unlike other Raman imaging work only showing the total lignin content, a Raman characterization study was performed to assign peaks for lignin compositions in terms of p-hydroxyphenyl, guaiacyl and syringyl units. Our imaging results have shown that significant amount of p-hydroxyphenyl units are present in the tracheids of corn stover stem, but not in the tracheids of the Eucalyptus globulus stem, which was corroborated by literature data (Galletti et al., 1996; Pinto et al., 2005). Polarized Raman is another very useful tool for plant cell wall research by including polarizers in the optical path. Cao et al. (2006) has demonstrated a Raman study on the net orientation of biomacromolecules in the outer epidermal walls of mature wheat stems by comparing spectra collected with Raman light polarized perpendicular or parallel to the longitudinal axis of the cell. By changing the laser polarization direction in 3° steps, Gierlinger et al. (2010) investigated the dependency between cellulose and laser orientation direction and determined cellulose microfibril angle in S1 and S2 layers of wood samples, which was validated by X-ray diffraction measurement.

Coherent anti-Stokes Raman scattering (CARS) microscopy and stimulated Raman scattering (SRS) microscopy are two types of nonlinear Raman techniques that have become important in plant cell wall research. CARS microscopy exhibits significantly higher sensitivity, faster image acquisition, and reduced susceptibility to fluorescence interference when compared to spontaneous Raman microscopy. Additionally, CARS microscopy possesses the inherent potential to do three-dimensional sectioning due to its nonlinear characteristics. The utilization of CARS imaging to see lignin in cell walls was successfully demonstrated using corn stover as the sample material (Evans and Xie, 2008). Nevertheless, a CARS spectrum diverges from its equivalent spontaneous Raman spectrum as a result of a nonresonant background, leading to challenges in interpreting images. SRS has significant benefits such as producing the same response as spontaneous Raman scattering, having a direct relationship with the concentration of the analyte, and enabling rapid image collection. Saar et al. (2010) achieved the real-time monitoring of the delignification reaction in maize stover using the acid chlorite method by employing stimulated Raman scattering (SRS) with a high level of precision in both space and time. Nevertheless, nonlinear Raman techniques pose significant challenges in terms of cost and restricted availability of non-commercial instruments, as well as the need for system optimization to ensure everyday usability, which restricts operation to highly skilled individuals.

7. CELL WALL ANALYSIS BY 2D AND 3D ELECTRON MICROSCOPY

Plant cell walls consist of intricate networks composed of sugars, lignins, and certain proteins. In addition to the intricate nature of individual plant cell walls, these walls can exhibit significant diversity in terms of composition and structure. This diversity can be observed not just between different groups of plants and species, but also within the same plant, organ, tissue, and cell type. The cell walls are also subject to dynamic changes in their ultrastructure as they grow and differentiate (Carpita and Gibeaut, 1993; Niklas, 2004; Popper, 2008; Sarkar et al., 2009). To gain a comprehensive understanding of the organization of plant cell walls, it is necessary to employ molecular resolution imaging techniques. This knowledge is crucial for the purpose of intelligently modifying cell walls to enhance the production of biofuels from biomass. Electron microscopy enables the examination of molecular structures with extremely high resolution, while optical microscopy techniques are generally restricted by the diffraction limit of the optical microscope and the signal-to-noise ratios observed in autofluorescent samples.

Transmission electron microscopy (TEM) enables a comprehensive examination of the intricate details of cellular ultrastructure. It has been employed in the study of plant cell walls since the 1940s, providing the initial high-resolution insights into their structure (Preston et al., 1948). In order to do TEM examination, it is necessary for the samples to be thin enough for the electron beam to pass through, ideally around 100-300 nm. Additionally, the samples must be viewed in a vacuum, which requires them to be embedded in resin and then thinly sliced. The chemical sample preparation steps used in a conventional protocol have limited preservation effects. These steps include chemical fixation, heavy metal postfixation, staining, and organic solvent dehydration. However, they can result in fixation staining and dehydration artifacts, such as denaturation, aggregation, and extraction of biological material, as well as uneven or preferential staining. Furthermore, contemporary sample preparation processes, which involve targeted staining techniques, are primarily designed to enhance the visualization of cell membrane lipids, nucleic acids, and proteins. However, these protocols are not extensively tuned for carbohydrates and lignin. Histochemical staining methods can be employed to stain specific components of cell walls. For instance, cellulose can be stained using negative staining with uranyl acetate, hemicelluloses can be stained using the PATCO method (Periodic acid - Thiocarbohydrazide - Silver Proteinate), pectins can be stained using ruthenium red, and lignins can be stained using potassium permanganate (Krishnamurthy, 1999). Nevertheless, these stains typically lack high specificity and instead stain several components of the cell wall to varying extents. Accurately distinguishing between cell wall components at high resolutions is challenging using only differential labeling.

Most of the early high-resolution TEM imaging of plant cell walls was done on samples prepared by metal shadowing and surface replication after freeze-fracturing or freeze-etching (Preston et al., 1948, McCann et al., 1990). While in principle no chemical fixatives, no dehydrating agents and no stains are used in this method, thus potentially retaining the samples closer to their native state, the images although potentially of high resolution are restricted to topological structural information in two-dimensions. Moreover, no chemical information can be obtained from this method as the imaging is done on the metal replica and not the biological sample. In recent years, sophisticated cryo-methods have been developed to minimize or completely overcome the limitations of conventional TEM sample preparation methods. Sophisticated cryo-methods such as high pressure freezing, followed by freeze-substitution and resin embedding typically display a superior quality of sample preservation in a much closer to native state (McDonald, 1999; McDonald & Muller-Reichert, 2002, McDonald & Auer 2006). High-pressure freezing followed by vitreous sectioning and cryo-TEM imaging offers preservation of biological samples closest to their native state (Al-Amoudi et al., 2004a, 2004b). These samples not only provide high-resolution structural information, but can also provide chemical information by specific staining or by immunolabeling with target-specific gold-conjugated antibodies. Several monoclonal antibodies and carbohydrate-binding modules (CBM) are being developed against different cell wall components in many laboratories around the world (Knox, 2008; Pattathil et al., 2010), which can be used with TEM analysis to localize the various chemical components of plant cell walls with high precision. All TEM sample preparation methods mentioned above are usually time-consuming and labor intensive, although automatic microwave tissue processors are now commercially available for rapid chemical fixation, dehydration, resin embedding and polymerization.

Apart from sample preparation issues, conventional TEM imaging runs into a few other problems. Due to a small field of view in a TEM, only small representative areas of any sample can be imaged at a time. The sections used in TEM are also very thin (under a micron), which makes imaging larger cellular structures in their entirety an almost impossible task. Use of high-resolution wide-field imaging (montaging) and imaging serial sections can help in covering relatively larger sample areas. Aligning serial sections is specimen-dependant due to the characteristics of each section. Individual sections in a series may have differences in scaling and/or may have non-linear deformation because of sectioning, folding, drying, specimen tilt, and optical distortions of the microscope (Stevens and Trogadis, 1984). A more complicated limitation of conventional TEM is that the images obtained are 2D projections of a 3D volume, which means multiple molecular layers of the sample contribute to the same layer of the image. Such images can be difficult to interpret if the structures of interest are only a few nanometers in dimension and are very closely packed. Plant cell walls are a good example of this imaging problem. They are made up of a tightly packed network of cellulose microfibrils, each microfibril being ~3 nm in diameter (Ohad & Danon, 1964; Frey-Wyssling, 1968; Heyn, 1969; Somerville et al., 2004; Ding and Himmel, 2006). The cellulose microfibrils are tightly surrounded by hemicelluloses such as xyloglucans or arabinoxylans that form hydrogen bonds with the cellulose microfibrils and form cross-links between two neighboring cellulose microfibrils. The matrix space in between cellulose and hemicelluloses is crowded with complex nano-scale molecules of pectins and/or lignins (McCann et. al., 1990; Carpita and Gibeaut, 1993; Somerville et al., 2004). It is extremely challenging to resolve the ultrastructure of plant cell walls in situ by most imaging techniques available currently, including conventional TEM. Atomic force microscopy (AFM) has been successfully used to image plant cell wall ultrastructure at high resolutions (Ding and Himmel, 2006) but the information available is two-dimensional and only topological.

Electron tomography overcomes some of the limitations faced by conventional TEM and AFM. In this method, several hundred two-dimensional TEM images of the same sample are collected by rotation of the sample along the central axis in small increments, which leads to a 3D volume of data, which can be visualized, segmented out to develop realistic models and quantitatively analyzed in 3D. Individual layers within the 3D volume can be separately visualized and analyzed in different planes using sophisticated image analysis software. Different types of algorithms for automated segmentation are being developed, though the currently available algorithms are only reliable for relatively simple image sets. Since individual components of the plant cell walls are typically spaced close to the resolution limit of the data set, manual segmentation is still the most reliable method for analyzing plant cell wall tomograms. EM tomography data of plant cell walls can be used to measure dimensions, orientations and spacing of the different cell wall components. This method paired with biochemical analysis methods like Raman imaging and immunolabeling has the potential to develop precise, comprehensive 3D ultrastructural cell wall model(s) at molecular resolution.

EM tomography has been applied to study the 3D organization of the cellulose microfibrils in the S2 layer of the secondary cell wall in Pinus wood tissue (Xu et al., 2007, 2011), however the samples used in this study had been harshly chemically treated. Cryo electron tomography of vitrified sections of plant tissue can provide preservation closest to their native state. The processing of tomographic datasets (reconstruction, filtering and segmentation) of plant cell wall cryo-sections and faithful model development however, is highly challenging, as the images obtained by cryo-tomography are extremely low in contrast. Furthermore, since plant cell walls contain densely packed nano-scale components, cryo-tomography is not suitable for high-throughput imaging of plant cell walls. Instead high pressure freezing, followed by freeze-substitution and resin embedding is a way to go and will be the key to obtain high-throughput realistic cell wall models. EM tomography of high-pressure frozen, freeze-substituted, resin embedded samples has been used to study at unprecedented resolution the dramatic structural changes during cytokinesis and the assembly process of cell plates during the final stage of cell division, (Seguí-Simarro et al., 2008).

8. ANALYSIS OF BIOMASS USING ACANNING AND TRANSMISSION ELECTRON MICROSCOPY

Apart from an in-depth look at the molecular ultrastructure of different plant cell walls, Scanning Electron Microscopy (SEM) can be utilized to swiftly characterize biomass samples from different pretreatment protocols and while 2D in nature provides a 3D impression of the effect such protocols had on plant cell walls. SEM uses backscattering properties of heavy atom-coated surfaces to determine surface topologies. SEM is strictly a surface technique, and while not yielding information about the inside of an object, it provides fast overviews of large areas at fairly high resolutions with relatively simple sample preparation, allowing rapid screening of many samples. Focused Ion Beam (FIB)/SEM is a promising new method that can provide continuous 3D information over a large depth range though it requires a much more sophisticated sample preparation as compared to conventional SEM and will be discussed further below.

Regarding sample preparation, usually, the best approach is to work with cross sections of plant stems (e.g. Arabidopsis or Brachopodium) that can be cut using a vibrating blade microtome (e.g. Leica VT1000S, Leica Microsystems), resulting in reproducible cross sections. In order to ensure reproducible results multiple sections of each stem are typically cut in water, picked up with a brush or a pair of very fine tweezers and placed on prepared SEM sample stubs. The sample can then be dehydrated and critical-point dried or directly mounted on sample stubs and sputter coated, which allows for better conductivity of the sample and therefore better SEM imaging results.

TEM with plant stem cross sections requires significantly more preparation than SEM (see section above), but yields more quantitative information regarding cell wall thickness and can discriminate between primary and secondary cell wall. We (Li et al. 2010, Çetinkol et al. 2010) have used the described SEM techniques to visualize the effects of different pretreatment techniques designed to break up plant parts before they can be subjected to either enzymes or microbes thus converting them into sugars and subsequently second or third generation biofuels. Both TEM and SEM were used (Yin et al. 2011) to visualize the isolation of plant organelles or changes of the plant structure due to mutations, affecting cell wall properties and susceptibility to deconstruction for second or third generation biofuels synthesis. Furthermore, SEM analysis can be used to study the fracture pattern in samples derived from the mechanical stress screening in order to determine whether the fracture pattern indicate ductile or brittle behavior.

9. ANALYZING MICROBIAL POPULATIONS USING SOPHISTICATED ELECTRON MICROSCOPY TECHNIQUES

A variety of microbial communities are thought to have unique lignocellulolytic capabilities making them of great interest for the development of second and third generation biofuels. Among the natural degraders of lignocellulose are microbial communities that are found e.g. in the gut of wood-degrading insects such as termites, in ruminants, in compost or soil. Electron microscopy with its ability to conduct an analysis at a spatial resolution of about a nanometer allows the investigation of the interplay between the different bacterial cells that constitute a community, its association with the plant material as well as their macromolecular inventory and degradation strategies.

SEM can only be conducted on surfaces, e.g. plant part from the digestive system of a cellulose degrading animal containing the microbial community, or microbes filtered out of a microbial culture suspension. Since SEM imaging typically requires sample fixation, dehydration and critical point drying, followed by sputter coating, the fine details tend to get lost. The microscopy on those samples in a high-resolution

SEM (typically, 2 – 10 kV accelerating voltage are used) can visualize the shape of microorganisms, thus sometimes enabling an identification of certain species, and show both interaction between microbes and between microbes and the surface (e.g. plant material). It can thus be a valuable first insight into the composition and functioning of a microbial community, facilitated by the fact that the technique is rather quick and does not require extensive preparations. However, the technique does not enable a more detailed analysis of the microbial community and is limited to the surface of the samples, thus possibly representing only a small fraction of the sample that might not be typical for the sample in general.

If a more in-depth analysis of the microbial community is the goal, one has to use TEM instead of additional SEM. The analysis of a microbial community with TEM can visualize a lot of interesting traits in this community, including frequent cell-cell interactions, either directly or via a variety of microstructures such as pili and flagella. Even more interesting in this case are the interactions between microbial cells and the plant biomass, which is present in the sample. In studies of such communities (Knierim et al., in preparation) we have found different strategies of cell wall attachment and biomas digestion. An unresolved challenge is the direct identification of species in TEM or SEM. There have been recent advances in this field, combining TEM with Catalized Reporter Deposition Fluorescence In-Situ Hybridization (CARD-FISH), which can identify either groups of bacteria or even certain species (Knierim et al., submitted). However, this technique is extremely tedious and requires some compromises of the TEM imaging quality in order to enable the identification via CARD-FISH.

The described electron microscopy techniques are not only applicable to the understanding of microbial communities but can also be utilized for the analysis of bacteria that have been engineered for certain capabilities. Such capabilities that are developed by synthetic biologists can be the increased production of fuels, the production of different fuels (especially those with longer carbon chains which are more valuable than ethanol) or the assessment of toxicity of the produced biofuels. When combined with tag-based labeling TEM also allows a rational monitoring of protein expression levels, cell-to cell variation and subcellular localization. Thus the application of SEM and TEM to those samples can facilitate the re-engineering of microorganisms that are needed for the production of second and third generation biofuels.

A natural limitation of TEM usually is its coverage of very small sample volumes covered by thin sections. To overcome this problem, FIB/SEM can be employed, which provides a three dimensional view of large volumes of a microbial community at a resolution comparable to electron tomography. For FIB/SEM a similar sample preparation protocol as described above for TEM can be used, but one has to ensure a high internal contrast inside the plastic blocks. We have had very good experience with tannic acid for this purpose. During FIB/SEM imaging one must pay attention to the area of interest careful, as this technique is very time intensive and hence currently can only be applied to a limited number of volumes. Typically, volumes of 10 x 10 x 5 µm can be covered at an estimated resolution around 10–15 nm. The datasets that are produced this way easily get very large (in the several Gigabyte range) and require careful 3D reconstruction and analysis using software packages with those capabilities such as UCSF Chimera (http://www.cgl.ucsf.edu/chimera) or Amira (http://www.amira.com). We are currently drafting a manuscript on the FIB/SEM 3D analysis of the termite hindgut microbial community, where we classify and quantify constituent microbial community member according to size, shape and internal density characteristics, and map out their distribution with respect to the biomass (Knierim et al., in preparation).

An understanding of the functioning of microbial communities that are capable of lignocellulose degradation will ultimately lead to the development of better techniques for lignocellulose degradation in an industrial setting such as a biorefinery for second or third generation biofuels. We are still in the early steps of this understanding due to the high complexity and variability of these microbial communities, but if we can reduce the complexity by identifying a small set of microbes with valuable capabilities, we may be able to speed up this process. As the principles that are present in these microbial communities have been developed by evolution over billions of years, we can assume that they are very energy efficient, thus providing a maximal energy output while taking up a small amount of energy themselves – an important challenge for the design of industrial processes for the production of biofuels

10. CONCISE SUMMARY

Enhancing the characteristics of feedstocks and improving every stage of the deconstruction process and the manufacture of fuel synthesis will result in a reduction in production costs. This is crucial for the substitution of fossil fuels with biofuels. In order to achieve the necessary technological advancements, it is necessary to employ a range of biophysical techniques, usually performed by experts. These techniques can measure the impact of experimental interventions and provide a comprehensive understanding of the physical, chemical, and biological processes involved in the production of lignocellulosic biofuels.

REFERENCES

Adapa P.K., Karunakaran C., Tabil L.G., and Schoenau G.J. 2009. Potential applications of infrared and raman spectromicroscopy for agricultural biomass. Agr Eng Inte: CIGR J 11:1-25

Agarwal U. 2006. Raman imaging to investigate ultrastructure and composition of plant cell walls: Distribution of lignin and cellulose in black spruce wood (Picea mariana) Planta 224:1141-1153

Agarwal U.P., and Ralph S.A. 1997. Ft-raman spectroscopy of wood: Identifying contributions of lignin and carbohydrate polymers in the spectrum of black spruce (Picea mariana). Appl Spectrosc 51:1648-1655

Akiyama K., Chikayama E., Yuasa H., Shimada Y., Tohge T., Shinozaki K., Hirai M.Y., Sakurai T., Kikuchi J., and Saito K. 2008. PRIMe: a Web site that assembles tools for metabolomics and transcriptomics. In Silico Biol 8: 339-345

Al-Amoudi A., Chang J., Leforestie A., McDowall A., Salamin L. M., Norlen L., Richter K., Blanc N. S., Studer D., and Dubochet, J. 2004. Cryo-electron microscopy of vitreous sections. EMBO J 23: 3583–3588

Al-Amoudi A., Norlen L., and Dubochet, J. 2004. Cryo-electron microscopy of vitreous sections of native biological cells and tissues. J Struct Biol 148: 131–135

Alves A., Schwanninger M., Pereira H., and Rodrigues J. 2006. Calibration of nir to assess lignin composition (h/g ratio) in maritime pine wood using analytical pyrolysis as the reference method. Holzforschung 60: 29-31

Atsumi S., Hanai T., and Liao, J.C. 2008. Non-fermentative pathways for synthesis of branched-chain higher alcohols as biofuels. Nature 451: 86-89

Baskin TI, Beemster GTS, Judy-March JE, Marga F. 2004. Disorganization of cortical microtubules stimulates tangential expansion and reduces the uniformity of cellulose microfibril alignment among cells in the root of arabidopsis. Plant Physiol 135(4):2279-2290.

Baucher M., Bernard-vailhé, M.A., Chabbert, B., Besle, J. M., Opsomer, C., Van Montagu M., and Botterman J. 1999. Down-regulation of cinnamyl alcohol dehydrogenase in transgenic alfalfa (Medicago sativa L.) and the effect on lignin composition and digestibility. Plant Mol Biol 39: 437-447

Bertran, M. S., and Dale B. E. 1986. Determination of cellulose accessibility by differential scanning calorimetry. J Appl Polymer Sci 32: 4241-4253

Boudet A. M., Goffner D., Marque C., Teulières C., and Grima-Pettenati J. 1998. Genetic manipulation of lignin profiles: a realistic challenge towards the qualitative improvement of plant biomass. Ag Biotech News Info 10: 295N-304N

Burk D. H., Liu B., Zhong R., Morrison W. H., and Ye Z. H. 2001. A katanin-like protein regulates normal cell wall biosynthesis and cell elongation. Plant Cell 13: 807-828

Cao Y., Shen D., Lu Y., and Huang Y. 2006. A raman-scattering study on the net orientation of biomacromolecules in the outer epidermal walls of mature wheat stems (Triticum aestivum). Ann Bot 97:1091-1094

Carpita N. C., and Gibeaut D. M. 1993. Structural models of primary cell walls in flowering plants: consistency of molecular structure with the physical properties of the walls during growth. Plant J 3: 1–30

Carr G. L. 1999. High-resolution microspectroscopy and sub-nanosecond time-resolved spectroscopy with the synchrotron infrared source. Vib Spectrosc 19: 53-60

Cavalier DM, Lerouxel O, Neumetzler L, Yamauchi K, Reinecke A, Freshour G, Zabotina OA, Hahn MG, Burgert I, Pauly M and others. 2008. Disrupting two Arabidopsis thaliana xylosyltransferase genes results in plants deficient in xyloglucan, a major primary cell wall component. Plant Cell 20(6):1519-1537.

Çetinkol, Ö. P., Dibble D. C., Cheng G., Kent M. S., Knierim B., Auer M., Wemmer, D. E., Pelton J. G., Melnichenko Y.B., Ralph J., Simmons B. A. and Holmes, B. M. 2010. Understanding the impact of ionic liquid pretreatment on eucalyptus. Biofuels 1: 33- 46

Chen L., Carpita N. C., Reiter W. D., Wilson R. H., Jeffries C., and McCann M. C. 1998. A rapid method to screen for cell-wall mutants using discriminant analysis of fourier transform infrared spectra. Plant J 16: 385-392

Chen F., and Dixon R. A. 2007. Lignin modification improves fermentable sugar yields for biofuel production. Nature Biotech 25: 759-761

Chiang V. L., and Funaoka M. 1990. The difference between guaiacyl and guaiacyl-syringyl lignins in their responses to kraft delignification. Holzforschung 44: 309-313

Chu L. Q., Masyuko R., Sweedler J. V., and Bohn P. W. 2010. Base-induced delignification of Miscanthus x giganteus studied by three-dimensional confocal raman imaging. Bioresour Technol 101:4919-4925

Coops H. and Van der Velde G. 1996. Effects of waves on helophyte stands: mechanical characteristics of stems of Phragmites australis and Scirpus lacustris. Aquatic Botany 53: 175-185

Couchman P. R. 1981. The effect of degree of polymerization on glass-transition temperatures. Polym Eng Sci 21: 377-380

Czichos H., Saito T., and Smith L. 2006. Measurement methods for mechanical properties. In Materials measurement methods, Springer, 1: 283-387

Dai C. C., Tao J., Xie F., Dai Y., and Zhao M. 2007. Biodiesel generation from oleaginous yeast Rhodotorula glutinis with xylose assimilating capacity. African J Biotech 6: 2130-2134

Davison B. H., Drescher S. R., Tuskan G. A., Davis M. F., and Nghiem N. P. 2006. Variation of S/G ratio and lignin content in a Populus family influences the release of xylose by dilute acid hydrolysis. Appl Biochem and Biotechnol 130: 427-435.

De Micco V, Aronne G. 2007. Combined histochemistry and autofluorescence for identifying lignin distribution in cell walls. Biotech Histochem 82(4):209-216.

D'Haeze W, Gao M, De Rycke R, Van Montagu M, Engler G, Holsters M. 2007. Roles for azorhizobial nod factors and surface polysaccharides in intercellular invasion and nodule penetration, respectively. Mol Plant Microbe Interact 11(10):999-1008.

Ding, S. and Himmel, M. E. 2006. The maize primary cell wall microfibril: A new model derived from direct visualization. J Agric Food Chem, 54: 597-606

Dokken K. M., Davis L. C. 2007. Infrared imaging of sunflower and maize root anatomy. J Agric Food Chem 55:10517-10530

Dokken K. M., Davis L. C., and Marinkovic N. S. 2005. Use of infrared microspectroscopy in plant growth and development. Appl Spectrosc Rev 40: 301 – 326

Ehara K., Takada D., and Saka S. 2005. GC-MS and IR spectroscopic analyses of the ligninderived products from softwood and hardwood treated in supercritical. J Wood Sci 51: 256-261.

Emmel A., Mathias A. L., Wypych F., and Ramos L. P. 2003. Fractionation of Eucalyptus grandis chips by dilute acid-catalysed steam explosion. Bioresour Technol 86: 105- 115.

Evans C. L., and Xie X. S. 2008. Coherent anti-stokes raman scattering microscopy: Chemical imaging for biology and medicine. Annu Rev Anal Chem 1:883-909

Flynn J. H., and Wall L. A. 1966. A quick, direct method for the determination of activation energy from thermogravimetric data. J Polym Sci Part B: Polym phys 4: 323-328 Frey-Wyssling (1968) The ultrastructure of wood. Wood Sci Tech 2: 73-83

Fromm J, Rockel B, Lautner S, Windeisen E, Wanner G. 2003. Lignin distribution in wood cell walls determined by tem and backscattered sem techniques. J Struct Biol 143(1):77-84.

Galletti G. C., Bocchini P., Smacchia A. M., and Reeves J. B. 1996. Monitoring phenolic composition of maturing maize stover by high performance liquid chromatography and pyrolysis/gas chromatography/mass spectrometry. J Sci Food Agric 71:1-9

Galletti G. C., Reeves J. B., and Bocchini P. 1997. Analytical pyrolysis as a tool to determine chemical changes in maize stovers during growth. J anal and appl pyrolysis 39: 105-114

Ghetti P., Ricca L., and Angelini L. 1996. Thermal analysis of biomass and corresponding pyrolysis products. Fuel 75: 565-573

Gierlinger N., Goswami L., Schmidt M., Burgert I., Coutand C., Rogge T., and Schwanninger M. 2008a. In situ ft-ir microscopic study on enzymatic treatment of poplar wood cross-sections. Biomacromolecules 9:2194-2201

Gierlinger N., Luss S., König C., Konnerth J., Eder M., and Fratzl P. 2010. Cellulose microfibril orientation of picea abies and its variability at the micron-level determined by raman imaging. J Exp Bot 61:587-595

Gierlinger N., Sapei L., and Paris O. 2008b. Insights into the chemical composition of Equisetum hyemale by high resolution raman imaging. Planta 227:969-980

Gierlinger N., and Schwanninger M. 2006. Chemical imaging of poplar wood cell walls by confocal raman microscopy. Plant Physiol 140:1246-1254

Gierlinger N., and Schwanninger M. 2007. The potential of raman microscopy and raman imaging in plant research. Spectrosc-Int J 21:69-89

Goddard T.D., Huang C.C., and Ferrin T.E. 2004 Visualizing density maps with UCSF Chimera, J. Struct Biol. 157: 281-287

Grethlein H. E., and Converse A. O. 1991. Common aspects of acid prehydrolysis and steam explosion for pretreating wood. Bioresour Technol 36: 77-82

Grünwald C, Ruel K, Kim YS, Schmitt U. 2002. On the cytochemistry of cell wall formation in poplar trees. Plant Biol 4(1):13-21. Guo D., Chen F., Wheeler J., Winder J., Selman S., Peterson M., and Dixon R. A. 2001. Improvement of in-rumen digestibility of alfalfa forage by genetic manipulation of lignin O-methyltransferases. Transgenic Research 10: 457-464

Hall M., Bansal P., Lee J. H., Realff M. J., and Bommarius A. S. 2010. Cellulose crystallinity – a key predictor of the enzymatic hydrolysis rate. FEBS Journal 277: 1571-1582

Harley C. D. G., and Bertness M. D. 1996. Structural interdependence: An ecological consequence of morphological responses to crowding in marsh plants. Functional Ecology 10: 654-661

Heraud P., Caine S., Sanson G., Gleadow R., Wood B. R., and McNaughton D. 2007. Focal plane array infrared imaging: A new way to analyse leaf tissue. New Phytologist 173:216-225

Heyn, A. N. J.1969 The Elementary Fibril and Supermolecular Structure of Cellulose in Soft Wood Fiber. J. Ultra. Res. 26: 52-68

Himmel M. E., Ding S., Johnson D. K., Adney W. S., Nimlos M. R., Brady J. W., and Foust T. D. 2007. Biomass recalcitrance: Engineering plants and enzymes for biofuels production. Science 315: 804 -807

Himmelsbach D. S., and Akin D. E. 1998. Near-infrared fourier-transform raman spectroscopy of flax (Linum usitatissimum l.) stems. J Agric Food Chem 46:991-998

Hu Y., Zhong R., Morrison W. H. III, and Ye Z. 2003. The Arabidopsis RHD3 gene is required for cell wall biosynthesis and actin organization. Planta 217: 912-921

Huyen T. L., Ramond C., Dheilly R.M., and Chabbert B. 2010. Effect of harvesting date on the composition and saccharification of Miscanthus x giganteus. Bioresour Technol 101: 8224-8231

Ibarra D., Chávez M. I., Rencoret J., Río J. C. D., Gutiérrez A., Romero J., Camarero S., Martínez M. J., Jiménez-Barbero J., and Martínez A. T. 2007. Lignin modification during eucalyptus globulus kraft pulping followed by totally chlorine-free bleaching: A two-dimensional nuclear magnetic resonance, fourier transform infrared, and pyrolysis-gas chromatography/mass spectrometry study. J Agric Food Chem 55: 3477-3490

Ibrahim M. N. M., and Pearce G. R. 1983. Effects of chemical treatments combined with high-pressure steaming on the chemical composition and in vitro digestibility of crop by-products. Agricultural Wastes 7: 235-250

Igathinathane C., Womac A. R., Sokhansanj S., and Narayan S. 2008. Knife grid size reduction to pre-process packed beds of high- and low-moisture switchgrass. Bioresour Technol 99: 2254-2264

Jackson L. A., Shadle G. L., Zhou R., Nakashima J., Chen F., and Dixon R. A. 2008. Improving saccharification efficiency of Alfalfa stems through modification of the terminal stages of monolignol biosynthesis. Bioenerg Res 1: 180-192

Jaffe M., Collins G., and Mencze J. 2006. The thermal analysis of fibers in the twenty first century: From textile, industrial and composite to nano, bio and multi-functional. Thermochim Acta 442: 95-99

Jung S., Foston M., Sullards M. C., and Ragauskas A. J. 2010. Surface characterization of dilute acid pretreated Populus deltoides by ToF-SIMS. Energy Fuels 24: 1347-1357

Kaloustian J., El-Moselhy T. F., and Portugal H. 2003. Chemical and thermal analysis of the biopolymers in thyme (Thymus vulgaris). Thermochim Acta 401: 77-86

Keppler B. D., and Showalter A. M. 2010. IRX14 and IRX14-LIKE, Two Glycosyl transferases involved in Glucuronoxylan biosynthesis and drought tolerance in Arabidopsis. Molecular Plant 1: 1-8

Kim U., Eom S. H., and Wada M. 2010. Thermal decomposition of native cellulose: Influence on crystallite size. Polym Degrad Stab 95: 778-781

Knierim B, Lin M., Desai M, van Leer B., Goddard T.D., Hugenholtz P., McDonald K.L., Webb R.I., Auer M. (2011) Multiscale three-dimensional Organization of the Termite Hindgut Elucidated by FIB/SEM, in preparation Knierim B, Luef B., Wilmes P., Webb R.I., Auer M., Comolli L.R., Banfield J.F. (2011) Correlative microscopy for phylogenetic and ultrastructural characterization of microbial communities, submitted to ISME J. Knox J.P. 2008 Revealing the structural and functional diversity of plant cell walls. Curr Opin Plant Biol 11: 308-313

Krishnamurthy KV. 1999. Methods in cell wall cytochemistry. Boca Raton, FL: CRC Press. pp: 190-220

Krongtaew C., Messner K., Ters T., and Fackler K. 2010a. Characterization of key parameters for biotechnological lignocellulose conversion assessed by ft-nir spectroscopy. Part i: Qualitative analysis of pretreated straw. BioResources 5: 2063-2080

Krongtaew C., Messner K., Ters T., and Fackler K. 2010b. Characterization of key parameters for biotechnological lignocellulose conversion assessed by ft-nir spectroscopy. Part ii: Quantitative analysis by partial least squares regression. BioResources 5: 2081-2096.

Kubo S., and Kadla J. F. 2005. Hydrogen Bonding in Lignin: A fourier transform infrared model compound study. Biomacromolecules 6: 2815-2821

Lee C., Teng Q., Huang W., Zhong R., and Ye Z. 2009. Down-regulation of PoGT47C expression in Poplar results in a reduced Glucuronoxylan content and an increased wood digestibility by cellulase. Plant Cell Physiol 50: 1075 -1089

Lee C., Teng Q., Huang W., Zhong R., and Ye Z. 2010. The Arabidopsis family GT43 Glycosyltransferases form two functionally nonredundant groups essential for the elongation of Glucuronoxylan backbone. Plant Physiol 153: 526-541

Li C., Knierim B., Manisseri C., Arora R., Scheller H. V., Auer M., Vogel K. P., Simmons B. A., and Singh S. 2010. Comparison of dilute acid and ionic liquid pretreatment of switchgrass: Biomass recalcitrance, delignification and enzymatic saccharification. Bioresour Technol 101: 4900-4906

Li Y., Qian Q., Zhou Y., Yan M., Sun L., Zhang M., Fu Z., Wang Y., Han B., Pang X., Chen M., and Li J. 2003. BRITTLE CULM1, which encodes a COBRA-Like protein, affects the mechanical properties of rice plants. Plant Cell 15: 2020-2031

Lin K. W., Ladisch M. R., Voloch M., Patterson J. A., and Noller C. H. 1985. Effect of pretreatments and fermentation on pore size in cellulosic materials. Biotechnol Bioeng 27: 1427-1433

Lu F., and Ralph J. 1997. DFRC method for lignin analysis. 1. New method for β-aryl ether cleavage: lignin model studies. J Agric Food Chem 45: 4655-4660

Lu F., and Ralph J. 1998. The DFRC method for lignin analysis. 2. Monomers from isolated lignins. J Agric Food Chem 46: 547-552

Lu F., and Ralph J. 1999. The DFRC method for lignin analysis. 7. Behavior of cinnamyl end groups. J Agric Food Chem 47: 1981-1987

Lu F.C., and Ralph J. 1997. Derivatization followed by reductive cleavage (DFRC method), a new method for lignin analysis: Protocol for analysis of DFRC monomers. J Agric Food Chem 45: 2590-2592

Lu X., Vora H., and Khosla C. 2008. Overproduction of free fatty acids in E. coli: Implications for biodiesel production. Metabolic Eng 10: 333-339

Marita J. M., Ralph J., Hatfield R. D., and Chapple C. 1999. NMR characterization of lignins in Arabidopsis altered in the activity of ferulate 5-hydroxylase. Proceedings of the National Academy of Sciences of the United States of America 96: 12328 -12332

McCann M, Stacey N, Wilson R, and Roberts K. 1993. Orientation of macromolecules in the walls of elongating carrot cells. J Cell Sci 106:1347-1356

McCann, M.C., Wells, B. and Roberts, K. 1990. Direct visualization of cross-links in the primary plant cell wall. J Cell Sci 96: 323–334

McDonald, K. 1999. High-pressure freezing for preservation of high resolution fine structure and antigenicity for immunolabeling. Electron Microscopy Methods and Protocols (ed. by N.Hajibagheri). Humana Press, Totowa, NJ. pp. 77–97.

McDonald Kent L, and Auer Manfred 2006. High-pressure freezing, cellular tomography, and structural cell biology. Biotechniques. 41:137-143

McDonald, K. and Müller-Reichert, T. (2002) Cryomethods for thin section electron microscopy. Meth. Enzymol. 351: 96–123

Mills T. Y., Sandoval N. R., and Gill R. T. 2009. Cellulosic hydrolysate toxicity and tolerance mechanisms in Escherichia coli. Biotechnol for Biofuels 2: 1-11

Mouille G., Robin S., Lecomte M., Pagant S., and Höfte H. 2003. Classification and identification of arabidopsis cell wall mutants using fourier-transform infrared (ftir) microspectroscopy. Plant J 35: 393-404

Niklas K. J. 2004. The cell walls that bind the tree of life. BioScience 54: 831–841 Nuopponen M., Willför S., Jääskeläinen A. S., Sundberg A., and Vuorinen T. 2004. A uv resonance raman (uvrr) spectroscopic study on the extractable compounds of scots pine (pinus sylvestris) wood: Part i: Lipophilic compounds. Spectrochim Acta A 60: 2953-2961

Ohad, I. and Danon, D. (1964) On the dimensions of cellulose microfibrils. J Cell Biol 22: 302-305

Papatheofanous M. G., Billa E., Koullas D. P., Monties B., and Koukios E. G. 1995. Two-stage acid-catalyzed fractionation of lignocellulosic biomass in aqueous ethanol systems at low temperatures. Bioresour Technol 54: 305-310

Pattathil S., Avci U., Baldwin D., Swennes A. G., McGill J. A., Popper Z., Bootten T., Albert A., Davis R. H., Chennareddy C., Dong R., O'Shea B., Rossi R. Leoff C., Freshour G., Narra R., O'Neil M., York W. S. and Hahn M. G. 2010. A comprehensive toolkit of plant cell wall glycan-directed monoclonal antibodies. Plant Physiol 153: 514–525

Paul S. A., Oommen C., Joseph K., Mathew G., and Thomas S. 2010. The role of interface modification on thermal degradation and crystallization behavior of composites from commingled polypropylene fiber and banana fiber. Polym Compos 31: 1113-1123

Pecina R., Burtscher P., Bonn G., and Bobleter O. 1986. GC-MS and HPLC analyses of lignin degradation products in biomass hydrolyzates. Fresenius' J Anal Chem 325: 461-465

Pena M. J., Zhong R., Zhou G., Richardson E. A., O'Neill M. A., Darvill A. G., York W. S., and Ye Z. 2007. Arabidopsis irregular xylem8 and irregular xylem9: Implications for the Complexity of Glucuronoxylan Biosynthesis. Plant Cell 19: 549-563

Peng J., Lu F., and Ralph J. 1998. The DFRC method for lignin analysis. 4. lignin dimers isolated from DFRC-degraded Loblolly Pine wood. J Agric Food Chem 46: 553-560

Persson S, Caffall KH, Freshour G, Hilley MT, Bauer S, Poindexter P, Hahn MG, Mohnen D, Somerville C. 2007a. The arabidopsis irregular xylem8 mutant is deficient in glucuronoxylan and homogalacturonan, which are essential for secondary cell wall integrity. Plant Cell 19(1):237-255.

Persson S, Paredez A, Carroll A, Palsdottir H, Doblin M, Poindexter P, Khitrov N, Auer M, Somerville CR. 2007b. Genetic evidence for three unique components in primary cell-wall cellulose synthase complexes in arabidopsis. Proc Natl Acad Sci USA 104(39):15566-15571.

Petterson E.F., Goddard T.D., Huang C.C., Couch G.S., Greenblatt D.M., Meng E.C. Ferrin T.E. 2004 UCSF Chimera – a visualization systems for exploratory research and analysis.

Pinto PC, Evtuguin DV, Neto CP. 2005. Effect of structural features of wood biopolymers on hardwood pulping and bleaching performance. Ind Eng Chem Res 44: 9777-9784

Popper Z. A. 2008. Evolution and diversity of green plant cell walls. Curr Opin Plant Biol 11: 286–292

Prat R., and Paresys G. 1989. Multiple use apparatus for cell wall extensibility and cell elongation studies. Plant Physiol Biochem 27: 955-962

Preston, R. D., Nicolai, E., Reed, R. and Millard, A. 1948. An electron microscope study of cellulose in the wall of Valonia ventricosa. Nature 162: 665–667

Ralph J. 2007. Perturbing ligninfication. In The Compromised Wood Workshop 2007, K. Entwistle, P.J. Harris and J. Walker, Eds., Wood Technology Research Centre, University of Canterbury, New Zealand, Canterbury, pp. 85-112

Ralph J., and Hatfield R. D. 1991. Pyrolysis-GC-MS charecterization of forage ,aterials. J Agric Food Chem 39: 1426-1437

Ralph J., and Lu F. 1998. The DFRC Method for Lignin Analysis. 6. A Simple Modification for identifying natural acetates on lignins. J Agric Food Chem 46: 4616-4619

Reh U., Kraepelin G., and Lamprecht I. 1987. Differential scanning calorimetry as a complementary tool in wood biodegradation studies. Thermochim Acta 119: 143-150

Reiter W., Chapple C., and Somerville C. R. 1993. Altered growth and cell walls in a fucosedeficient mutant of arabidosis. Science 261: 1032-1035

Ryden P., Sugimoto-Shirasu K., Smith A. C., Findlay K., Reiter W., and McCann M. C. 2003. Tensile properties of Arabidopsis cell walls depend on both a Xyloglucan crosslinked microfibrillar network and Rhamnogalacturonan II-Borate complexes. Plant Physiol 132: 1033-1040

Saar B. G., Zeng Y., Freudiger C. W., Liu Y. S., Himmel M. E., Xie X. S., Ding S. Y. 2010. Label-free, real-time monitoring of biomass processing with stimulated raman scattering microscopy. Angew Chem 122: 5608-5611.

Saariaho A-M, Jääskeläinen A-S, Nuopponen M, Vuorinen T. 2003. Ultra violet resonance raman spectroscopy in lignin analysis: Determination of characteristic vibrations of p-hydroxyphenyl, guaiacyl, and syringyl lignin structures. Appl Spectrosc 57: 58-66

Sarkar P., Bosneaga E. and Auer M. 2009. Plant cell walls throughout evolution: towards a molecular understanding of their design principles. J Exp Bot 60: 3615-3635

Samuel R., Pu Y., Raman B., and Ragauskas A. J. 2010. Structural characterization and comparison of Switchgrass ball-milled lignin before and after dilute acid pretreatment. Appl Biochem Biotechnol 162: 62-74

Schmidt M., Schwartzberg A., Perera P., Weber-Bargioni A., Carroll A., Sarkar P., Bosneaga E., Urban J., Song J., Balakshin M., Capanema E. A., Auer, M., Adams P. D., Chiang V. L. and Schuck P. J. 2009. Label-free in situ imaging of lignification in the cell wall of low lignin transgenic populus trichocarpa. Planta 230: 589-597.

Schmidt M., Schwartzberg A. M., Carroll A., Chaibang A., Adams P. D., and Schuck PJ. 2010. Raman imaging of cell wall polymers in Arabidopsis thaliana. Biochem Biophys Res Commun 395: 521-523.

Seguí-Simarro, J.M., Otegui, M. S., Austin, J. R. and Staehelin, A. L. 2008. Plant cytokinesis – Insights gained from electron tomography studies. In: Verma DPS, Hong Z (eds) Cell division control in plants. Springer, Berlin/Heidelberg, pp 251–287

Serapiglia M. J., Cameron K. D., Stipanovic A. J., and Smart L. B. 2008. High-resolution thermogravimetric analysis for rapid characterization of biomass composition and selection of shrub willow varieties. Appl Biochem Biotechnol 145: 3-11

Sewalt V. J. H., Glasser W. G., and Beauchemin K. A. 1997. Lignin impact on fiber degradation. 3. reversal of inhibition of enzymatic hydrolysis by chemical modification of lignin and by additives. J Agric Food Chem 45: 1823-1828

Shao S., Jin Z., Wen G., and Liyama K. 2009. Thermo characteristics of steam-exploded bamboo (Phyllostachys pubescens) lignin. Wood Sci Technol 43: 643-652

Singh S, Simmons BA, Vogel KP. 2009. Visualization of biomass solubilization and cellulose regeneration during ionic liquid pretreatment of switchgrass. Biotechnol Bioeng 104(1):68-75.

Smith E, Dent G. 2005. Modern raman spectroscopy: A practical approach. Chichester, England: John Wiley & Sons Ltd.

Soares, S., Cammino G., and Levchick S. 1995. Comparative study of the thermal decomposition of pure cellulose and pulp paper. Polym Degrad Stab 49: 275-283

Somerville, C., Bauer, S., Brininstool, G., Facette, M., Hamann, T., Milne, J., Osborne, E., Paredez, A., Persson, S., Raab, T., Vorwerk, S., and Youngs, H. (2004) Toward a systems approach to understanding plant-cell walls. Science 306: 2206-2211

Stevens, J.K. and Trogadis, J. 1984. Computer-assisted reconstruction from serial electron micrographs: a tool for the systematic study of neuronal form and function. Advan Cell Neurobiol 5: 341–369

Stewart D, Wilson H. M., Hendra P. J., and Morrison I. M. 1995. Fourier-transform infrared and raman spectroscopic study of biochemical and chemical treatments of oak wood (Quercus rubra) and barley (Hordeum vulgare) straw. J Agric Food Chem 43: 2219-2225

Sun L., and Simmons B. A, Singh S. 2011. Understanding tissue specific compositions of bioenergy feedstocks through hyperspectral raman imaging. Biotechnol Bioeng 108: 286-295 Tang, Y., Pingitore F., Mukhopadhyay A., Phan R., Hazen T. C., and Keasling J. D. 2007. Pathway confirmation and flux analysis of central metabolic pathways in desulfovibrio vulgaris hildenborough using gas chromatography-mass spectrometry and fourier transform-ion cyclotron resonance mass spectrometry. J Bacteriol 189: 940-949

Tirumalai V, Agarwal U, Obst J. 1996. Heterogeneity of lignin concentration in cell corner middle lamella of white birch and black spruce. Wood Sci Technol 30(2):99-104.

Tsujiyama S., and Miyamori A. 2000. Assignment of DSC thermograms of wood and its components. Thermochim Acta 351: 177-181

Vailhé M. A. B., Besle J. M., Maillot M. P., Cornu A., Halpin C., and Knight M. 1998. Effect of down-regulation of cinnamyl alcohol dehydrogenase on cell wall composition and on degradability of tobacco stems. J Sci Food Agric 76: 505-514

Vinzant T., Ehrman C., Adney W., Thomas S., and Himmel M. 1997. Simultaneous saccharification and fermentation of pretreated hardwoods. Appl Biochem Biotechnol 62: 99-104

Wilkinson J.M., and Santillana R. G. 1978. Ensiled alkali-treated straw. I. Effect of level and type of alkali on the composition and digestibility in vitro of ensiled barley straw. Anim Feed Sci Technol 3: 117-132

Wilson RH, Smith AC, Kačuráková M, Saunders PK, Wellner N, Waldron KW. 2000. The mechanical properties and molecular dynamics of plant cell wall polysaccharides studied by fourier-transform infrared spectroscopy. Plant Physiol 124: 397-406

Ximenes E., Kim Y., Mosier N., Dien B., and Ladisch M. 2010. Inhibition of cellulases by phenols. Enzyme Microb Technol 46: 170-176

Ximenes E., Kim Y., Mosier N., Dien B., and Ladisch M. 2011. Deactivation of cellulases by phenols. Enzyme Microbl Technol 48: 54-60

Xu, P, Donaldson, L. A., Gergely, Z. R, and Staehelin, L. A. 2007. Dual-axis electron tomography: a new approach for investigating the spatial organization of wood cellulose microÞbrils. Wood Sc and Tech 41: 101–116

Xu, P., Liu, H., Donaldson, L. A., and Zhang, Y. 2011. Mechanical performance and cellulose microfibrils in wood with high S2 microfibril angles. J Mater Sci 46: 534-540

Yin, L., Verhertbruggen Y., Oikawa A, Manisseri C, Knierim B, Prak L., Krüger Jensen J., Knox J.P., Auer M., Willats W.G.T. Scheller H. V. (2011). "The Cooperative Activities of CSLD2, CSLD3 and CSLD5 are Required for Normal Arabidopsis Development." Molecular Plant, in press. Yu M., Womac A. R., Igathinathane C., Ayers P.D., and Buschermohle M.J. 2006. Switchgrass ultimate stresses at typical biomass conditions available for processing. Biomass Bioenergy 30: 214-219

Yu P, McKinnon JJ, Christensen CR, Christensen DA, Marinkovic NS, Miller LM. 2003. Chemical imaging of microstructures of plant tissues within cellular dimension using synchrotron infrared microspectroscopy. J Agric Food Chem 51: 6062-6067

Zhong R., Taylor J. J., and Ye Z. H. 1997. Disruption of interfascicular fiber differentiation in an Arabidopsis mutant. Plant Cell 9: 2159-2170

Zhong R., Burk D. H., Morrison W. H., and Ye Z. 2002. A kinesin-like protein is essential for oriented deposition of cellulose microfibrils and cell wall strength. Plant Cell 14: 3101-3117

Zhong R., Burk D. H., Morrison W. H., and Ye Z. 2004. FRAGILE FIBER3, an Arabidopsis gene encoding a Type II Inositol Polyphosphate 5-Phosphatase, is required for secondary wall synthesis and Actin organization in fiber cells. Plant Cell 16: 3242-3259

Zhong R., Richardson E., and Ye Z. 2007. Two NAC domain transcription factors, SND1 and NST1, function redundantly in regulation of secondary wall synthesis in fibers of Arabidopsis. Planta 225: 1603-1611

CHAPTER-22

INVESTIGATION OF THE RATE OF DECOMPOSITION OF PALM OIL WASTE

1. INTRODUCTION: AN OVERVIEW

Malaysia holds the title of being the foremost producer of palm oil, accounting for a significant 43% of global production (Shuit et al., 2009). In addition to palm oil, the palm oil sector produced 169.72 million metric tons of solid waste, accounting for 85.5% of the total biomass waste generated in the country (Khan et al., 2010). The substantial quantity of trash can be transformed into valuable chemical raw materials and fuels, as a result of the environmental issues linked to traditional fossil fuels. Lignocellulosic biomass primarily comprises hemicellulose, cellulose, and lignin, which is widely recognized. The typical composition (wt%) typically ranges from 40-50% cellulose, 20-60% hemicellulose, and 10-25% lignin (Yang et al., 2007). The thermal decomposition of these substances is significant as they impact the fundamental aspects of thermochemical conversion processes, including pyrolysis, combustion, and gasification. The decomposition of these components is extensively examined in the literature. Demirbas et al. (2001) found that the decomposition of lignocellulosic biomass components follows the order of hemicellulose being easier to decompose than cellulose, and lignin being the most difficult to decompose. Yang et al. (2007) suggested distinct temperature ranges for the decomposition of hemicellulose, cellulose, and lignin. Specifically, they proposed decomposition areas of 220-300 °C, 300-340 °C, and >340 °C for hemicellulose, cellulose, and lignin, respectively. Lignin has a prolonged decomposition process as a result of its intricate and robust cross-linked structure (Guo & Lua, 2001). Various methodologies exist for investigating the kinetics of biomass degradation. Thermogravimetric analysis (TGA) is the most widely used and straightforward technique for observing the loss of sample mass over time or temperature, often at a particular heating rate (Luangkiattikhun et al., 2008). TGA offers exceptional accuracy (Várhegyi et al., 2009), rapid data gathering, and excellent reproducibility (Yang et al., 2004) within a clearly defined domain of kinetic control. Only a limited number of studies have been conducted to investigate the kinetics of empty fruit bunch (EFB) and palm shell (PS) using thermogravimetric analysis (TGA). Guo and Lua (2001) examined the impact of sample particle size and heating rate on the pyrolysis process and kinetic parameters for polystyrene (PS).A first-order reaction mechanism was determined for the degradation of PS at various heating speeds.. They also recommended increasing the heating rates to expedite and facilitate the thermal degradation of PS.In their study, Yang et al. (2004) investigated the activation energy required for the degradation of hemicellulose and cellulose in EFB and PS. They examined several temperature ranges to analyze the first-order kinetic reaction.The researchers assessed the mean activation energy and pre-exponential factor by analyzing the single-step breakdown of hemicellulose and cellulose.Luangkiattikhun et al. (2008) examined how the heating rate and particle size of palm oil shell, fibre, and kernel affect their thermogram behavior and kinetic parameters.Upon observation, it was determined that the particle size of palm oil shell does not have a notable impact on the behavior of the thermogram at temperatures below 320 °C.The researchers additionally suggested an nth order reaction mechanism to assess the kinetic parameters using various models.Previous studies on the kinetics of EFB and PS focused on a single heating rate, where the activation energy was only dependent on temperature. The current study assesses the kinetic parameters using a method that necessitates a minimum of three sets of experimental data obtained at varying heating rates.This approach enables the investigation of how the activation energy varies with temperature and conversion under a specific heating rate (Vyazovkin& Wight, 1999).Furthermore, the process of lignin degradation in EFB and PS has not been extensively investigated under conditions of elevated heating rates.The current study investigates the breakdown of lignin in empty fruit bunches (EFB) and palm shells (PS) in order to determine how the lignin content affects the kinetic parameters and rate of decomposition.In addition, the degradation of pure lignin is investigated by the determination of its thermogram and kinetic characteristics.The study examined the kinetics of biomass degradation, specifically focusing on EFB, PS, pure cellulose, and lignin. The investigation was conducted using TGA (thermogravimetric analysis) under non-isothermal circumstances.A detailed thermogram study was presented to comprehend the degradation of cellulose, hemicellulose, and lignin, which are the primary constituents of lignocellulosic biomass.The breakdown kinetics of cellulose and lignin were analyzed using a single-step first order kinetic model.. Meanwhile, the degradation of empty fruit bunches (EFB) and polystyrene (PS) was documented using a single-step nth order kinetic model.The activation energy, pre-exponential factor, and order of reaction were determined and compared to the values given in the literature.There is no text provided.

2. MATERIALS AND METHODS

2.1. Materials preparation and experimental procedure

The fibrous powder form of cellulose and the dark alkali powder form of lignin were acquired from Sigma Aldrich Sdn. Bhd., Malaysia. EFB and PS were obtained from the local palm oil business in Perak, Malaysia. The biomass samples were subjected to drying at a temperature of 105°C, and their weight was continuously monitored at one-hour intervals until the measurements reached a steady state. The samples were subsequently ground to a particle size of 150-250μm. The earlier work (Abdullah et al., 2010) provided the procedure for drying, characterisation, and analysis. The characteristics of biomass, pure cellulose, and lignin are provided in Tables 1 and 2. The tests on biomass breakdown were conducted using the EXSTAR TG/DTA 6300 instrument manufactured by SII in Japan. Nitrogen gas (N2) was employed as an inert gas throughout the whole range of studies, with a consistent flow rate of 100 ml/min. The initial weight utilized in all tests ranged from 3 to 6 mg. Thermogravimetric (TG) studies were conducted using heating rates of 10, 30, and 50 °C/min. The samples underwent an initial heating process, starting at 50 °C and reaching a continuous temperature of 150 °C for 10 minutes to eliminate moisture. Subsequently, they were further heated to a final temperature of 800 °C. The experiments were conducted twice to ensure repeatability. The second experimental measurements did not show any notable deviations.

2.2. Kinetic parameters determination

The biomass decomposition rate under non-isothermal condition is described (Cai & Bi, 2009).

$$\frac{d\alpha}{dT} = \frac{A}{\beta}\exp(-\frac{E}{RT})f(\alpha) \tag{1}$$

f (α) depends on the reaction mechanism as listed in Table 3 and α is the mass fraction reacted.

$$\alpha = \frac{w_0 - w}{w_0 - w_f} \tag{2}$$

Where, 0 and f shows initial and final sample weight

$$g(\alpha) = \int_0^\alpha \frac{d\alpha}{f(\alpha)} = \frac{A}{\beta}\int_0^T \exp(-\frac{E}{RT})dT = \frac{AE}{\beta R}p(\frac{E}{RT}) \tag{3}$$

p(E/RT) function has no exact analytical solution, and therefore different approximations are reported to evaluate the function (Budrugeac et al., 2000). The method developed by Flynn-Wall (Flynn & Wall, 1966), Ozawa (1965) using Doyle's approximation (1961) is the most popular and commonly used by several researchers for biomass decomposition (Cai & Bi, 2009; Hu et al., 2007; Zhouling et al., 2009).

$$p(\frac{E}{RT}) = 0.0048e^{-1.0516(\frac{E}{RT})} \tag{4}$$

(3) is then rearranged for β

$$ln\beta_i = ln(\frac{A_\alpha E_\alpha}{Rg(\alpha)}) - 5.331 - 1.0516\frac{E_\alpha}{RT_{\alpha,i}} \tag{5}$$

Where; w = sample weight (mg); β = heating rate (K/min); R = universal gas constant (8.314 × 10-3 kJ mol-1 K-1)

To determine the activation energy, lnβi vs. 1/Tα,i is plotted for different α values and heating rates (i) to give a straight line and the slope of which gives the activation energy (Doyle, 1961; Ozawa, 1965; Zsakó, &Zsakó, 1980; Flynn & Wall, 1966).

Table 1. Biomass analysis (wt% dry basis)

Analysis	EFB	PS
Volatiles	84.61	81.03
Ash	5.50	4.10
Fixed Carbon (by difference)	9.89	14.87
C	40.73	49.65
H	5.75	6.13
N	1.40	0.41
S	0.22	0.48
O (by difference)	51.90	43.33
Cellulosea	38.30	20.8
Hemicellulosea	35.30	22.7
Lignina	22.10	50.7
a Kelly-Yong et al. (2007)		

Table 2.Pure cellulose and Lignin properties (wt % dry ash free)

Analysis	Cellulose	Lignin
C	43.09	47.71
H	5.96	4.53
N	0.13	0.04
S	0.14	4.24
O (by difference)	50.67	43.40

Table 3.Different *f*(α) and *g*(α) values based on kinetic control regime (Ahmad et al., 2009)

Function	f(α)	g(α)
First order reaction	1-α	-ln(1-α)
Second order reaction	(1-α)2	(1-α)-1-1
Third order reaction	1/2(1-α)3	(1-α)-2-1
nth order reaction	(1-α)	1-(1-α)1-n/1-n

2.3. Kinetic parameter determination model\

The following assumptions are considered for the decomposition of EFB, PS, pure cellulose and lignin.

- Reaction is purely kinetic controlled.
- The decompositions follow single-step processes.
- First order reaction kinetics is considered for pure cellulose and lignin and PS and EFB kinetics are assumed to be nth order.
- No secondary reaction takes place among the gaseous products.

3. RESULTS AND DISCUSSIONS

3.1. Thermogram analysis

The TG and DTG curves for cellulose, lignin, EFB and PS at different heating rates are shown in Figures 1-4. The effect of different heating rate can be described by a lateral shift appeared at high heating rates. These lateral shifts are due to the thermal lag effect between surrounding and biomass particles (Yang et al., 2004; Luangkiattikhun et al., 2008). As a result, conversions are delayed at high heating rates. Thermal lag effect is due to the smallheat conductive property of biomass particles (Zhang et al., 2006).

In the DTG curves (Fig. 1-4, b) for all samples, high decomposition rate was observed at 50 °C/min, which shows the increase of thermal decomposition rate of biomass at high heating rates.

The investigated EFB exhibited the decomposition rate corresponds to -41 wt%/min which is higher than -33 wt%/min of PS at 50 °C/min (see Fig. 5). The high decomposition rate for EFB and PS appeared at 342 and 382 °C, respectively. It is important to consider that 60 wt% of EFB and PS is decomposed at 400 and 429 °C for 50 °C/min. These results depict relatively easy and fast decomposition for EFB as compared to PS. This fast decomposition of EFB may be attributed to the comparatively high volatiles matter and low lignin content present in EFB as compared to PS. Conversely, pure cellulose and lignin decomposition rate is the highest and lowest among all species which is -124 and -19 wt%/min at 50 °C/min, respectively. Furthermore, the highest decomposition rate for cellulose and lignin is observed at 386 and 418 °C.

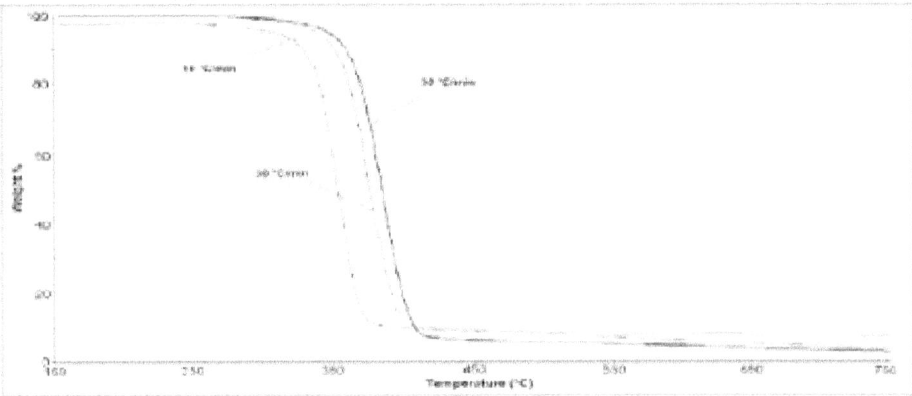

(a)

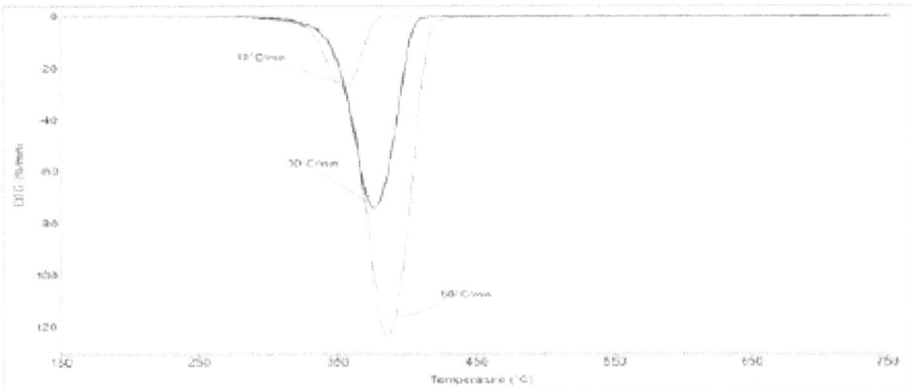

(b)

Figure 1.Cellulose (a) TG and (b) DTG curves

The TG and DTG curves for EFB and PS are given in Figures 3-4. In these figures, the first peak represents the decomposition of hemicellulose. The second peak, which is sharper, gives the highest rate corresponds to the cellulose decomposition. The decomposition range of hemicellulose and cellulose of EFB is between 240-300 °C and 300-340 °C, respectively, at heating rate of 10 °C/min. Decomposition rate of hemicellulose in PS falls almost in the same temperature region as for EFB but higher decomposition range for cellulose (340-370 °C). It is important to consider that the cellulose decomposition rate in PS is in the same temperature region as pure cellulose (340-370 °C at 10 °C/min). The tail at high temperature shows lignin decomposition as found by Yang et al. (2004) and Luangkiattikhun et al. (2008). In the present study, at 10 °C/min, no lignin decomposition was observed for EFB and PS. Similar observation is reported by Yang et al. (2004) for heating rate of 10 °C/min at temperature >340 °C. At higher heating rates, there is some small lignin decomposition observed for EFB and PS which is in the range of 450-530 °C and 680-750 °C, respectively. Different region for lignin decomposition in EFB and PS may be due to different lignin structure and composition in both species.

Among all species, lignin decomposition produced highest residual fraction of ~40% followed by ~27% of EFB and PS and <7% for cellulose, respectively. High residual fraction for lignin shows its high resistance to thermal decomposition which can be seen by its lowest decomposition rate.

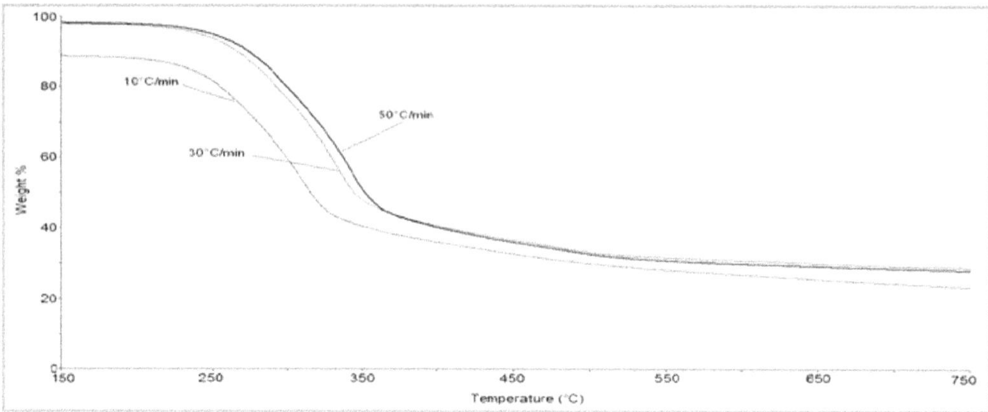

(a)

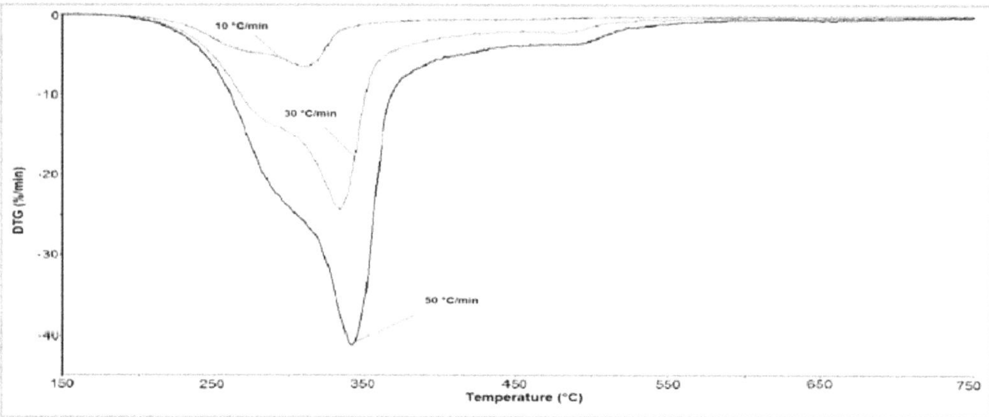

(b)

Figure 2.Lignin (a) TG and (b) DTG curves

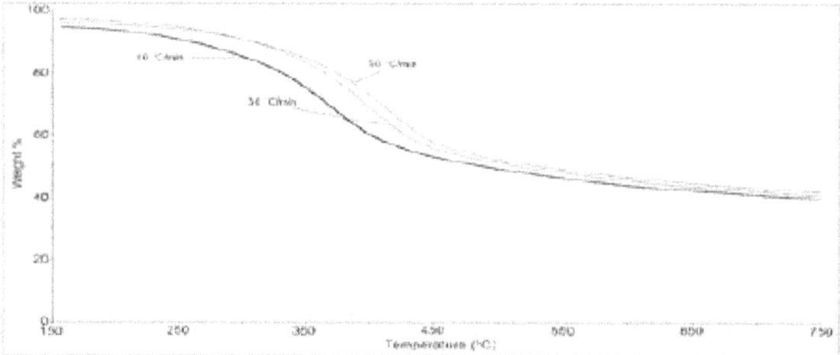

(a)

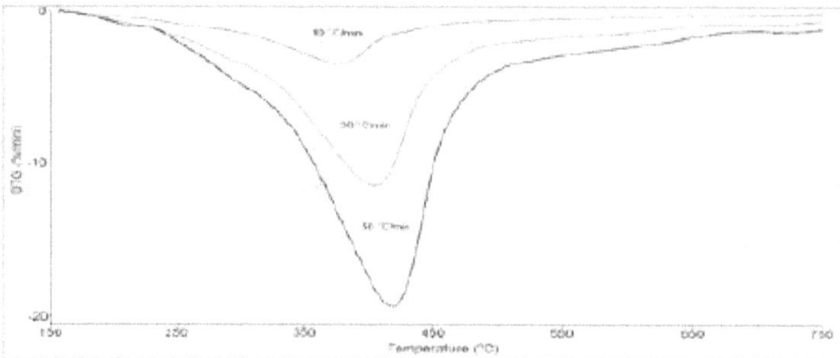

(b)

Figure 3. EFB (a) TG and (b) DTG curves

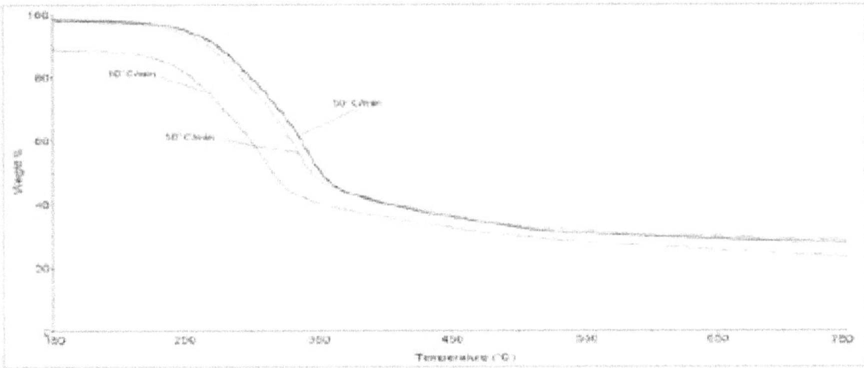

(a)

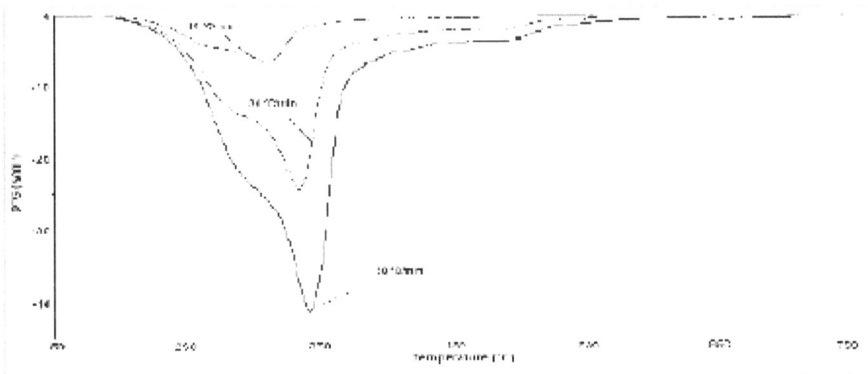

(b)

Figure 4.PS (a) TG and (b) DTG curves

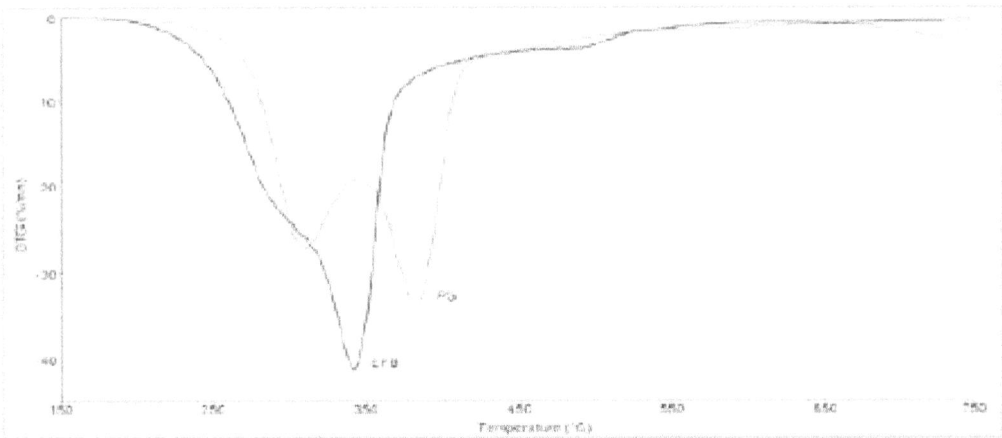

Figure 5.EFB and PS DTG curves at 50 °C/min

3.2. Kinetic parameters

Kinetic parameters are evaluated using equation (5) at a heating rate of 10, 30 and 50 °C/min. Figures 6-9 shows the plot of lnβi against 1/T at different mass fraction reacted (α) for cellulose, lignin, EFB and PS. The higher lnβi values shows high heating rate (50 °C/min) followed by 30 and 10 °C/min. Kinetic parameters evaluated at each α are given in the Tables 4-7. For pure cellulose and lignin, the kinetic parameters are determined at α=0.1 to 0.8 and α=0.1 to 0.6, respectively. EFB and PS kinetic parameters are evaluated at α=0.1 to 0.7. Among all samples, pure lignin produced highest residual fraction and hence kinetic parameter determined up to α=0.6. The correlation coefficients (R2) determined are higher than 0.991 for all cases.

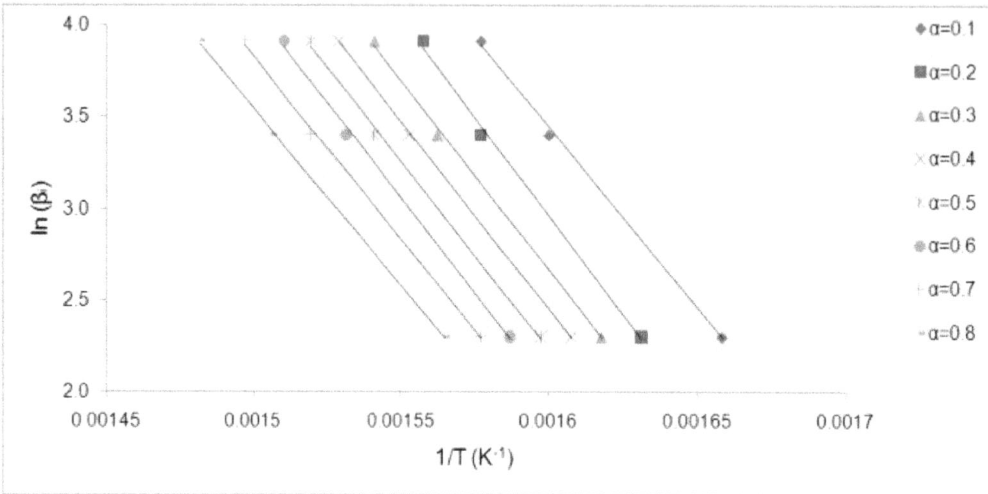

Figure 6.The dependence of ln(βi) on 1/T at a different α values of cellulose (solid lines shows linear fitting)

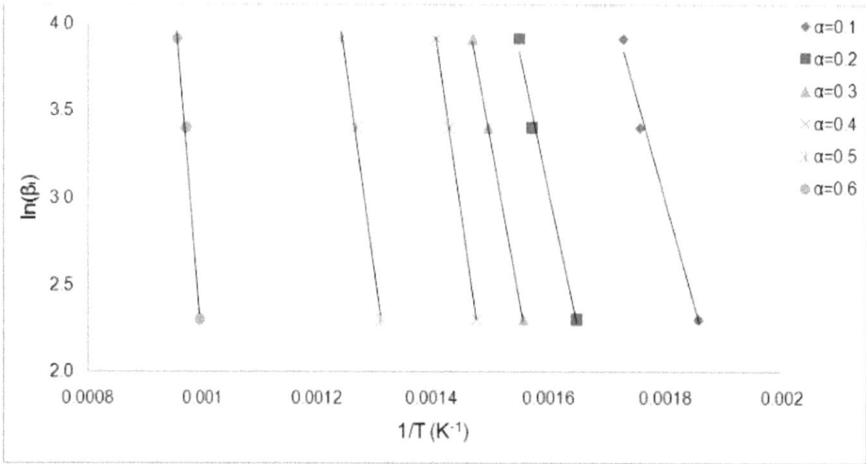

Figure 7.The dependence of ln(βi) on 1/T at a different α values of lignin (solid lines shows linear fitting)

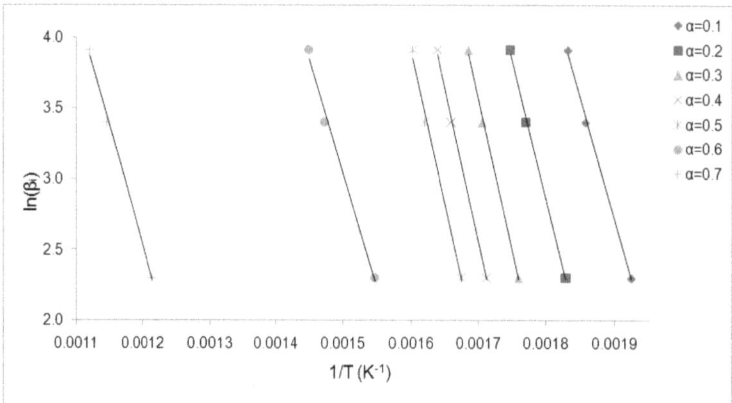

Figure 8.The dependence of ln(βi) on 1/T at a different α values of EFB (solid lines shows linear fitting)

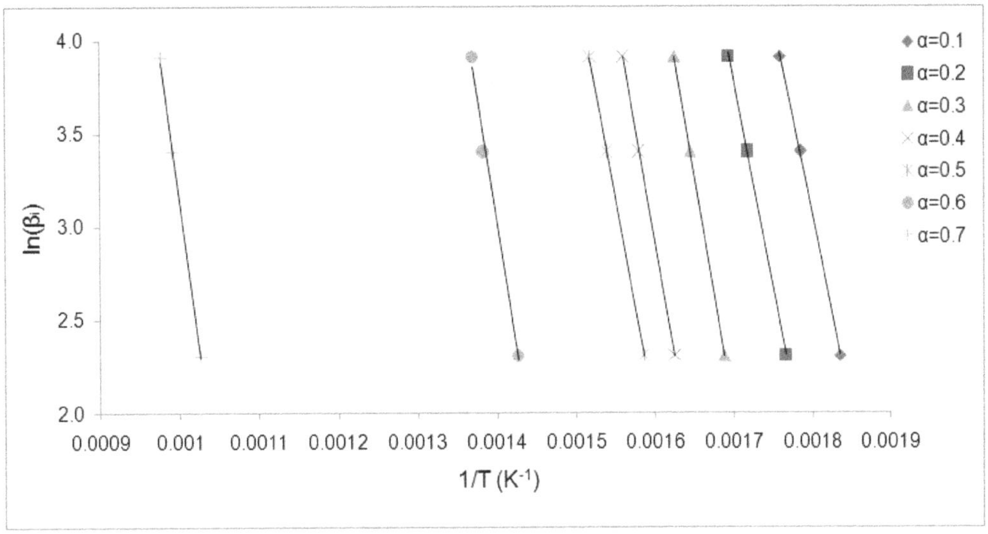

Figure 9.The dependence of ln(βi) on 1/T at a different α values of EFB (solid lines shows linear fitting)

Table 4.Kinetic parameters of cellulose at different α values

α	E (kJ/mol)	R^2	A (m^{-1})
0.1	155.31	0.996	3.9×10^{13}
0.2	170.14	0.996	3.9×10^{13}
0.3	162.95	0.998	1.1×10^{13}
0.4	161.12	0.999	9.2×10^{12}
0.5	161.64	0.997	1.1×10^{13}
0.6	163.66	0.997	1.8×10^{13}
0.7	157.03	0.998	5.3×10^{12}
0.8	151.53	0.991	1.9×10^{13}
Average	160	0.997	1.2×10^{13}

Table 5.Kinetic parameters of lignin at different α values

α	E (kJ/mol)	R2	A (m-1)
0.1	95.01	0.990	9.1E+107
0.2	123.28	0.990	4.4E+109
0.3	143.67	0.999	7.9E+1010
0.4	183.89	0.999	3.8E+1013
0.5	184.22	0.995	1.2E+1012
0.6	318.48	0.992	1.2E+1016
Average	175	0.994	2.06E+1015

Table 6.Kinetic parameters of EFB at different α values

α	E (kJ/mol)	R^2	A (m^{-1})
0.1	135.88	0.998	3.9×10^{13}
0.2	154.59	0.996	6.0×10^{14}
0.3	169.01	0.998	4.5×10^{15}
0.4	171.96	0.998	4.7×10^{15}
0.5	169.92	0.993	9.1×10^{13}
0.6	125.60	0.997	9.4×10^{10}
0.7	131.69	0.997	4.0×10^{9}
Average	151	0.997	1.4×10^{15}

Table 7.Kinetic parameters of PS at different α values

α	E (KJ/mol)	R2	A (m-1)
0.1	165.20	0.999	6.5×1015
0.2	176.23	0.999	1.9×1016
0.3	200.13	0.999	6.2×1017
0.4	194.15	0.998	5.4×1016
0.5	186.27	0.999	7.6×1015
0.6	214.97	0.992	5.9×1016
0.7	255.68	0.998	4.0×1014
Average	199	0.998	1.1×1017

The kinetic parameters for EFB, PS, pure cellulose and lignin are determined as listed in the Table 8 and compared with experimental works reported. The average activation energy of cellulose is 160 kJ/mol and pre-exponential factor is 1.2×1013 m-1 for first order kinetic model. Several researchers found that first order kinetics fit well for cellulose decomposition reaction (Várhegyi et al., 1997; Gronli et al., 1999; Hu et al., 2007).

Activation energy evaluated for cellulose in the present study is in good agreement with the work reported by Zhang et al. (2009) for a first order kinetic model. Nevertheless, the value is comparatively low corresponding to values reported by Varhegyi et al. (1997), Gronli et al. (1999) and Yang et al. (2004). The reason may be different source of cellulose and different method followed by the authors. Secondly, this may be due to relatively high heating rates used in the present study. Meanwhile, Hu et al. (2007) reported high activation energy (233 kJ/mol) using Flynn-Wall-Ozawa method at low heating rates of 2.5, 5, 10 °C/min. Gronli et al. (1999) observed the effect of different heating rates on activation energy for cellulose and found low activation energy at high heating rates. Similarly, much lower activation energies were found by Milosavljevic et al. (1995) at high heating rates. The same effect was also observed for pre-exponential factor. This is due to the heat transfer limitation between sample particles and the surroundings at high heating rates. Varhegyi et al. (1997) suggested utilization of low sample mass to minimize the heat transfer limitations at high heating rates.

Table 8.Comparison of kinetic parameters

Biomass	Kinetic Parameters E (kJ/mol)	A (m-1)	n (-)	R2	Reference
Cellulose	160	1.2×1013	1	0.997	This study
	175	7.16×1012	1	0.999	Zhang et al. (2009)
Lignin	175	2.06×1015	1	0.994	This study
	171	2.73×1025	1	-	Morugun et al. (2008)
EFB	151	1.4×1015	5.3	0.997	This study
	61	3.14×102	1	0.991	Yang et al. (2004)
PS	199	1.1×1017	5.0	0.998	This study
	111	5.27×107	2.54	-	Luangkiattikhun et al. (2008)

Pure lignin is decomposed with average activation energy of 175 kJ/mol and pre-exponential factor of 2.06×1015 m-1 for first order kinetics model. These kinetic parameters are in good agreement with work reported by Murugan et al. (2008).

EFB and PS are decomposed with the average activation energy of 151 and 199 kJ/mol and reaction order of 5 and 5.3, respectively. The corresponding average pre-exponential factors evaluated for EFB and PS are 1.4×1015 and 1.1×1017 m-1, respectively. These values are somehow larger than those reported by Yang et al. (2004) for first order reaction kinetics. Luangkiattikhun et al. (2008) and Guo & Lua, (2001) observed lower activation energy and pre-exponential factor based on single-step nth order and first order kinetic model for PS. However, in these studies, comparatively high values were obtained using two step kinetic models.

Based on kinetic parameters, it is easy to decomposed EFB as compared to PS, pure cellulose and lignin. The order of decomposition from fast to slow is EFB > cellulose > lignin > PS.

4. CONCISE SUMMARY

A study on biomass decomposition was conducted to examine various decomposition regions and evaluate kinetic parameters for EFB and PS. The degradation kinetics of pure cellulose and lignin, which are significant components of biomass, were also investigated. An in-depth analysis was conducted on TG and DTG curves to get insight into the primary decomposition area in EFB, PS, cellulose, and lignin. The kinetic parameters of EFB and PS exhibit greater values in comparison to those reported in the literature. The variation could be attributed to the distinct approaches employed for determining kinetic parameters and the substantially elevated heating rates utilized in the current investigation. PS exhibited greater resistance to decomposition compared to EFB, as indicated by the kinetic parameters. The potential cause may be the considerably elevated lignin level found in PS. The increased lignin concentration was the cause of the low breakdown rate of PS in comparison to EFB. The breakdown rate of pure lignin was the slowest compared to all other species. Furthermore, the lignin component in PS was degraded at a greater temperature compared to EFB due to its higher heating rates.

REFERENCES

Abdullah, S.S. & Yusup, S. (2010). Method for Screening of Malaysian Biomass Based on Aggregated Matrix for Hydrogen Production through Gasification. Journal of Applied Sciences, Vol.10, No.24, pp. 3301-3306, ISSN 1812-5654

Ahmed, I.I. & Gupta, A.K. (2009). Hydrogen Production from Polystyrene Pyrolysis and Gasification: Characteristics and Kinetics. International Journal of Hydrogen Energy, Vol.34, No.15, pp. 6253-6464, ISSN 1520-5029

Budrugeac, P.; Homentcovschi, D. & Segal, E. (2000). Critical Analysis of the Isoconversional Methods for Evaluating the Activation Energy. I. Theoretical Background. Journal of Thermal Analysis and Calorimetry, Vol.63, No.2, pp. 457-463, ISSN 1388-6150

Cai, J. & Bi, L. (2009). Kinetic Analysis of Wheat Straw Pyrolysis using Isoconversional Methods. Journal of Thermal Analysis and Calorimetry, Vol.98, No.1, pp. 325-330, ISSN 1388-6150

Demirbas, A. (2001). Biomass Resource Facilities and Biomass Conversion Processing for Fuels and Chemicals. Energy Conversion and Management, Vol.42, No.11, pp. 1357- 1378, ISSN 0196-8904

Doyle, C.D. (1961). Kinetic Analysis of Thermogravimetric Data. Journal of Applied Polymer Science, Vol.5, No.15, pp. 285-292, ISSN 0021-8995

Flynn, J.H. & Wall, L.A. (1966). A Quick, Direct Method for the Determination of Activation Energy from Thermogravimetric Data. Journal of Polymer Science, Part B: Polymer Letter, Vol.4, No.5, pp. 323-328, ISSN 1542-6254

Gronli, M.; Antal, M.J. & Varhegyi, G. (1999). A Round-Robin Study of Cellulose Pyrolysis Kinetics by Thermogravimetry. Industrial Engineering and Chemistry Research, Vol.38, No.6, pp. 2238-2244, ISSN 1520-5045

Guo, J. & Lua, A.C. (2001). Kinetic Study on Pyrolytic Process of Oil-Palm Solid Waste Using Two-Step Consecutive Reaction Model. Biomass and Bioenergy, Vol.20, No.3, pp. 223-233, ISSN 0961-9534

Hu, S.; Jess, A. & Xu, M. (2007). Kinetic Study of Chinese Biomass Slow Pyrolysis: Comparison of Different Kinetic Models. Fuel, Vol.86, No.17-18, pp. 2778-2788, ISSN 0016-2361

Kelly-Yong, T.L.; Lee, K.T.; Mohamed, A.R. & Bhatia, S. (2007). Potential of Hydrogen from Oil Palm Biomass as a Source of Renewable Energy Worldwide. Energy Policy, Vol.35, No.11, pp. 5692-5701, ISSN 0301-4215

Khan, Z.; Yusup, S.; Ahmad, M.M.; Chok, V.S.; Uemura, Y. & Sabil, K.M. (2010). Review on Hydrogen Production Technologies in Malaysia. International Journal of Engineering & Technology, Vol. 10, No.02, pp. 111-118, ISSN 2077-1185

Luangkiattikhun, P.; Tangsathitkulchai, C. & Tangsathitkulchai, M. (2008). Non-isothermal Thermogravimetric Analysis of Oil-Palm Solid Wastes. Bioresource Technology, Vol.99, No.5, pp. 986-997, ISSN 0960-8524

Milosavljevic, I. & Suuberg, E.M. (1995). Cellulose Thermal Decomposition Kinetics: Global Mass Loss Kinetics. Industrial Engineering and Chemistry Research, Vol.34, No.4, pp. 1081-1091, ISSN 1520-5045

Murugan, P.; Mahinpey, N.; Johnson, K.E. & Wilson, M. (2008). Kinetics of the Pyrolysis of Lignin Using Thermogravimetric and Differential Scanning Calorimetry Methods. Energy & Fuels, Vol.22, No. pp. 2720-2724, ISSN 0887-0624

Ozawa, T. (1965). A New Method of Analyzing Thermogravimetric Data. Bulletin of the Chemical Society of Japan, Vol.38, No.11, pp. 1881-1886, ISSN 0009-2673

Shuit, S.H.; Tan, K.T.; Lee, K.T. & Kamaruddin, A.H. (2009). Oil Palm Biomass as a Sustainable Energy Source: A Malaysian Case Study. Energy, Vol.34, No.4, pp. 1225-1235, ISSN 0360-5442

Várhegyi, G.B.; Chen, H. and Godoy, S. (2009). Thermal Decomposition of Wheat, Oat, Barley, and Brassica carinata Straws. A Kinetic Study. Energy and Fuels, Vol.23, No.2, pp. 646-652, ISSN 0887-0624

Várhegyi, G.; Antal, M. J.; Jakab, E. & Szabó, P. (1997). Kinetic Modeling of Biomass Pyrolysis. Journal of Analytical and Applied Pyrolysis, Vol.42, No.1, pp. 73-87, ISSN 0165-2370

Vyazovkin, S. & Wight, C.A. (1999). Model-Free and Model-Fitting Approaches to Kinetic Analysis of Isothermal and Nonisothermal Data. Thermochimica Acta, Vol. 340-341, pp. 53-68, ISSN 0040-6031

Yang, H.; Yan, R.; Chin, T.; Liang, D.T.; Chen, H. & Zheng, C. (2004). Thermogravimetric Analysis–Fourier Transform Infrared Analysis of Palm Oil Waste Pyrolysis. Energy and Fuels, Vol.18, No.6, pp. 1814-1821, ISSN 0887-0624

Yang, H.; Yan, R.; Chen, H.; Lee, D.H. & Zheng, C. (2007). Characteristics of Hemicellulose, Cellulose and Lignin Pyrolysis. Fuel, Vol.86, Vol.12-13, pp. 1781-1788, ISSN 0016-2361

Zhang, X.; de Jong, W. & Preto, F. (2009). Estimating Kinetic Parameters in TGA using Bspline Smoothing and the Friedman Method. Biomass and Bioenergy, Vol. 33, No.10, pp. 1435-1441, ISSN 0961-9534

Zhang, X.; Xu, M.; Sun, R. & Sun, L. (2006). Study on Biomass Pyrolysis Kinetics. Journal of Engineering for Gas Turbines and Power, Vol.128, No.3, pp. 493-496, ISSN 0742-4795

Zhouling.; Chang-qing, S.; Chao-yu, L.; En-cheng, J. & Xi-wei, X. (2009). Pyrolysis Characteristics and Kinetics of Cotton Stalk, International Conference on Energy and Environment Technology, pp. 768-772, ISBN 978-0-7695-3819-8, Guilin, China, October 16-18, 2009

Zsakó, J. & Zsakó, J. (1980). Kinetic Analysis of Thermogravimetric Data XIV. Three integral methods and their computer programs. Journal of Thermal Analysis and Calorimetry, Vol.19, No.2, pp. 333-345, ISSN 1388-6150

CHAPTER-23

THE COLOMBIAN EXPERIENCE WITH BIOFUELS AND ACHIEVING ENERGY SELF-SUFFICIENCY

1. INTRODUCTION:AN OVERVIEW

The finite nature of fossil fuels, along with the significant contribution of the transportation sector to overall primary energy consumption and atmospheric pollution, has led to increased research on alternative sources for vehicles, particularly those derived from biomass. As a consequence, there has been a rise in environmental awareness that aims to substitute fossil fuels or develop mixtures that decrease their overall usage. I am mostly looking for sources in the field of agriculture. It is important to note that tropical countries, particularly Colombia, play a significant role in this area. Colombia is known for having a wide diversity of plant species and favorable environmental conditions for their production. The global energy crisis necessitates an examination of the extent, potential, and risks associated with the utilization and partial substitution of traditional fossil fuels with biofuels or agrofuels in order to foster the growth of a country. This analysis will focus on Colombia as a case study. The increasing significance of alternative energy sources, which can be obtained from various crops, and the need for large amounts of biomass, raises concerns about the potential negative impacts of deforestation, the destruction of forests, and the substitution of essential crop fields for human nutrition (food security). In addition to the difficulties posed by climate, geography, and physical factors, such issues also have significant impacts on the economics of entire nations (Cortés et al., 2009). Today, these emerging energy sources have become the latest financial, political, and even environmental policies. Currently, over 30 raw materials are undergoing testing on a global scale due to their significant importance. However, despite this significant increase, they have not yet offered a resolution to the worldwide energy dilemma (Cortés & Álvarez, 1998). The potential utilization of biofuels in the advancement of automobiles and engines has been contemplated since the inception. However, it is only due to the prevailing energy and environmental circumstances that the conditions have emerged for the establishment of a worldwide biofuels sector. The advancement of alternative energy sources has led to the increasing significance of biofuels and energy crops in the agriculture and energy policy of both industrialized and developing nations. The driving forces behind this phenomenon include the clear exhaustion of fossil fuels, the recurring oil crises, and the greenhouse impact resulting from the accumulation of CO_2 in the atmosphere. However, it is crucial to acknowledge that biofuels alone will not eliminate the need of developed nations on oil, as there is insufficient land and water resources to fulfill the energy demands of the automotive sector. Consequently, numerous countries, including Colombia, have formulated programs to mitigate the effects of greenhouse gases on Earth's climate and prevent permanent alterations by diversifying energy production through the utilization of renewable sources. The initial approach involves substituting oil-based fuels with biofuels, resulting in a decrease in CO_2 emissions produced by mobile sources. It is crucial to start utilizing alternative energies, specifically ones that are regarded as clean and renewable. Therefore, biofuels could be a viable option. Hence, the adoption of renewable energy sources as a substitute for fossil fuels is a crucial approach to mitigating greenhouse gas emissions (Consejo Nacional de Política Económica y Social (CONPES, 2008). Recently, there has been a discourse, debate, and confrontation about biofuels, which has aided the development of new technologies and refineries for their production. The significance of this matter stems not solely from a sudden advancement in scientific understanding, although such progress has indeed occurred. Rather, it is mostly due to a substantial increase in government financing, driven by concerns over escalating oil costs and the geopolitical implications of dependence on them. If funding persists, a forthcoming era of biofuels may become accessible, regardless of the underlying cause. Amidst the excitement, marketing, and support, a fundamental question arises: can biofuels be considered a feasible and economically sustainable energy and environmental alternative to replace future gasoline imports? Simultaneously, the implementation of promotional measures and incentives, encompassing legal, regulatory, fiscal, and financial frameworks, for alcohol fuels and bio-oils, will have a favorable effect on employment rates in agricultural regions. It is imperative to not only promote the development of biofuels but also establish programs that assist the biomass requirements of new refineries. This will ensure that the cost of raw materials, which serve both as food and biofuel, remains unaffected. Colombia must diversify its raw materials portfolio for the production of anhydrous alcohol and biodiesel. It should also encourage research and development of proprietary technology programs that can produce biofuels at competitive levels to meet domestic demand in the short and medium term. In the long term, Colombia should consider exporting biofuels. The policies aimed at replacing energy sources in the Colombian biofuel industry present a valuable opportunity to mitigate the depletion of reserves, mitigate the increasing cost of imports, reduce gas emissions, and minimize the impact of particulate matter on the atmosphere. This is particularly advantageous given the escalating price of oil, as it helps to decrease the risk of energy vulnerability. The text seeks to demonstrate the production and substitution of fossil fuels with bioenergy, specifically ethanol and biodiesel. It discusses the advancements, uncertainties, and challenges associated with these processes in generating and utilizing new forms of energy. The focus is primarily on the potential impact on food security and environmental consequences in developing nations. Specifically, it outlines the political, regulatory, and legal structure through which the Colombian government encourages the production and utilization of biofuels.

2. ENERGY REPLACEMENT AND CONSUMPTION

For operating the machines, during this century and part from the previous: man has used greater amounts of energy from fuels. Fuels coming from forest resources like wood and natural energy coming from sun have not been enough to satisfy the energy appetite that this modern civilization has (Silvestrini, 2000).

This growing demand has created an energy crisis that has led to create programs, proposals, funding and research, to find processes for greater efficiency in energy generation, consumption reduction and uncover alternative sources that allow a diverse offering and increase life span of existing resources.

As long as societies develop, it is clear that energy dependence increases. As a result, the search for more efficient, cleaner, more economic new energy sources and fuels becomes a social need. Therefore, in many processes and business it is possible and feasible to replace oil with natural gas; electric power with solar power; firewood with biogas; tractors with animal traction, etc. At the moment, there is an increase and trend for the conversion of internal combustion engines (ICE) that use gasoline by replacing them with CNG (compressed natural gas). Several CNG dispensers have been installed in the main cities of Colombia. The introduction and claimed benefits have been linked with reductions in pollution and operation costs.

Mankind's welfare has always been associated with high-energy availability; unfortunately a society's progress is directly related to its level of consumption. Man needs energy to survive, manage his environment and produce goods. In distant times, at least, he needed energy to keep from starving. Over millions of years, the time it took to get from primitive forms to its present forms; mankind's evolution is closely linked to the different forms and quantities of energies that were available during each period. The energy consumption growth per capita and energy control has been a constant feature. Evolution changed Man from a gathering society, to a hunting society, an agriculture society, until today becoming a technological society, where per capita consumption in the U.S.A is close to 450 MBTU/inhab. Perhaps primitive man during his tenure in a gathering society consumed no more than 10 BTU/inhab. This growth in consumption is an indication of progress and risks as well as the complexities of the social organizations in which we live.

Although the safety and quality of our energy supply is of real concerns, we must avoid viewing the energy problem from a local perspective, surrounding countries, and in short term perspectives, now and in the near future. A realistic and deep approach regarding the energy issue must consider that a third of humanity lacks supply of electricity and any other form of advanced energy. It must take into account the safety of supplies for future generations and must be aware of the consequences of the environmental impact that energy production and consumption are having on the planet being passed on to our descendants. The amount of energy needed in the future mainly depends on its efficient production and use. With the purpose of evaluating energy efficiency, energy intensity may be applied as an indicator, i.e. energy consumption per GDP unit in each country. Along the way a trend in energy intensity reduction can be observed alongside the increase in economic development (Pérez, 2002).

Since the 70s it has been common to correlate the Gross Domestic Product (GDP) with countries' energy consumption. In the beginning it was an indicator of a developing country to see their energy consumption growing: more infrastructures and vehicles, appliances, heating and air conditioning systems, more power and better quality demand. It went from a traditional economy based on biomass combustion to an economy that makes a general use of electricity and fossil fuels, both for automotive and industrial use. On the other hand, increased consumption means more man-made fibers to fulfill clothing demands, larger workplaces and leisure trips, more housing space and in general more services. A country's welfare for better or worse is associated with increased income and this leads to an energy consumption increasing per capita (Valero, 2004).

In the last thirty years, global energy use has increased almost 70%, but this growth is uneven, because developing countries have almost tripled their energy consumption, while industrialized countries increased by 21%. The world's energy consumption grows faster than wealth or population because developing countries hold the 77% of the world's population. On the other hand, since 1973 the countries belonging the Organization for Economic Co-operation and Development (OECD) have reduced their energy intensity or energy use by 24%per GDP unit. Energy demand has annually increased more than 2% since the 1973 crisis until today, and if current trends persist, this rate will continue over the next fifteen years (Brown et al., 2000).

Because the population is increasing and countries are getting industrialized, the global energy demand grows. That growth's projection classified by economic sectors can be seen in Figure 1. In addition, fossil fuels that are used are the main source of pollution in the atmosphere. The important role that energy plays in all human activities is widely recognized. Energy is not only transformed but it also transforms society. With its many developments and uses, energy has substantially changed modern life, by creating new services related to technological progress, and becoming the principal supply for the development and progress of any society.

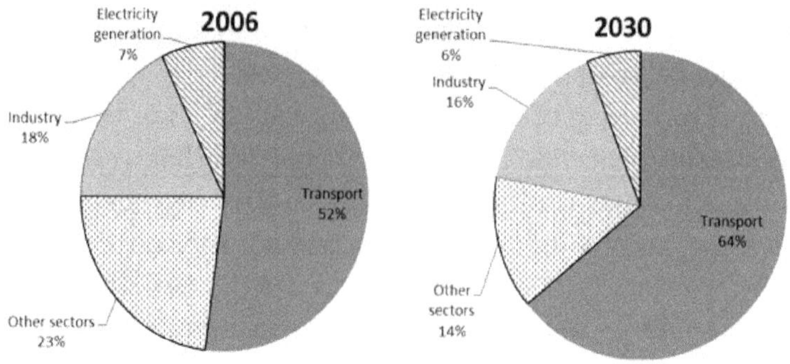

Figure 1.World oil consumption participation by sectors 2006 and 2030

Source: *Perspectives énergétiquesmondiales à long terme-le casspécifique des transports. AgenceInternationale de l'Energie. July 2007*

According to the International Energy Agency (2006), dynamic growth in the biofuels market is associated with the evolution of the world's demand for primary energy, where fossil fuels have increased participation. Energy demand depends on factor's performance such as: a) world population increasing b) economic growth c) technological developments which allow maximize efficiency production and use d) implementation of measures towards climate change such as the development of alternative energy sources.

Thus, the IEA estimates that by 2020 energy demand will be 16,000 Mtoe (Million Tonnes of Oil Equivalent), it means an annual growth rate of 1.7%.

The World Energy Council considers that in the next twenty years, world energy consumption shall approximately increase by 50%, which means it could be possible to provide energy to 4,000 million people (2,000 million that currently do not have it, plus the other 2,000 expected during this period).

But at the moment, energy has six considerable issues that must not be ignored in the scope any global economic analysis (Casilda, 2002).

First problem: production and consumption in unequal distribution around the world: large areas of primary energy production are different from the major consumption areas.

Second problem: has to do with the limited energy sources currently in use. Still nowadays, about 80% of the world's primary energy production comes from fossil fuels (coal, oil and natural gas), i.e. non-renewable sources and limited reserves.

Third problem: is the dominant role that oil plays within energy supply, oil where there is huge gap between production and consumption.

Fourth problem: comes from the relationship between energy and its development. The current energy consumption per capita is varied, as are the geographic levels of development. If in the coming years part of the developing world were come close to the level of energy consumption of industrialized countries, the world would face a long energy crisis, and Latin America would have an uneven behavior, as countries like Mexico and Venezuela would be favored, the two largest oil producers in the region and others would not, well most of them, due to their dependence. Likewise, developed countries would also be seriously affected.

Fifth major problem: production and energy consumption pose serious environmental problems that affect other productive resources worldwide, which could lead to climate change of irreparable consequences.

Sixth problem: is the ability to increase energy supply, which directly depends on the capital to be allocated for this purpose. In many Latin American countries capital is restricted due to regulations and low energy prices.

The use of different types of (electricity, motion, light or heat) and forms of energy (fossil fuels like coal, oil and natural gas, hydropower, nuclear energy or alternative energy) necessary for technologic and economic advancement, has produced the energy crisis that, since 1970, questions the possibility of keeping the current development model, in addition to other harmful effects, both because of the uneven development and the environmental consequences (pollution, global warming, etc.)

The nature of non-renewable fossil fuels and the high level of participation within transport sector of the total primary energy consumption and air pollution, have become the force behind promoting research of alternative sources for vehicles, mainly those sources derived from biomass. This has resulted in an increasing environmental consciousness, and seeks to replace fossil fuels or provide blends that reduce their consumption, mainly searching for those sources in agribusiness. Tropical countries play lead roles because it is where the greatest variety of plant species can be found and where the environmental conditions make production of these more advantageous.

Add to this the geopolitical crisis existing with the U.S.A and Iran, Iraq and Venezuela, and the decrease in global reserves that have influenced oil prices (over 75 U.S.D $/barrel), threatening the stability of many economies that do not have energy self-sufficiency. Under these circumstances, biofuels have become a strategic matter for the U.S.A: in 10 years they are seeking to replace 20% of oil consumption, and so does the European continent.

A technological solution to this problem is the use of biofuels; however, the price competitiveness of these compared to liquid fuels is still disputable. One of biofuels' restrictions is the cost of raw material and its transformation; so low-cost substrates are required to reach competitive price levels.

In practice, different raw materials can be used for industrial alcohol and biodiesel production; however, it is fair to consider that production cost of each liter or gallon strongly depends on geopolitical crisis the characteristics of the raw materials used and the type of process or technology used for their production.

Mainly in industrialized countries, biofuels supply behavior has been due to: biomass availability for their production, production costs and subsidies, and incentives for their production and use. In such respect, demand has been associated with fossil fuels dynamics and the growing interest for the reduction of greenhouse gases (GHGs), from major fossil fuel consuming countries, the latter being an opinion that is under discussion worldwide and must be evaluated in the country.

The contribution of biomass, wood, agricultural, livestock and urban-world residues to energy consumption is currently limited to traditional use as a fuel, especially in Least Developed Countries. However, advanced technologies like gasification, fermentation and anaerobic digestion for biomass exploitation, are increasing its important role as a sustainable energy source like liquid fuel or electricity production. When biomass is grown for burning, there are no net CO2 emissions in the whole process.

Current challenge with biomass is its sustainability management. By 2050 the calculated potential of energy production from biomass is about 10 times the current production, which would be enough to meet the current global energy needs. However, there are several factors that restrict this potential, among them the principal being water availability (UNDP, UNDESA, WEC, "World Energy Assessment, United Nations Development Program, 2000).

According to the growth projections from the International Energy Agency, it is estimated that biofuels participation within the energy market will be 4% by 2030 in comparison with the current 1%. The United States, the European Union and Brazil will be the leading countries in biofuels demand. On the other hand, developing countries will have an important role in such energy supply, so that it is estimated Asian and African countries participation for ethanol production, and Malaysian and Indonesian for biodiesel. Considering this, the expected biofuels demand by 2020 is estimated in 50 Mtoe, which is an annual increasing of 6.3%. Thus, it is estimated that ethanol production will be 524 mmba and biodiesel will be mmba 397.

It becomes important to reconsider current energy consumptions, given the evidence that free energy flows will be renewed as long as earth is habitable, unlike stored energy (oil, coal, gas), which is finite, or in constant decrease with increasing cost. This reason justifies the systematic capture of free energy flows (solar, wind, hydraulic, geothermal) or the use of biomass for generating biofuels. Thus, as long as man and society dominate and know the physical phenomena and natural contributions, the more energy that can be exploited. Therefore, it is not whether there is or is not enough energy to sustain human development, but the physical limits of energy use are going to be or have already been overcome.

Recently, within the scenario of energy generation from diversified sources, rises the clash of the feasibility for replacing crops for human consumption or for biofuels production in the automotive industry. The burden on agriculture for energy production seems to be very hard and biofuels promotion measures in many countries may trigger a very serious food conflict, with still unknown impacts on the poorest countries. In short term they could bring higher food costs.

The world faces complex challenges, it can not be considered the solution to life survival on Earth based on the alternative of renewable energy from biofuels, as it would increase food crops replacement by monocultures, deforestation for energy crops, would boost diversity extinction, cause reduction of fertile lands and water; plus social consequences caused by population displacement.

The use of a specific type of energy depends on two factors: availability for potential energy and technological capacity to turn it into heat or work. Of these two conditions, Colombia has enough natural energy resources (coal, oil, gas, water falls, solar radiation, wind, biomass), being researched and developed in very early stages for its use. Therefore, replacing some conventional energy sources by renewable energies, with endless reserves, aims to reduce pollutant loads and is more environmentally friendly.

The kind or type of energy source used together with the application or purpose, varies over time within the countries, because the use of one or another depends on demand, pricing, availability in the same country or importation, technical reasons, international political situations, etc. In Colombia it can be said that most of current energy comes from burning fossil fuels, which are petroleum products processed in refineries.

Fossil fuels' finite and non renewable nature, plus the high air pollution from major population centers, automotive sector which has a high proportion in the total primary energy consumption; all this has promoted new energy or biofuels from organic resources (plants or animals). Colombia is a vulnerable country in regards to oil due to limited exploratory programs, lack of important new discoveries and decreasing reserves. As a result, the country is about to lose its oil self-sufficiency in the short term (2020), in the midst of high prices unfavorable situation. In addition there it is of poor quality due to its high sulfur content, especially in diesel which exceeds the international standards. Diversification of energy sources, preventing degraded climate change, and promoting social rejecting of polluting energy sources, constitute the challenges faced in order for science and technology to make a contribution for a development that is sustainable, recognized and supported by citizens

A condition for sustainable economic development is to ensure self-sufficiency in energy supply, which must be supported in a flexible and diversified energy structure; being at the same time, energy policy components. This policy should consider the agricultural sector's characteristics, pressed by other energy consumption patterns. Energy as a transformation, transference and accumulation process, should lead those communities to a more efficient use of the different available sources.

It must be a national goal to plan actions for replacing and optimizing the use of different energy forms. External energy inputs replacement by original sources in agricultural production units is one of the self-sufficiency energy objectives through the use of digesters, animal traction, solar, wind, mini stations and alternative fuels (alcohol, vegetable oils, natural gas, etc.)

The prospects resulting from energy optimization and replacement, will allow greater coverage of efficiency and economy of incorporated social benefits. Promoting this energy replacement, due mainly to higher oil prices in recent years, an increase that represents an onerous burden on the world's poorest countries' economies.

An energy program must be an integrated set of studies or research projects within biomass production, processing and the areas of energy consumption, linked to socio-economic studies and systems analysis, aimed to develop techniques and technologies on various energy sources (biogas digesters, gasifiers, solar cells, windmills, micro stations), bioenergy systems in rural areas, use of waste and design systems. Productivity and profitability paradigms that favor: high production values per unit of time, area and, invested money, must be accompanied by processes that also favor energy efficiency, in terms of consumption and savings.

Bioenergy is a rather broad term covering all energy products obtained from agricultural commodities or animal waste conversion processes, mainly required to meet automotive sector's demand in developed countries. Biofuels are a reality in several countries, including Colombia. Biofuels have become an agro-industrial development model, which attempts to take part in the economic landscape and for this it has stimulus and a broad policy framework as well as multiple advocates and detractors.

Given the biofuels impressive growth (biodiesel and bioethanol) as a renewable energy source, it is necessary to consider in a more analytical and balanced way its options; benefits and limitations in a society that also requires increased food production. In any case, the use of these alternative fuels has negative effects both in natural and socio-economic levels. Despite some partial benefits, from the incorporation of new lands for the production of energy crops and agro-industrial development processes with the collateral effects on employment and possible exports as an added tax product. It is also necessary to face the nearly inevitable competition possibility between food and biofuels production, that within a high poverty levels society, it is worrying that production displacement in our food and agricultural system.

Biofuels present both opportunities and risks. The outcome would depend on the specific context of the country and the policies adopted. Current policies tend to favor producers in some developed countries over producers in most of developing countries. The challenge is to reduce or manage the risks while sharing the opportunities more widely. Biofuel production based on agricultural commodities increased more than threefold from 2000 to 2007, and now covers nearly two percent of the world's consumption of transport fuels. The growth is expected to continue, but the contribution of liquid biofuels (mostly ethanol and biodiesel) to transport energy, and even more so, for global energy use will remain limited. Despite the limited importance of liquid biofuels in terms of global energy supply, the demand for agricultural feedstocks (sugar, maize, oilseeds) for liquid biofuels will continue to grow over the next decade and perhaps beyond, putting upward pressure on food prices (FAO, 2008).

Today Colombia seeks different alternatives for solving the increasing difficulties set out by its development, mainly its population's diet in a natural sources panorama, especially when related to water and soil which may be eroded and contaminated. In order to prevent irreversible changes and reduce the impact of greenhouse gases on Earth's climate, many countries have decided to put their hopes on diversification of energy production strategies by using renewable sources. The first strategy has been replacing petroleum fuels with biofuels, achieving thereby a reduction in CO2 emissions generated by mobile sources (1). Therefore it is urgent to start using alternative energies, i.e. clean and renewable. For this reason biofuels can be a choice, not without question.

It is becoming more urgent to use other fuels in vehicles, mainly for replacing gasoline and diesel, diesel being the one that is gaining preference because of the costs (2). Therefore, it is advisable to research diesel replacements as opposed to gasoline.

The feasibility of using alcohol or biodiesel as automobile fuel in greater proportions than those currently stipulated is a proven fact. Even though there are great expectations, regarding hydrocarbon reserves existence underground in Colombian, the possibility that in a few decades the country has used all its oil must be. This consideration makes it necessary to pay attention to Brazil's experience, which aims to replace oil with other fuels. With optimism of our economy recovery, if new options are not found, the country will: increase its fossil fuels consumption of high priced products within the international market, that in the near future could consume our currency. No matter what, its price tends to increase as a result of the world demand and conflicts with governments in the Middle East and Venezuelan. For that reason, it is advisable to have a replacement from organic sources available, such as alcohol and biodiesel.

Unlike the oil industry, the new agro-energy industry involves a productive chain that impacts different economic sectors more directly, especially with regard to employment generation and agricultural and agribusiness development. The addition of Biofuels to blends helps to mitigate oil import requirements. This supports the national biofuel policy facing the balance of energy trade, and to some extent, defines security settings on the supply level.

Colombia has a great potential for establishing a big biofuel industry. Developing this industry gives the country an opportunity for exploiting its comparative advantages, as a tropical country with an agricultural tendency (biomass production) and suitable soils for feedstocks. Moreover, it becomes a technologically scientific challenge for research groups to focus their efforts on achieving proprietary technological developments by working together and interdisciplinary with public and private sectors. Unfortunately the substitution of conventional fuels with biofuels (ethanol and biodiesel) is also having ecological consequences.

Because of the current high demand, which is expected to increase in the coming years, many countries are cutting down large areas of tropical forests to cultivate biofuels. Although this is not the case in Colombia, as long as a favorable competition for energy feedstocks are shown (given the high government incentives and subsidies), against food without any government protection and also competing with the different free trade agreements. This is worrying given the large expanses of land necessary for feedstocks.

In conclusion, to really consolidate a coherent policy on new energy matter within the Colombian bucket, the following remarks must be considered:

- Environmental Ethics.
- The physical boundaries (finite resources).

- Climatic and geographical conditions.
- Crop Yields (kg/ha, l/ha, l/t).
- Energy intensity and energy return rate.
- Water requirements.
- Self Sufficient Process.
- Technology, return on investment (ROI) and profitability.

Not forget that compressed natural gas – natural gas vehicle (CNG-NGV) is another strong rival among the already diverse energy supply.

In that context, the National Government has promoted research and of new renewable energy sources that are sustainable with the increasing pace of life, while oil and its derivatives are partially replaced in different applications, mainly in transport sector. This promotion must also consider the implications when allocating millions of hectares for bioenergy production. This reality highlights the urgent need of fulfilling food needs or allocating those lands and feedstocks to meet the automotive industry energy requirements.

Energy is, undoubtedly, both a solution and a problem for sustainable development. As such, the hope is focused on the dilemma for fulfilling growing consumption needs while minimizing the social impact and ensuring resources. The world energy problem, fossil fuel reserves reduction, the air pollution problem and global warming, are matters of great importance for humanity without a global solution until now.

3. PRECEDENTS

The public sector supports the biofuel industry due to its potential environmental and social benefits, such as mitigating climate change and contributing to energy security. The industry has experienced rapid expansion. Therefore, in recent years, there has been a global increase in the manufacture of fuel from renewable resources as an alternative to fossil fuels. This is because the global output of biofuels has doubled and is likely to continue growing strongly in the near future due to increasing demand. Some of the motivations include the eventual depletion of fossil resources, recurring oil crises, and the greenhouse effect resulting from the accumulation of $CO2$ in the atmosphere.

In particular, in our country, diesel is in higher demand when in comparison to the quantities supplied by the crude oil processed in the refinery. For that reason it must be partially imported already manufactured. In addition to high oil prices, the agriculture crisis and low international oil trade rates are some of the factors that have contributed to give an additional role to biodiesel.

Liquid biofuels also known as biofuels, which are derived from agricultural raw materials, are products that are being used as replacements for gasoline and diesel in vehicles, they can also be used as blends. At the moment many countries encourage the idea of planting their own fuel, to lessen dependency on imports or exhaustible reserves, while generating stable and top-quality jobs. Biofuels are alcohols, ethers, esters and other chemical products made from cellulosic biomass such as grasses and woods, agricultural and forest waste and most of the municipal and industrial waste; as wells as vegetable oils. The term Biofuel refers both to fuels for electricity or transportation. Unlike oil which is a nonrenewable resource, biofuels are renewable and represent an inexhaustible fuel source. Both commercial and noncommercial, these fuels should always be considered as valuable products which can meet the growing demand.

The most important biofuels are ethanol (from sugar cane, sugar beet and corn) and biodiesel (from oil palm, soybeans and other oilseeds). Biofuel promoters highlight the ecological character of these fuels: they are renewable and are apparently environmentally friendly and produce less greenhouse gas emissions (GHG) in comparison to petroleum fuels.

It has been extensively spoken about biofuel models for more than a decade, and the opportunities and challenges that these oil substituting fuels can offer. This potential is not only related to environmental improvement, but it also includes economical, cultural and social aspects (Cortés, 2007).

In many countries (e.g. Brazil, USA and some European), including Colombia, have implemented policies favoring biofuels. Now these actions are having an effect: more and more agricultural commodities are for biofuel production. For this reason demand grows and causes these agricultural commodities in world market have high prices.

Biofuels apparently help to reduce greenhouse gas emissions, but most of the time it is not considered that for biofuels production, fossil fuels are used (diesel for machinery and products transportation, inputs production, etc.) It is also left aside that the expansion of agricultural frontiers (caused by the increasing demand of agricultural commodities) does not reduce greenhouse gas emissions. In the contrary, the forest is a carbon reserve and turning it into cropland releases this carbon as CO2 (the most important GHG). However, with the production of biofuels there are side effects that disturb an apparent prosperity. There are several scientists who defend that for true CO2 reduction re-forestation is vital as opposed to de-forestation for farming.

We must also consider that in large scale certain that biofuel inputs are produced by an agro industrial agriculture. This type of agriculture is supported in large monocultures; abuse of agrochemicals and the soil fertility overexploitation provokes water pollution (with pesticides), soil erosion, air pollution and loss of biodiversity.

On a national level, within the context of energy replacement policies, a delay in exhaustion of reserves, prevention in the rise of import costs and reduction in the impact of gas emissions and particulate matter into the atmosphere, present a great opportunity for the biofuel industry due to in part to the rise in oil prices. This opportunity is supported by a regulatory and legal framework of agro-energy production, including Act 693, of 2001 that proposes an initial 5% replacement of gasoline with alcohol. Later increased to 10% by 2010 and 12% by 2012; with similar proportions for replacing diesel with biodiesel. The same for the use of suitable lands for energy crops such as: sugar cane, cassava, sugar beet, oil palm, castor oil plant and jatropha, all of them with studies and different productivity levels (ton/ha, l/ton). Without denying the possibility of cellulosic biofuels from different sources.

In addition to the aforementioned, the National Government has promoted the development and search of new renewable energy sources, sustainable with the increasing pace of life, by partially replacing oil or its derivatives in different uses, especially within the transport sector. This promotion must also consider the implications when allocating millions of hectares for bioenergy production. This reality shows the urgent need for meeting food demands or allocating lands and feedstocks to meet the energy requirements of the automotive industry(Cortés, 2007). Within this context it is proposed that it will promote competition between the different biofuels, with criteria for financial sustainability and energy supply. For these purposes the feasibility and advisability of releasing biofuels prices and promoting removal of import duties on these products. Notwithstanding the aforementioned, the National Development Plan states that in any case the current pricing scheme based on opportunity costs of such energy, their replacements and raw materials used in its production must be considered. While simultaneously promoting strategies for prevention and control of air contamination, by promoting cleaner fuels, including those derived from crops with production potential for biodiesel and alcohol fuels.

The different types of biodiesel from the oil palm, castor oil plant, jatropha and sachainchi, considered by the Ministerio de Agricultura y Desarrollo Rural (English: Ministry of Agriculture and Rural Development) as strategic for Colombia, have a wide range of chemical compositions and qualities. Oil palm biodiesel given its highly saturated chemical nature has excellent ignition and chemical stability qualities, but it has limitations as for its ease of flow at low temperatures. Oil palm biodiesel cloud point, i.e. the temperature at which crystals formation is visualized, is around 16°C. Crystals emerging and further agglomeration can clog fuel filters preventing the fuel to reach the engine. Castor oil biodiesel is characterized by high alkyl esters content from ricinoleic acid (about 90%), which are monounsaturated and have a hydroxyl group in its structure, giving the biodiesel a high viscosity. Jatropha biodiesel holds 80% of unsaturated alkyl esters from which 34.8% are di-unsaturated. Sacha inchi oil can reach an unsaturated level up to 94%, and it is the most unsaturated oil according to technical literature. Therefore, with jatropha biodiesel and, the oil obtained from sachainchi, there could be problems in its chemical stability (Ministerio de Minas y Energía, 2007; Mesa, 2006).

Crops' biggest problem is that they can be expensive as raw materials, which makes the final product price high; so the State must allocate considerable tax resources to make these energies competitive. Aside from being expensive, these raw materials are crops' primary products, and only recently has the use of waste for biofuel production been taken advantage of. Faced with this adversity, many countries are researching and developing methods for producing ethanol from agricultural, forest and industrial waste, which are abundant and very cheap.

In this scenario, sugars will be derived from cellulose found in plant waste, such as banana or lumber industry waste. Currently, Brazil has developed methods that utilize husks through fermentation processes. Are biofuels a technically, economically, and environmentally viable energy source for the nation, with the goal of replacing future petroleum imports? After the implementation of rules pertaining to alcohol fuels and biological oils, it may be inferred that employment rates in developing areas will be positively impacted. In order to ensure the competitiveness of efficient biofuels in the global market, it is imperative for the country to not only promote their production, but also establish programs that facilitate the establishment of new biomass refineries. These programs should aim to prevent any negative impact on the price of raw materials that serve a dual purpose, both for food and biofuel production (Cortés, 2007). Additionally, it is imperative to establish precise agricultural land management techniques to ensure the preservation of forest regions and prevent their conversion into biomass cultivation zones. Based on past observations, the agricultural sector of the country can be connected to the energy sector without causing any negative impact on the food business in terms of higher prices for raw materials, as has been the case with sugar, wheat, and maize.

3.1. Different energy issues

The relationship between development and energy usage is widely recognized. It is appropriate to view the total energy consumption of a society as a means to enhance human effort. Similarly, technical advancements enable significant progress while only requiring a slight increase in the use of primary energy. The availability and utilization of energy have a significant impact on and are influenced by factors such as population increase, urbanization, development opportunities, and poverty reduction for a large portion of the population. For instance, the energy consumption patterns of around one-third of the global population who rely solely on biomass as their energy source likely to perpetuate their dire socioeconomic position. Countless individuals, particularly children and women, dedicate numerous hours each day to the arduous task of procuring firewood or transporting water over significant distances. Consequently, they are deprived of valuable possibilities for education and more fruitful endeavors.

The existing development and consumption paradigm, along with the escalating energy wastage by affluent individuals and the consumption patterns of the most marginalized, contribute to pollution and destruction, resulting in poverty. Simultaneously, this poverty further exacerbates pollution and destruction. This represents the pernicious cycle: consumption leads to pollution, which in turn leads to poverty. This is an intricate network of relationships, which is not always apparent, where certain phenomena are both the source and effect simultaneously, and no element can be regarded as distinct or isolated. One of the primary obstacles that humanity must overcome is the discovery of methods to generate and utilize energy in a manner that fosters long-term human progress across all aspects of society, economy, and the environment (Pérez, 2002).

It is expected that in the course of this century, the use of oil for producing electricity will be replaced by gas, clean burning coal, energy from renewable sources, and nuclear energy. However, the largest contributors to oil consumption and increased pollution is the transport sector, where in the medium term there is not yet a replacement, since current petroleum products are characterized by their high calorific value, they are also easy to store, carry and use.

The creation of liquid biofuels has been one of the ways that science has developed for replacing gasoline and diesel and preserving the environment. Modern bioenergy technologies are renewable energy sources that produce transportation fuels and are advancing very fast, mainly towards ethanol made from maize or sugar cane, which is blended with gasoline in order to reduce both oil consumption and pollution. To use ethanol as fuel by blending it with gasoline, it is necessary to remove water for purity close to 99.9%, which requires special distillation methods.

Bioenergy competitiveness is associated with oil price, if price keeps current trends, there will be options for those trends. It must be considered if benefits and efficiency of these new fuels could survive without stimuli (subsidies) that now favor them. In a realistic framework it is necessary to avoid ambiguous positions that require a choice between biofuels or food production. It is important to combine both processes and add technology that improves production. But food safety issues cannot be jeopardized.

Security of supply is synonymous with the availability of all the energy needed, at an affordable price and for a long time, in fact indefinitely so it can be sustainable. If supply security is considered from a national perspective; dependence on external resources and the uncertainty of this non native supply becomes an important aspect.

"Opportunities for developing countries to take advantage of biofuel demand would be greatly advanced by the removal of the agricultural and biofuel subsidies and trade barriers that create an artificial market and currently benefit producers in OECD countries at the expense of producers in developing countries "said the director general of FAO, Jacques Diouf.

In the coming decades several challenges must be faced because of the energy and environmental problems arisen from it:

- **Energy efficiency:** It will be necessary to radically increase the energy efficiency of processes and systems.
- **New technologies:** It is necessary to develop and incorporate new technologies that achieve the above goal.
- **Diversify energy sources**: At the moment we strongly depend on hydrocarbons as a primary energy source. It is necessary to incorporate new energy sources: natural gas, biomass, wind power, micro-hydro, solar energy and others.

- **Cogeneration:** cogeneration as a means for energy efficiency and savings is not new. But incorporating it into the system on a grand scale may have a very large positive impact.
- Then it is important to recognize that biofuels will not end industrialized countries' dependency on oil -not even Colombia-, because there will not be enough land and water to meet the high energy demand. In spite of this, the development of a national biofuel industry is an opportunity for the country. There are a number of technological, regulatory, economic and environmental restrictions or challenges that may affect critical links in the chain of biofuel production, and if they are not superseded they could lead to failure.

3.2. Energy Impact on the environment

Undoubtedly, development and energy use are intricately interconnected. In the future, a crucial issue will be how to ensure that energy sources remain cost-effective and dependable enough to enable a sufficient level of development. Limited energy supply poses a hindrance to development, while potential environmental repercussions might also impede or jeopardize development. Nevertheless, this does not encompass the entirety of the situation. Evidently, every activity will inevitably exert an influence on the environment. The problem arises when this influence is detrimental or perhaps irreversible. Throughout history, numerous instances can be found when societies have caused the collapse of their environment, leading to their own downfall. At a global scale, it is crucial to carefully assess the influence of energy-related activities on climate change, commonly referred to as the greenhouse effect. While there are other worldwide threats to environmental sustainability, climate change is widely regarded as the most significant. Climate change is considered a paradigm due of its vastness, intricacy, and direct correlation with energy-related activities. For instance, the degree of success or failure in implementing the Kyoto Protocol remains a reliable measure of the global community's and each country's dedication to sustainable development (Pérez, 2002). The effects of energy generation and utilization have been documented over an extended period. The widespread clearing of forests and the contamination resulting from industrial activities are widely recognized instances. However, despite their seriousness, their focus was primarily on local effects. Over the past century, localized impacts have evolved into worldwide perils. The acknowledgment of the correlation between energy and global environmental issues is a recent development that is impacting both human well-being and the quality of life, particularly for future generations. Without a doubt, human activity exerts a significant influence on the environment. Today, there is a significant amount of discussion surrounding Sustainable Development; yet, it can be argued that this word is inherently contradictory. Indeed, if we consider that development, which entails the continual augmentation of resource utilization, must always result in growth. Inevitably, both in the short and long term, we will encounter significant limitations to progress due to the unavoidable scarcity of resources. Consequently, it is crucial to understand the distinction between growth and development. Undoubtedly, a nation can experience robust economic expansion while also attaining a significant level of advancement, albeit at a more gradual rate. Significant development advancements can be achieved even with minimal rates of growth. Distinguishing between these two categories is crucial as it provides an alternative viewpoint on a country's development over time. Over the past decade, it is evident that Colombia has had significant economic expansion. However, our level of development has remained considerably lower. It is evident that finite expansion is not feasible due to the requirement of unlimited resources, which are not available on our planet.

During the last century, impacts of human activity have been higher had taken place since the beginning of civilization. Footprints of human activity are changing the world at a rapid pace and energy has a lot to do with this impact.

Instead of detesting the technological development, it is necessary that some philosophical, ethical and social principles guide it. Not everything is good, cars have not given humans more freedom: because now travel time has increased, people need to work longer hours to pay it, cities have become uninhabitable, it is literally asphalting the living space, social conflicts are increasing because of the lack of communication, and due to noise and urban stress increasing, we move away from downtown and then our dependence on the car and our isolation increases. From that overview, it doesn't look like the car is synonymous with social development and welfare (Valero, 2004).

3.3. Biofuel, a sustainable energy?

In Colombia due to the act (693/2001), production of biofuel is it is expected to increase for use it in transportation as a measure to reduce greenhouse gas emissions. A target by 2010 is to replace 10% of fuels, used today for transportation, with biofuels that will increase up to 20%. With that, reduce CO2 emissions to the atmosphere that cause the greenhouse effect.

One of the main problems is that it causes higher prices for basic foods. Biofuels are produced from products like maize (corn), sugar cane, wheat or soybeans; and when the availability of them decreases in the food market, it drives up prices. At the same time, available area for food crops also decreases because most of these crops are for biofuels, which results in increasing the agricultural commodity's price in general.

Economic effects from this rising food prices are being felt in recent years, especially in some poor countries. For example the crisis that took place in Mexico because of the maize tortillas rising price, which has doubled in recent years (similar to what is happening in our country with the rising price of bread and milk).

Replacement of conventional fuels with biofuels is also generating adverse ecological consequences. Most of the feedstocks needed for processing take place in developing countries, mainly in Latin America and Asia; most of these countries are cutting down large areas of tropical forests for growing biofuels.

For the production of clean fuels it is necessary to use dirty fuels as energy source. For instance, intensive sugar cane crops (for ethanol) or other oil crops (for biodiesel) will need petroleum products: fertilizers, insecticides, fuel pumps, transport and industrial processing. Therefore it is possible that pollution levels increase by using dirty energy sources for producing and exporting clean energy sources.

So far it is clear that bio in biofuels must have a question mark. Then, it can not be neither justified nor adopted policies for biofuels promotion and support, based on ecological arguments, or in industrialized countries (where people want to use agro-energy) or in developing countries (where people want to produce them).

To classify biofuels as bio, it would necessary to grow in degraded and poor soils that are unsuitable for food production (the so-called second generation biofuels). This prevents the rising prices of food and deforestation. An international certification scheme could ensure the sustainability of agricultural practices for the production of raw materials for biofuels.

In order to reduce possible impacts caused by biofuel production, certification for procedures of its production have been developed around the world; this is how the Dutch government, among others, aims that imported biofuels are certified according to environmental and social criteria. Certification of the entire process shall be necessary to ensure the world sustainability production and the use of biofuels (Testing framework for sustainable biomass, 2007).

Likewise, one of the most important factors for defining biofuel production feasibility is energy balance (the comparison between the energy used for producing biofuels and energy production). From the energy perspective, it is not enough to take into account the energy generated by a fuel, but it also must be considered the global energy balance, considering energy expenditures for fuel production and energy derived from it. Undoubtedly, for the production of that fuel to be profitable, the balance must be positive, i.e. it must generate more energy than consume.

Again the usefulness of biofuels as potential replacement for fossil fuels in the reduction of greenhouse gas emissions has been questioned. Several specialists have shown that the cultivation and use of, is not as efficient as a measure to slow down climate change as their advocates say.

Specifically deforestation, caused because of these feedstocks expansion, can have devastating effects in terms of climate, as well as from the ecological perspective. According to studies, forests from a particular area can reduce CO2 emissions nine times more than a biofuel feedstock with the same area, as well as the subsequent use of those biofuels for transportation.

In addition, during the acquisition process of these fuels, which involves cultivation, processing, transportation, and distribution, a greater amount of CO2 is emitted than the crops can absorb during their growth.Thisisduetothesubstantialquantitiesoffossilfuelsrequired,whichleadstothereleaseofgreenhouse gases. In the case of bioethanol, these plants are unable to fully absorb these emissions. Additionally, the significant water use associated with their production, particularly in the case of biodiesel (which requires 12 liters of water per liter), renders them an unsustainable alternative when compared to fossil fuels. In light of the numerous issues presented by initial biofuels, a technological remedy is being proposed: the development of liquid biofuels (BtL, Biomass to Liquid) derived from lignocellulosic biomass, such as straw or wood chips. These include bio-ethanol created through the fermentation of hydrolyzed biomass and biofuels derived through thermochemistry, such as bio-oil obtained from pyrolysis (carbonization), gasoline, and diesel oils produced using Fischer-Tropsch Synthesis, among others.

3.4 Objectives of biofuel programs in Colombia

It is mainly to promote the diversification of the energy basket through the use of biofuels, with the following criteria (Mesa, 2006):

- Environmental sustainability.
- Favor lignocellulosic crops replacement.
- Agricultural employee maintenance and development.
- Energy self-sufficiency.
- Agro-industrial development.
- Improving the quality of country's fuels, as a result of a blending between biofuels and fossil fuels.

To achieve these goals, Colombia faces the challenge of moving into strategic areas, among them are: a) consolidation of an institutional framework for the formulation of actions related to the handling of biofuels; b) reduction in the production of biofuels in the most critical points of the production chain, c) increasing the productivity of biofuels throughout all the production chain, d) research and development looking towards increasing biomass crop yields, develop new varieties adapted to different growing conditions and resistant to plagues, and develop changing processes of first and second generation e) price regulation in order to encourage the efficient production of biofuels, and f) differentiation of the Colombian product in order make easier the access to international markets by adding strategic environmental and social variables, besides food safety protection measures (Consejo Nacional de Política Económica y Social (CONPES, 2008).

As stated by the Consejo Nacional de Política Económica y Social (CONPES) (in English: National Council of Economic and Social Policy) of the Colombian government: This will enable the ability to take advantage, in a competitive and sustainable way, of economic and social development opportunities offered by biofuels emerging markets. At the same time it will allow: increasing competitive sustainable biofuels production by contributing to employment generation, rural development and population welfare; promoting an alternative productive development to the reliable rural land occupation; contributing to the formal employment generation within the rural sector; diversifying the country's energy basket throughout biofuels efficient production, by using current and future technologies; ensuring an environmentally sustainable performance throughout the addition of environmental variables when making decisions in the chain of biofuel production.

4. THE MOST COMMON RAW MATERIALS

Energy crops are those developed only for fuel. These crops include fast growing trees, shrubs and grasses. These can be grown in agricultural land not needed neither for food, nor pasture nor fibers. In addition, farmers can grow energy crops along the banks of rivers, around lakes or in farms areas including, natural forests or swamps, for creating habitat for wildlife, renewing and improving soil biodiversity. Trees can be grown for a decade and then being cut down for energy.

Thus, bioenergy covers all forms of energy derived from organic fuels (biofuels) form biological origin used for producing energy. It includes both crops intended to produce energy which are particularly grown and multipurpose crops and by-products (residues and wastes). The term By-products includes solid, liquid and gaseous byproducts derived from human activities. It can be considered biomass as a sort of converted solar energy.

It can be said that biodiesel production tends to come mainly from oils extracted from oilseeds plants, but any material containing triglycerides can be used for biodiesel production (sunflower, rapeseed, soybean, oil palm, castor oil, used cooking oils, animal fat). Here are the main raw materials for biodiesel production (Mesa, 2006).

Conventional vegetable oils: raw materials traditionally used for biodiesel production have been: oils from oilseeds such as sunflower and rapeseed (in Europe), soybeans (in The United States) and coconut (in The Philippines), and oils from oilseeds fruits such as oil palm (in Malaysia, Indonesia and Colombia).

Alternative vegetable oils: in addition to traditional vegetable oils, there are other species adapted to the conditions from the country where they are developed and better positioned within the field of energy crops: Jatropha curcas oil (Ministerio de Minas y Energia, 2007). Biofuels have become very important because of the variety of crops from which they can be derived, but this energy supply demands a high production of them. This would have harmful effects because of the destruction of forests and jungles and replacement of crops that are essential to human diet; besides the drawbacks shown in the following fields: climatic, geographical and physical. The main supply sources of raw materials for biofuels production are shown in Table 1 and Figures 2, 3 and 4.

Table 1.Raw materials for biofuel production: Source: Ministerio de Agricultura y Desarrollo Rural, MADR (English: Ministry of Agriculture and Rural Development); Portafolio: Goldaman Sachs (2007)

Crop	Efficiency (l/ ha/year)	Efficiency (ton/ ha)	Estimated barrel price (US $)
Sugar Cane	9	100	45
Cassava	4,5	25	NA
Sugar Beet	5,000	NA	100
sweet sorghum	1,189	NA	NA
Cellulose	NA	NA	305
Maize	3,2	10	83
Oil palm	5,55	NA	NA
Coconut	4,2	NA	NA
Castor oil	2,6	NA	NA
Avocado	2,46	NA	NA
Jatropha	1,559	NA	43
Rapeseed	1,1	NA	NA
Peanut	990	NA	NA
Soybeans	840	NA	122
Rapeseed	NA	NA	125
Wheat	NA	NA	125
Sunflower	890	NA	NA
Oil	NA	NA	70-80

But not all the questions are clear and therefore the UN declares: if growing fields for biofuels production increase disproportionately, food and the environment could be at risk. Increased logging. Also food prices could increase.

For major producing countries, costs of ethanol production range between 32 and 87 USD/barrel (International Energy Agency, 2006). According to the available information, about 47% and 58% of this cost is raw materials, about 13% and 24% for inputs, about 6% and 18% for operation and maintenance costs and, about 11% and 23% to capital costs. It can be said that production costs widely vary between countries due to agro-climatic factors, land availability and labor cost that affect the kind of biomass used as raw material; this factor affects transformation technologies selection.

Figure 2 shows sources of raw materials sources for alcohol and biodiesel production and the corresponding efficiency. Figure 3 and 4 show ethanol efficiency from biomass sources in countries outstanding in their production. There is higher ethanol efficiency from sugar beet, in comparison with sugar cane and corn.

For every ton of cassava, 200 liters of ethanol can be obtained, when making the cassava calculations as a yield base of 25 ton/ha it can be obtained a yield of 5000 liters/ha can be obtained which is lower in comparison to sugar beet but higher compared to corn and sugar cane. With fertilization programs and cassava crops mechanization, yields can be increased to values of 70 ton/ha, which will triple cassava yield in liters/ha (Altin et al., 2001).

Another important factor is that biofuels do not work as well as petroleum fuels. In order to increase their production most of the fertile lands would have to be assigned for farming them, which could be counterproductive in a world where hungry and desertification are two problems with difficult solution.

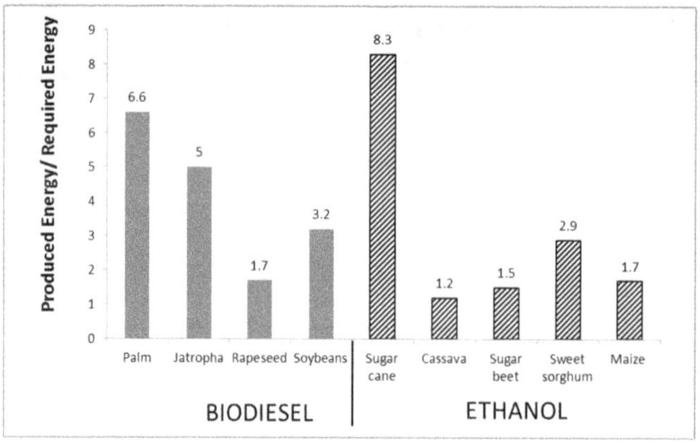

Figure 2.Energy efficiency in biofuel production

Source: *Ministerio de Minas y Energía (English: Ministry of Mines and Energy), based on Goldman Sachs and LMC*

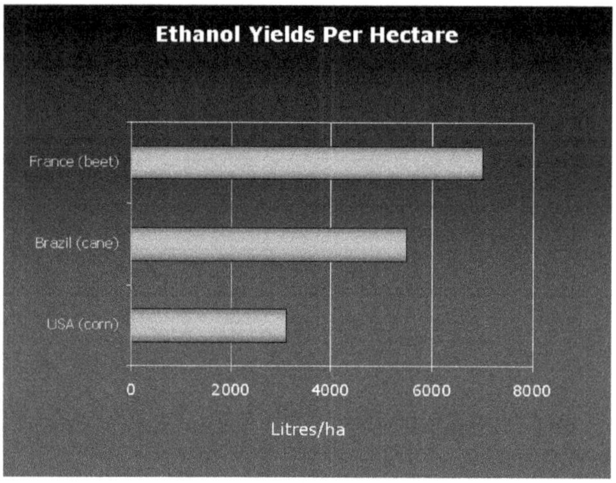

Figure 3.Ethanol yields from biomass (Source: FAO, 2008)

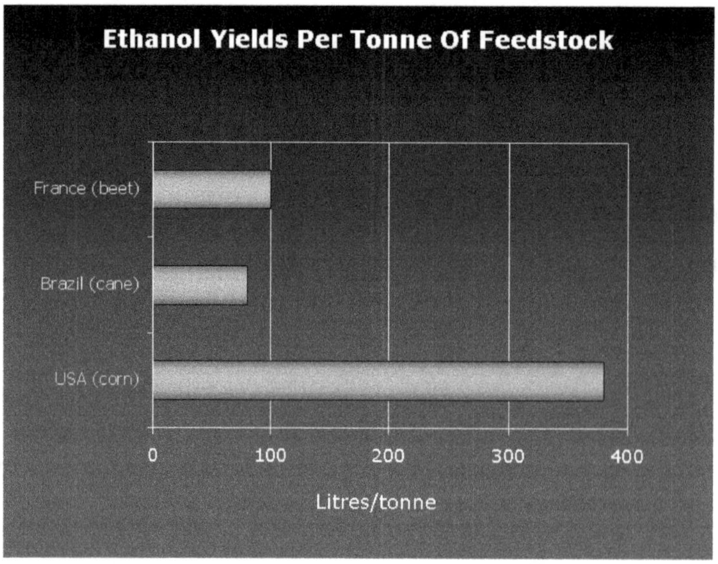

Figure 4.Ethanol yields in liters per Tone of Feedstock. (Source: FAO, 2008)

5. TECHNICAL CONSIDERATIONS

Biodiesel use in diesel engines is more limited. As well as ethanol, biodiesel is produced by fast pyrolysis of lignocellulosic biomass and mainly fermentation, because fast pyrolysis is a more expensive way (Bridgwater et al., 2002), it is a renewable oxygenated fuel with low cetane components (Ikura et al., 2003). Its heating value is about 60% of ethanol, but its high density makes up for its percentage. When using biodiesel in machines and engines there are some problems (Lopez & Salva, 2000) because of its higher viscosity and acidity, tar and fine particles resulting during working hours and solid residues during the combustion.

Following the direction of ethanol research, attempts have been made to overcome these problems by blending bio-oil with diesel to form an emulsion (Chiaramonti et al., 2003). In some success these efforts solve the operation with these fuels, however it is necessary to prove the feasibility and the additional cost of surfactant required to stabilize the blending which is a barrier for using it.

It must be considered that the blending of biodiesel and ethanol makes a stable blend and a fast pyrolysis, without using additives and surfactants. Current research on these blends is limited to gas turbines (López & Salva, 2000) and their use in these engines has shown positive results. Biodiesel blended with ethanol shall not exceed the problems of direct ethanol use in diesel engines without modification. However, using modified engines to use ethanol blends of ethanol/biodiesel could overcome the problems related to pure biodiesel combustion. As all new fuels, it is necessary to solve technical problems such as fuel storage, material compatibility, and procedures for turning engines on and off and long operation periods (Nguyen &Honnery, 2008).

5.1. Benefits

However, in Colombia, promotion of biofuels production may represent several benefits:

Energy sustainability: it will help to reduce the use of fossil fuels, thus protecting oil reserves. That is, a decreased risk of energy vulnerability. According to the Ministerio de Minas y Energía (English: Ministry of Mines and Energy) estimates show if new deposits are not found, known reserves will support the demand only for a few years. In this context, adding 10% of ethanol to gasoline helps to support fuel needs. Furthermore, Colombia has set the goal of increasing that percentage to 25% by 2020, which requires the new projects for ethanol production and the use of biomass sources other than sugar cane. In the short term the national program for Biofuels, seeks to improve fuel trade balance, and thus avoid wasting foreign reserves and spending at high prices by importing oil and petroleum products, that now tare close to 100 USD/barrel).

Environmental: biofuels are biodegradable, 85% is degraded in about 28 days.

Ethanol is a compound free of aroma, benzene and sulfur components, so the blending produces less smoke (particulates) and generate lower emissions (Stern, 2006). By using a 10% ethanol blending there is a reduction in CO emissions between 22% and 50% in carbureted vehicles, and a decrease of total hydrocarbons between 20% and 24% (Lopez & Salva, 2000).

With only a 10% blending of ethanol with gasoline, in new cars, 27% of carbon monoxide emissions decrease. In typical Colombian cars with 7-8 years of use it decreases 45%, and there is 20% reduction in hydrocarbons emissions. The effects of these reductions shall be reflected in the environmental emissions indices (Kumar, 2007), and in improve the citizens' living conditions, for example Bogotá where acute respiratory diseases are public health problems. Diesel blending decreases vehicle emissions such as particulate matter, polycyclic aromatic hydrocarbons, carbon dioxide and sulfur dioxide (U.S. Environmental Protection Agency, 2003).

Biodiesel is biodegradable, nontoxic and sulfur and aromatic components free, no matter the source of the oil used in its production. It reduces the soot emission in 40%-60%, and CO between 10% and 50%. Biodiesel can replace diesel (diesel fuel) without changes in ICE. Emissions with primary pollutants; with the exception of nitrogen oxides NOx. Despite these obvious benefits, there is not enough information about the solution to by-products and waste generated from biethanol-vinasses-and biodiesel-glycerin production processes, which are a source of future contamination if they are not properly disposed.

Agricultural development: biofuels production from agricultural raw materials, can guarantee both jobs growth and the possibility of crops diversification, including those for biofuel production. Export expectation, if there is pipe dream with Free Trades Agreement implementation, where Colombia supposedly is able to export bioenergy to poor energy countries, or that require large amounts of fuel for supporting economic growth.

Advantages of Colombia: As a reference, the abundance and variety of raw materials could be pointed out; several regions suitable for cultivation in all the country; guaranteed domestic market; government incentives and appropriate legal framework; high-yield crops, uninterrupted interest in research and development.

5.2. Regulations

Colombia, in order to reduce gasoline and diesel consumption, has implemented policies to encourage domestic production of biofuels. This purpose is economically boosted compared to fuels consumption reduction by the automotive industry and the best environmental indicators of mobile source emissions given the oxygenating effect of biofuels in combustion. For that reason in 2001 it is passed the Act Nº 693 and in 2004 the Act Nº 939, which states regulations on alcohol fuels and vegetable oils in the country, and creates incentives for their production, marketing and consumption.

In this regard, the Government has promoted development of biofuels through different measures to encourage their production and use. In this matter there is a broad regulatory and incentives for bioenergy production in Colombia, namely (Ministerio de Minas y Energía, 2007; Cala, 2003):

Act 693/2001: the regulations about the use of alcohol fuels are thereby stated; Incentives are created for their production, marketing and consumption. This act makes obligatory the use of oxygenated components in fuels for vehicles from cities with more than 500,000 inhabitants. A deadline of 5 years was established for gradual implementation of this regulation.

Act 788/2002: tax reform where exemptions were introduced to the Value Added Tax (VAT), the income tax and surcharge on alcohol fuel blended with gasoline engine.

Act 939/2004: defines the legal framework for the use of biofuels, by which the production and commercialization of biofuels of plant or animal origin, are thereby encouraged for use in diesel engines and other purposes. Exempts biodiesel from VAT and the income tax and establishes a net income exemption for 10 years to new oil palm plantation. This exemption applies to all plantations to be developed before 2015.

- **Act 1111/2006:** establishes a 40% income tax deduction of investments in real productive fixed assets of industrial projects, including financial leasing.
- **Act 1083 2006:** some regulations on sustainable urban planning and other provisions are thereby stated.
- **Resolution 1289/2005:** establishes biofuels criteria quality for their use in diesel engines, states the date of January 1st 2008 as a blending start of 5% of biodiesel with diesel fuel.

- **Resolution No. 180127/2007:** the heading "MD" in Act 4 from Resolution 82439 from December 23th, 1998 is thereby amended and amends Act 1st from Resolution 180822 from June 29th, 2005 and, states the provisions relating to Diesel Fuel pricing structure.
- **Decree 383/2007:** Amends the Foreign-Trade Zones Decree 2685 of 1999, regulates the set up of Special Foreign-Trade Zones for high economic and social impact.
- **Decree 3492/2007:** Act 939 of 2004 is thereby regulated.
- **Decree 2328/2008:** The Intersectoral Commission for Biofuels Management is thereby created.
- **Decree 4051/2007:** Permanent Foreign-Trade Zones area requirements is thereby stated; requirements for stating the existence of a Special and Permanent Foreign-Trade Zone and Industrial User recognition.
- **Resolution No. 180158/2007:** clean fuels are stated thereby in accordance with the Paragraph in Article 1, Act 1083.
- **Resolution No. 180782/2007:** biofuels quality criteria for use in diesel engines as a component of the blending with fossil diesel fuel in combustion processes are thereby amended.
- **Resolution No. 180212/2007:** Resolution 181780 December 29th, 2005 is thereby partially amended, regarding the pricing structure of diesel fuels blended with biofuel for their use in diesel engines.
- **Decree 2629/2007:** provisions for promoting the use of biofuels in the country are thereby stated, as well as applicable measures for vehicles and other motorized devices that use fuels. From January 1st, 2010 timetable is thereby set up for extending the mandatory blending of biofuels of 10% and, 20% from 2012 as well as the requirement that from January 1st 2012, new vehicle parc and other new motorized devices should be Flex-fuel at least 20%, for both E-20 blending (80% of gasoline from fossil fuel, with 20% of alcohol fuel) and B-20 (80% of diesel fuel with 20% of biofuels).
- **Decree 1135/2009:** In connection with the use of alcohol fuels in the country and applicable measures to motor vehicles using gasoline, decree 2629, 2007 is thereby amended. And which states in its article 1: from January 1st, 2012 motor vehicles up to 2000 cm3 manufactured, assembled, imported, distributed and marketed in the country and requiring gasoline to operate, must be soup up so that their engines run Flex-fuel system (E85), i.e. they can work normally by using either basic gasoline or blends composed of basic fossil fuel with at least 85% alcohol fuel. To meet the above, each brand shall sell vehicles in the Colombian market according to the following schedule and provisions:
- **From January 1st, 2012:** 60% of its annual supply must support E85.
- **From January 1st, 2014:** 80% of its annual supply must support E85.
- **From January 1st, 2016:** 100% of its annual supply must support E85.
- **From January 1st, 2013:** vehicles with engine cubic capacity greater than 2000 cm3 from all brands and models shall bear E85.

It is worth mentioning CONPES-3510/2008 document (in English: National Council for Economic and Social Policy document 3510/2008), where a policy to promote the production of sustainable biofuels in Colombia is thereby established, by taking advantage of economic and social development opportunities which are offered by biofuels emerging markets. Thus, it intends to expand the known biomass crops in the country and diversify the energy basket within a framework of production that is financially, socially and environmentally efficient and sustainable, that makes possible to compete in domestic and international markets.

Likewise the promotion of biofuels is also done through: the National Development Plan (NDP), the establishment of a regulatory framework and the development of financial and tax incentives. Also, the National Government has policy guidelines in areas such as: agriculture, research and development, infrastructure and environment that influence biofuels development.

There are also other complementary policy developments in the form of decrees and ministerial decisions that define the technical regulations, quality standards, as well as pricing, margins and rate parameters for fuel ethanol and biodiesel transport. There is an applicable regime in the Foreign-Trade Zone and several soft loan sources for agricultural development (González, 2008).

Among them, in the framework of AgroIngreso Seguro Program (AIS), financial instruments that provide soft loan sources for growing crops that produce biomass for ethanol and biodiesel production have been implemented. In addition, through the Incentivo a la Capitalización Rural, ICR (in English: Rural Capitalization Incentive) it is promoted, among others, oil palm crops establishment and renewal, and the construction of infrastructure for biomass processing (Consejo Nacional de Política Económica y Social (CONPES, 2008)

Despite this broad regulatory framework, there is uncertainty about changes in: regulation, raw material prices and emerging new technologies. In particular, with gallon prices as defined by state intervention (subsidies), that generates the discussion about how much does it mean for the national treasury, and whether it is advisable or heavy subsidies is fair to benefit a minority that supply biofuels, for even small domestic market and one that is difficult to be exported.

As shown, the Colombian Government has a fairly strong policy and information that allows for investment in projects, sustainable energy and biofuels plans and programs through a set of tools, studies and institutional strengthening.

Therefore, the Colombian Government has promoted assessments that seek to: a) study the implications of the biofuel industry, from planting crops for biofuel production to the final consumers of ethanol or biodiesel (flex-fuel or normal vehicles); b) analyze the current infrastructure requirements for the expansion of the biofuel market; c) know the sector current status, as well as the economic instruments, regulatory elements, policies and tax incentives required or recommended for promoting renewable energy, energy efficiency and biofuels; d) analyze the renewable energy potential, energy efficiency and carbon credits through the Clean Development Mechanism.Likewise, institutional strengthening assessment required by the Ministerio de Minas y Energía (English: Ministry of Mines and Energy) (MME), in energy efficiency, renewable energy, bioenergy and carbon financing.

This set of measures that promote the enthusiasm for liquid biofuels such as the mandatory blending of biofuels with fossil fuels and tax incentives, have created a fast artificial growth in biofuel production. These incentives have broad social impacts, as they are resources that do not come into the State, and are taken for solving important issues such as health, education and basic sanitation.

These measures entail high economic, social and environmental costs and should be monitored promptly.

5.3. Current projects under construction

Ethanol: In compliance with the provisions of Act 693/01, the country began to implement initiatives for alcohol fuel from sugar cane. At the moment 5 ethanol plants are running: Incauca, Providencia, Manuelita, Mayaguez and Risaralda refineries that produce about 1,050,000 liters of alcohol fuel a day and this production is mainly to supply the domestic market. It is estimated a domestic demand close to 1,500,000 liters per day to cover the 10% of blending needs.

Likewise, in the country several alcohol production projects are being implemented in several departments: Antioquia, Boyacá, Santander and the coast, derived from different raw materials such as sugar cane, sugar beet, banana and cassava.

Unfortunately, due to the economic crisis there is absence of new plants. Projects are standstill, and Ecopetrol plant would only come into operation in 2011, starting with a production of 385,000 liters a day. At the moment there is another project being developed in Magdalena, where an international company sowed a very large sugar cane area for producing an average of 300,000 liters a day. With this, the 20% blending could be reached by in 2012 without any problem.

Biodiesel: At the moment there are five projects under construction for producing biodiesel from oil palm (Oleoflores – already in production, Odin Energy, BiocombustiblesSostenibles del Caribe) and two in the eastern region (Biocastilla, Bio D. SA). In addition, they are other projects under development, one in the central region (ECOPETROL), one in the eastern region (Manuelita), one in the west region and another in the north region. In 2008 it is expected they shall enter into production, with a total amount of 400,000 t/year (19).

How are investments for biodiesel production doing? Construction of the Ecopetrol plant in Barrancabermeja is almost over. With this in total there will be seven plants in the country. A total installed capacity of 526,000 biodiesel tonnes a year may be achieved.

6. CONCISE SUMMARY

It is necessary to acknowledge that the contemporary individual faces the same obstacle that our predecessors successfully resolved many centuries ago, which is the realization that life is not concluded. The availability and utilization of natural resources necessitate the development of technological innovations and social coexistence. In this context, the prioritization of ethical considerations over monetary concerns becomes increasingly important, particularly in our interconnected global society, which demands new economic, lifestyle, consumption, and value paradigms. Society need energy for its progress, however, progress does not inherently entail an excessive consumption of energy. During any productive process, there is a possibility of wastage of resources and water. However, it is definite that energy will be consumed, and this energy consumption will have a tangible environmental impact. Should energy production bear the full burden of costs, it would result in much higher expenses. Emerging energy sources have become a significant economic, political, and environmental strategy. Currently, more than 30 raw materials are undergoing global testing due to their significant importance. Although they have experienced significant growth, they have not yet offered a resolution to the worldwide energy challenges.

Biofuels should not be regarded as a definitive answer to the energy and environmental predicament, but rather as a component of a multifaceted human and energy endeavor, in which major nations still hold divergent views on the optimal resolution. When Bioenergy is utilized effectively, it presents a significant chance to support the development of several impoverished nations worldwide. It is important to highlight a fact: alcohol fuel is pricier than both gasoline and biodiesel. It is not conducive to excellent business for a market economy to operate in isolation and without intervention. In order for other options to be feasible, the market must be regulated, as competing fuel sources are more cost-effective. Oil is found in the reservoir, whereas crops such as cassava, sugar cane, oil palm, or other raw materials need to be cultivated, typically on costly land. By definition, a feasible project is one that can only be operated outside the market framework with the intervention of the State. The world is confronted with intricate problems, and the sustainability of life on Earth cannot be sustained solely by relying on biofuels as a renewable energy alternative. This approach would lead to the substitution of food crops with monocultures, deforestation for energy crops, and an increase in the extinction of diverse species. Additionally, it would result in the depletion of fertile lands and water resources, as well as social consequences such as population displacement. The FAO has stated that it is necessary to promptly reassess biofuel policies and subsidies in order to maintain the objective of global food security, safeguard the interests of impoverished farmers, stimulate comprehensive rural development, and guarantee environmental sustainability. Additionally, it is mentioned that the increasing demand for biofuels and the subsequent rise in agricultural commodity prices present significant prospects for certain developing nations. Agriculture has the potential to serve as the primary driver for reducing hunger and alleviating poverty. By providing help to impoverished small farmers, particularly in terms of expanding their production and accessing markets, the development of biofuel feedstocks can generate income and employment opportunities. Additionally, a certification mechanism is necessary to guarantee that biofuels will only be sold if they meet the required environmental standards. Colombia cannot ignore or remain unaffected by the worldwide market trend for crude oil and its byproducts. This feature presents the possibility of producing items, such as biofuels, which contribute to diversifying the range of energy sources in the home market and can also be exported to global markets. However, an essential prerequisite for participating in the global market is the presence of effective production conditions for these items. Colombia possesses ample land suitable for cultivating biofuels. Out of the total land area of 14 million hectares designated for agriculture and 20 million hectares for livestock, only 5 million hectares are currently utilized. The remaining land is primarily utilized for extensive cattle ranching. However, a more beneficial utilization of this land would be for biofuel production, which would not only provide plant coverage but also create opportunities for rural income. Additionally, it exhibits significant efficiency in sugar production derived from sugar cane. However, this activity is primarily concentrated within agribusiness models, with production being dominated by a few number of corporations affiliated with well-established economic conglomerates. While Colombia has been a trailblazer in the use of ethanol as a biofuel, biodiesel projects are now gathering momentum and have the potential to make a more significant impact across the country. In rural areas, there is a suboptimal utilization of natural resources and a significant reliance on them. There is a lack of consensus regarding the appropriate allocation and sustainable utilization of these resources. The productivity paradigm emphasizes predatory models and prioritizes economic efficiency and profitability as the main metric. This approach overlooks the social and environmental advantages of productive operations. Next, it is necessary to establish a long-term sustainable agriculture that is in harmony with the environment. The objective of this is to conduct a thorough evaluation of the existing modernization model, considering that various technical options, in conjunction with a range of socio-economic and environmental issues, necessitate distinct technological remedies. Hence, while making judgments on biofuels, it is imperative to take into account not only the food safety issue, but also the availability of land and water resources. Energy is intricately connected to the three dimensions of sustainability (economic, social, and environmental). This means that it needs to be integrated, harmonized, and optimized. The services provided by energy are essential for fulfilling various fundamental needs, including water supply, lighting, healthcare, food production, transportation, and information access. Therefore, access to advanced forms of energy such as electricity, liquid fuels, and gaseous fuels should be considered as inherent human rights in the 21st century. Ensuring the security of energy supply and maintaining stable energy prices are essential factors for promoting economic growth. However, it is evident that numerous methods of production and consumption can diminish environmental sustainability. It is imperative to inquire whether the present energy production and consumption are sustainable. One of the primary difficulties confronting humanity is to discover methods for generating and utilizing energy in a manner that fosters long-term human progress across all aspects, including social, economic, and environmental dimensions. In order to achieve a balanced perspective, it is imperative to thoroughly analyze the economic, social, and environmental impact of bioenergy before determining the desired pace of development. Additionally, significant consideration should be given to the selection of appropriate technologies, legislation, investment, and research initiatives.

REFERENCES

Altin, Recep, Çetinkaya, Selim & Serdar Yücesu, Hüseyin. (2001). The potential of using vegetable oil fuels as fuel for diesel engines, Energy Conversion and Management, Vol. 42, No. 5, pp. 529-538, ISSN 01968904

Bridgwater, A.V., Toft A.J. & Brammer, J.G. (2002). A techno-economic comparison of power production by biomass fast pyrolysis with gasification and combustion, Renewable Sustainable Energy Reviews, Vol. 6, No. 3, pp. 181–246, ISSN 13640321

Brown, Lester R., Renner, Michael & Halweil, Brian. (2000). Signos vitales 2000: Las tendencias que guiarán nuestro futuro, GAIA Proyecto 2050, ISBN 9788493023225, Spain.

Cala, David F. (2003). Proyecto para producción de biodiesel a partir de palma africana en Colombia, Corporación para el Desarrollo Industrial de la Biotecnología y Producción limpia (CORPODIB), Bogotá, Colombia

Casilda, Béjar Ramón. (2002). Energía y desarrollo económico en América Latina. Boletín ICE Económico, No.2750, (December 2002), pp.31-43, ISSN 02102633

Chiaramonti, D., Bonini, M., Fratini, E., Tondi, G., Gart, K., Bridgwater, A.V., Grimm, H.P., Soldaini, I., Webster, A. & Baglioni, P. (2003). Development of emulsions from biomass pyrolysis liquid and diesel and their use in engines, part 1: Emulsion production, Biomass and Bioenergy, Vol. 25, pp. 85-99, ISSN 09619534

Consejo Nacional de Política Económica y Social (CONPES 3510). (2008). Lineamientos de política para promover la producción sostenible de biocombustibles en Colombia, Bogotá, Colombia

Cortés, Elkin Marin, Suarez, Héctor Mahecha & Pardo, Sandra Carrasco. (2009). Biocombustibles y autosuficiencia energética, Dyna, Vol.76, Nro. 158, pp. 101-110, ISSN 00127353

Cortés, M. E. (2007). Biocombustibles: ¿alternativa para la agricultura colombiana?, Memorias Agroexpo, Bogotá, Colombia

Cortés, Elkin & Álvarez, Fernando. (1998). Consumo energético y desarrollo agrícola sostenible, Ponencias de la III semana técnica Nacional de Ingeniería Agrícola, Medellín, Colombia, October 1998.

Food and Agriculture Organization of the United Nations (FAO). (2008). The state of food and agriculture, FAO, Retrieved from ftp://ftp.fao.org/docrep/fao/011/i0100e/i0100e.pdf

Gonzáles, M. César. (2008). Peor el remedio, In: Portafolio, 10.12.2010, Available from: http://www.portafolio.com.co/archivo/documento/MAM-2833067

Ikura M., Stanciulescu, M. & Hogan, E. (2003). Emulsification of pyrolysis derived bio-oil in diesel fuel, Biomass and Bioenergy, Vol. 24, No. 3, pp. 221 – 232, ISSN 09619534

International Energy Agency. (March 2006). Word Energy outlook 2006 In: International Energy Agency, 08.09.2010, Available from: http://www.iea.org/textbase/nppdf/free/2006/weo2006.pdf

Kumar, A.A. (2007). Biofuels (alcohols and biodiesel) applications as fuels for internal combustion engines, Progress in Energy and Combustion Science, Vol. 33, No.3, pp. 233–271, ISSN 0360-1285

López, Juste G. & Salva Monfort, J. J. (2000). Preliminary test on combustion of wood derived fast pyrolysis oils in a gas turbine combustor, Biomass Bioenergy, Vol. 19, No.2, pp. 119–128, ISSN 09619534

Mesa, Jens. (2006). Biocombustibles y Agricultura, Primer Congreso Grupo Empresarial del Campo, Bogotá, Colombia, November 15-17, 2006.

Ministerio de Minas y Energía-Unidad de Planeación Minero Energética. (2007). Desarrollo y consolidación del mercado de biocombustibles en Colombia, Bogotá, Colombia.

Nguyen, D. & Honnery, D. (2008). Combustion of bio-oil ethanol blends at elevated pressure, Fuel, Vol. 87, No. 2, pp. 232–243, ISSN 00162361

Pérez A., José Ignacio. (2002). In: Energía y desarrollo sostenible, 10.12.2010, Available from: http://www.oei.es/decada/portadas/MedAmbPDF4.pdf

Silvestrini, Vittorio. (2000). Qué es la entropía?, Colección Milenio/Norma, ISBN 9580438757, Colombia

Stern, D.I. (2006). Reversal of the trend in global anthropogenic sulfur emissions, Global Environmental Change, Vol. 16, No. 2, pp. 207–220, ISSN 09593780

Testing framework for sustainable biomass. (February 2007). In: Final report from the project group"Sustainable production of biomass", 16.11.2010, Available from: http://www.lowcvp.org.uk/assets/reports/070427-cramer-finalreport_en.pdf

U.S. Environmental Protection Agency. (2003). National Air Quality and Emissions Trends Report: 2003 Special Studies Edition, North Carolina, USA

Valero, Antonio. (March 2004). In: Energía y Desarrollo Social, 9.10.2010, Available from: http://habitat.aq.upm.es/boletin/n32/aaval.html.

CHAPTER-24

MICROFLUIDIC BIOFUEL CELL UTILIZING ENZYMES FOR MICROPOWER GENERATION

1. INTRODUCTION: AN OVERVIEW

Enzymatic biofuel cells (BFCs) utilize enzymes to facilitate chemical processes, thereby substituting the usual electrocatalysts found in standard fuel cells. These systems produce electricity by oxidizing renewable energy sources under mild circumstances (Calabrese Barton et al., 2004). At the anode, the fuel undergoes oxidation, resulting in the release of electrons. These electrons are then utilized to reduce the oxidant at the cathode (see Figure 1).

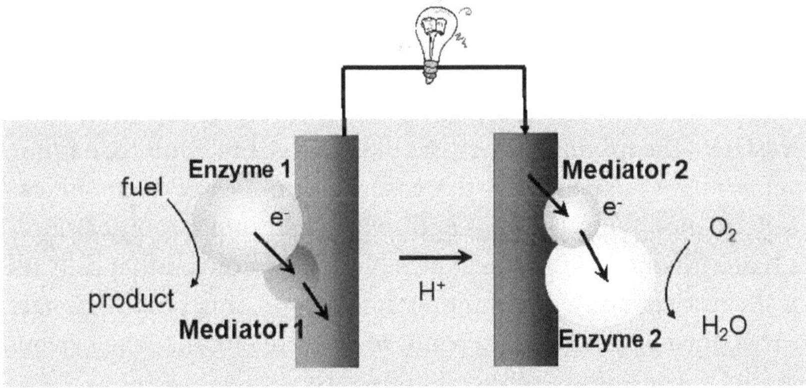

anode-oxidation Cathode-reduction

Figure 1.Enzymatic biofuel cell principle.

An efficient connection is established by employing suitable redox mediators, which are commonly dyes or organometallic complexes. These mediators play a crucial role in transporting electrons from the enzymes to the electrode surface. Biocatalysts have several advantages over precious metal catalysts, including the capacity to selectively react with specific reactants, high activity even at room temperature under physiological settings, and ease of manufacturing. Various plentiful organic raw materials, such as sugars, low aliphatic alcohols, and organic acids, can be utilized as substrates for the oxidation process. The primary substrate being reduced in this process is predominantly molecular dissolved oxygen. The concept of a biological fuel cell was introduced in 1964 by Yahiro and his colleagues (Yahiro et al., 1964). Their paper detailed the development of a methanol/O2 cell. During the 1990s, there was a rise in the importance of biofuel cells (BFCs) due to the development of new electrode chemistries by Katz and Willner (Wilson, 2002), as well as Heller (Degani et al., 1987). The biofuel cell that has been extensively researched utilizes glucose as the fuel and oxygen as the oxidant (Service, 2002). Glucose is oxidized to gluconolactone at the anode by the enzymes glucose oxidase (GOx) or glucose dehydrogenase. At the cathode, dioxygen is reduced to water by the enzymes laccase or Bilirubin oxidase (BOD), which belong to the group of multicopper oxidases. Enzymatic fuel cells often exhibit power outputs ranging from microwatts to milliwatts. Nevertheless, the experiments are frequently conducted under varying parameters such as concentration, temperature, pH, and mass transit conditions, making it challenging to compare diverse setups described in the literature. Biofuel cells (BFCs) have lower power densities and shorter lifetimes compared to traditional fuel cells due to issues with enzyme stability and electron transfer rate (Bullen et al., 2006). To enhance performance, it is necessary to optimize the components. The literature describes many options, such as catalysts, enzymatic electrode assemblies, and design, which have been studied by Davis et al. (2007) and Ivanov et al. (2010). In recent times, the rapid expansion of portable, wireless consumer electronics and biomedical equipment has significantly contributed to the advancement of novel micro power sources capable of providing sustained power for extended durations. Miniature battery fuel cells (BFCs) are seen as a potentially advantageous substitute (Gellett et al., 2010) for powering wireless sensor networks (WSN). Nevertheless, the process of making these devices smaller has considerable technical obstacles related to manufacturing methods, expenses, device design, and material properties. These devices must exhibit comparable efficiency and power density to bigger biofuel cells, while utilizing fewer reagents, occupying less space, and requiring less time. The literature primarily refers to a limited number of tiny biofuel cells. These devices encompass both tiny versions of standard devices and micro-devices that employ innovative ways of energy conversion. This chapter discusses the latest developments in the miniaturization of biofuel cells (BFCs). The initial presentation will feature miniature traditional BFCs. However, our discussion will primarily concentrate on the advancement of microfluidic enzymatic BFCs, in which microfluidics plays a crucial and direct role. These little devices function within the structure of a microfluidic chip. They harness the laminar flow of fluids, which restricts the mixing of fuel and oxidant within a microchannel, hence removing the requirement for a membrane. Consequently, the reaction kinetics of both the cathode and anode can be enhanced separately by modifying the fuel and oxidant stream's composition. This text proposes a debate on the factors that influence the performance of microfluidic biofuel cells (BFCs). The focus of the discussion is on both theoretical and experimental research in this field. Ultimately, this presentation highlights the factors that must be taken into account in order to enhance the performance of microfluidic devices for optimal results in the energy conversion process.

2. MINIATURE BFCS

This section provides a concise overview of the miniature traditional BFCs. The literature mostly refers to a limited number of miniaturized BFCs that operate using glucose and O2. These devices have mainly been developed by lowering the size of the electrodes and the volume of the cells. Various approaches have been employed to design smaller BFCs. Heller and his colleagues have effectively showcased the efficacy of novel small membraneless biofuel cells operating in physiological settings. A novel small gadget has been created, consisting of two components: a 7-μm diameter anode and a 2-cm long cathode comprised of carbon fibers. These components are housed within a polycarbonate framework. The anode was altered using glucose oxidase (GOx) while the cathode was adjusted using either laccase or biological oxygen demand (BOD). These modifications were facilitated using redox osmium-based hydrogels, as described by Mao et al. (2003). In 2001, researchers successfully created the initial compact fuel cell without a membrane, which produced a power density of 140 μW cm-2 at 0.4V (Chen et al., 2001). This prototype demonstrated the feasibility of a small, self-contained sensor and transmitter system (Bullen et al., 2006; Heller, 2004). The devices achieved greater power densities of 431 μm cm-2 at 0.52 V (Mano et al., 2002) and 440 μm cm-2 at 0.52 V (Mano et al., 2003) by utilizing a better redox polymer to connect the reaction centers of enzymes to the electrodes. These results were obtained at pH 7.2, 37 °C, and with a glucose concentration of 15 mM. The high power density of these devices is achieved through the efficient cylindrical mass transport at the carbon fibers and the utilization of effective redox polymers for electron transportation. Additionally, they demonstrated that when this device was implanted in a living organism, namely near the skin of a grape, it generated a power density of 240 μm cm-2 at 0.52V. Subsequently, by the substitution of carbon fibers with manufactured porous microwires composed of oriented carbon nanotubes, researchers were able to create the most effective glucose/O2 biofuel cell (BFC) to date (Gao et al., 2010). This innovative design achieved a remarkable power density of 740 μW cm-2 at a cell voltage of 0.57 V. The experiment's success likely leads to an augmentation in the transfer of substrates' mass. Currently, there are alternative small devices that have inferior performance, which are referred to as stacked biofuel cell designs. A study detailed a chip-based stacking structure consisting of six cells connected in series, as outlined by Nishizawa et al. in 2005. The structure was comprised of a glucose oxidase (GOx) anode and a polydimethylsiloxane-coated platinum (Pt) cathode. The arranged cells on the chip exhibited a power output of 40 μW cm-2 when tested in an air-saturated buffer solution with a glucose concentration of 5 mM. Another study documented the creation of a compact biofuel cell (BFC) that occupies an area of 1.4 cm2. This was achieved by implementing a design based on stackable proton exchange membrane (PEM) fuel cells (Fischback et al., 2006). The gadget was comprised of a cathode that takes in air and an anode that contains clusters of crosslinked GOx enzymes on the surface of carbon nanotubes. This work demonstrated the crucial significance of buffer solution in establishing the effectiveness and durability of tiny biofuel cells (BFCs). The initial performance of the buffered fuel solution was significantly high, measuring 371 μW cm-2. However, it rapidly declined as a result of the deactivation of the proton exchange membrane. Nevertheless, in an unbuffered solution, the initial performance exhibited a lower value of 117 μW cm-2, primarily attributed to the low pH state. However, the performance remained remarkably steady for a duration of 10 hours. This study indicated that the utilization of a compact system and fuel solution without buffering would be advantageous for real-world applications. Presently, in these small-scale devices, the density of electric current and the amount of power produced are primarily restricted by the process of fuel diffusion to the surface of the electrode. An intriguing solution for optimizing transport efficiency involves utilizing hydrodynamic flow and pumping the fuel directly to the electrode.

3. MICROFABRICATED DEVICES

3.1. Advantages of microfluidics

Microfabrication techniques offer an alternate method for reducing the size of energy conversion devices. Microchemical systems include intrinsic advantages compared to macrosystems, such as enhanced rates of mass transfer, minimal reagent usage, improved safety due to smaller volumes, and the ability to connect many microreactors. Microfluidic techniques are well-suited for reducing the size of devices that have channels with a height of less than a millimeter and exhibit smooth flow. The application of microfluidics in fuel cells has experienced significant advancements since the early 2000s (Ferrigno et al., 2002; Choban et al., 2004). All functions and components associated with the delivery and removal of fluids, reaction sites, and electrode structures are contained within a microfluidic channel in these devices. In the channel, depicted in Figure 2, the fuel (pink) and oxidant (blue) streams flow in close proximity, maintaining a near-parallel arrangement. This arrangement minimizes the diffusion of the two streams. Molecular diffusion is the sole mechanism by which molecules in opposing fluid streams can intermingle across their contact. The absence of convective mixing facilitates the smooth flow of fluids in a laminar manner.

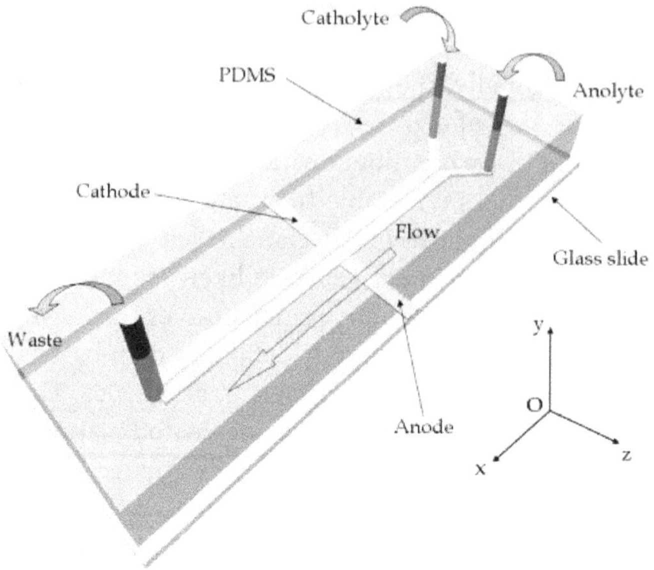

Figure 2.Laminar flow of streams in a microfluidic channel.

The electrochemical reactions take place at the anode and cathode located within the respective streams, without needing a membrane to minimize the ohmic drop, what maximises the current density. Protons diffuse through the liquid-liquid interface created by the contacting streams of fuel and oxidant. The cathode and the anode are connected to an external circuit. The technique to force the fluid through microchannels is the pressure driven flow, in which the fluid is pumped through the device via positive displacement pumps, such as syringe pumps.

As summarized by authors (Luo et al., 2005; Gervais et al., 2006; Sun et al., 2007), the limiting factors in laminar flow-based microfluidic fuel cells that influence the performances are (i) cross-diffusional mixing of fuel and oxidant at the interface between the two streams, and (ii) the formation of depletion boundary layers at the surface of the electrodes as the result of the reaction of fuel and oxidant.

Interesting papers have presented theoretical and experimental works to describe how to prevent or reduce these phenomena by concentring research efforts on designs, electronic and ionic conductivity, and electron-transfer kinetics in microfluidic fuel cells (Lee et al., 2007). The role of flow rate, microchannel geometry, and location of electrodes within microfluidic systems was also studied (Choban et al., 2005; Sun et al., 2007; Amatore et al., 2007; Chen et al. 2007).

Similarly to microfluidic fuel cells, advanced microfabrication techniques can be applied to build components of microfluidic enzymatic BFCs. The number of devices presented to date is limited. The devices have been developed based on both laminar flow within a microchannel and biological enzyme strategies. Indeed, the advantage of the co-laminar flow is to choose the composition of the two oxidant and fuel streams independently for optimum enzymatic activity and stability to improve reaction rates and current density (Zebda et al., 2009a).

3.2. Microfluidic biofuel cell fundamentals

In a microfluidic channel, the relationship between the fluid velocity and the absolute pressure for an incompressible viscous liquid is given by the classical fluid dynamics theory and the well-known Navier-Stokes equation:

$$\frac{\partial \vec{v}}{\partial t}+\left(\vec{v}\cdot\vec{\nabla}\right)\vec{v}=-\vec{\nabla}\left(\frac{P}{\rho}\right)+\mu\,\Delta\vec{v} \tag{1}$$

Where stands for the fluid velocity vector with components (u, v, w), each expressed for a set of Euler components (x, y, z, t), P is the absolute pressure, ρ is the relative density, and μ is the kinematic viscosity.

In the case of a microfluidic horizontal straight channel (x-direction), the flow is always laminar under low pressure drop (typically a few bar), leading thus to a unidirectional flow and a uniform absolute pressure in the cross-section. For a fixed pressure drop ΔP between the inlet and the outlet of the channel, Eq. 1 simplifies to:

$$\frac{\partial u}{\partial t}=-\frac{1}{\rho}\frac{\Delta P}{L}+\mu\left(\frac{\partial^2 u}{\partial y^2}+\frac{\partial^2 u}{\partial z^2}\right) \tag{2}$$

Where L is the length of the microchannel. When the permanent flow is reached, the time derivative term becomes zero and Eq. 2 simplifies to:

$$\frac{-\Delta P}{\eta L}+\frac{\partial^2 u}{\partial y^2}+\frac{\partial^2 u}{\partial z^2}=0 \tag{3}$$

Where η is the dynamic viscosity (10-3 Pa.s for water at 20°C), defined as the product of the dynamic viscosity and the relative density ρ. Due to the very large aspect ratio of the rectangular cross-section of the microchannel, a 2D approach is usually considered that leads to a pseudo infinite-plate flow (except in the borders). The directions along the length and height of the microchannel are indicated as x and y coordinates, respectively (see fig. 2). A typical parabolic rate profile is obtained for pressure driven flow:

$$u(y)=\frac{\Delta P}{2.\eta.L}\left(\frac{h^2}{4}-y^2\right) \tag{4}$$

Where h is the height of the microchannel and y is defined as y=0 at the middle of the microchannel and y= ± h/2 at the upper and under walls. By considering a rectangular microchannel (with l the width of the microchannel) in Eq. 4, the flow rate, Q, in laminar regime, is deduced and is proportional to the applied pressure (Eq. 5):

$$Q=\frac{\Delta P.l.h^3}{12.\eta.L} \tag{5}$$

In electrochemical laminar flow systems, the mass transport is achieved by both diffusion and convection transport. In the case of Y-shaped microchannel, the mixing between the two laminar streams occurs by transverse diffusion. Microscale devices are generally characterized by high Péclet number, Pe, (Pe = U av h/D, with U av the average velocity of the flow, h the height of the microchannel and D the diffusion coefficient of the molecule). In this condition, the transverse diffusion is much lower than the convection, and the diffusive mixing of the co-laminar streams is restricted to a thin interfacial width, δ mix , in the center of the channel (Fig. 3) that grows as a function of the downstream position (x) and the flow rate, determined from Eq. 6 (Ismagilov et al., 2000):

$$\delta_{mix} = \left(\frac{(Dhx)}{U_{av}} \right)^{\frac{1}{3}} \quad (6)$$

Where D is the diffusion coefficient for ions of type i and U av is the average flow velocity defined as:

$$U_{av} = \frac{\Delta P.h^2}{12.\eta.L}$$

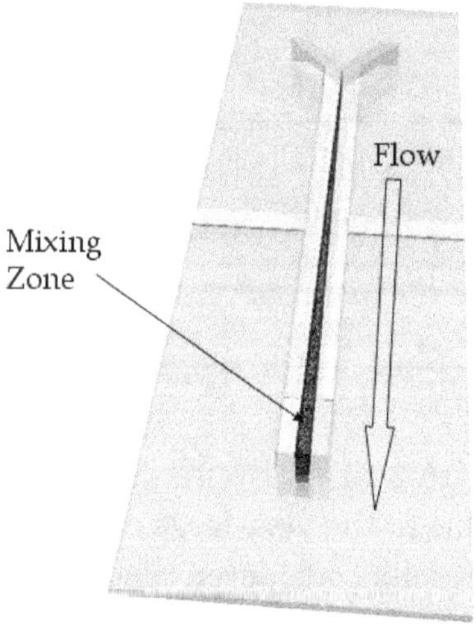

Figure 3.Schematic of a laminar flow in a microchannel with the formation of the diffusion region during operation of a microfluidic BFC.

For fast electron transfer and in excess of supporting electrolyte, the kinetics of a simple electrochemical redox reaction is controlled by diffusion and convection. The concentration profiles of the chemical species involved in the reaction are determined by solving the convective diffusion equation (Eq. 8):

$$\frac{\partial c_i}{\partial t} + \overline{\nabla}\left(-D_i \overline{\nabla} c_i\right) + \overline{v}\overline{\nabla} c_i + R_i = 0 \quad (8)$$

Where ci is the concentration of species i, D i its diffusion coefficient, t the time, the fluid velocity vector (given by Eq. 1) and R i a term describing the rate of net generation or consumption of species i formed by homogeneous chemical reaction.

In the case of a microfluidic biofuel cell as described in this work, Eq. 8 can be simplified into a 2-dimensionnal cartesian steady state (Eq. 9):

$$-D_i \left(\frac{\partial^2 c_i}{\partial x^2} + \frac{\partial^2 c_i}{\partial y^2} \right) + u(y) \frac{\partial c_i}{\partial x} = 0 \quad (9)$$

The boundary conditions associated are usually: (i) ci=c°i (bulk concentration) at the inlet of the microchannel,Ci=0 at the electrode surface andno flux at the other walls (no electrochemical reaction).

Those simulations were exploited in order to calculate the diffusive flux at the electrode, defined as:

$$J_{diff}(x) = D \left. \frac{\partial C}{\partial y} \right|_{electrode} \quad (10)$$

And, therefore, the total current is expressed as:

$$I_{tot} = \int_{electrode} n.F.J_{diff}(x)dx = n.F.D. \int_{electrode} \left(\frac{\partial C}{\partial y}\bigg|_{y=0}\right)dx \tag{11}$$

Where n is the number of electron exchanged, and F is the Faraday constant.

The pumping power Wpumping , required to sustain steady laminar flow in the microchannel by the syringe pump, is estimated on laminar flow theory (Bazylak et al., 2005) as the pressure drop multiplied by the flow rate:

$$W_{pumping} = \frac{8.\eta.L.Q^2}{l.h^3} \tag{12}$$

One can note that the contributions from inlet and outlet feed tubes are not included, because they are negligible.

The fuel utilization (FU) is estimated by the following Eq. 13, defined as the current output divided by the flux of reactant entering the channel (Bazylak et al., 2005):

$$FU = \frac{I}{n.F.C.Q} \tag{13}$$

The fuel utilization is maximized for the lowest flow rate and decreases with flow rate. Typical fuel utilization for microfluidic fuel cell is ~ 1% (Hayes et al., 2008).

3.3. Manufacturing technology

3.3.1. Fabrication of the microchannel

Microchannel architecture typically represents a T- or Y-channel configuration. There are mainly two fabrication techniques. The first one utilizes conventional chip-manufacturing techniques of semiconductor industrics. Silicon wafers are patterned by lithography step followed by etched step in order to get the desired form of the channel (Moore et al., 2005; Lee et al., 2007). The second technique allows the fabrication of microchannels by rapid prototyping using standard soft lithography procedure to build the channel in poly(dimethylsiloxane) (PDMS) (Duffy et al., 1998). PDMS is relatively inert and compatible with most solvents and electrolytes (Kjeang et al., 2008). Besides it is permeable to gases, which is essential for biofuel cells working from enzymes with oxygen as the cofactor. Typically, a pretreated microscope glass slide or a silicon wafer is coated with a thin layer of photoresist by spin-coating and exposed directly to UV light through a photomask that defines the desired channel structure. Several thick photoresist layers are sequentially laminated on the first layer to get the desired channel depth, and then exposed to UV light. The structure is then developed by spraying an aqueous solution of sodium carbonate (1% wt) and hardened by a final irradiation. The result is a master with a positive pattern defined by the master. The channel structure is thus obtained by pouring PDMS monomer over the master, followed by curing at 70 °C during 2 h. After cooling, the PDMS slab is peeled off from the master, and holes are punched to provide fluid access (Stephan et al., 2007).

3.3.2. Fabrication of the microelectrodes

Most of the microfluidic devices employ patterned electrodes positioned in parallel on the bottom wall or on sides of the channel (Kjeang et al., 2008). Electrodes, with varying length and wide, are patterned by coating glass slides with conductive materials such as gold, graphite over an adhesive layer (often chromium or titanium) by standard sputtering techniques (Zebda et al., 2010) or by photolithography and sputtering (Moore et al., 2005; Lim et al., 2007; Togo et al., 2008). The inter-electrode gap varies between 0.2 mm and 1.4 mm (Lee et al., 2007; Togo et al., 2008; Zebda et al., 2010).

3.3.3. Fabrication of the microfluidic devices

The microfluidic device is finally obtained by physically clamping the PDMS slab with the glass substrate that accommodated the electrode pattern. This approach works well with elastomer polymer like PDMS. Alternatively, an irreversible seal may be achieved between both parts by oxygen-plasma treating prior to improve the adhesion (Lim et al., 2007; Lee et al., 2007). Alignment of the flow channel over the microelectrodes is often aided by a microscope. As an example, a device, consisted of a Y-shaped channel with two inlets and two outlets, is presented in Fig. 4. The pressure-driven laminar flow required for injection of fuel and oxidant is typically driven by a syringe pump via polyethylene tubing.

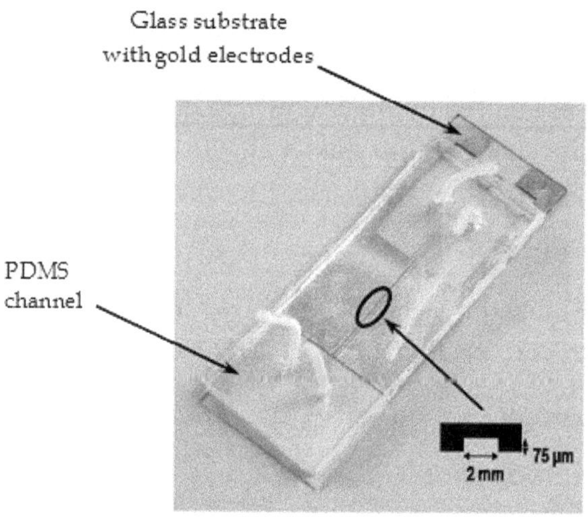

Figure 4.Schematic microfluidic BFC based on a Y-shaped channel with two inlets and two outlets.

3.4. Performances of microfluidic biofuel cells

This paragraph mainly describes microfluidic BFCs involving mediated monoenzymatic systems, which are capable of only partial oxidation of the fuel. Devices have been developed either with diffusional enzymes flowing through the microchannel, or with immobilized enzymes on electrode surface. This paragraph also includes preliminary works on devices allowing improvement of fuel utilization by complete oxidation, which have been designed with multienzymatic systems.

The performances of microfluidic BFCs are evaluated from cell voltage and current density. The cell voltage of the biofuel cell reflects both the open circuit voltage (OCV), partially controlled by the formal potential of the two redox mediators and the overpotential losses. The delivered current density reflects the rate of catalytic turnover and transport processes as a function of the surface area of the electrode. The power density is the product of cell voltage and cell current density. As already mentioned, the performances of the microfluidic BFCs are limited (i) by cross-diffusional mixing (δ_{mix}) of fuel and oxidant at the interface between the two streams, (ii) by formation of depletion boundary layers at the surface of the electrodes as the result of the reaction of fuel and oxidant, and (iii) by low concentration and low diffusion coefficient of oxygen.

These factors depends on geometric and process parameters such as the microchannel dimensions, the electrode parameters (number of electrodes, electrode surface area, electrode spacing), and operating conditions (electrolyte, flow rate, pH, concentration of species). The influence of these parameters on open circuit voltage, current density and power density, have been evaluated both experimentally and theoretically in literature.

3.4.1. Microfluidic BFCs with soluble enzymes

3.4.1.1. Strategies to limit the cross-diffusional mixing

In order to restrict fuel and oxidant mixing to a thin interfacial width **δ** mix sufficiently far from the electrodes (see Fig. 3), the flow rate should be increased to an optimal value to provide little to no fuel crossover, while yielding high reactants consumption (Lee et al., 2007), and besides, the electrodes must have sufficient separation distance within the microchannel (Kjeang et al., 2009). Generally, to confirm that the diffusive crossover doesn't contribute to the loss of current, the width of the mixed region **δ** mix is calculated using Eq. 6. (Zebda et al., 2009b). Another strategy to prevent the direct contact and the reaction between oxidant and fuel was proposed in the case of a microfluidic fuel cell working from formic acid as fuel (Sun et al., 2007). A three-stream laminar flow fuel cell was developed that consisted to introduce a third stream containing only electrolyte solution between fuel and oxidant streams.

3.4.1.2. Strategies to reduce the depletion layer effect

Rapid transport of reactants to the electrodes is essential to provide high power densities. When a heterogeneous reaction occurs at electrode surface, depletion of the reactant results in formation of a depletion zone near the electrodes surface where lower conversion rates occur as the reactant concentration is lower than in the bulk region. The thickness of the depletion layer increases usually in the direction of the convective flow (Fig. 5), thus resulting in the decrease of the current density along the electrode length.

Concentration profiles of reactive species are usually described computationally by resolving the convection-diffusion Eq. 9 and by setting appropriate boundary conditions. Modelling results in targeting optimal electrodes configuration in microchannel, fuel utilization and flow rate. Fig. 6 describes the 2D profile concentration during the operation of a glucose/O2 biofuel cell based on Y-shaped microfluidic channel of height, h, and with electrodes length L. As observed, the concentration of the active species decreases near the gold electrode surface that generates a depletion zone gradually increasing. The thickness of the depletion zone is a function of the distance from the inlet edge of the microchannel, and decreases with high flow rates. According to the Fick's law (Eq. 11), it results that the simulated current density decreases sharply because of the concomitant increase of the depletion zone.

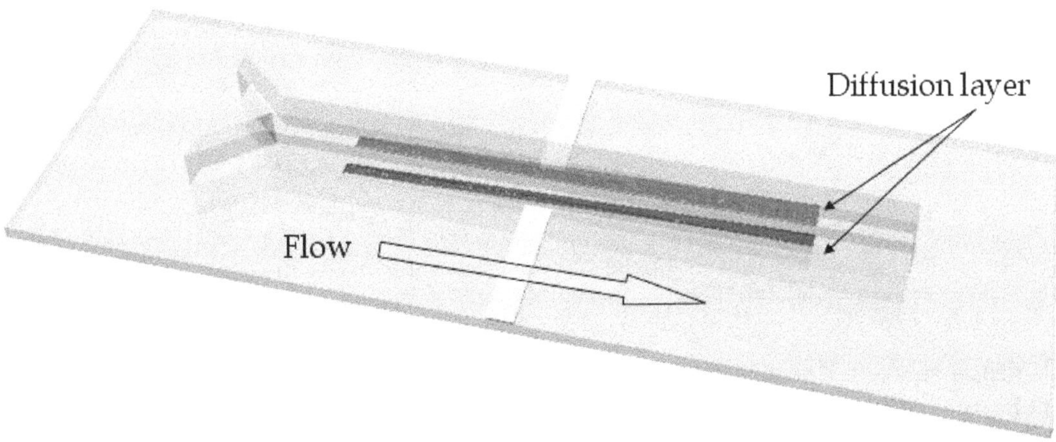

Figure 5.Schematic of the formation of the depletion zone near the electrodes surface.

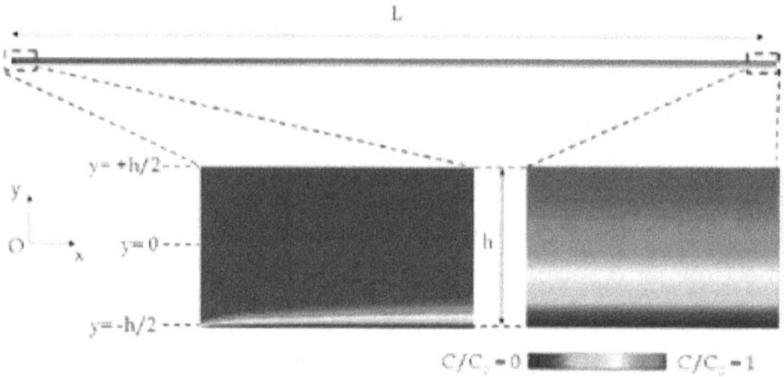

Figure 6.profile of ABTS concentration during the operation of a glucose/O2 biofuel cell based on Y-shaped microfluidic channel (Q=100 μL min-1).

Reduction of the thickness of the depletion boundary layer is necessary to increase mass transport since the diffusion distance will be shorter and the diffusional flux will be higher.

Several approaches have been developed to control the transport rate of reactants towards the electrode surface and thus the current density. It was demonstrated experimentally that the adjustment of flow rates controls the electrochemical processes that take place at the electrodes and regulate the depletion layer thickness. As observed in Fig. 7, the delivered current densities are influenced by flow rate of streams containing GOx for the anolyte and laccase for the catholyte during the operation of a glucose/O2 biofuel cell based on Y-shaped microfluidic channel.

The shape of the voltage-current density curves indicates that the current density is maximal when the voltage is almost zero, due to the consumption of all the fuel instantaneously at the electrode. Maximal current densities increase with flow rate from ~ 0.4 to 0.7 mA cm-2 as the impact of mass transport limitations is reduced from the bulk solution to the electrode surface. The maximum current density is thus limited by the diffusion of fuel and oxygen to their respective electrodes. The performance of the biofuel cell was evaluated at the operating flow rate of 1000 μL min-1. With an oxidant stream under oxygen at pH 3 and a fuel stream under nitrogen gas at pH 7, the maximum power density delivered by the biofuel cell is 110 μW cm-2 at 0.3 V (Fig. 7). Moreover, the pumping power to sustain the necessary flow in the microchannel was evaluated, according to Eq. 13, and compared with the delivered power by the cell. By varying the flow rate, it was found that the ratio of the input power to the output power increased from 1.5 % at 100 μL min-1 to 76 % at 1000 μL min-1. This experiment pointed out the importance of the flow rate on the power output delivered by the microfluidic BFC.

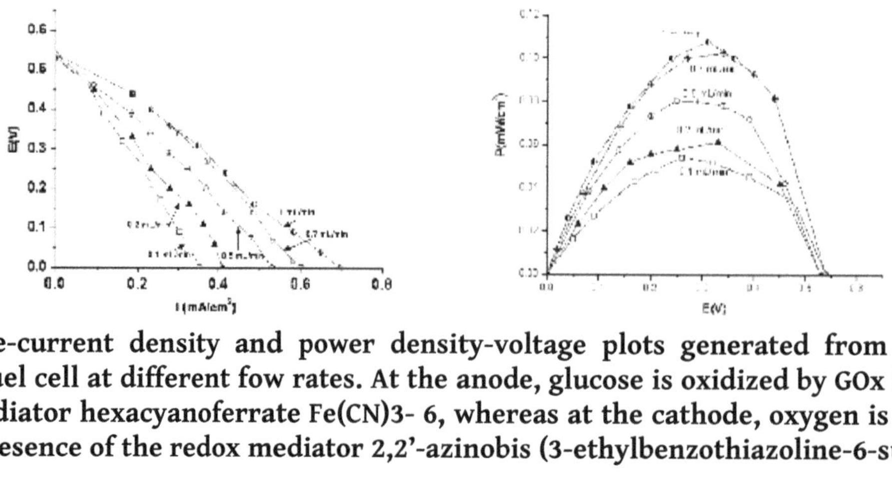

Figure 7.Voltage-current density and power density-voltage plots generated from a microfluidic glucose/O2 biofuel cell at different fow rates. At the anode, glucose is oxidized by GOx in the presence of the redox mediator hexacyanoferrate Fe(CN)3- 6, whereas at the cathode, oxygen is reduced by the laccase in the presence of the redox mediator 2,2'-azinobis (3-ethylbenzothiazoline-6-sulfonate) ABTS.

Another way to reduce depletion layer limitation and to enhance the transport rate of reactants towards the electrode surface lies in optimization of electrode geometry. Detailed experiments and simulations have revealed that current density decreases with increasing length of electrodes in the direction of the convective flow (Lim et al., 2007). To promote uniforme current density across the entire electrode assembly, Palmore and co-workers have demonstrated that splitting electrodes into smaller units separated by a gap in a microfluidic cell decreased the diffusion layer and improved the delivered power density by about 25% (Lee et al., 2007). In this work, the microfluidic fuel cell was built from a biocathode operating with laccase and ABTS as mediator to perform oxygen reduction, and from an anode operating with ABTS under N2. When operated at pH 4, this microfluidic cell exhibited a maximum power density of 26 µW cm-2 at the open-circuit voltage 0.4 V with a flow rate of 100 µL min-1. However, since the geometrical surface area required for the gaps did not contribute to any net current, the overall current density was not improved.

The depletion layer can be also manipulated to overcome mass transfer limitations by the development of original microfluidic configurations. This strategy was pointed out by simulation and experiments in the case of a microfluidic fuel cell working from formic acid (Yoon et al., 2006). These configurations featured, along the electrodes, either multiple periodically located outlets to remove consumed species or multiple periodically located inlets to add fresh reactants in the microfluidic channel. Such devices require controlling the volumetric flow rate trough each segments of the fluidic network. For both configurations, mass transfer was enhanced and reactant conversion at the electrodes was increased from 10 to 100 %.

3.4.1.3. Strategies to overcome oxygen limitation

A few studies report the analysis of oxygen limitation. However, the use of dissolved oxygen in enzymatic fuel cells is one of the main limitations of these systems due to low concentration (~ 0.2 mM) and low diffusion coefficient of oxygen (1.97 10-5 cm2 s-1 at 20 °C) (Barton, 2005). Usually, an exponential decay in the availability of oxygen at the cathode is observed along the length of microchannel (Bedekar et al., 2008). As a result, the oxygen flux is very low, limiting the generation of current. In the middle portion of the channel, the oxygen remains unconsumed and can still diffuse to the electrode. However, due to pressure-driven convective flow, the oxygen is still not consumed. To increase the availability of oxygen at the cathode surface, one strategy consists of designing a branched-microchannel configuration with several electrodes that allows periodically full contact of the electrolyte with the electrodes (Bedekar et al., 2008).

Another promising approach developed until now and only for fuel cells (methanol or acid formic/O2), is to incorporate cathodes that access the surrounding air with higher diffusivity and O2 concentration. Porous gas diffusion electrodes allow gaseous reactants to pass. Devices were developed either with flow-through porous electrodes able to increase delivered power densities with near complete fuel utilization (Kjeang et al., 2008), or with an air-breathing porous cathode structure (Jayashree et al., 2005). In the latter case, the use of air breathing cathode showed that the rate of oxygen reduction is enhanced with corresponding increase in current densities. However this approach requires precise control of pore size of the electrode to maintain constant rate of air delivery to the cathode.

3.4.2. Microfluidic BFCs with immobilized enzymes

A few works to date present enzyme-based microfluidic BFCs with immobilized enzymes. Enzymes in solution are only stable for a few days, it is why immobilization of active enzymes is of interest in order to improve lifetime (Bullen et al., 2006) and to develop integrated microfluidic BFC designs. Besides, full selectivity of both enzymatic half-cells allows microfluidic BFC operation in a single, combined fuel and oxidant channel with mixed reactants at constant concentrations favourable for stability studies. Additionally, the close proximity of the enzymes with the electrodes reduces ohmic losses because electrons are harvest from reaction sites with lower electrical resistance.

In microfluidic BFCs described in literature, enzymes with their respective redox mediators are immobilized on electrodes surface by encapsulation in polymer film. Currently, Nafion® is commonly used as it possesses surfactant properties interesting to immobilize enzymes in micellar structures when treated by quaternary ammonium salt (Moore et al., 2004). Such matrix provides an optimal enzyme environment where the enzyme retains its activity for greater than 90 days. Based on this strategy, a microchip-based bioanode with alcohol dehydrogenase enzymes immobilized in treated Nafion® associated with an external platinum cathode delivered a power density of 5 µW cm-2 (Moore et al., 2005). The low value was attributed to the thick coating of polymer, casting by hydrodynamic flow and thus difficult to control in the microchannels.

Another polymer to entrap enzymes is poly-L-lysine. This polymer mixed with an enzyme solution can stand on electrode surface after drying in air. According to this technique, a microfluidic glucose/O2 BFC was developed (Togo et al., 2007; Togo et al., 2008). The originality of this study concerned the location of the electrodes in a single flow channel. This device has been developed to generate electric power from glucose oxidation with an anode coated by immobilized glucose deshydrogenase and a bilirubin oxidase-adsorbed O2 cathode (Togo et al., 2008). The device was featured with a specific electrode-arrangement within a microchannel to prevent dissolved O2 to react at the anode as an interfering substance. A large biocathode (10 times the anode size) was placed strategically upstream of a bioanode to pre-electrolyseO2 to protect the anode vicinity from interfering oxygen. The maximum cell current was increased by 10% with this cell configuration at pH 7. The experimental results showed the influence of the channel height that should be in the same order of the depletion layer thickness for optimal operating of the device. However, in such design, the composition of fuel and oxidant could not be adjusted independently for optimum enzymatic activity and stability.

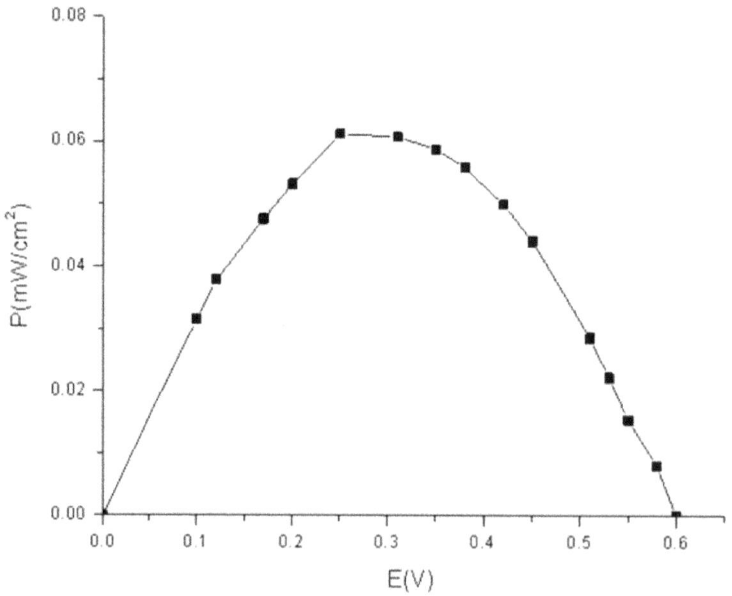

Figure 8.Experimental power density vs voltage plot generated from a microfluidic glucose/O2 biofuel cell at 23 °C and under flow rate 300 µL min−1. At the cathode, the couple laccase/ABTS and at the anode the couple GOx/Fe(CN)6 3− are immobilized in a poly-L-lysine film.

In order to choose independently the composition of the two streams, the configuration based on Y-microfluidic single channel is required. This approach has been developed for a microfluidic glucose biofuel cell working from GOx and laccase immobilised on gold electrodes in a poly-L-lysine matrix. The immobilization process was realized by mixing the enzymes and their respective redox mediators in poly-L-lysine solution. After drying, the device was tested with a phosphate buffer pH 7 solution with 50 mM glucose for the anolyte, and with a citrate buffer pH 3 solution saturated with oxygen for the catholyte. Different redox mediators were tested for efficient electron transfer at the anode. Among the mediators hexacyanoferrate (Fe(CN)6 3-), ferrocene and 8-hydroxyquinoline-5-sulfonic acid hydrate (HQS), the higher power density (60 μW cm-2) was obtained at a cell voltage 0.25 V with Fe(CN)6 3- (Fig. 8).

3.4.3. Microfluidic BFCs with immobilized multi-enzyme systems

In general there are numerous combinations of enzymes and mediators that have been employed in biofuel cells but the respective studies typically involve monoenzymatic systems, which are capable of only partial oxidation of the fuel. Improvement of fuel utilization can be achieved by complete oxidation, which can be realized by introduction of enzyme cascades to increase the overall efficiency of the fuel cell (Nick et al., 2005; Sokic-Lazic et al., 2010; Addo et al., 2010). By this way, the overall performance of the BFC is increased as well.

In a theoretical work described by Kjeang and co-workers, a concept of an enzymatic fuel cell is proofed, without presenting experimental system (Kjeang et al., 2006). This work describes how to optimize the structure of a microfluidic enzymatic fuel cell, involving three-step-catalyzed methanol oxidation (Fig. 9). Different enzyme patterning strategies are tested, e.g., spatially distributed- or evenly-mixed enzymes along the electrode surface. The model predicts high fuel utilization at low flow rates, i.e., in the diffusion dominated and mixed mass, transfer conditions. According to the model, the investigated theoretical concept is reaction limited, which means that the system performance can be improved by improving enzyme turnover numbers. This work also demonstrates that the power required for pumping of the fuel is negligible in comparison to predicted power of the fuel cell.

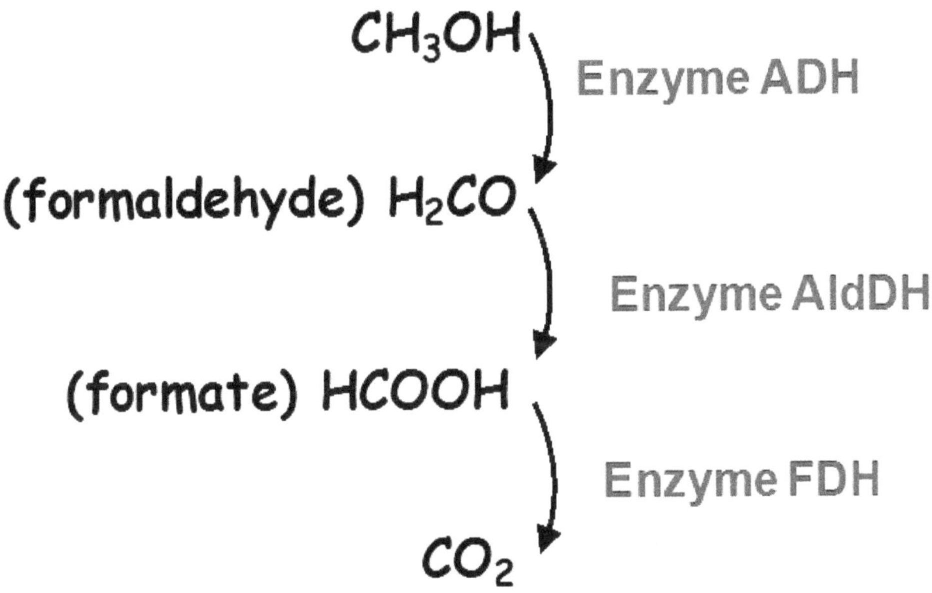

Figure 9. Schematic of the three-step-catalyzed methanol oxidation by the enzymes alcohol dehydrogenase (ADH), aldehyde dehydrogenase (AldDH) and formate dehydrogenase (FDH).

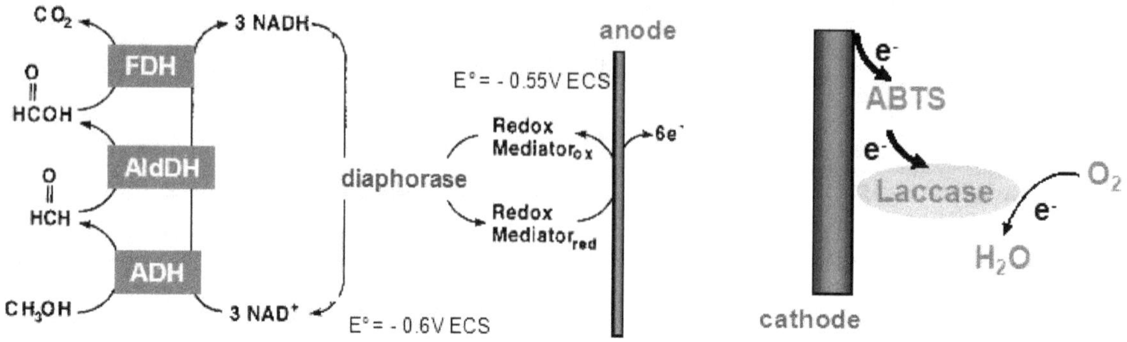

Anode: $CH_3OH + H_2O \rightarrow 6H^+ + 6\ e^- + CO_2$ Cathode: $3/2\ O_2 + 6H^+ + 6e^- \rightarrow 3H_2O$

$(NAD^+ + 2e^- + 2H^+ <=> NADH + H^+)$

Figure 10.Electron transfer steps at the biocathode and the bioanode.

Based on this work, we built a methanol/O2 microfluidic BFC able to completely oxidize methanol, according to the electron steps described for each electrode in Fig. 10. The enzymes and the mediators were immobilized in poly-L-lysine.

These enzymes are NADH-dependant. The electrochemical connection and regeneration of NADH at the anode is achieved using the enzyme diaphorase and the redox mediator benzylviologen as already described (Palmore et al., 1998). At the anode, a mixture of the three dehydrogenase enzymes was deposited along a gold electrode surface according two configurations (Fig. 11). When enzymes are mixed along the anode, optimal power density is achieved as observed in Fig. 12. The Y-channel device delivers a power density of 70 µW cm-2 at a cell voltage 0.25 V.

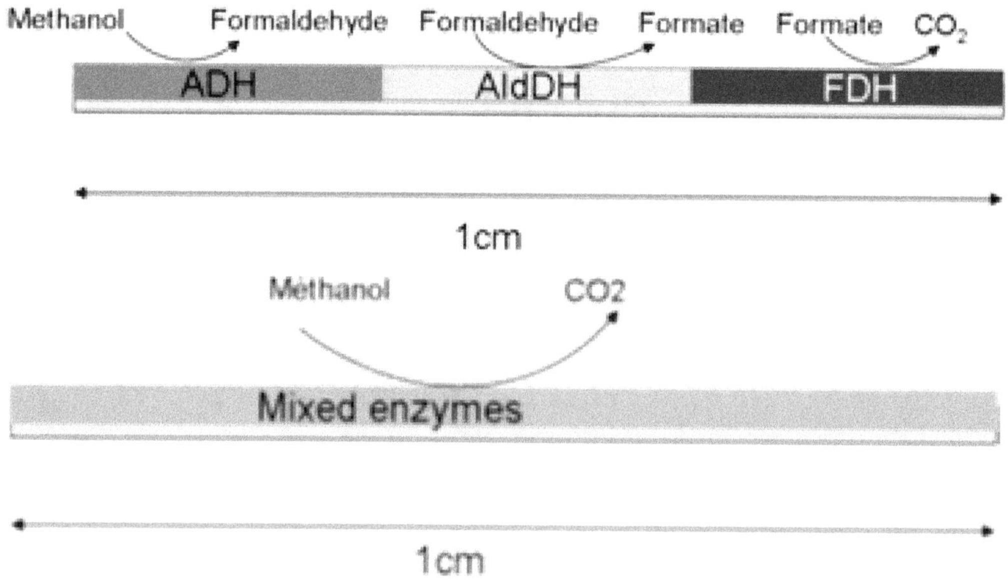

(A) (B)

Figure 11. (A) enzymes are distributed along the anode surface in three distinct, equally sized zones of 10 mm in length, (B) enzymes are mixed randomly along the anode surface of 10 mm in length.

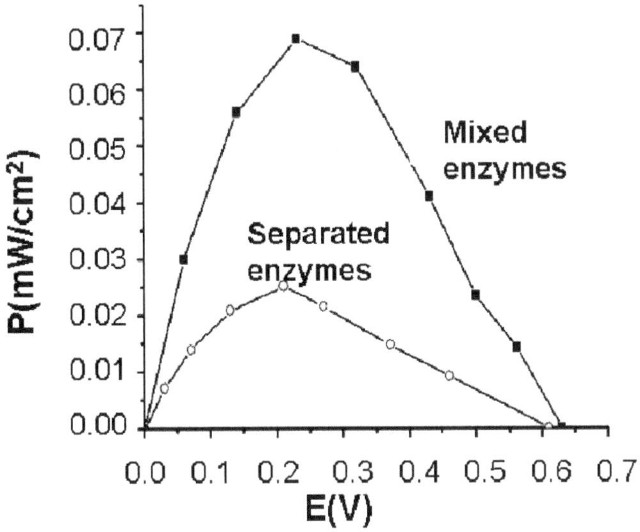

Figure 12.Power density-voltage plots generated from a microfluidic methanol/O2 biofuel cell for 2 cases: separated and mixed enzyme patterns at the anode (1:1:1 molecular basis).

4. CONCISE SUMMARY

Microfluidic biofuel cells (BFCs) offer a promising alternative for powering small devices including biological sensors, implantable medical devices, and portable electronics. Nevertheless, substantial research endeavors are necessary for the implementation of practical applications. Research should focus on discovering the most resilient and dynamic enzymes, optimizing the immobilization conditions for enzymes and mediators in a microfluidic environment, and enhancing the longevity of enzymes. In order to achieve greater power densities, the primary obstacles lie in the realms of energy density and fuel consumption. Microfluidic biofuel cells (BFCs) should have sophisticated immobilization arrangements in order to boost the efficiency of enzymes, as well as high surface area electrodes to improve the rates of convective-diffusive transport of reactants. Further nanofluidic research and development are necessary to fully showcase the actual capabilities of this type of energy conversion technology. In addition, the power generated by a single microfluidic biofuel cell is insufficient for the majority of practical uses. The expansion of a solitary cell is limited by fuel/oxidant crossover and increased ohmic losses as the geometrical electrodes area and microchannel are enlarged. To generate sufficient power, it is possible to arrange many autonomous cells in a stacked configuration, similar to conventional fuel cells.

REFERENCES

Addo, P.K., Arechederra, R. L. & Minteer S.D., 2010. Evaluating enzyme cascades for methanol/air biofuel cells based on NAD+-Dependent Enzymes. Electroanalysis, vol. 22, No. 7-8 (April 2010) pp 807-812, ISSN 1040-0397.

Amatore, C.; Da Mota, N.; Della C. & Thouin L. (2007). Theory and experiments of transport at channel microband electrodes under laminar flows. 1. Steady-state regimes at a single electrode, Analytical Chemistry, vol. 79, (November 2007), pp. 8502-8510, ISSN 0003-2700.

Barton, S. C. (2005). Oxygen transport in composite mediated cathode. Electrochimica Acta, vol. 50, No. 10, (March 2005) pp 2145-2153, ISSN 0013-4686.

Bazylak, A.; Sinton, D. & Djilali, N. (2005). Improved fuel utilization in microfluidic fuel cells: A computational study, Journal of Power Sources, vol. 143, No. 1-2, (April 2005), pp 57-66, ISSN 0378-7753.

Bedekar, A.S.; Feng, J.J.; Krishnamoorthy, S.; Lim, K.G.; Palmore, G.T.R. & Sundaram, S. (2008) Oxygen limitation in microfluidic biofuel cells, Chemical Engineering Communication 195 pp. 256–266, ISSN 0098-6445.

Bullen, R.A.; Arnot, T.C.; Lakemanc, J.B. & Walsh, F.C. (2006). Biofuel cells and their development. Biosensors & Bioelectronics, vol. 2, No. 11, (January 2006), pp. 2015– 2045, ISSN 0956-5663.

Calabrese Barton, S.; Gallaway, J. & Atanassov, P. (2004). Enzymatic Biofuel Cells for Implantable and Microscale Devices. Chemical Reviews, Vol. 104, No. 10, (October 2004), pp. 4867-4886, ISSN 0009-2665.

Chen, T.; Barton, S.C.; Binyamin, G.; Gao, Z.; Zhang, Y.; Kim, H. & Heller, A. (2001). A Miniature Biofuel Cell, Journal of the American Chemical Society, vol. 123, No.35, (august 2001), pp 8630-863, ISSN 0002-7863.

Chen, F.; Chang, M.H. & Lin M.K. (2007). Analysis of membraneless formic acid microfuel cell using a planar microchannel, Electrochimica acta Vol. 52, No. 25, (September 2007), pp. 2506-2514, ISSN 0013-4686.

Choban, E.R.; Markoski, L.J.; Wieckowski, A. & Kenis, P.J.A. (2004). Microfluidic fuel cell based on laminar flow, Journal of Power Sources, vol. 128, No. 1, (march 2004), pp. 54-60, ISSN 0378-7753.

Choban, E.R., Waszczuk, P. & Kenis P.J.A. (2005). Characterization of limiting factors in laminar flow-based membraneless microfuel cells. Electrochemical and Solid-State letters, vol. 8, No. 7, pp. A348-A352, ISSN 1099-0062.

Davis, F. & Higson, S.P.J. (2007). Biofuel cells—Recent advances and applications. Biosensors & Bioelectronics, vol. 22, No. 7,(February 2007) pp 1224–1235, ISSN 0956-5663.

Degani Y.& Heller A. (1987). Direct electrical communication between chemically modified enzymes and metal- electrodes. 1. Electron-transfer from glucose oxidase to metal electrode via electron relays, bound covalently to the enzymes, Journal of Physical Chemistry, vol. 91, (March 1987), pp. 1285-1289, ISSN 0022-3654.

Duffy D. C.; Cooper McDonald J.; Schueller O. J. A & Whitesides G. M. (1998) Rapid Prototyping of Microfluidic Systems in Poly(dimethylsiloxane), Analytical Chemistry 70, pp. 4974-4984, ISSN 0003-2700.

Ferrigno, R.; Stroock, A.D.; Clark, T.D.; Mayer M. & Whitesides M. (2002). Membraneless Vanadium Redox Fuel Cell Using Laminar Flow, Journal of American Chemical Society, vol. 124, No. 7, (October 2002), pp. 12930-12931, ISSN 0002-7863.

Fischback, M.B.; Youn, J.K.; Zhao, X.; Wang, P.; Park, H.G.; Chang, H.N.; Kim, J. & Haa, S. (2006). Miniature biofuel cells with improved stability under continuous operation, Electroanalysis, vol. 18, No.19-20 (july 2006), pp 2016-2022, ISSN 1040-0397.

Gao, F.; Viry, L.; Maugey, M.; Poulin, P. & Mano, N. (2010). Engineering hybrid nanotube wires for high-power biofuel cells. Nature Communications, Vol. 1, No. 1, pp. 1-7, ISSN 2041-1723.

Gellett, W.; Kesmez, M.; Schumacher, J.; Akers N. & Minteer, S.D. (2010). Biofuel Cells for Portable Power. Electroanalysis , vol. 22, No. 7-8, (April 2010) pp 727-731, ISSN 1040-0397.

Gervais, T. & Jensen, K.F. (2006) Mass transport and surface reactions in microfluidic systems, Chemical Engineering Science, vol. 61, (February 2006), pp. 1102-1121, ISSN 0009-2509.

Hayes, J.R.; Engstrom, A. M. & Friesen, C. (2008). Orthogonal flow membraneless fuel cell. Journal of Power Sources, vol. 183, No. 1, (august 2008), pp. 257-259, ISSN 0378-7753.

Heller, A. (2004). Miniature biofuel cells. Physical Chemistry Chemical Physics, Vol. 6, No. 2, (January 2004) pp. 209-216, ISSN 1463-9076.

Ismagilov R.F.; Stroock A.D.; Kenis P.J.A.; Whitesides G. & Stone H.A. (2000). Experimental and theoretical scaling laws for transverse diffusive broadening in two-phase laminar flows in microchannels. Applied Physics Letters, Vol. 76, (April 2000), pp. 2376-2378, ISSN 0003-6951.

Ivanov, I., Vidaković-Koch, T. & Sundmacher, K. (2010). Recent Advances in Enzymatic Fuel Cells: Experiments and Modeling. Energies, Vol. 3, No. 4, (April 2010), pp. 803-846, ISSN 1996-1073.

Jayashree R. S.; Gancs L.; Choban E. R.; Primak A.; Natarajan D.; Markoski L. J. & Kenis P. J. A. (2005) Air-Breathing Laminar Flow-Based Microfluidic Fuel Cell, Journal of the American Chemical Society, 127, 48 (December 2005) pp 16758–16759, ISSN 0002-7863.

Kjeang E.; Sinton D. & Harrington D.A. (2006) Strategic enzyme patterning for microfluidic biofuel cells. Journal of Power Sources vol.158, No. 1, (July 2006) pp. 1–12, ISSN 0378- 7753.

Kjeang E.; Michel R.; Harrington D. A.; Djilali N. & Sinton D. (2008) A Microfluidic Fuel Cell with Flow-Through Porous Electrodes, (2008) Journal of the American Chemical Society 130 (March 2008) pp. 4000-4006, ISSN 0002-7863.

Kjeang, E.; Djilali, N. & Sinton, D. (2009). Microfluidic fuel cells: a review, Journal of Power Sources, vol. 186, (October 2008), pp 353-369, ISSN 0378-7753.

Lee, J.; Lim, K.G.; Palmore, G.T.R. & Tripathi, A. (2007). Optimization of microfluidic cells using transport principles, Analytical Chemistry, vol. 79, No. 19, (October 2007), pp 7301-7307, ISSN 0003-2700.

Lim K.G. & Palmore G. T. R. (2007) Microfluidic biofuel cells: the influence of electrode diffusion layer on performance, Biosensors & Bioelectronics 22 (January 2007) pp. 941-947, ISSN 0956-5663.

Luo, T.J.M.; Fei, J.F.; Lim, K.G. & Palmore, G.T.M. (2005). Membraneless fuel cells: an application of microfluidics. In: ACS Symposium Series 890, 2005.

Mano, N.; Mao, F. & Heller, A. (2002). A Miniature Biofuel Cell Operating in A Physiological Buffer. Journal of the American Chemical Society, Vol. 124, No. 44, (November 2002) pp. 12962-12963, ISSN 0002-7863.

Mano, N.; Mao, F. & Heller, A. (2003). Characteristics of a miniature compart-mentless glucose-O2 biofuel cell and its operation in a living plant, Journal of the American Chemical Society, vol.125, No.21, (april 2003), pp 6588-6594, ISSN 0002-7863.

Mao, F.; Mano, N. & Heller, A. (2003). Long Tethers Binding Redox Centers to Polymer Backbones Enhance Electron Transport in Enzyme "Wiring" Hydrogels. Journal of the American Chemical Society, Vol. 125, No. 16, pp. 4951-4957, ISSN 0002-7863.

Moore, C.M.; Akers, N. L.; Hill, A. D.; Johson, Z. C. & Minteer, S.D. (2004). Improving the environment for immobilized dehydrogenase enzymes by modifying Nafion with tetraalkylammonium bromides. Biomacromolecules, vol. 5, No. 47, (July-August 2004), pp 1241-1247, ISSN 1525-7797.

Moore, C.M.; Minteer, S.D. & Martin, R.S. (2005). Microchip-based Ethanol/Oxygen. Biofuel Cell, Lab on Chip, vol. 5, No. 2,pp 218-225, ISSN 1473-0189.

Nick, A.; Akers L.; Moore C. M. & Minteer S. D. (2005) Development of alcohol/O2 biofuel cells using salt-extracted tetrabutylammonium bromide/Nafion membranes to immobilize dehydrogenase enzymes, Electrochimica Acta, Volume 50, Issue 12, (April 2005) pp. 2521-2525, ISSN 0013-4686.

Nishizawa, M.; Togo, M.; Takamura, A. & Abe, T. (2005). Enzyme-based fuel cells for biomedical microdevices, Proceedings of the fifth international workshop on micro and nanotechnology for power generation and energy conversion applications, Takeda Hall, University of Tokyo, November 2005.

Palmore G. T. R.; Bertschy H.; Bergens S. H. & Whitesides G. M. (1998). A methanol/dioxygen biofuel cell that uses NAD+-dependent dehydrogenases as catalysts: application of an electro-enzymatic method to regenerate nicotinamide adenine dinucleotide at low overpotentials . Journal of Electroanalytical Chemistry, 443 (February 1998),pp.155–161, ISSN 0022-0728.

Service, R.F. (2002). Shrinking fuel cells promise power in your pocket, Science, 296 (May 2002), pp 1222-1224, ISSN 0036-8075.

Sokic-Lazic D.; Arechederra R. L.; Treu B. L. & Minteer S. D. (2010). Oxidation of Biofuels: Fuel Diversity and Effectiveness of Fuel Oxidation through Multiple Enzyme Cascades, Electroanalysis 22, No. 7-8, (April 2010) pp. 757 – 764, ISSN 1040-0397.

Stephan K.; Pittet P.; Renaud L.; Kleimann P.; Morin P.; Ouaini N. & Ferrigno R. (2007) Fast prototyping using a dry film photoresist: microfabrication of soft-lithography masters for microfluidic structures, J. Micromech. Microeng. 17 (October 2007) N69–N74, ISSN 0960-1317.

Sun, M.H.; Casquillas, G.V.; Guo, S.S.; Shi, J.; Ji, H.; Ouyang, Q. & Chen, Y. (2007). Characterization of microfluidic fuel cell based on multiple laminar flow. Microelectronic Engineering, vol. 84, (May 2007), pp. 1182-1185, ISSN 0167-9317.

Togo M.; Takamura A.; Asai T.; Kaji H. & Nishizawa M. (2007) An enzyme based microfluidic biofuel cell using vitamin K3-mediated glucose oxidation, Electrochimica Acta, Volume 52, Issue 14, (April 2007) pp. 4669-4674, ISSN 0013- 4686.

Togo, M.; Takamura, A.; Asai, T.; Kaji, H. & Nishizawa, M. (2008). Structural studies of enzyme-based microfluidic biofuel cells, Journal of Power Sources, vol. 178, (December 2007), pp 53-58, ISSN 0378-7753.

Wilson, G.S. (2002). Bioelectrochemistry, Encyclopedia of electrochemistry, vol 9, John Willey, 2002, ISBN: 3527304010, Weinheim.

Yahiro A. T.; Lee S. M. & Kimble D. O. (1964). Bioelectrochemistry .I. Enzyme Utilizing Bio- Fuel cell studies, Biochimica Biophysica Acta, vol. 88, pp. 375, ISSN 0006-3002.

Yoon S.K.; Fichtl G. W. & Kenis P.J.A. (2006) Active control of the depletion boundary layers in microfluidic electrochemical reactors, Lab On a Chip 6 (December 2006) pp. 1516- 1524, ISSN 1473-0197.

Zebda, A.; Renaud, L.; Tingry, S; Cretin, M.; Pichot, F.; Ferrigno, R. & Innocent, C. (2009a). Microfluidic Biofuel Cell for Energy Production. Sensor letters, vol. 7, No. 5, (October 2009), pp. 824-828, ISSN 1546-198X.

Zebda, A.; Renaud, L.; Cretin, M.; Pichot, F.; Innocent, C.; Ferrigno, R. & Tingry, S. (2009b). Electrochemical performance of a Glucose/Oxygen microfluidic biofuel cell., Journal of Power Sources, vol. 193, No. 2, (may 2009), pp 602–606, ISSN0378-7753.

Zebda, A.; Renaud, L.; Cretin, M.; Innocent, C.; Ferrigno, R. & Tingry, S. (2010). Membraneless microchannel glucose biofuel cell with improved electrical performances, Sensors and Actuators B, vol. 149, No. 1, (august 2010), pp 44–50, ISSN 0925-4005.

CHAPTER-26

ENERGY PATHWAYS RESULTING FROM THE BLUE TOWER PROCESS

1. INTRODUCTION: AN OVERVIEW

Decarbonisation has gained global prominence as a means to address the environmental issue. Specifically, hydrogen energy, derived from renewable energy sources such as wind power, hydropower, photovoltaic, solar thermal power, geothermal power, and biomass, holds the greatest potential. The biomass feedstock can be converted into electrical and/or heat energy using a traditional incineration system such as a gas-engine or gas-turbine. Alternatively, it can be transformed into secondary energy sources such as hydrogen, methanol, and/or dimethyl-ether (DME), which are known as "biomass to liquid" fuels (BTL). The biomass energy system is an environmentally friendly energy technology that helps mitigate global warming. However, there may be certain challenges when promoting eco-friendly solutions in the target location. It is widely recognized that the resource can potentially compete with biomass as a food source. Furthermore, despite the presence of ample material resources, the products produced by the system would be more expensive compared to the conventional one due to the poor energy density features. Based on recent interviews, the cost issue is highly critical, and the market capabilities of items are also crucial for sustainable business management. Prior to implementing the project, it is crucial to thoroughly evaluate the operational state of the eco-business, the technological obstacles, and the impact on environmental contribution when considering the promotion of an eco-friendly system. In order to accomplish this, it is necessary to establish the appropriate energy infrastructure and develop a viable business strategy. In our earlier research, we examined a biomass gasification system that is not typically utilized in the chemical industry. We determined that it is necessary to consider various factors, including practical examples of how the system may be applied. Therefore, it is imperative to have an appropriately sized plant in order to accommodate the low energy density of biomass fuel. In order to successfully operate a plant operation business, it is necessary to have public support from the central or local government in the early stages of promoting new technology. Additionally, a sound business scheme based on emission trading or the increased awareness of consumers' willingness to pay for eco-products is also crucial. In this section, we present various energy pathways for biomass consumption, considering the perspectives of "energy," "environment," and "economy." namely, we are focusing on the biomass energy system, namely the Blue-Tower process. This process has the capability to generate both electrical and heat energy through a co-generation unit, such as a gas-engine, gas-turbine, or fuel cell. Additionally, it can purify Bio-H2 using a PSA (pressure swing adsorption) unit. Currently, the emphasis has primarily been on the utilization of biomass energy systems, specifically the biomass gasification process, which has predominantly been advanced in Europe. Primarily, we believe that the Blue Tower (BT) process, created by D.M.2 Projekt GmbH, is nearing the point of being a commercially operational facility. The Dowaki laboratory at Tokyo

University of Science, in collaboration with Japan Planning Organization Inc., has utilized the patent on the basic BT process to develop advanced technology. This has been achieved through the analysis of results from the demo-plant, lab-scale experiments, and comprehensive investigations.

Since its implementation in 2005, the Kyoto agreement has led to substantial advancements in technology for mitigating CO_2 emissions. The biomass energy system, which utilizes organic materials like wood, organic waste, sewage, and similar substances, is a highly promising energy method for reducing CO_2 emissions. Gasification technologies have garnered global interest as a promising future technology. However, their energy efficiency vary significantly due to differences in energy conversion processes. In such conditions, the cost of energy production is expected to remain high, even with the implementation of subsidies, environmental laws, or policies aimed at providing support. Initially, it was believed that the biomass energy system, utilizing gas-engine cogeneration, would become widely used. However, because to a disparity between the demand and supply, namely because the generated energy cannot be stored, there would be an excess of electricity and/or heat energy. Therefore, in such a system, it would not be possible to achieve reductions in CO_2 emissions, and the cost of energy generation would be high. In light of this context, the Japanese government has endeavored to advance the biomass energy system as a means to address the issue of global warming. For instance, the Ministry of Economy, Trade and Industry (METI) forecasts that the utilization of biomass energy will reach 39.1 petajoules in the future energy demand and supply plan (METI, 2007). In the 2010 energy demand and supply plan, the Ministry of Agriculture, Forestry and Fisheries (MAFF) forecasts that biomass energy usage will range from 110.5 to 210.1 petajoules (PJ). Specifically, MAFF is undertaking a project to manufacture 500 million liters of BTL gasoline for use in the transportation industry (MAFF, 2006). In contrast, Japan utilized approximately 1.5 to 2.0×1010 Nm^3 of hydrogen gas in 2005. According to the K.K. Gas Review in 2007, 50% of the hydrogen energy was generated from fossil fuel. In 2015, numerous Japanese automobile manufacturers made significant efforts to market fuel cell vehicles (FCVs). Therefore, harnessing hydrogen energy from sustainable energy sources is crucial for reducing CO2 emissions. Given the aforementioned factors, we recognized a strong likelihood of being able to create Bio-H2 fuel. Nevertheless, it is imperative to identify the effective business model in the initial phase to facilitate the growth of the eco-enterprise through the utilization of this gasification process. It may be necessary to streamline the concrete application system using the Business Transformation approach. Subsequently, we assessed the suitability of the associated amenities in the agricultural or forestry sector. The adoption of environmentally friendly systems, such as PV systems or fuel cell co-generation systems, is currently insufficient for the promotion of these facilities. There is potential to integrate the environmentally friendly biomass energy system with agriculture-related facilities. Furthermore, MAFF plays a role in mitigating global warming by considering the carbon footprint of agricultural products. The ministry offers several subsidy programs to promote the system. In addition, there are institutions that require energy companies to purchase extra electricity and pay additional fees for it. By utilizing the aforementioned organizations and subsidy programs under the guidance of the Japanese government, we have examined the implementation of a biomass gasification (BT) process with a gas-engine or fuel cell for the agricultural products gathering system. In this section, we present multiple energy pathways. We conducted a comprehensive analysis of CO_2 emissions and energy intensities for the complete system design using the Life Cycle Assessment (LCA) technique. This assessment adheres to the requirements set forth by ISO 14041. We established the system boundary for each path. The system boundary encompasses the complete life cycle of any energy input, such as electricity or thermal energy, including the pre-processing and energy conversion processes. Specifically, we conducted process design for the given energy system using a simulator and relied on experimental data to estimate the energy efficiency, CO2 intensity, and energy cost, including the consideration of the business model.

This is the schematic design of the section shown in Figure 1. In this text, we will discuss the biomass energy system of Blue Tower technology and provide an explanation of the environmental system analysis in relation to the eco-business operation.

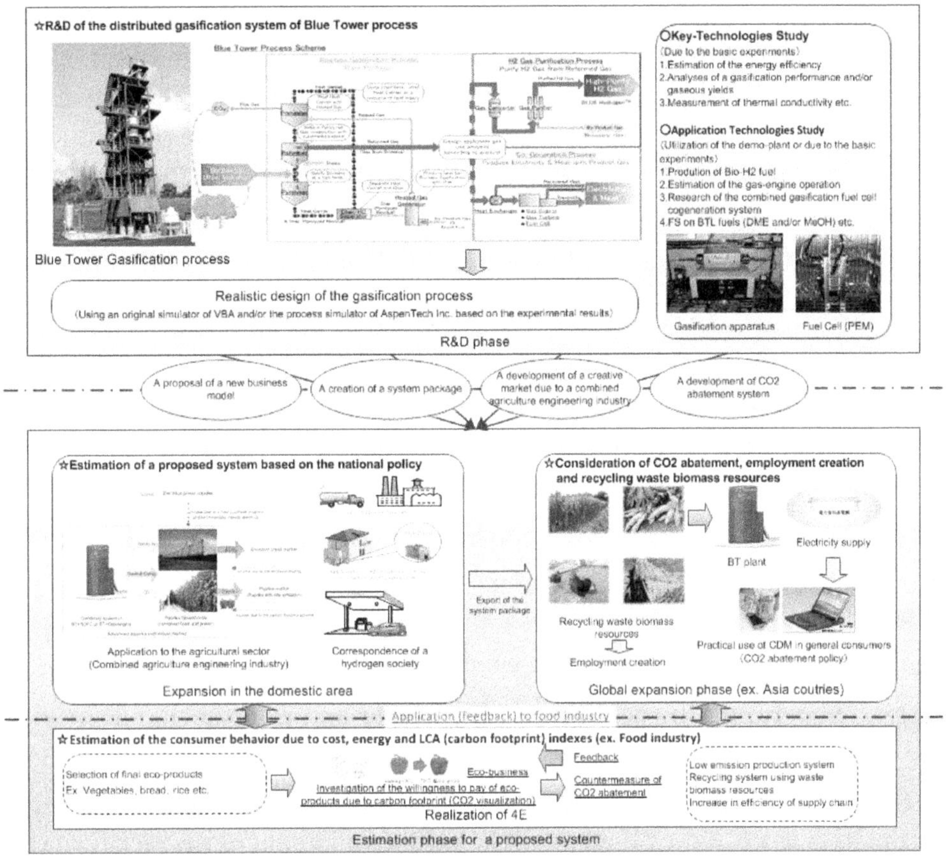

Figure 1.Outline of biomass energy system through Blue Tower process

2. BLUE TOWER TECHNOLOGY

The characteristics of the BT process are similar to the indirect gasification process with a pyrolyzer. This gasifier has three reactors for the preheating process, the reforming process and the pyrolysis process. The gasifier has the following characteristics; 1) pyrolysis reactions take place in the pyrolyzer, 2) pyrolysis gases are reformed by H_2O (steam), 3) heat required for their reactions is supplied by combustion of off-gas, Tar and Char.

Additional heat through a chip boiler might be required in order to accelerate the reactions well. With regard to the gasification performance, the gaseous yield and concentration are dependent upon the kinds of materials, the operating temperature, and the inner pressure (Mühlen et al., 1999, Mayer et al., 2004). Here, in order to evaluate the reliability of each fuel production, we fabricated an apparatus, which included the concept of the BT process (a pyrolyzer and a reformer), and we executed the basic chemical experiments using the biomass samples with the size of 2-3 mm. In the experiments, we measured the syngas components and estimated the equilibrium constants, adjusting the temperature condition and/or steam-carbon ratio (S/C). Also, we ensured the reliability of our simulator which was available for some analyses on the energy cost and/or CO_2 intensity etc. comparing our calculation to the demo-plant data. Assuming that the materials chopped at the size of 20-30 mm are fed into the reactor, H2 of 54.4 vol.% and CO of 24.4 vol.% were generated at 950ºC and Steam/Carbon=1.0 due to our simulator (Dowaki et al., 2007). On the other hand, we executed the studies in order to confirm the absolute proof of the chemical equilibrium reactions, and/or the heat balance in use of the experimental apparatus and/or the demo-plant (1t/d scale) at Izumo, shimane prefecture in Japan. Likewise, the studies on the handling of equipment (the plant operation) have been done at Izumo. In the previous studies, we made the simulator of BT process in order to estimate the operational performance. This simulation program used the parameters estimated by the experimental results of a room condition. Also, the estimation accuracy due to the simulator was analysed. For instance, Kameyama et al. compared the operational result of the demo-plant to that of the simulator. Accordingly, we made sure that the simulated data were corresponding with the practice data to some extent (Kameyama et al., 2010).

Here, we describe the system outline and the performance characteristics of this process which was evaluated through the demonstrated operation with 1t/d plant (see Fig. 2) of the Blue Tower gasification process.

2.1. BT process simulator due to the basic experiments

Based on the basic experiment on the pyrolysis and the reforming, we have developed the simulator by which the gaseous yields and/or the energy efficiency through BT process can be estimated. Here, we compared the practice data through the demo-plant with the result of the simulator.

In general, there would be somewhat deviation between the practice data and the estimated one due to the simulator. That is, it would be extremely significant to identify the deviation from the viewpoint of the reliable plant operation.

The calculation logic of the simulator which we developed is as follows (Dowaki et al. 2007):

- The reaction temperature in each furnace (pyrolyzer and reformer) and the steam feeding rate are fixed.
- Based on the gaseous yields in the pyrolysis and/or the reforming reactions, which were analysed by gas chromatograph (GC-8A, Shimadzu), the gaseous components in each furnace were estimated due to the following two equilibrium reactions. In our experiments, Shincarbon-ST was used as a measure column, and the six kinds of gases which can be measured include H2, CO, CO2 CH4, C2H4 and C2H6, respectively.

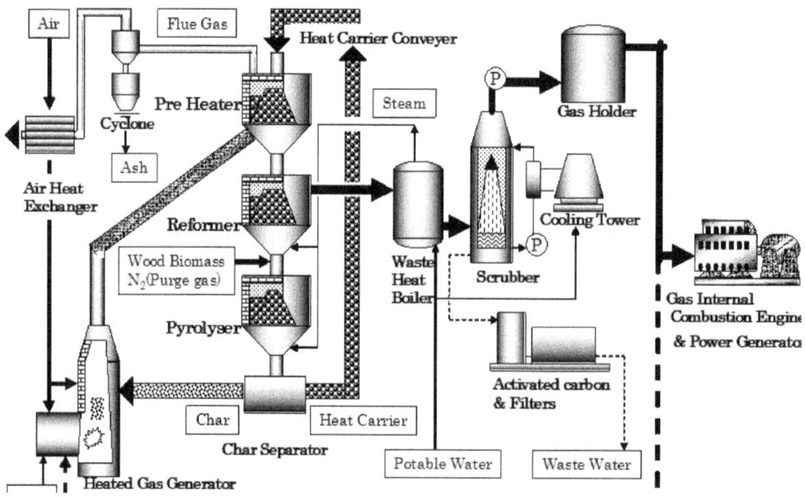

Figure 2.Schematic design of 1t/d Blue Tower plant

Note that we considered the approach temperature difference between the theoretical one and the reaction one on these equilibrium reactions. Here, in the gasification process after the pyrolysis reaction, it is usually thought that the following two reactions take place.

$$CH_4 + H_2O \leftrightarrow CO + 3H_2 \quad (1)$$

$$CO + H_2O \leftrightarrow CO_2 + H_2 \quad (2)$$

Using our experimental apparatus, the approach temperature differences between theoretical temperature and the actual temperature was measured. This temperature difference is known as the approach temperature. In Eqs. (1) and (2), the equilibrium constant of shift reaction and that of methanation reaction are represented as follows:

$$K_s = \exp\left(\frac{-\Delta G^{\circ}(T - \Delta T)}{RT}\right) = \frac{P_{CO_2}P_{H_2}}{P_{CO}P_{H_2O}} \quad (3)$$

$$K_m = \exp\left(\frac{-\Delta G^{\circ}(T - \Delta T)}{RT}\right) = \frac{P_{CH_4}P_{H_2O}}{P_{H_2}{}^3 P_{CO}} \quad (4)$$

Where, DT, R and Pi are the approach temperature [K], gas constant [J/molK] and the partial pressure of i-component [Pa], respectively.

Next, we compared the measurement data in Izumo plant (1t/d) with the estimated results using the simulator (see Table 1). Here, the tasks of Run 11 and Run 11.2 in which the decomposed reactions are assumed to be completed in the pyrolyzer are described. Both tasks were executed on March, 2009 (Kameyama et al. 2010).

Table 1.Comparison of the measured data and the simulated ones

	Deviatoin [-]	Temparature at Refomer [°C]	
		Measured	Estimated
Run 11.1	0.134	820	801
Run 11.2	0.367	750	774

In this verification, we focused on the molar fractions of H2, CO, CH4 and CO2, and found the average reaction temperature so that the total deviation on molar fraction between the measurement data and the estimated one due to Eqs. (3) and (4) is a minimum. That is, we investigated if the gaseous components based on the temperature which was measured in the plant corresponded to the estimated ones due to the simulator. The reason why we verified the gaseous components using the temperature as a variable is as follows; in the demo-plant, we did not know the temperature profile on the vertical and/or horizontal directions precisely since the sampling point of the temperature in reformer is one position. Thus, assumed that the estimated temperature based on the measured gaseous components would represent the average one of reformer, we made sure that the process simulator would be more suitable.

Next, based on the process simulator which has a precision to some extent, we describe the example due to the biomass feedstock of waste Japanese cedar. With regard to the gasification performance, since gaseous yields and concentrations are dependent upon the kind of materials, the operating temperature, and the inner pressure, they were examined using the gasifier apparatus which has a reformer and a pyrolyzer. Here, Table 2 shows the ultimate analysis of the waste Japanese cedar.

Table 2.Ultimate analysis of the waste Japanese cedar

C*	46.660%	wt.%
H*	5.480%	wt.%
O*	47.351%	wt.%
S*	0.000%	wt.%
N*	0.120%	wt.%
Cl*	0.000%	wt.%
Ash*	0.389%	wt.%
HHV*	18,348	kJ/kg
Moisture Content	20.0	wt.%
Volatile Matter	86.21%	wt.%
Bulk density	0.14	t/m3
*Dry-Base		

Through the tests, the syngas components and the equilibrium constants were obtained. For instance, Fig.3 illustrates the gaseous yields on the pyrolysis at 550 °C with variation of S/C =0.14 to 0.98, and the reforming reaction at S/C=1.0 with variation of 800 to 950 °C, respectively. Here, a steam carbon ratio is defined as the following equation.

$$S/C\left[mol-H_2O/mol-C\right]=\frac{\text{Added Steam [mol/s]}+\text{Moisture [mol/s]}}{\text{Carbon Content of Material [mol/s]}} \tag{5}$$

Note that the gaseous components are modified at 20% moisture content. Also, the approach temperature for each reaction is shown as Table 3.

Table 3.Approach temperature for each reaction (estimated)

Reaction	ΔT	Unit
Pyrolysis	78.3	°C
Reforming	252.0	°C

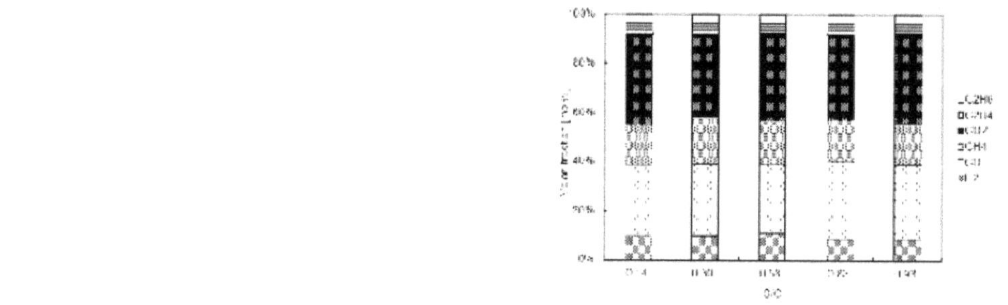

a) Pyrolysis (550 °C, S/C=0.14-0.93)

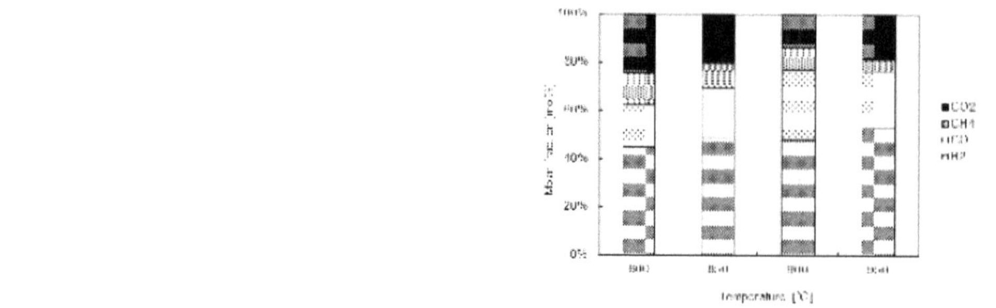

b) Reforming reaction (800-950 °C, S/C=1.0)

Figure 3.Gaseous yields of pyrolysis (a) and reforming reaction (b).

Based on the above experimental results, we estimated the following material balance:

$$\begin{aligned}&C_{31.078}H_{43.495}O_{23.677}N_{0.069} + 31.078H_2O \rightarrow \\ &24.052H_2 + 11.682CO + 2.435CH_4 + 9.442CO_2 + 23.228H_2O \\ &+2.92\times10^{-2}N_2 + 4.40\times10^{-5}NH_3 + C_{2.686}H_{0.482}O_{0.343}N_{0.004}\end{aligned} \tag{6}$$

C2.686H0.482O0.343N0.004 is the chemical component of char, and its heating value was 32.0 MJ/kg. In our simulator, the energy performance would be solved so that the input and the output on heat and materials would be balanced.

Next, using ϕ=9.5mm ball, we measured the temperature profiles at the surface of ball and the center of it. In the phase of absorption of heat, the ball was kept at each designed temperature between 200 and 950 ºC. At the time, there was difference between the surface temperature and the center one, and the temperature differences were measured. Inversely, in the phase of heat radiation, the ball was heated up to 1,000 ºC in the furnace, and it was put in a room temperature. Simultaneously, the temperature differences were measured. Note that these temperature profiles are time series data.

As a result, the thermal conductivities can be obtained. Also, since the thermal circulation time has to be the same as the reacting time on a pyrolysis and a steam reforming reaction, the optimal size of the ball is decided. Thus, the adequate auxiliary power for the circulation of HC would be obtained. Due to this result, we can estimate the suitable residence time in each reactor for the temperature profile which would be led by the simulator. Based on the above concept, we could estimate the syngas through BT process (Dowaki et al., 2008a, Dowaki, 2011a).

2.2. Process design of energy production system through BT process

Next, we introduce the examples of process design through BT process. As we mentioned before, there would be many energy paths through BT process. Here, as the examples, H2 production and Cogeneration system (CGS) would be concentrated. The purpose of each process design would be due to the energy analysis and/or the environmental one using LCA methodology.

2.2.1. Case study of Bio-H2 production system

Through a reaction process based on superheated steam, the biomass is converted to the syngas with a high concentration of H2. In the BT process, pyrolysis gases are reformed with H_2O (steam), and Tar and Char are generated as co-products. Since Tar contents pass through the higher temperature zone, the residual volume would be negligible. Also, due to the recycling of the sensible heat of syngas, the total efficiency of the entire system would be improved.

Here, the process design of Bio-H2 was executed by the consideration of basic experimental results. The capability of the biomass gasification plant is 12 t/d, and the annual operation days are 300 day/yr. In the process design, the heat energy generated from the gasifier was assumed to be utilized as the energy for materials dryer. Due to the recycling of thermal energy, the energy of dryer can be reduced at most. For instance, the moisture content can be compensated up to 42 wt.% against the initial moisture content of 50 wt.%. The syngas generated through BT gasifier is transferred to the shift-reaction convertor, and then is fed into PSA (Pressure Swing Adsorption). In the PSA, the high concentrated H2 gas was purified to 99.99Vol.% (4N) of H2 gas.

Here, Tables 4 shows the performance of Bio-H2 production system. In Tables 4, the cold gas efficiency is defined as follows:

$$\eta_{Cold} = \frac{Syngas\ [\mathrm{MJ/h}]}{Feedstock[\mathrm{MJ/h}] + Char[\mathrm{MJ/h}] + Offgas[\mathrm{MJ/h}]} \tag{7}$$

$$\eta_{Total} = \frac{\text{Bio-H}_2\ [\mathrm{MJ/h}]}{Feedstock[\mathrm{MJ/h}]} \tag{8}$$

Also, the total efficiency

Table 4.Performance of Bio-H2 production system (estimated)

BT Process (15 t/d)	Feedstock	635.9	kg/h
		8,415	MJ/h
	Syngas (For Bio-H2)	678.5	Nm3/h
		4,544	MJ/h
	Cold-Gas Eff. ηCold	62.0%	LHV-%
	Auxiliary Power	247.4	kW
PSA (4N-H2)	Bio-H2	303.5	Nm3/h
		3,275	MJ/h
	Total Eff. ηTotal	38.9%	LHV-%

2.2.2. Case study of Cogeneration system

Next, we explained about the co-generation system by which electricity and thermal energy can be generated. In the case of BT-CGS, due to the heat balance, the reaction energy in the furnace might be shortage. Thus, the additional feedstock would be necessary. In the case of Bio-H2 production system, since off-gas through PSA is available, the additional biomass material is not required.

Also, from the viewpoint of the economic condition, the case that the additional one is fed into BT would be much better in comparison to the case without any feedstock. That is, more products (i.e. electricity and/or thermal energy) can be generated. Consequently, the economic condition of BT-CGS operation would be improved by a lot of energy products.

Thus, we consider BT-CGS case in which the additional feedstock is required.For the operation of gas-engine due to the low calorific heating value of bio-gas which means the syngas of BT gasifier, although there are sometimes problems on the heating value of fuel, we executed the process design using the practice parameters which were analysed by the engine manufacturing maker.

Table 5 shows the performance of Bio-CGS. In Tables 5, the cold gas efficiency nCold, the net power efficiency nPow, the heat recovery efficiency nHeat and the net total efficiency nTotal are defined as follows:

$$\eta_{Cold} = \frac{Syngas\ [\mathrm{MJ/h}]}{Feedstock[\mathrm{MJ/h}] + Char[\mathrm{MJ/h}] + Add.\ Feedstock[\mathrm{MJ/h}]} \tag{9}$$

$$\eta_{Pow} = \frac{Net\ \text{Power}\ (=\text{Power-Auxiliary})[\mathrm{MJ/h}]}{Feedstock[\mathrm{MJ/h}] + Add.\ Feedstock[\mathrm{MJ/h}]} \tag{10}$$

$$\eta_{Heat} = \frac{Steam[MJ/h] + Hot\ water[MJ/h]}{Feedstock[MJ/h] + Add.\ Feedstock[MJ/h]} \tag{11}$$

$$\eta_{Total} = \eta_{Pow} + \eta_{Heat} \tag{12}$$

Table 5.Performance of BT-CGS (estimated)

(9pt)BT Process (18 t/d)	Feedstock	625.0	kg/h
		8,278	MJ/h
	Additional Feedstock	139.5	kg/h
		1,846	MJ/h
	Syngas (For Gas-engine)	1,021	Nm3/h
		6,922	MJ/h
	Cold-Gas Eff. η Cold	59.0%	LHV-%
	Auxiliary Power	111	kW
Gas-Engine	Power (Net)	459	kW
	Steam	1,344	MJ/h
	Hot water	1,551	MJ/h
	Power Eff.	16.3%	LHV-%
	Heat Recovery Eff.	28.6%	LHV-%
	Total Eff.(Net)	44.9%	LHV-%

3. CONCEPT OF THE BIOMASS LIFE CYCLE ASSESSMENT

So far, the biomass Life Cycle Assessment (LCA) analyses, in which the pre-processing process of chipping, transportation and drying of biomass materials are included, and in which the energy conversion process of a production energy of electricity and/or heat through an integrated gasification combined cycle (IGCC) power system or a co-generation system (CGS) is included, were analysed (Dowaki et al. 2002, Dowaki et al. 2003).

In this section, we describe on the BT-CGS and the production system of Bio-H2. At the beginning, in this section, we defined the system boundary of the biomass LCA. A target is to estimate a life cycle inventory (CO2 emissions and/or energy intensities) of the entire system with a biomass gasification system and/or a purification one. That is, we refer to the environmentally friendly system, such as the biomass energy system, considering CO2 emissions and/or energy intensities from the entire system based on LCA methodology. In the case of BT-CGS or Bio-H2, due to the shortage of reaction heat in the furnace or the larger auxiliary power output of PSA, the specific CO2 emission might be affected. That is, the process design and the energy analysis on basis of the process simulation would be extremely significant.

3.1. System boundary

Following ISO 14041 guidelines, we define the system boundary in the biomass energy system. The system boundary includes the entire life cycle of Bio-H2 fuel or electricity and thermal energy products, including the pre-processing process and the energy conversion process. (See Fig. 4). In the pre-processing process, there are sub-processes of chipping, transportation by trucks, and drying. In the energy conversion process, there are sub-processes of the gasification through the BT plant with a purification process or a CGS unit. Also, in our estimation, we focused on "well to tank (WtW)" analysis.

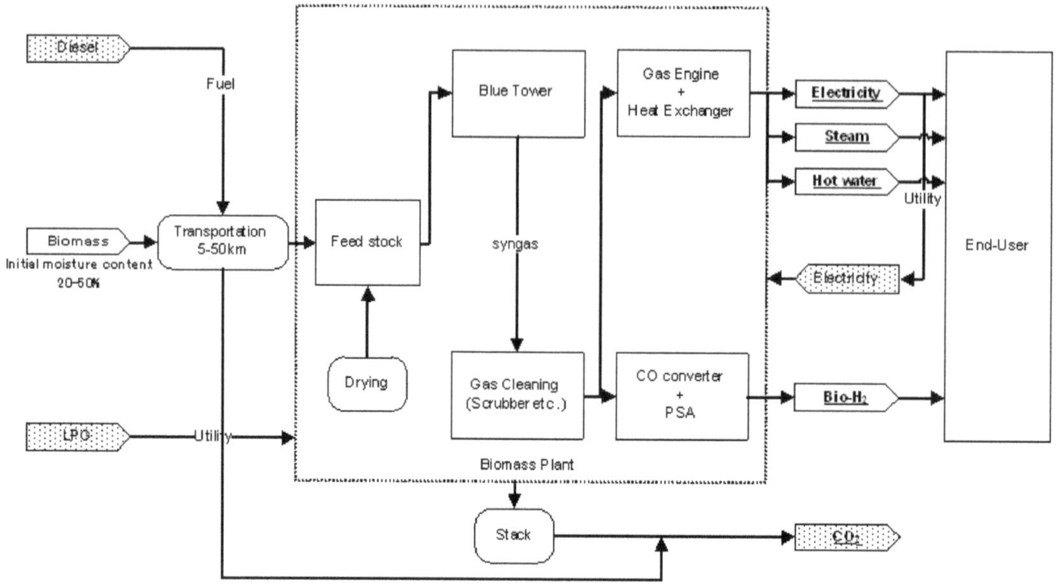

Figure 4.System boundary of biomass LCA.

3.2. Functional unit

The target product is Bio-H2 and CGS products of electricity and/or heat energy. Thus, the functional unit is assumed to be the unit per a produced energy. The lower heating values of H2 and electricity are 10.8 MJ/Nm3 and 3.6 MJ/kWh, respectively.

3.3. Pre-processing process

In the pre-processing process, there are sub-processes of chipping, transportation, and drying of biomass materials. In particular, within the sub-processes of transportation and drying, we have to consider uncertainties. To date, there are few studies considering these uncertainties. CO2 emissions and energy intensities in the biomass LCA would be affected by the moisture content of biomass materials, and the transportation distance from the cultivation site, or the site of accumulating waste materials, to the energy plant. Table 6 shows heating values, and that of CO2 emissions, for each fuel with biomass materials, respectively. Also, CO2 emissions and energy intensities were estimated using the Monte Carlo simulation in order to consider these uncertainties (Dowaki and Genchi, 2009).

Table 6.Data of the specific CO2 emissions

Fuel	CO2	Note
Feedstock	0.0 g-CO2/MJ-Fuel	at 20 wt.% (moisture content), Japanese Cedar, HV:13.23 MJ/kg
Diesel	74.4 g-CO2/MJ-Fuel	Chipping, Transportation, HV: 35.50 MJ/L
Electricity	123.1 g-CO2/MJ-Fuel	Auxiliary power of the plant (Primary Energy)

3.3.1. Sub-processes of chipping, transportation and drying

The energy consumption of chipping, transportation and drying is as follows:

- **Chipping:** The energy consumption of the chipping process is due to electricity and diesel. The specific units of energy consumption are 13.6 kWh/material-t (122.4 MJ/material-t) and 1.23 L-diesel/material-t (43.7 MJ/material-t), respectively (Hashimoto et al., 2000).

- **Transportation:** The chopped biomass materials are delivered to the plant by 10 ton diesel trucks. CO_2 emissions and/or energy intensities on a given transportation run would be affected by the weight of biomass materials. That is, the weight of which the materials can be carried is restricted to bulk density. We measured the bulk density (=0.14 t/m3) in the atmosphere. The bulk density is dependent upon the moisture content. Thus, assuming that the bulk density is at a moisture content of 15 wt.% is

$$\rho_{MC} = \frac{0.85}{1-MC}\rho_{15} \quad (13)$$

Next, the loading platform of 10t-trucks is to be approximately 24.7 m3 (Suri-Keikaku Co. Ltd., 2005). Consequently, even a truck with 10 ton's volume cannot always carry that in full weight. Here, CO_2 emissions and/or energy intensities are assumed to be due to the fuel consumption of truck, which is indicated as a function of the loading rate of weight. That is, using the loading rate of λ, the fuel consumption rate of a 10t-truck $f_{FC}(\lambda)$ is

$$f_{FC}(\lambda) = a\lambda + b \quad (14)$$

where, a(=714 g-CO_2/km) and b(=508 g-CO_2/km) are constants on the fuel consumption of the truck (Dowaki et al., 2008b).

The definition of the loading rate of λ is as follows: Assuming that the plant scale is Ps dry-t/d, and that the annual operating time is 300 days, the annual material balance on the feed materials is 300Ps t-dry/yr. Since the throughput per year at MC wt.% is 300Ps/1-MC), the total number of transportation by 10 t trucks at MC wt.% (N_{mat}) is

$$N_{mat} = \left[\frac{300Ps/(1-MC)}{24.7\rho_{MC}}\right] + 1 \quad (15)$$

Where, [α] is represented as the maximum integer, so as not to exceed a. Thus, theaverage loading rate of a 10 t-truck (λ_{ave}) is

$$\lambda_{ave} = \frac{300Ps/(1-MC)}{10N_{mat}} \quad (16)$$

Providing the average loading rate, and multiplying $f_{FC}(\lambda)$ by the transportation distance and the specific CO_2 emissions or the energy consumption of diesel, we can estimate CO_2 emissions or fuel consumption in the transportation sub-process. In this paper, the transportation distance is the range between 5 (Distmin) and 50 km (Distmax), because the wooden materials in Japan are distributed widely. That is, it is assumed that the feed materials are collected within a radius of 50 km.

Drying: Next, on the sub-process of drying, the energy consumption was estimated under the condition that the moisture content of feed materials would decrease to 20 wt.%. Here, assuming that the initial moisture contents are from 20 (MCmin)) to 50 wt.% (MCmax), the raw materials are dried by a boiler. Also, the auxiliary power of a pump in a boiler is included in the energy consumption of the sub-process. The operational specification of a wood-chip dryer (boiler) is the energy efficiency of 80 %, and the auxiliary power of a pump of 0.195 kWh/t-water (1.75 MJ/t-water). Note that the moisture content of feedstock can be reduced by the hot exhausted gas to some extent.

Monte Carlo simulation on the uncertainties: As the above, in this paper, we estimated CO2 emissions and/or energy intensities, considering the uncertainties of the transportation distance and the moisture content. In this paper, the following two uncertainties of the distance and the moisture content were considered by the Monte Carlo simulation.

That is, the uncertainties on the transportation distance (Dist km) and the moisture content (MC wt.%)are represented by uniform random numbers Rndibetween 0 and 1 in Eqs. (17) and (18). Note that Rnd1 andRnd2are independent and identically distributed.

$$Dist = Dist_{min} + Rnd_2 \left(Dist_{max} - Dist_{min} \right) \tag{17}$$

$$MC = MC_{min} + Rnd_1 \left(MC_{max} - MC_{min} \right) \tag{18}$$

An iteration count in the simulation was executed up to 10,000. The range within a 95 % significant level was adopted as the uncertain data on the distance and the moisture content, in order to estimate CO2 emissions. In this case, the gross distributions on CO2 emissions would be normal distributions.

3.4. CO2 emissions on CGS and Bio-H2 fuel

Based on the above data, CO2 emissions of CGS (electricity and/or thermal energy) and Bio-H2 fuels are shown in Fig. 5.

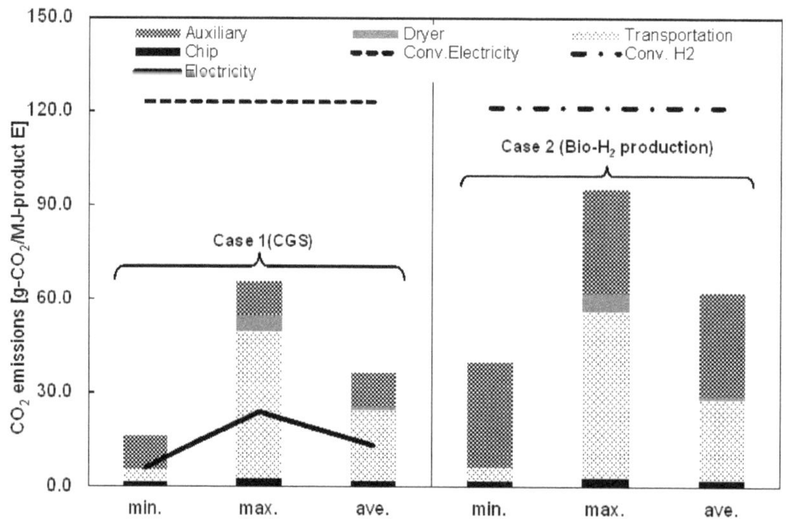

Figure 5.CO2 emission in each case (Case 1:CGS, Case 2 Bio-H2 production).

According to Fig. 5, the entire CO2 emissions are 16.3-65.7 g-CO2/MJ of CGS and 39.6-95.3 g-CO2/MJ of Bio-H2, respectively. Especially, in the CGS case, the specific CO2 emissions of electricity are 5.9-23.9 g-CO2/MJ, and the reduction percentages in comparison to the conventional electricity in Japan are 80.6-95.2%. In the case of Bio-H2 case, the reduction percentages against the conventional H2 production (121.3 g-CO2/MJ, Natural gas origin) are 21.4-67.3%.

CO2 emissions at the material drying and at the auxiliary power of a purification process of PSA occupy a large portion of the entire CO2 emission. Especially, the influence due to the compression power of H2 purification would be significant. In the case of Bio-H2, the amount of 35.1% to 84.4 % of the total CO2 emissions would be emitted from the auxiliary power including the power for BT operation. Also, in the case of CGS, that of 16.5% to 66.6 % would be emitted from the auxiliary power origin, even if the PSA operation is not equipped.

The deviations of CO_2 emissions (the maximum value – the minimum one) due to the uncertainties on the moisture content and the transportation distance would be within 49.5 g-CO_2/MJ of CGS and 55.7 g-CO_2/MJ of Bio-H_2, respectively.

That is, the range of collection of biomass feedstock would be extremely significant from the viewpoint of CO_2 emission reduction on basis of LCA methodology.

4. FUTURE APPLICATION OF BIO-FUEL

As we mentioned before, the renewable energy source, especially, the biomass energy source would be promising for global warming protection. Using the biomass feedstock, there are many fuels which can be converted through the gasification, the fermentation or another process. Here, we concentrated to the biomass gasification process by which electricity and thermal energy or Bio-H_2 fuel are produced. Also, the CO_2 emission due to LCA methodology, which is estimated in order to understand the impact of Global warming numerically, was estimated. As a next step, we have to create the countermeasure for promotion of our proposed system. However, there is not example in which the relationship between the supply and the demand is argued enough. Based on the sequential and entire system, we have to judge the effects and/or the benefits such as CO_2 emission etc. (See Fig. 1).

Here, as a good example, we introduce the following system. However, that might be difficult to promote our proposed system due to the cost barrier against a conventional system at the present time. The combined system in which the renewable energy such as Bio-H_2 can be available would have a significant meaning in the future utilization for Global warming protection. Simultaneously, we have to create the new business model which would be suitable for the end users.

Now, there is the proposal to install an advanced cell phone (a smart phone) with a PEFC unit so as to get CO_2 benefit. A smart phone is an electronic device used for two-way radio telecommunication over a cellular network of base stations known as cell sites. The sale of mobile phones has been one of the fastest growing markets in the world today. For instance, the cell phone users of Japan were approximately 107 million in 2005 (Infoplease, 2005). At present, around 85% people in America have used cell phone. In addition, new technology of a mobile communication is being developed very quickly. A few years ago, people used their cell phone just for making a call or sending a short mail through a SMS function. However, at the current time, there are a lot of features of a smart phone such as music player, video player, game, chatting, internet browsing and email, etc. These factors should increase energy consumption and increase CO_2 emission.

The current power supply system in a smart phone is dominated by a Li-ion battery, which has some advantage such as wide variety of shapes/sizes without a memory effect. In addition, the rapidly advancing needs for mobile communication are increasing the consumer demand for portable application with even higher power output, longer operation time, smaller size, and lighter weight. A Li-ion and other rechargeable battery system might not be suitable for high power and long time span portable devices due to their lower energy density, shorter operational time, and safety. Li-ion batteries are well established as a power supply for portable devices. Recently, since the power demand has been increasing faster than battery capabilities, the fuel cells might become a promising alternate for niche applications. A fuel cell is an electrochemical device which continuously converts chemical energy into electricity and thermal energy by feeding H_2 fuel and oxygen into it. A fuel cell power supply can be higher energy per a unit mass than conventional batteries. Also, the using of fuel cell system is not harmful to the environment, if compared with a Li-ion battery (Hoogers, 2003). Also, there are the following two types of fuel cell: 1) Polymer Electrolyte Fuel Cell (PEFC) and2)Direct Methanol Fuel Cell (DMFC), which are operated in low temperature.

These two systems are almost same, the difference is only in fuel, that is, the PEFC is operated by H2 (gas) and DMFC is done by methanol (liquid). Here, we focused on the PEFC into which H2 fuel is fed. The reason why we concentrate the system is that the fuel for a PEFC can be produced by the renewable resources such as biomass feedstock with a lower CO2 emission in comparison to the conventional production system. In the area where there is plenty of biomass feedstock (e.g. Indonesia and Malaysia etc.), there is a good potential to install that. A PEFC is applied to replace a Li-ion battery. A comparison of CO2 emission between a Li-ion battery cell phone and a PEFC cell phone was calculated using Life Cycle Assessment (LCA) methodology, in consideration of the user's behaviour.

4.1. A case study on a smart phone due to LCA methodology

The goal of this study is to compare the CO2 emission of the conventional Li-ion cell phone and the PEFC cell phone. The functional unit is the specific CO2 emission per a life cycle (LC) of kg-CO2/LC. Fig. 6 shows the life cycle stage on the schematic design of system boundary, in which a pre-processing of raw materials, a manufacture, a transportation and distribution, an energy consumption of end users and a disposal process are included. Also, in this study, we referred to the duration time of each operation of cell phone (Dowaki et al., 2010a).

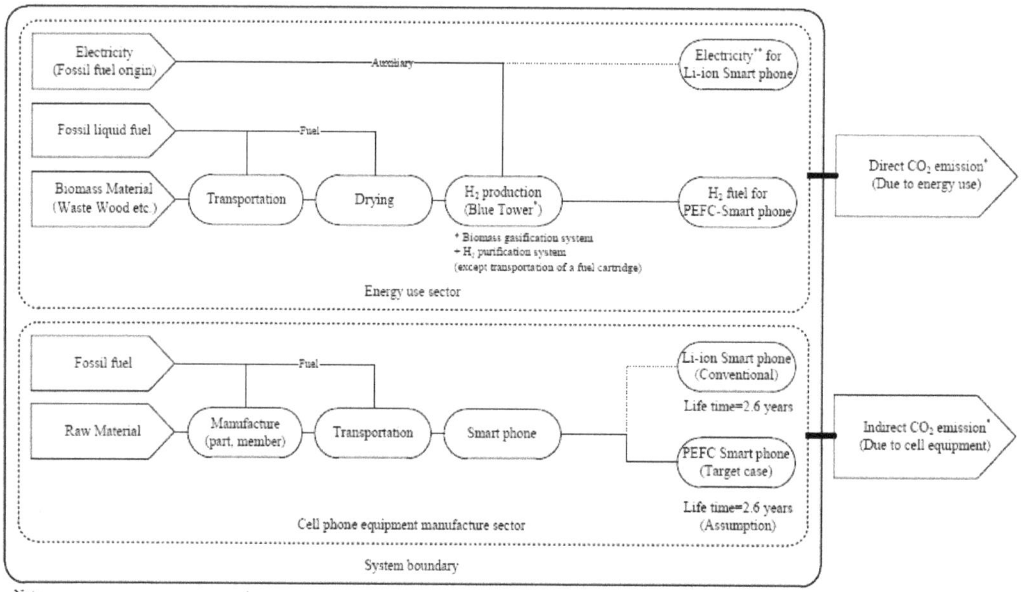

Note:
* Estimated period=2.6 years
** This value is based on the wheel to tank, that is, the input energy for producing the fuel besides raw energy source (primary energy) is considered, too.

Figure 6.System boundary of a cell phone analysis.

In the system boundary, as we described the prior section, we think about the availability of Bio-H2 through BT process. For the purpose, we executed the questionnaire on the way to use a smart phone firstly. Also, we executed the performance of a PEM cell which is based on a PEFC unit using the electric power measurement device.

The difference between a Li-ion and a PEFC cell phone is in electrical energy sources. The Li-ion cell phone is supplied by conventional electricity, whereas a PEFC cell phone is done by Bio-H2 as an energy input. The battery charge due to the conventional electricity emits CO2 of one of the greenhouse gases. On the other hand, since the Bio-H2 would be carbon neutral, the CO2 emission is equivalent to zero in a combustion process. However, the production process of a renewable fuel is accompanied with the conventional energy inputs (i.e. fossil fuels). Thus, it is extremely important to estimate the energy system based on LCA methodology.

4.1.1. A questionnaire for the smart phone users

In order to investigate the way to use a smart phone in each user, we executed the questionnaire between February 17 and February 24, 2011. 200 respondents in Japan participated in this research. Also, target respondents are the users who use a smart phone, with their ages between 15 and 65 years, respectively. In the questionnaire content, the duration time of a talking, a SMS, music (MP3), a game, a web-site (internet), an e-mail checking, and an idle time were estimated for each age category. Fig. 7 shows the result of the duration time of each function. The checking time of internet would be larger in both weekdays and holyday (Dowaki et al., 2011b).

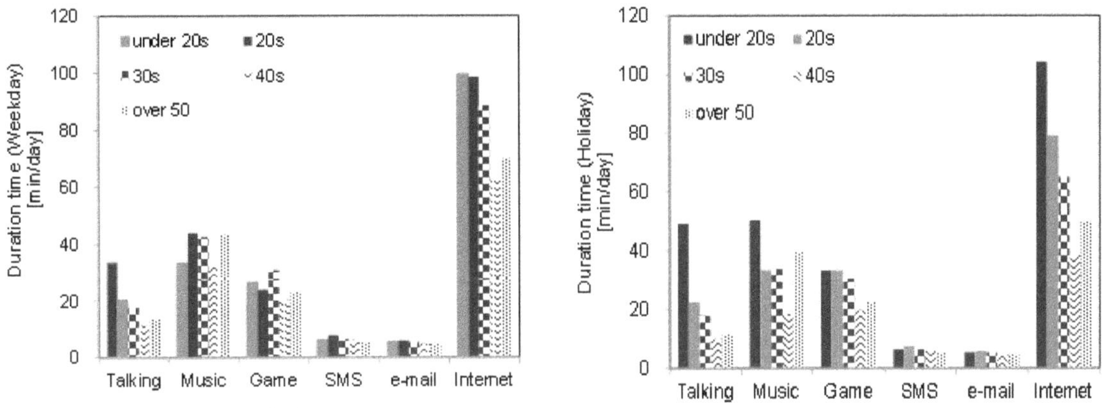

a) Weekdays b) Holiday

Figure 7.Duration time of each function in a smart phone.

4.1.2. Measurement of the performance of a PEFC unit

Based on the duration time, we measured the performance of a PEM-cell which is based on a PEFC unit for a smart phone. That is, using the result of output capability of a PEM-cell and the maximum duration time, we designed the cell area of a PEFC, and estimated the energy consumption for each function.

Here, in order to have a good reliability, 10 times experiments have been done for the following tasks: a talking, a SMS, music (MP3), a game, a web-site (internet) and an e-mail checking, respectively. In our experiments, we used the electric measurement device (AC/DC POWER HiTESTER 3334, HIOKI E.E. Corp.) to measure the voltage and the current, and the power which is obtained by these factors.

Next, for the purpose of estimating the cell performance, we measured the potential of a PEM-cell in varying currents. The apparatus consists of a PEM-cell (Micro Inc.) and a potentiostat (HAB-151, Hokuto Denko Corp.). The size of the cell with low platinum loading electrodes (1.0 Pt mg/cm2) and Nafion® 115 is 4 cm2×3 cells. Using a potentiostat, 1) an open circuit voltage V0 [Volt] was measured and 2) the relationship of current density J vs. cell potential V was evaluated between approximately 200 mV/cell and an open circuit potential V0. Also, H2 flow in anode was up to 20 ml/min and the concentration of H2 was 100 vol.% at a constant percentage (see Fig.8). Note that each parameter on the performance of a PEM-cell is decided at the condition which is not rate-limiting. In this case, we adopted the flow rate condition of 20 ml/min. The cathode was stayed at the atmospheric condition. The conditions in both the anode and cathode sides were not saturated by steam (Dowaki et al., 2010a, Dowaki et al., 2011b).

Next, the relationship between a cell's potential and current density, in the low and intermediate current density region of a PEM-cell, has been shown to obey the following Eq. (19) (Kim et al., 1995). Note that the result was shown by the condition of a single PEM- cell.

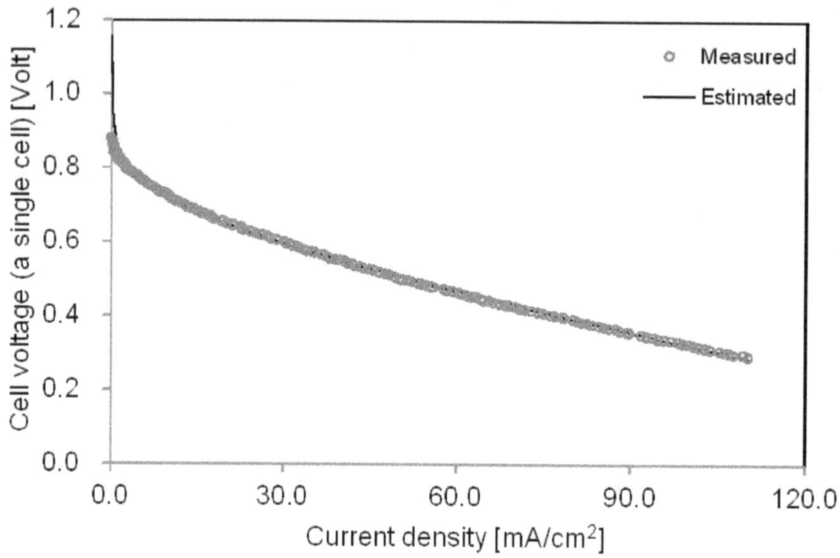

Figure 8.Relationship between current density and a cell voltage in the single PEM-cell.

$$V = V_0 - b\log J - R_{cell}J - m\exp(nJ) \tag{19}$$

Where, J , b , Rcell ,m and n are current density [mA/cm2], Tafel slope [mV/decade], a cell resistance [ohm-cm2] and constant parameters, respectively.

Based on the experimental result, we analysed each parameter by the approximation formula of Eq. (19). Consequently, the open circuit voltage V0, the Tafel slope b, a cell resistance Rcelland constant parameters of mandnwere 1.16 Volt, 54.0 mV/decade, 2.98 ohm-cm2, 9.08 and -2.53×10-3 were obtained. Using these parameters, we designed the PEFC unit of a smart phone as follows: the surface area is 22 cm2, and the stack number of cell is 3. These conditions would be suitable of a conventional smart phone size and satisfy the maximum output among each function. Also, the stoichiometric ratio is assumed to be 1.00. This means the supplied H2 would be fully consumed.

Next, due to the questionnaire for smart phone users (see Fig. 7), the energy consumption for each function and the performance of PEFC using a PEM-cell experimental result, we estimated CO2 emission on basis of LCA methodology. Using Eq. (19), the H2 flow rate in practice use is able to be calculated. Here, a PEFC would be operated between 1.17 and 2.63 Volt. The specific energy consumption in each function was between 0.39 and 37.7 Nml/min.

4.1.3. Specific CO2 emission of a smart phone

Based on the above analysed results, we estimated the CO2 emission of a smart phone use. Here, we considered the indirect and the direct CO2 emissions. The direct CO2 emission is equivalent to the fuel consumption origin. On the other hand, the indirect one is mainly on the device of a smart phone. In this study, we focused on HTC Desire X06HT made in Taiwan as a model phone. The indirect CO2 emission is calculated by Input-Output (IO) table, and this emission referred to the prior result. Also, we estimated the conventional smart phone including Li-ion battery in order to compare to the new one. Assuming that the holding time (life time: LT) when one user has a smart phone until he or she change the new one is 2.6 years, the indirect CO2 emission of HTC Desire X06HT including Li-ion battery would be 15.32 kg-CO2/unit. The emission of a smart phone with a PEFC unit would be 15.30 kg-CO2/unit. Although there are uncertainties on the storage tank of H2 to some extent, referring to the data of DMFC storage tank which has already developed, we estimated the emission as almost same as the conventional case (Dowaki et al., 2010a).

Next, the direct CO2 emission is affected by the specific CO2 emission of each fuel. Here, the CO2 emissions of conventional electricity, H2 fuel of natural gas origin (on-site) and Bio-H2 are assumed to be 123.1 g-CO2/MJ, 121.3 and 39.6 g-CO2/MJ-H2, respectively. The CO2 emission per one life cycle is shown in Fig. 9. Note that the specific emission of Bio-H2 is a minimum level (see Fig. 5).

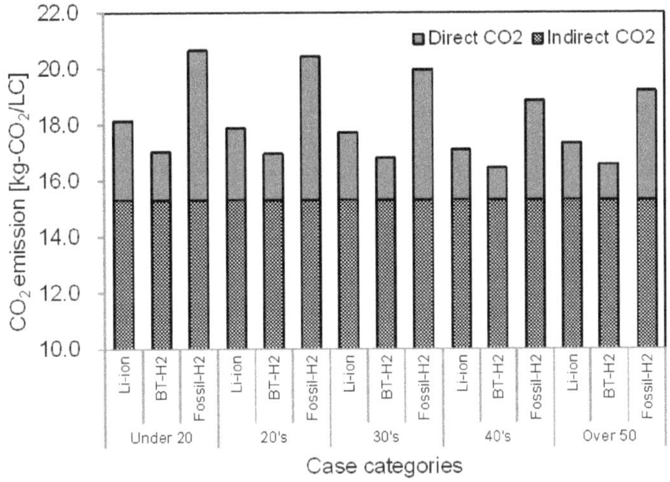

Figure 9.Life Cycle CO_2 emission for a smart phone use.

According to this result, due to application of a PEFC unit to the smart phone, we would be able to reduce CO_2 emissions of 3.9% to 6.1 % in comparison to the conventional phone. Especially, in the category of younger generation, the CO2 reduction benefit would be effective.

4.2. A case study on a greenhouse facility due to LCA methodology

Next, we propose the advanced greenhouse system for paprika cultivation with the combined biomass gasification process of BT with SOFC (Solid Oxide Fuel Cell). The BT gasifier which is a biomass gasification process has a characteristic of generating hydrogen gas of high concentration in syngas. Here, we considered the acceptability for the related facilities in agriculture field. Because the environmentally friendly system such as a PV system or a fuel cell co-generation system is still not enough to be promoted for those facilities. That is, there would be potential to combine the biomass energy system which is environmentally friendly with the agriculture related facilities. In addition, MAFF contribute to the global warming protection through the carbon-footprint of agricultural products. The ministry has a few subsidy menus on the promotion of the system. Also, on the surplus energy of electricity and/or thermal energy, there are institutions by which the energy companies are obliged to purchase them with additional fees.

Using the above institutions and/or subsidy menus under the leadership of the Japanese government, we considered the following concrete paprika harvesting system in which the biomass gasification (BT) process with SOFC is assumed to be introduced (Dowaki et al., 2010b).

First, our model site is the paprika harvesting facility in Miyagi of Japan, whose area is 4.6 ha. In our study, through interviews from the owner company, we used the data of not only the energy consumption of electricity and oil, but also the supply of CO2 gas which is fed into the greenhouse as a growth promoting agent. That is, in the model we proposed, the electricity, the thermal energy and the CO2 gas which is included in exhausted gas through BT plant are assumed to be available for the greenhouse facility of paprika harvesting. In addition, due to the combination of the advanced power generation such as SOFC, additional benefit of CO2 emission reduction would be obtained. This may be advantageous from the profit aspect since the surplus electricity would be able to be sold to the commercial energy companies. Also, from the viewpoint of thermal energy use, the combined BT with SOFC units would be advantageous since the exhausted gas with a high temperature (ca.700 °C) is generated. Although the operation of SOFC has been in a developmental stage, we used the published parameters. The initial cost of SOFC unit seems to be costly in comparison to the conventional power system. However, it is said that the commercial stage of SOFC is close. Thus, the initial cost was assumed to be equivalent to the target price as of 2015. The thermal energy for the greenhouse is supplied by the heat pump equipment. This would bring to the benefit of cost and/or CO2 emission reduction, since there is little waste thermal energy (Dowaki et al., 2011c).

On the other hand, MAFF tries to introduce the carbon-footprint for the agricultural products. It is difficult to estimate the monetary values of CO2 emissions of agricultural products. For instance, Kikuchi et al. investigated the willingness to pay for CO2 emission reduction of vegetables (Kikuchi and Itsubo, 2009). They found out that the consumers have a willingness to pay for an additional cost of approximately 5% up against a conventional price. Although this is only a limited effect, there would be a potential to earn income due to the carbon footprint. That is, with regard to income in our system, revenues to the plant owner would include the related subsidy, the processing fee of waste material, the sale of surplus electricity and the paprika sale with low CO2 emission. The carbon-footprint of agricultural product might be important one of income sources.In this study, we analysed the CO2 emission due to LCA methodology.

4.2.1. A LCA for the paprika cultivation

In the LCA concept of this paper, the direct factors and the indirect ones have to be considered. In our definition, fossil fuel energy inputs (primary energy basis) and the electricity of fossil fuel origin are included in the direct factors. Also, chemical fertilizers are included in the indirect ones. Here, note that another greenhouse gases such as N2O and CH4 are not taken into consideration.

So far, in the biomass LCA analyses, the pre-processing process of chipping, transportation and drying of biomass materials, and the energy conversion process of a production energy of electricity and/or heat, through an energy system are included. This time, the paprika harvesting process has to be added to the entire life cycle stage. Using the chemical experimental data, the design of BT plant with SOFC units would be extremely significant in the biomass LCA. A target is to estimate a life cycle inventory of the entire system with BT gasifier and SOFC.

Here, we describe on the system boundary in this study. Following ISO 14041 guidelines, we define the system boundary in the biomass energy system (see Fig. 10) (Dowaki et al., 2010b).

The system boundary includes the entire life cycle of each energy input (electricity/thermal energy), including the pre-processing process, the energy conversion process and the paprika harvesting process. In the pre-processing process, there are sub-processes of chipping, transportation by trucks, and drying. In the energy conversion process, there are sub-processes of the gasification through the BT plant (19 t/d) with the four units of SOFC (200 kW/unit) process. In the paprika harvesting process, it is assumed that the exhausted gas of CO2 is available as a growth agent. Here, the target product is a paprika. Thus, the functional unit is assumed to be the unit per a produced paprika (Dowaki et al., 2011c).

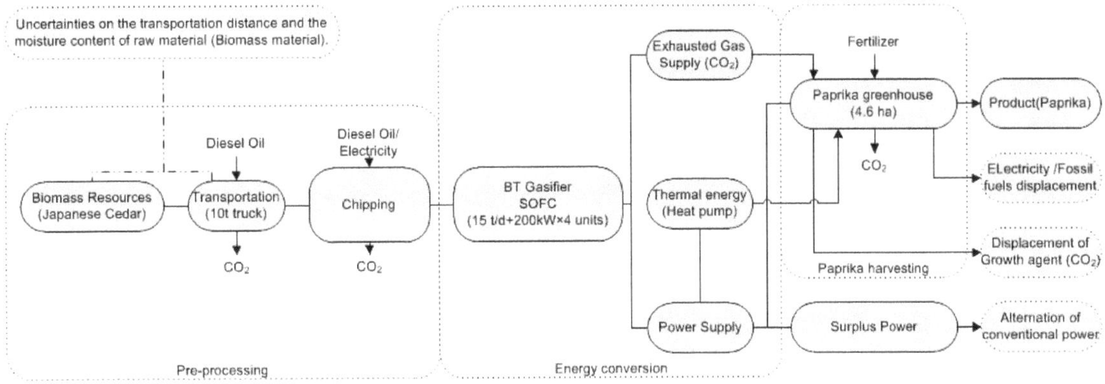

Figure 10.System boundary of a paprika production system.

Next, in the pre-processing process, there are sub-processes of chipping, transportation, and drying of biomass materials. In particular, within the sub-processes of transportation and drying, we have to consider uncertainties (see section 3.3.1). To date, there are a few studies considering these uncertainties. CO2 emissions in the biomass LCA would be affected by the moisture content of biomass materials, and the transportation distance from the cultivation site, or the site of accumulating waste materials, to the energy plant. Hence, it would be extremely significant to consider these factors. Table 7 shows the specific CO2 emissions, for each fuel with biomass materials, respectively.

Table 7.Data of the specific CO_2 emissions.

Item	CO_2	Note
Feedstock	0.0 g-CO_2/MJ-Fuel	at 20 wt.% (moisture content), Japanese Cedar, HV:13.23 MJ/kg
Diesel	74.4 g-CO_2/MJ-Fuel	Chipping, Transportation, HV: 35.50 MJ/L
Bunker A	76.9 g-CO_2/MJ-Fuel	Paprika production (Boiler)
Kerosene	73.6 g-CO_2/MJ-Fuel	Paprika production (Boiler)
Electricity	123.1 g-CO_2/MJ-Fuel*	Paprika production (Ventilation and lightning)
Fertilizer (N)	5.67 kg-CO_2/kg	Indirect CO_2 emission
Fertilizer (P2O5)	0.88 kg-CO_2/kg	Indirect CO_2 emission
Fertilizer (K2O)	1.85 kg-CO_2/kg	Indirect CO_2 emission

On the energy conversion process, assuming that the 19 t/d BT plant and 4×200 kW SOFC (BT-SOFC system) were installed, we estimated the CO_2 emission in the paprika production system. Here, the operational condition of SOFC unit is assumed to be almost full load operation. Also, the specification of SOFC unit is shown in Table 8.

Table 8.Specification of SOFC unit.

Unit Scale	[kW]	200
Number of unit	-	4
Operating Temp.	[°C]	900
Current density	[mA/cm2]	612
Stoichiometric ratio	-	1.25
Tafel slope	[mV/dec.]	2.2
Cell Resistance	[ohm]	0.52
Open Circuit Voltage	[mV]	950
DC/AC converter Eff.	[%]	95

Due to the specification data in each system, the performance of BT-SOFC system is obtained as Table 9. Also, the thermal energy supply to the facility is assumed to be due to the heat pump (COP: 5.5).

Table 9.Performance of BT-SOFC system.

BT Process (19t/d)	Feedstock	781.3	kg/h
		10,338	MJ/h
	Cold-Gas eff.(Eq. (7))	56.2	LHV%
	Auxiliary Power	127.3	kW
SOFC (4×200 kW)	Power eff.vs. syngas	45.5	LHV%
	Power eff. vs. feed	25.0	LHV%
	Net eff. vs. feed	20.6	LHV%
	Net power scale	590	kW

4.2.2. Paprika cultivation facility

In this study, we investigated the greenhouse facility at Miyagi of Japan where paprika is brought into cultivation. In this facility, the annual product yields are around 200 t/yr. The energy of electricity, kerosene and bunker A for lighting and a heater, and the input of CO2 gas as a growth agent are consumed. Here, since the energy data of time series was necessary, the boiler fuels of kerosene and/or bunker A were assumed to be in proportion to a difference between the minimum temperature for growing and the atmospheric one. Also, electricity was assumed to be consumed for 12 hours per a day.

Next, the consumption of CO2 gas as a growth agent would be analysed statistically. In a plant such as paprika, CO2 is consumed through photosynthesis. That is, this volume would be proportional to the duration of bright sunshine and an intensity of radiation. Fig.11 shows the statistically estimated CO2 consumption.

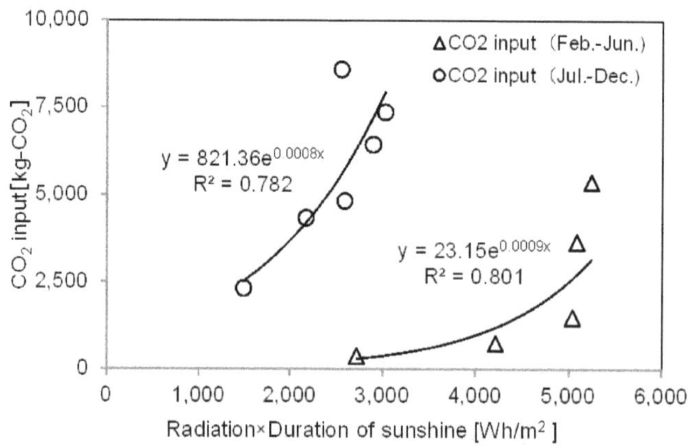

Figure 11.CO_2 supply volume as a growth agent.

On the other hand, fertilizers of N, P205 and K_2O only were considered, however another chemical inputs were ignored (Dowaki et al., 2010b).

4.2.3. Specific CO_2 emission of a paprika production

Based on the above results, we estimated the CO2 emission in conventional case and that of BT-SOFC case (see Fig. 12). In this study, the CO_2 intensities in BT-SOFC case are included on the uncertainties of moisture content and a transportation distance (See section 3.3.1).

In the conventional case, the specific CO_2 emission of 622.6 g-CO2/paprika was estimated. On the other hand, in the BT-SOFC case, the specific CO_2 emission of 38.1 to 218.4 g-CO2/paprika was analysed, and CO_2 reduction rate was 64.9% to 93.9%, respectively. Also, since the surplus electricity of 4,137 MWh/yr would be generated through this system, the much CO_2 reduction benefit might be obtained due to the alternation with the conventional electricity.

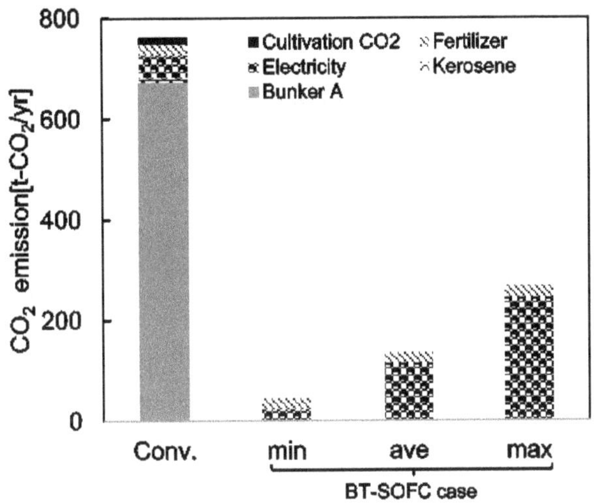

Figure 12.Specific CO2 emission of a paprika.

5. CONCISE SUMMARY

As previously said, there is significant potential for implementing a renewable energy system, such as a biomass energy system. This section was dedicated to examining the Blue Tower gasification process. In the foreseeable future, when contemplating the advancement of environmentally conscious enterprises, it is imperative to acknowledge the implementation of a sustainable business model that can function under favorable cost conditions and/or a decrease in CO_2 emissions. It is necessary to take into account both the technological barrier and the CO_2 reduction effect. Given the circumstances, the utilization of LCA technique would be both logical and essential. Undoubtedly, the commercial concept would be really significant. In this section, we presented two scenarios that are centered around the biomass gasification system of the BT process. If we take advantage of the subsidies provided by the national and/or local governments at the beginning, or if there is regulation of feed-in tariff, the proposed business scheme will become more competitive compared to the traditional company model. Furthermore, there is currently a significant level of public interest in the carbon footprint, as measured by the Life Cycle Assessment (LCA) technique. This indicates that there is the possibility to acquire the product with a low level of carbon emissions. In order to reduce greenhouse gas emissions in the future, it may be necessary to evaluate and implement appropriate technological systems and effective eco-socio systems.

REFERENCES

Ministry of Economy, Trade and Industry, Japan (METI). (2007). Report on Next-Generation Vehicle and Fuel Initiative, 11.07.2007, Available from: www.meti.go.jp/press/20070528001/initiative-torimatome.pdf

The Ministry of Agriculture, Forestry and Fisheries, Japan (MAFF). (2006). 11.07.2007, Available from: http://www.maff.go.jp/biomass/h18_point.pdf

K.K. Gas Review. (2007). Demand and Supply for industrial gas, 2007 Gas Georama in Japan (in Japanse). p.65

M .Mayer, H. J. Mühlen and C. Schmid. (2004). Blue Tower hydrogen from biomass, European Conference on Renewable Energies and Cooperation Exchange, Vienna, Austria H. J. Mühlen and C. Schmid. (1999). Staged Reforming-A Thermal Conversion Process Specifically Developed for Biomass, Proceedings 2nd Olle Lindström Symposium, pp. 75-82, Stockholm, Sweden

K. Dowaki, T. Ohta, Y. Kasahara, M. Kameyama, K. Sakawaki and S. Mori. (2007) An Economic and Energy Analysis on Bio-hydrogen Fuel using a Gasification Process. Renewable Energy, Vol. 32, Issue 1, pp. 80-94

M. Kameyamaa, S. Ogura, H. Kamiuchi and K. Dowaki. (2010). Performance evaluation of 1t/d scale plant through Blue Tower gasification process (in Japanese). Journal of Japan Society of Energy and Resources, Vol. 31, No. 2, pp.43-50

K. Dowaki, Y. Kon, and M. Kameyama. (2008a). A Modified Design of CGS Operation or Bio-H2 Production through Blue Tower (BT) Process. The 8th International conference on Eco balance, B-07, Tokyo, Japan

K. Dowaki, M. Nakazono, M. Yamanari, M. Kameyama and H. Kamiuchi. (2011a). An Analysis of Heat and Flow Characteristic of Heat Carrier in BT Process (in Japanese). Proceedings of the 27th conference on energy, economy, and environment, pp.271-274, Tokyo, Japan

K. Dowaki, H. Ishitani, R. Matsuhashi and N. Sam. (2002). A comprehensive life cycle analysis of a biomass energy system. Technology, Vol. 8, pp. 193–204

K. Dowaki, S. Mori, H. Abe, P. Grierson, M. Adams, N. Sam and P. Nimiago. (2003). A life cycle analysis of a biomass energy system taking sustainable forest management into consideration. Proceedings of 6th International Conference on Greenhouse Gas Control Technologies, pp. 1383–1388, Kyoto, Japan

K. Dowaki and Y. Genchi. (2009). Life Cycle Inventory Analysis on Bio-DME and/or Bio- MeOH products through BLUE Tower Process. International Journal of LCA, Vol.14, Number 7, pp.611-620

S. Hashimoto, T. Obara and Y. Terashima. (2000). Evaluation of demolition wood waste recycling from environmental aspect (in Japanese), Journal of environmental systems and engineering VII, Vol. 643, Issue VII-14, pp.37–48

Suri-Keikaku Co. Ltd. (2005). FY2004 Survey on estimation and investigation of exhaust emission factors and their total amount (in Japanese).

K. Dowaki, T. Eguchi, R. Ohkubo and Y. Genchi (2008b) A life cycle assessment on a fuel production through distributed biomass gasification process (in Japanese). Transaction of the institute of Electrical Engineers of Japan C, Vol. 128, Issue 2, pp.168– 175

Infoplease. (2005). Cell Phone Usage Worldwide, by Country. 01.04.2011, Available from: http://www.infoplease.com/ipa/A0933605.html

G. Hoogers. (2003). Fuel Cell Technology Handbook, In: Portable Applications, G. Hoogers, (Ed.), Capter 9, CRC Press LLC, ISBN 0849308771

K. Dowaki, T. Izawa and K. Sakamoto. (2010a). An analysis of CO2 emission on the cell phone with PEFC using Bio-H2 fuel (in Japanese). Journal of Life Cycle Assessment, Japan, Vol. 6, Issue 3, pp. 259-269

K. Dowaki, K. Yasumoto, M. Yamanari, Y. Iwai and H. Nishiyama. (2011b). A system analysis on the mobile device in consideration of the characteristics of cell phone users (in Japanese). Proceedings of The 6th meeting of the institute of Life Cycle Assessment, Japan, pp.170-171, Sendai, Japan

J. Kim, S. M. Lee and S. Srinivasan. (1995). Modeling of Proton Exchange Membrane Fuel Cell Performance with an Empirical Equation. Journal of The Electrochemical Society, Volume 142, Issue 8, pp. 2670-2674

K. Dowaki, M. Yamanari, Y. Takada and J. Kamihirata. (2010b). An indication of a carbonfootprint of Paprika due to the greenhouse facilities with an installation of biomass energy system, The 9th International conference on Eco balance, A2-1510, Tokyo, Japan

K. Dowaki, K. Fukumoto, M. Shimoguchi and M. Yamanari. (2011c). Proposals of the agricultural products cultivation system due to Blue Tower Gasification combinedcycle systems to reduce CO2 emission. Proceedings of The Life Cycle Management conference (LCM 2011), Submitted, Berlin, Germany

K. Kikuchi and N. Itsubo. (2009). Utility of Applying Carbon Footprint at Farmers' Market Aimed at Promotion of Local Production for Local Consumption (in Japanese), Journal of Life Cycle Assessment, Japan, Vol. 5, No. 4, 2009, pp.446-455

CHAPTER-26

ADVANCEMENT OF BIOETHANOL DEVELOPMENT

1. INTRODUCTION: AN OVERVIEW

Henry Ford, the progenitor of the contemporary automobile, fabricated his Model T during the initial years of the 20th century with the intention of powering it using ethanol derived from cereals. Due to Ford's strong belief, the utilization of this fuel was encouraged to such an extent that by 1938, ethanol production in Kansas reached 18 million gallons year (about 54,000 metric tons per year). However, the demand for ethanol decreased following the Second World War due to the abundant supply of natural gas and oil. In the late 1970s, in response to the initial oil crisis, several oil firms started selling a type of gasoline called gasohol, which contained 10% ethanol. They did this to benefit from the tax discounts offered on ethanol. Despite its merits, bioethanol initially had challenges in gaining widespread acceptance due to the presence of competing alternatives on the market, such as methyl tert-butyl ether (MTBE). MTBE was considered superior to ethyl tert-butyl ether (ETBE) in terms of both economic viability and performance. Following its introduction, MTBE was shown to be highly contaminating, leading to its prohibition. As a result, bioethanol resurfaced as a highly promising option for mitigating CO2 emissions. Another contributing element in the resurgence of bioethanol was the increasing recognition that we are approaching a critical juncture, sometimes referred to as the tipping point, where the demand for oil intersects with its diminishing availability. An ethical question arises in relation to bioethanol, which also extends to other biofuels. Biofuels are mostly derived from agricultural crops and grains that would otherwise be used for food production. In order to address this issue, current studies have focused on a non-edible perennial herbaceous plant known as Miscanthus giganteus. This plant possesses a high calorific content of around 4200 kcal/kg of dry matter. By utilizing lignocellulosic materials, such as municipal solid waste or the annual waste from wheat (about 5% of the total), it is possible to produce around 9.3 Gigaliters of bioethanol. This approach can also help address ethical concerns. Bioethanol may be utilized in different manners: it can be blended with diesel oil in diesel engines at a ratio of 5-10%; it can be mixed with petrol at a ratio of 10-85% for internal combustion engines; or it can entirely replace petrol in flexible fuel vehicles (FFV). The prevalence of FFV on the roads is steadily rising. In Brazil, annual sales have reached 400,000 vehicles, and there are roughly 1,500,000 FFVs, primarily used as public vehicles, now in circulation in the USA. In Europe, Sweden boasts a fleet of over 15,000 FFVs powered by E85, which consists of 85% ethanol. Ongoing research is being conducted to enhance engines powered by bioethanol and to develop fuel cells that utilize internal reforming of bioethanol to produce hydrogen.

1.1. Ethyl tertiary butyl ether

Ethyl tertiary butyl ether (ETBE) is a bioethanol product with a high octane rating. It is mostly produced by reacting ethanol with isobutylene, a byproduct of oil refining, using heat and various catalysts. Therefore, it is regarded as somewhat renewable. ETBE possesses technological and functional characteristics that closely resemble and significantly surpass those of the alcohol from which it is derived. Furthermore, it does not have the same issues of instability or compatibility with gasoline, and it possesses a high octane rating. As an ether, it possesses oxygen inside its molecular structure, which allows it to enhance the vehicle's ability to reduce emissions of pollutants. In a recent study conducted by Da Silva et al. (2005), the effects of different additives on the anti-detonating properties and Reid vapor pressure (RVP) of petrols were examined. The study concluded that the addition of ETBE enhances the anti-detonating properties of the mixture and decreases the vapor pressure, while still maintaining the necessary volatility to initiate a cold engine. BioETBE, derived from bioethanol, provides the same advantages as bioethanol, such as reduced pollutant emissions, a higher octane rating, and decreased reliance on crude oil imports. Unlike bioethanol, however, bioETBE does not present the technical and logistical challenges associated with its alcoholic nature. BioETBE additionally facilitates the proliferation of biofuels in the transportation industry.

1.2. Diesel and bioethanol mixtures (e-diesel)

The development and increasing use of diesel and bioethanol mixtures in diesel engines has been driven mainly by the European countries needing to comply with the European Union directive 2003/30/CE (which establishes that at least 5.75% of the fuels market must consist of biofuels by the year 2010), as well as the need to dispose of a petrol surplus in the refineries due to the greater demand for diesel vehicles. The drawbacks of the so-called e-diesel mainly concern a reduced viscosity and lubrication issues, a lower cetane number and injection capacity, a greater volatility (which can lead to an increase in the emissions of uncombusted hydrocarbons) and a lesser miscibility (Marek & Evanoff, 2001; Hansen et al., 2005; Lapuerta et al., 2007). In particular, Lapuerta et al. studied different diesel-bioethanol mixtures in different conditions of temperature, water content and additives, developing level maps that give a precise idea of the mixtures' areas of stability and of kinetic separation, that prompt the following conclusions:

- The presence of water in the mixture facilitates the separation of the ethanol phase;
- When its temperature increases, the mixture becomes more stable and the solubility of the ethanol in the diesel also increases;
- The mixture's sensitivity to the effects of water content and additives is higher, the higher the temperature of the mixture;
- Mixtures with a bioethanol content up to 10% (v/v) can be used in diesel engines in regions where temperatures in winter rarely drop below -5°C;
- Using stability-improving additives can increase the range of ethanol proportions in the mixtures, or the geographical extension of their applicability, enabling any phase separation to be avoided.

1.3. Research projects and bioethanol promotion

To succeed in demonstrating the feasibility of replacing petrol and diesel oil with bioethanol, the European Union developed the BEST project (BioEthanol for Sustainable Transportation) (European Union, 2001), involving six European countries (Sweden, the Netherlands, the United Kingdom, Ireland, Spain and Italy), and also Brazil and China: the global aims of the project are to introduce at least 10,500

FFV and 160 bioethanol-fueled buses, as well as to build 148 service stations, 135 to provide E85 and 13 to provide E95.

The NILE project (New Improvements for Lignocellulosic Ethanol) (Eurec, 2011) focuses instead on proposing the best processes for an economically effective production of bioethanol from lignocellulose biomass, suitable for use in internal combustion engines. The main goal of the NILE project is to reduce the cost of producing bioethanol from this type of raw material so as to make the technology commercially competitive. The NILE project brings together 21 industrial and research organizations from 11 member states, with complementary professional backgrounds and expertise so as to cover the whole cycle of bioethanol production and usage. On a technical level, the problems that remain to be solved concern reducing the cost of the enzymatic hydrolysis process by developing new artificial enzymatic systems, eliminating the current drawbacks intrinsic in converting fermentable sugars into ethanol, and validating the artificial enzymatic systems and yeast strains in a fully-integrated pilot plant.

Finally, there is the European LAMNET research program (Latin America Thematic Network on Bioenergy) (LAMNET, 2001), the main aim of which is to establish a trans-national forum to promote the sustainable use of biomass in Latin America and other emerging countries.

2. RAW MATERIALS

One of the great merits of bioethanol consists in the enormous variety of raw materials, and not only plants, from which it can be produced. The production methods vary depending on whether or not the raw material is rich in fiber.

The basic materials for producing biofuels must have certain features, including high carbon and hydrogen concentrations and low concentrations of oxygen, nitrogen and other organic components. The following is a brief description of some of the most important raw materials suitable for use in bioethanol production.

2.1. Alfalfa (medicago sativa)

This is a lucerne of the Fabaceae family that grows in cool subtropical and warm temperate regions. It demands no nitrogen-based fertilizers and its leaves are a precious source of protein in animal fodder. In a recent paper (Dien et al., 2006) it was observed that this plant has a low glucose yield due to a low-efficiency cellulose hydrolysis. The stems contain high concentrations of crude proteins and organic acids.

2.2. Switch grass (panicum virgatum)

This is a perennial herbaceous plant that grows mainly in the United States. Its ethanol yield per hectare is the same as for wheat. It responds to nitrogen fertilizers and can sequester the carbon in the soil. It is a highly versatile plant, capable of adapting easily to lean soils and marginal farmland (Heaton et al., 2004). Like maize, it is a type C4 plant, i.e. it makes an alternative use of CO2 fixation (a process forming part of photosynthesis). Most of the genotypes of Panicum virgatum have short underground stems, or rhizomes, that enable them with time to form a grassy carpet. Single hybrids of Panicum virgatum have shown a marked potential for increasing their energy yield (Bouton, 2007), but genetic engineering methods on this plant are still in a developmental stage and for the time being only their tetraploid and octaploid forms are known; we also now know that similar cell types (isotypes) reproduce easily.

2.3. Sweet sorghum (sorghum bicolor L)

The grains obtained from this plant are rich in starch and the stems have a high saccharose content, while the leaves and bagasse have a high lignocellulose content. The plant can be grown in both temperate and tropical countries, and it tolerates drought, flooding and alkalinity. Sorghum is considered an excellent raw material because the methods for growing and transporting it are well established. Ethanol can be obtained from it by exploiting both its starch and its sugar content. Research is currently underway on the use of hybrid or genetically modified species, although those obtained so far are weaker and need to be further refined and tested as concerns energy conversion efficiency (Rooney et al., 2007).

2.4. Cassava (manihoc esculenta)

This tuber is of considerable interest not only for ethanol production but also to produce glucose syrup, and it is available in tropical countries. The ethanol yield from the whole manioc is equivalent to the ethanol produced from cereals using dry milling methods. The only known lies in that the manioc has to be processed 3-4 days after it was harvested. To avoid such lengthy processing times, the manioc is first sliced and then left to dry in the sun. The waste water produced in the process can be treated by means of anaerobic digestion to produce bio gas.

2.5. Spruce (picea abies)

This tree has attracted a great deal of attention as a raw material for ethanol production because it is a lignocellulose material mainly composed of hexose sugars, which are more readily convertible than pentose sugars.

2.6. Willow (salix)

This is a member of the Angiosperm family and is consequently characterized by a hard wood. In this species, a fraction of the xylose units is acetylated. Some of the OH groups of the xylose carbons C2 and C3 are replaced by O-acetyl groups. With pretreatment, these groups release acetic acid that, in high enough concentrations, inhibits the yeasts involved in the fermentation process, according to some studies (Sassner et al., 2008a). It was recently demonstrated (Sassner et al., 2008b) that, by pretreating willow with sulfuric acid before the enzymatic hydrolysis process, and then simultaneously performing saccharification and fermentation, they succeeded in obtaining a global ethanol yield of 79%.

2.7. Reed canary grass (phalarisarundinacea)

This is a type C3 perennial herbaceous plant that grows in the cool season and has an excellent resistance to flooding. Its productivity is strongly influenced by high levels of nitrogen fertilizers, a feature that makes it very useful for the distribution of fertilizer from livestock.

2.8. Sugar cane (saccharum officinarum)

This plant only grows well in tropical and subtropical regions, which is why it is particularly common in Brazil. It has a 12-17% sugar content, 10% of which is glucose and the other 90% is saccharose. Milling can extract 95% of the total sugar content and the juice can subsequently be used to produce sugar or allowed to ferment to produce bioethanol. The bagasse (i.e. the solid residue remaining after milling) can be used as a source of energy and heat.

2.9. Sugar beet (beta vulgaris)

This plant generally grows in the cooler temperate regions, so it is abundant in Europe, North America and Asia. In the ethanol production process, the sugar beet is sliced and, while the juice is used to produce sugar or ethanol, the pulp is dried and used as animal feed or sold for pharmaceutical purposes.

2.10. Cereals

In order to obtain starch, it is necessary to grind these substances, which can then be used to produce bioethanol. Cereals with lower protein content and higher carbohydrate content are more desirable for distillation applications due to their superior bioethanol conversion rate. This implies that the nitrogen levels in cereals can be adjusted to promote the accumulation of starch rather than the synthesis of proteins. As a result, both the energy output and the quality of the fermentation process can be enhanced (Rosenberg et al., 2001). The principal cereals are:

2.10.1. Wheat

It grows mainly in temperate regions. The wheat treatment process is much the same as for the other cereals and it is best to use high-gravity fermentation to obtain the best performance in the fermentation process.

2.10.2. Barley

The most suitable is the so-called Winter variety, which is often underestimated as a foodstuff, despite the fact that it can tolerate drought and is highly adaptable.

2.10.3. Winter rye (*secale cereale* L.)

This cereal relies heavily on the availability of nitrogen in the soil; it has high contents of both glucan and xylan (40.8% and 22.3% respectively) (Petersson et al., 2007).

2.10.4. Corn stover

This is what remains on the ground after maize has been harvested. This raw material is abundantly available and demands no further investment in biomass, although not all of the corn stover can be removed - 30% of it must be left on the ground to prevent erosion (by facilitating water infiltration and reducing evaporation), and as the main source of soil organic carbon (SOC) in order to preserve the soil's productivity. Corn stover contains polymeric hemicellulose and cellulose, but their biodegradability by glycosidase is strongly inhibited by a small quantity (12-15%) of lignin (Gressel, 2008; Varvel et al., 2007).

2.11. Jerusalem artichoke (helianthus tuberosus)

This plant grows in summer, reaching its maximum height in July and dying in October. The tubers are rich in inulin (a fructose polymer), which can be used to obtain a syrup for use both in the foodstuffs industry and in the production of ethanol. It was demonstrated (Curt et al., 2006) that, towards the end of the season, the potential for bioethanol production of the stems of clones is 38% of that of the tubers.

2.12. Municipal solid waste (MSW)

The most suitable waste for converting into bioethanol is the waste from the fruit and vegetable industries, for instance, cotton fiber, milk whey from cheese-making, the waste products of coffee making, and so on. Generally speaking, such waste contains approximately 45% of cellulose (glucose polymer), which can be simultaneously hydrolyzed and fermented to produce ethanol. SSL (Spent Sulfite Liquor) is a byproduct of bisulfite "pulp" manufacturing that can also be fermented to produce ethanol. Waste varies considerably in content from one area to another, but the majority of the volume generally consists of paper (20-40%), gardening waste (10-20%), plastics, glass, metals and various other materials (Prasad et al., 2007).

2.13. Miscanthus

This is a perennial grass of the C4 type that spreads through rhizomes. Miscanthus x giganteus is primarily utilized for the production of biofuels. It is a sterile tetraploid hybrid derived from the combination of Miscanthus sinensis and Miscanthus sacchariflorus. This hybrid exhibits a high yield, reaching 30 tons per hectare in irrigated soils during fall and 10-25 tons per hectare in non-irrigated soils. Miscanthus sacchariflorus contributes to the genome of Miscanthus x giganteus by being well-suited to warm temperatures, while Miscanthus sinensis offers the necessary genetic resources for colder places. It is frequently employed for decorative purposes or as a ground cover, and it has the potential to reach a height of up to 4 meters. The process of achieving a consistent yield takes three years (or around 5 years in poor quality soils). During the first year of growth, the rhizomes are highly susceptible to low temperatures, whereas in the following years they can tolerate temperatures as low as -40°C. The rhizomes remain dormant during the winter and initiate growth once the soil temperatures reach 10-12°C. Regarding the plant's energy content, the dry matter possesses a net calorific value of around 17 MJ/kg. The energy content of 20 metric tons of dried Miscanthus is roughly equivalent to that of 8 metric tons of coal (Heaton et al., 2004; Sánchez & Cardona, 2008; DEFRA, 2001). When comparing Panicum virgatum and Miscanthus, both of which are type C4 plants with significant potential as energy sources, it is shown that Miscanthus produces a greater amount of biomass per unit than Panicum virgatum, specifically 12 Mg ha-1 (Heaton et al., 2004). Both plants are perennial, which eliminates the need for replanting and results in cost savings. In regions characterized by ample precipitation but facing issues of nitrogen pollution in the water supply, it is more advisable to cultivate Miscanthus as an energy crop. Conversely, in uncontaminated arid regions, cultivating Panicum virgatum with sufficient nitrogen fertilization undoubtedly results in a higher crop yield.

3. PRODUCTION PROCESSES

The production processes of bioethanol exhibit significant variations depending on the specific raw material used. However, many key phases in the process remain consistent, albeit under varying temperature and pressure settings, and occasionally using different microorganisms. The phases involved in this process are hydrolysis (which can be done through chemical or enzymatic means), fermentation, and distillation. Hydrolysis is an initial process that allows for the extraction of sugars from the raw materials, which can then be subjected to fermentation. For enzymatic hydrolysis to be successful, it is necessary to use effective pretreatments that enhance the vulnerability of lignocellulose materials to the enzymes. The subsequent paragraphs delineate the diverse production methodologies, categorizing them based on the specific type of raw material utilized.

3.1. Lignocellulose biomass

The biofuels obtained from wood cellulose and from organic materials in general offer considerable advantages over conventional biofuels. Burning ethanol obtained from cellulose produces 87% lower emissions than burning petrol, while for the ethanol from cereals the figure is no more than 28%. Ethanol obtained from cellulose contains 16 times the energy needed to produce it (Martinez et al., 2008), petrol only 5 times and ethanol from maize only 1.3 times. The problem is a matter of how to disrupt the bonds of this molecule in order to convert it into fermentable sugars.

In fact, this is unquestionably the type of raw material that is the most complicated to process. The starting material may be farming and forest waste, scrap woods, grassy crops grown for energy purposes or even municipal solid waste. Lignocellulose occurs in the walls of vegetable cells and consists of cellulose microfibers contained in the lignin, hemicellulose and pectin. The procedure to obtain ethanol consists first in depolymerizing the carbohydrates into their monomeric sugars, then fermenting the sugars with the aid of appropriate microorganisms. The lignocellulose biomass consists mainly of three basic polymers: cellulose, hemicellulose (such as xylane), lignin and other minor components (essential oils, acids, salts and minerals).

3.1.1. Pretreatments

These are used to modify the structure and dimensions of macroscopic and microscopic raw materials, and also their chemical composition. They have the effect of solubilizing the hemicellulose, reducing the crystallinity, and increasing the available surface area and porosity of the substrate. An effective pretreatment must meet following requirements: - it must increase sugar formation or facilitate the subsequent formation of sugars during hydrolysis, preventing any degradation or the loss of carbohydrates, and avoiding the formation of byproducts capable of inhibiting the subsequent processes of hydrolysis and fermentation, all at a competitive cost (Balat et al., 2008).

Pretreatments are particularly essential before enzymatic hydrolysis and may be of various types, i.e. physical, chemical, biological, steam explosion, and ammonia fiber explosion (AFEX).

Physical pretreatments may or may not be mechanical. The mechanical physical pretreatments include milling and grinding, that not only reduce the substrate, but also increase its surface area to volume ratio, thus making the cellulose easier to convert during hydrolysis. "Ball milling" could also be used to reduce the crystallinity of the cellulose, but this practice is not only very expensive, but also takes a long time (nearly a week) to complete, so it is hardly practicable on an industrial scale. The non-mechanical pretreatments feature a combination of high-power internal and external forces that decompose the lignocellulose.

Chemical pretreatments are used mainly to reduce the crystalline content of the cellulose. Using this type of pretreatment poses plant-related problems, however, since all the structural materials have to be capable of withstanding the severe working conditions imposed by the chemical agents.

The chemical pretreatments most often used are an alkaline treatment to delignify and solubilize the glycan, and an NaOH treatment that dissolves the lignocellulose biomass, destroying its lignin structure. Pretreatment with diluted sulfuric acid is also very important but this poses serious problems if it is associated with diluted acid hydrolysis, because the hydrolyzed end products become scarcely fermentable.

Other chemical pretreatments include: pretreatment with hydrogen peroxide (H2O2), which exploits oxidative delignification to separate and solubilize the lignin, and dissolve the lignocellulose matrices, thereby increasing the enzymatic digestibility of the mass; pretreatment with ozone, which degrades the lignin polymers; and pretreatment with liquid hot water (LHW), which is applied mainly to alfalfa. It was demonstrated (Laser et al., 2002) that, in ideal conditions, this method is as effective as diluted acid hydrolysis, without the need to use any acid or create any products of neutralization).

Biological pretreatments involve the use of enzymes, which are already useful in industrial processes on timber waste, in the processing of pulp and scraps. Several microorganisms studied years ago are the enzymes produced by the basidiomycetes Pleurotus ostreatus: these enzymes are homologous proteins characterized by different specifications, depending on which phenols are substituted. Another fungus in the basidiomycetes class that is effective in delignification is the Phanerochaetechrysosporium (Palmieri et al., 1997).

In the steam explosion process, saturated steam is used at very high temperatures and pressures to break up the chemical bonds in the cellulose, hemicellulose and lignin in order to break down the fibers and hydrolyze the biomass. The process consists in delivering steam under high pressure into a sealed chamber containing the lignocellulose material, then reducing the pressure and thus making the steam and matrix expand, and obtaining its explosive decompression through an orifice, which disrupts the cellular structure of the substrate, breaking up the acetyl groups of the hemicellulose. In some cases (e.g. Angiosperm), it is preferable to use acid catalysts, such as H2SO2 or SO2, to make the cellulose-rich components more accessible to the enzymes. SO2 gas is better able to attack the fibers (Shevchenko et al., 2000), but its use makes it necessary to carefully consider the working conditions in which the steam explosion takes place. In fact, it becomes necessary to find the best compromise between a strong enzymatic hydrolysis (obtainable in very severe conditions) and a good recovery of the components containing hemicellulose, that are in the form of monomeric sugars (which demand much less severe conditions) (Silverstein et al., 2007). That is why a severity indicator has been developed (Overend &Chornet, 1987), which correlates pretreatment temperatures and times, assuming that the pretreatment obeys Arrhenius's equation and has first-order kinetics. The indicator R 0 is:

$$R_o = t \cdot \exp\left[\frac{(T_r - T_b)}{14.75}\right] \tag{1}$$

where t is the duration of the pretreatment (min), T r is the reaction temperature (°C), T b is the baseline temperature (100°C) and the constant 14.75 is the conventional activation energy, assuming that the whole conversion is of the first order. If the version with sulfuric acid is being used, then the severity parameter, called M 0 in this case, is slightly modified:

$$M_o = t \cdot C^n \cdot \exp\left(\frac{T_r - T_b}{14.75}\right) \tag{2}$$

where C is the chemical concentration (wt%) and n is an arbitrary constant (Chum et al., 1990).Ammonia fiber/freeze explosion (AFEX) pretreatment involves the use of liquid ammonia and steam explosion: in this process, the previously-humidified lignocellulose material is placed in a vessel under pressure with liquid NH3 in proportions of 1-2 kg NH3/kg of dried biomass. This method is very effective for non-woody materials such as bagasse and newspaper, but less so in the case of "soft" wooden materials. This system does not release any sugars directly, but it does make the polymers (hemicellulose and cellulose) easier for the enzymes to attack. The ammonia can also be replaced with carbon dioxide because the latter is relatively less costly and also because the alcohol waste product contains traces of pollutants that would thus derive only from the lignin.

The most promising pretreatments for farming waste are AFEX and LHW, while pretreatment with steam affords a high output of sugars from both farming waste and forest waste.

3.1.2. The hydrolysis process

Hydrolysis is governed by the law:

$$(C_6H_{10}O_5)_n + nH_2O \rightarrow nC_6H_{12}O_6 \tag{3}$$

and can be mainly of two types: acid (using diluted or concentrated acids) or enzymatic. A lignocellulose biomass is more complicated to hydrolyze than pure cellulose because it contains components that are not glucose-based, such as hemicellulose and lignin.

A lignocellulose biomass undergoing acid hydrolysis mainly produces xylose, while the lignin and cellulose fractions remain unchanged. This is because xylan is more susceptible to hydrolysis in moderately acid conditions because of its amorphous structure, while cellulose demands more severe conditions because of its crystalline nature.

If hydrolysis is implemented using 1% diluted sulfuric acid, the hemicellulose is depolymerized at a lower temperature than the cellulose. This process is usually conducted in two consecutive stages.

One of the most important characteristics of this type of hydrolysis is the rate of the reactions involved, which facilitate the continuity of the process. To speed up the diffusion of the acid, the raw material is mechanically reduced to pieces a few millimeters in size.

Hydrolysis with concentrated acids (10-30%), on the other hand, rapidly and completely converts cellulose into glucose and hemicellulose into xylose, with some degree of degradation. The acids most often used are sulfuric and hydrochloric acid, and hydrogen fluoride.

This type of acid hydrolysis has the great advantage of recovering the sugars very efficiently (approximately 90% of hemicellulose and cellulose are depolymerized into monomeric sugars). From an economic standpoint, this process enables a reduction in production costs by comparison with the diluted acid solution, especially if the acids are retrieved and reconcentrated. The acids and sugars in solution are separated by ion exchange so the acid is reconcentrated by passing it through a series of multiple-effect evaporators. The remaining solid fractions, which are rich in lignin, are collected and can be made into pellets for use as fuel.

So, in short, we can divide concentrated acid hydrolysis into two stages: in the first stage, the concentrated acid (70%) destroys the crystalline structure of the cellulose, breaking up the hydrogen links between the cellulose chains; in the second stage, hydrolysis induces a hydrolytic reaction in the single isolated cellulose chains.

The enzymatic hydrolysis of natural lignocellulose materials is a very slow process, because it is hindered by several structural parameters of the substrate, such as its of cellulose and hemicellulose content, and the surface area and crystallinity of the cellulose. Pretreatments are consequently needed to make the biomass more susceptible to attack by hydrolysis. For the same reason, a cocktail of enzymes has to be used that is capable of breaking the links in the polymeric chains. This cocktail is usually a mixture of various hydrolytic enzymes, including cellulase, xylanase, hemicellulase and mannoxidase. Enzymatic cellulose degradation is a complex process because it takes place in limit conditions between the solid and liquid phases, where the enzymes are the mobile components. Generally speaking, degradation is characterized by a rapid initial phase followed by a slower second phase that can continue until all the substrate has been used up. The reason for this behavior is usually assumed to be because the accessible fraction of cellulose is quick to hydrolyze, followed by the slow activation of the absorbed enzyme molecules.

Chopping up the biomass increases the surface area accessible to the enzymes and reduces the polymerization and crystallinity of the cellulose, thus enabling a smaller quantity of enzymes to be used and the production costs to be contained.

Both bacteria and fungi can produce the cellulase for the hydrolysis of lignocellulose materials. The bacteria may be aerobic or anaerobic, mesophylic or thermophylic. The bacteria most often used are Clostridium, Cellulomonas, Bacillus, Thermomonospora, Ruminococcus, Bacteriodes, Erwinia, Acetovibrio, Microbispora and Streptomyces. The enzymes are usually classified according to their reaction site, so they may be intracellular (or cell-associated) or extracellular. The main function of extracellular enzymes is to convert the substrate into an external medium by taking effect on the cell mass constituents. Conversely, intracellular enzymes need the substrate to spread through the cellular mass before it can be converted.

The most widely accepted mechanism for the enzymatic hydrolysis of cellulose involves the synergic action of the enzymes endoglucanase (or endo-1,4-**β**-glucanase, EG), exoglucanase (or cellobiohydrolase, CBH), and **β**-glucosidase. Both EG and CBH are extracellular enzymes, while **β**-glucosidase is intracellular. EG randomly disrupts the cellulose chains, consequently inducing their strong degradation. It takes effect by hydrolyzing the **β**-1,4-glucoside bonds, creating new ends in the chains. Exoglucanase breaks up the ends of the chains, thus enabling the release of soluble cellobiose or glucose. BGL hydrolyses the cellobiose into glucose, thus eliminating the inhibitory cellobiose; then BGL completes the process by catalyzing the hydrolysis of cellobiose into glucose. Most cellulase and hemicellulase producers are microorganisms such as the filamentous fungi, e.g. Trichoderma sp., which can be used in their natural form or genetically modified (Trichoderma viride, Trichoderma reesei, Trichoderma longibrachiatum). CBH I and CBH II are the main enzymes of Trichoderma reesei, while EG I and EG II are the dominant endoglucanases.

Enzymatic activity is influenced by various parameters, such as temperature (a 20-30°C increase in temperature leads to a 3- to 5-fold increment in the end products). The crucial issue of temperature lies in the risk of an unwanted denaturation when the temperature is too high (Balat et al., 2008). Enzymatic hydrolysis, with or without the addition of catalysts, has generally proved capable of a high yield of both glucose (>90%) and xylose (>80%).

3.1.3. Fermentation

After hydrolysis, the hydrolyzed products must be fermented by means of microorganisms such as yeasts (Hahn-Hägerdal et al., 2006). Since the hydrolyzed products are composed mainly of glucose, xylose, arabinose and cellobiose, the microorganisms used must be capable of fermenting all of them efficiently for ethanol to be produced on a large scale. The reactions that involve glucose and xylose are respectively:

$$3C_5H_{10}O_5 \rightarrow 5C_2H_5OH + 5CO_2 \quad (4)$$

$$C_6H_{12}O_6 \rightarrow 2C_2H_5OH + 2CO_2 \quad (5)$$

The classic method used in the fermentation of the hydrolyzed biomass is separate hydrolysis and fermentation (SHF), in which the two processes are completed in different units. A commonly used alternative is simultaneous saccharification and fermentation (SSF), in which hydrolysis and fermentation are completed in the same unit. A last option is represented by consolidated bioprocessing (CBP).

When the SHF process is used, the solid fraction of the lignocellulose material undergoes hydrolysis and this process is called saccharification. The liquid fraction, on the other hand, goes first to the reactor for glucose fermentation, then it is distilled to extract bioethanol, leaving behind only the unconverted xylose, which is then fermented in a second reactor and then undergoes a second, final distilling phase.

The main advantage of this process consists in that separating the processes of hydrolysis and fermentation enables optimal working conditions to be adopted in each case. The enzymes are free to work at high temperatures, while the microorganisms can induce fermentation at more moderate temperatures.

Among the disadvantages, in addition to needing two twin reactors, there is the fact that the enzymes for hydrolyzing the cellulose are inhibited end products. The rate of hydrolysis progressively declines due to the accumulation of glucose and cellobiose.

This process has sometimes been used to produce ethanol from a mix of municipal solid waste: in this case, enzyme recycling was improved using micro- and ultra-filtering procedures, thus achieving the hydrolysis of 90% of the cellulose with a net enzyme load of 10 FPU/g of cellulose (where FPU stands for filter paper unit) (Sánchez & Cardona, 2008).

In the SSF procedure, enzymatic hydrolysis and fermentation take place simultaneously. Cellulases and microorganisms take effect in the same process, so the glucose produced by hydrolysis of the cellulose is immediately consumed by the bacterial cells that convert it into ethanol. SSF achieves the highest output of bioethanol at the lowest costs, since the lesser demand for enzymes is lower because the inhibitory effect of the cellobiose and glucose end products is alleviated by fermentation with yeast. This is a discontinuous type of process that uses natural heterogeneous materials containing complex polymers such as lignin, pectin and lignocellulose. The greatest advantages offered by SSF are a faster rate of hydrolysis thanks to the conversion of the sugars that inhibit cellulase activity, a low enzyme demand, a high product yield, the need for less sterile conditions, a shorter process time, and smaller overall reactor dimensions (Sun & Cheng, 2002).

This process also has far from negligible disadvantages, however, the most significant of which consists in the need to complete fermentation and hydrolysis in suboptimal conditions. That is why microorganism selection and preparation is so important for this process. The cocktail of enzymes for hydrolyzing the cellulose must likewise remain stable within a wide range of temperatures and pH. As for the Saccharomyces cerevisiae cultures, the typical working conditions in SSF involve a pH of 4.5 and temperatures of around 310 K.

Experiments have recently been conducted with a new variant of this process called simultaneous saccharification and cofermentation (SSCF), in which the five- and six-carbon sugars are fermented simultaneously. In SSCF, hydrolysis continuously releases hexose sugars that increase the rate of glycolysis, so that the pentose sugars can ferment more quickly and produce a higher yield.

In CBP, four biologically-mediated conversions take place in a single process, i.e. the production of glycolytic enzymes (cellulase and hemicellulase), hydrolysis of the carbohydrate component of the pretreated biomass to obtain sugars, fermentation of the six-carbon sugars (mannose, galactose and glucose), and fermentation of the five-carbon sugars (xylose and arabinose).

The main difference between CBP and the other processes consists in that there is no single process focusing on cellulase production. CBP, also known as direct microbial conversion (DMC), requires just one microbial community for both cellulase production and fermentation. The weakness of this approach lies in the difficulty of finding an organism sturdy enough to simultaneously produce cellulase and ethanol with a high yield. Wyman (Wyman, 1994) wrote that many studies on CBP involved the use of the bacterium Clostridium thermocellum for enzyme production, cellulose hydrolysis and glucose fermentation, while Clostridium thermosaccharolyticum enabled the simultaneous conversion of the pentose sugars obtained from hemicellulose hydrolysis into ethanol. Using Clostridium thermocellum in the system also induces a 31% higher conversion of the substrate than when Trichoderma reesei or Saccharomyces cerevisiae are used. Recent studies have focused on cellulase production combined with a high ethanol yield using strains of Escherichia coli, Klebsiella oxytoca and Zymomonasmobilis as well as the yeast Saccharomyces cerevisiae. The expression of cellulase in Klebsiella oxytoca increased the yield from microcrystalline cellulose hydrolysis and enabled an anaerobic growth in the amorphous cellulose. Various cellobiohydrolases have likewise been functionally expressed in the Saccharomyces cerevisiae. Genetic engineering and metabolic studies will enable the development of new stable strains of microorganisms capable of converting the cellulose biomass into bioethanol, leading to improvements in the industrial bioethanol production process (Lynd et al., 2005).

The microorganisms used during the fermentation process must be capable of working efficiently on both monosaccharide and polysaccharide sugars, so they have to be very versatile. The survival of these bacteria is only assured in controlled pH conditions and the majority of the microorganisms cannot tolerate bioethanol concentrations in excess of 10-15% (w/v).

Saccharomyces cerevisiae is one of the microorganisms most often used because it affords a high ethanol yield from hexose sugars, and it can tolerate bioethanol and inhibitory compounds very well. It has the great disadvantage, however, of being unable to assimilate C6 sugars.

The ethanol-generating bacteria that seem industrially most promising are Escherichia coli, Klebsiella oxytoca and Zymomonasmobilis. Zymomonas, in particular, has demonstrated an aptitude for rapidly and efficiently producing bioethanol from glucose-based raw materials and, by comparison with the other yeasts, it has demonstrated a 5-fold higher yield. The ethanol it produces in the fermentation of the glucose corresponds to a yield that is 97% of the theoretical yield and in concentrations up to 12% (w/v). This bacterium is also capable of producing bioethanol efficiently from fructose and saccharose (C5), but not from C6 sugars.

There are also yeasts that naturally ferment xylose, such as Pichia stipitis, Candida Shehatae and Candida parapsilopis, and they can do so through the action first of xylose reductase (XR), which converts xylose into xylitol, and then of xylitol dehydrogenase (XDH), which converts xylitol into xylulose. Bioethanol fermentation from xylose can also be achieved by recombinant Saccharomyces cerevisiae using the heterologues XR and XDH of Pichia stipitis and xylulose kinase (XK) of Saccharomyces cerevisiae.

Table 1 summarizes the commonly used bacteria and microorganisms (Balat et al. (2008)), highlighting the principal parameters used to assess the performance of the various types of fermentation.

Table 1.Commonly used bacteria and microorganisms (Balat et al. (2008)).

Species	Characteristics
Clostridium acetobutilicum	Useful in fermentation of xylose to acetone and Butanol.
Clostridium thermocellum	Capable of converting cellulose directly to ethanol and acetic acid.
Escherichia coli	Native strains ferment xylose to a mixture of Bioethanol.
Klebsiella oxytoca	Native strains rapidly ferment xylose and cellobiose.
Lactobacillus pentoaceticus	Consumes xylose and arabinose.
Latobacillus casei	Ferments lactose very well; particularly useful for bioconversion of whey.
Lactobacillus xylosus	Uses cellobiose if nutrients are supplied: uses nglucose, D-xylose, and L-arabinose.
Lactobacillus pentosus	Homolactic fermentation. Some strains produce lactic acid from sulfite waste liquors.
Lactobacillus plantarum	Consumes cellobiose more rapidly than glucose, xylose, or arabinose.
Zymomonasmobilis	Normally ferments glucose and fructose.

Fermentation can occur in various ways, i.e. discontinuously, continuously, with cells immobilized, and batch-fed (Chandel et al., 2007).

A problem encountered in enzymatic hydrolysis consists in the formation of inhibitors. The activity of the enzymes is strongly influenced by certain levels of cellobiose, glucose or products such as furfural and organic acids deriving from pretreatments.

Inhibitors form in relation to the conditions in which enzymatic hydrolysis takes place. Conditions can be selected that should provide maximum solubilisation and recovery of the hemicellulose component (low severity), optimum enzymatic hydrolysis of the water insoluble cellulosic component (high severity), and a compromise between the two conditions (medium severity). The combined severity (CS) links the severity factor (R0) to the ambient pH, and this index expresses the intensity of the previously-described factors. Its value is expressed as:

$$CS = \log R_0 - pH \tag{6}$$

When the CS increases beyond the value that generates the highest concentrations of mannose and glucose, the cellulose and hemicellulose break down and there is a drop in the concentration of fermentable sugars that coincides with the formation of furfural and hydroxy methyl furfural (HMF), which subsequently degrade into levulinic and formic acids. To achieve both the maximum fermentability and a high yield of fermentable sugars, the CS should be around 3 (Palmqvist & Hahn-Hägerdal, 2000)).

Inhibitors can come from various sources, e.g. equipment, carbohydrate degradation, lignin decomposition, wood extracts and their decomposition. They can be classified according to their structure as organic, acid, furanes and phenolic compounds. The fermentation inhibitors in particular include the furane derivatives, such as furfural and 5-hydroxy-methyl-furfural (5-HMF), the aliphatic acids, such as acetic acid, formic acid and levulinic acid, and the phenolic compounds. The furane derivatives can further react to form certain polymeric materials. The formation of inhibitory compounds makes it necessary to introduce changes in the production process, such as process water recirculation. It was demonstrated (Palmqvist et al., 1996), for instance, that unconcentrated hydrolyzed products have a moderately inhibitory action, while five-fold concentrations of nonvolatile components almost completely inhibit the fermentation of ethanol by Saccharomyces cerevisiae.

The formation of inhibitors and consequently of toxic compounds is a problem that has a negative fallout on the rate of both enzymatic hydrolysis and fermentation. The toxic compounds can form during steam explosion pretreatments and also during hydrolysis in the presence of low acid concentrations, and they are mainly the products of lignin degradation.

Four main groups of inhibitors have been identified in hydrolyzed lignocellulose products these are acetic acid from the hemicellulose fraction, products of lignin degradation, products of sugar degradation, and extracts that have been solubilized during the pretreatment.

The fermentation inhibitors, on the other hand, can be divided into various groups, depending on their origin:

- Substances released during pre-hydrolysis and hydrolysis: acetic acid and extracts including terpenes, alcohols and aromatic compounds (e.g. tannins);
- Inhibitors produced as a byproduct of pre-hydrolysis and hydrolysis, due to sugar degradation (furfural, 5-HMF);
- Products of lignin degradation, including sizable groups of aromatic and polyaromatic compounds with a great variety of constituents (cinnamaldehyde, p-hydroxybenzaldehyde, syringaldehyde);
- Products of the fermentation process, such as ethanol, acetic acid, glycerol and lactic acid;

- Metals released by equipment and additives, e.g. nickel, iron, copper and chrome.
- The compounds revealing the greatest inhibitory potential are acetic acid and the products of lignin degradation (Larsson et al., 1999).

A detoxification procedure can be used to improve the sugars' fermentability. Detox methods may be physical, chemical or biological, and they are impossible to compare directly with one another because the degree to which they can neutralize the inhibitors varies. The different microorganisms suitable for this purpose can tolerate the inhibitors to varying degrees. The choice of the most suitable method consequently depends on the raw materials involved and the composition of the hydrolyzed products. Figure 1 shows a flowchart of ethanol production from lignocellulose raw materials.

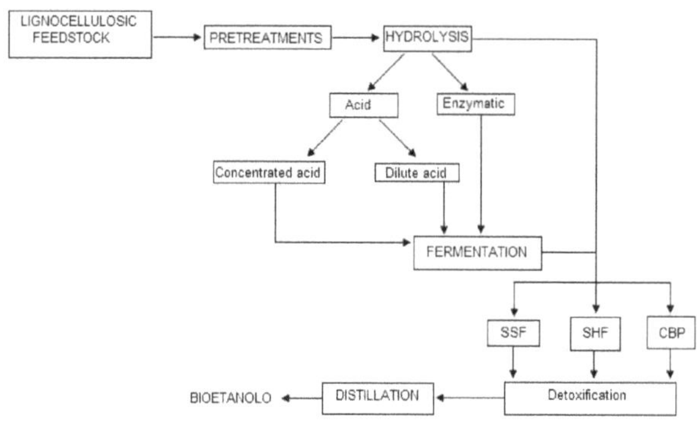

Figure 1.Flowchart of ethanol production from lignocellulose raw materials

3.2. Raw materials containing starch

Starch is a biopolymer defined as a homopolymer. The constituent monomers are molecules of glucose held together by bonds between the oxygen atom of one molecule and the carbon atoms of adjacent molecules. These particular bonds are described as glycosidic and may be type α or type **β**, depending on the stereoisomery of the anomeric carbon in the molecule. Starch in plants occurs in the form of granules containing two main constituents in variable proportions, depending on the resource, i.e. amylose (16-30%) and amylopectin (65-85%). These are both type α glucose polymers. Amylose is a glucose polymer held together by α-1,4 bonds in linear chains, while amylopectin is a highly-branched glucose polymer with type α-1,6 bonds. Inside the cell, the starch is in the form of granules located in the amyloplasts. The granules contain both amorphous and crystalline regions, in proportions of approximately 30-70, respectively.

The starch for ethanol production comes mainly from cereals, wheat or corm being at the top of the list in North America and Europe, and tubers such as manioc in the tropical regions. In order to produce bioethanol from starch, its carbohydrate chains have to be broken down to obtain glucose syrup, which can then be converted into bioethanol with the aid of yeasts.

3.2.1. Starch hydrolysis

Various microorganisms are capable of hydrolyzing starch, though a preliminary process called gelatinization is needed to ensure an efficient hydrolysis. During this preliminary process, the starch granules swell, particularly rupturing the hydrogen bonds in the crystalline regions. The long glucose chains comprising the starch must be converted into fermentable sugars by means of a process called the "hydrolysis technique", during which the starch reacts with the water normally used to break down the starch into its fermentable sugars. There are numerous microorganisms capable of hydrolyzing starch, but those involved in the starch degradation process are generally amylase, α-amylase, **β**-amylase and isoamylase. The most important for the purposes of the SSF process are certainly the first two. α-amylase is an endo-amylase that randomly attacks the α-1,4 bonds, rapidly reducing the starch molecule's dimensions and consequently also its viscosity, i.e. it liquefies the starch. α-amylase can be obtained by means of heat-resistant bacteria such as Bacillus licheniformis, or by means of new strains of Escherichia coli or Bacillus subtilis, used on the starch suspensions during the first hydrolysis stage. For amylase to succeed in attacking these suspensions, they must be brought up to high temperatures (90-110°C) to rupture the starch cell nuclei. The products of this preliminary hydrolysis phase, called liquefaction, is a solution containing dextrins and a small amount of glucose.

$$\text{Starch} \xrightarrow[\text{(liquefaction)}]{\alpha\text{-amylase}} \text{Dextrins} \xrightarrow[\text{(saccharification)}]{\text{glycoamylase}} \text{Glucose} \qquad (7)$$

At this point, the liquefied starch undergoes saccharification at low temperatures (60-70°C), induced by the action of glycoamylase generally obtained from Aspergillus or Rhizopus species. This enzyme is an exo-amylase capable of producing units of glucose from amylose and amylopectin chains.

The factors that influence starch hydrolysis include the substrate, enzyme activity and the reaction conditions (temperature, pH and other parameters) (Prasad et al., 2007). The microorganisms take effect more easily on gelatinized starch, but this process demands large amounts of energy so on an industrial level there has been a tendency to focus on using microorganisms capable of growing on ungelatinized starch. Various studies on this issue have considered certain species of fungi for producing enzymes capable of degrading starch in its natural state (Soccol et al., 1994). Liquefaction is followed by a saccharification stage under the effect of glycoamylase.

3.2.2. Milling

The milling phase enables the starch to be extracted from the biomass and it is very important for the purposes of analyzing the bioethanol production process as a whole because it strongly influences not only the subsequent stages but also the co-products obtained at the end of intermediate stages, which also vary according to the specific raw material entering the process (wheat, barley, corn, oats). The two main options are wet milling and dry milling.

Wet milling is the standard procedure generally used in the starch-based foodstuffs industry. Though this procedure demands more energy and more economic resources, and it delivers a smaller quantity of ethanol, it is still preferred at industrial level because its capacity to separate the grain into its components enables a purer form of starch to be obtained, along with more valuable byproducts. Wet milling can be used to obtain not only ethanol, but also products such as corn oil, gluten-based foods and flour, and corn steep liquor (CSL).

Dry milling means there is no need to pre-treat the raw material, which simply has to be ground before going through the other processing stages (hydrolysis, fermentation, distillation), which are identical to those following the wet milling process. Because dry milling does not break down the cereals into their various components, the unfermentable residue leaving the process that extracts the ethanol from the fermentation broth is rich in proteins, fibers, fats and sugars.

3.2.3. From hydrolysis to bioethanol

After the preparatory stage, the glucose solution can be fermented in ethanol. The temperature of the glucose is lowered to around 35°C, and then the yeast (usually Saccharomyces cerevisiae) is added and the anaerobic fermentation process begins, which converts the glucose into ethanol and carbon dioxide.

$$\text{Glucose} \xrightarrow[\text{Alcoholic fermentation}]{\text{Saccharomyces cerevisia}} 2\ \text{EtOH} + 2\ CO_2 \qquad (8)$$

As a rule, the preferred method is to conduct the saccharification and fermentation steps during the same stage of the production process. Fermentation can be completed in two stages (Verma et al., 2000) using starch treated with α-amylase and glycoamylase.

Fermentation may be continuous or discontinuous, it makes no difference. When the fermentation broth reaches an ethanol content of around 8-10% v/v (beyond which the yeast can no longer survive), the ternary mixture is distilled by adding benzene or cyclohexanone, or using molecular sieves. After distillation, the ethanol is 95% pure.

In 2006, a research group (Robertson et al., 2006) experimented with the so-called "cold hydrolysis" of starch, concluding that the potential use of this method relies on the discovery and characterization of more efficient enzymes and the development of processes with a high level of integration, such as simultaneous liquefaction, saccharification and fermentation, along with other factors. Figure 2 shows the flow chart for bioethanol production from materials containing starch.

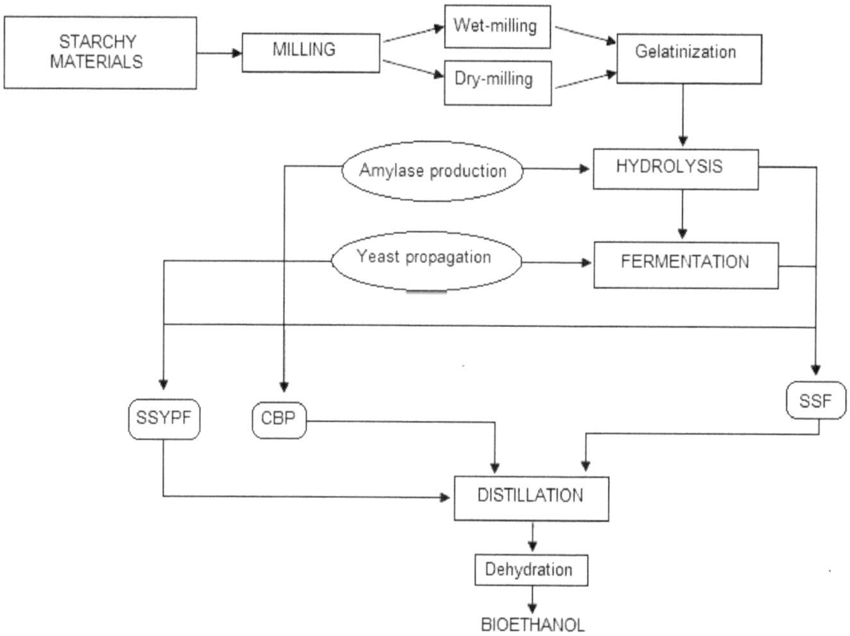

Figure 2.Flow chart for bioethanol production from materials containing starch

3.3. Raw materials containing saccharose

For the purposes of bioethanol production, the most important raw materials containing saccharose are unquestionably sugar cane and sugar beet. Two thirds of the world's sugar production derives from cane, the other third from beet.

3.3.1. Sugar cane (saccharum officinarum)

Sugar cane contains approximately 12-17% of total sugars, 90% of which are saccharose and 10% are glucose. Milling can extract approximately 95% of the sugar, leaving behind the solid residue. This cane residue goes by the name of bagasse. Sugar cane is washed in order to undergo a primary "crushing" process before milling. The cane juice obtained undergoes a clarification process in which the pH is balanced and cachaça is obtained, which can be sold as animal feed or as a component in mixtures. Fermentation is usually done with the aid of a yeast, Saccharomyces cerevisiae, which is separated in a continuous phase by centrifugation and reused in the fermenter. The fermentation process differs slightly, depending on whether all the juice is used to obtain bioethanol or whether part of it is drawn off to obtain sugar: in the former case, the juice is heated up to 110°C (to reduce the risk of bacterial contamination), then decanted and fermented; in the latter, the crystals formed by concentration are centrifuged, leaving a viscous syrup called molasses.

The extract leaving the fermenter must then be distilled to extract the hydrated ethanol (an azeotropic solution containing 95.5% v/v of ethanol and 4.5% v/v of water), which is dehydrated using molecular sieves or azeotropic distillation (i.e. with cyclohexanone or benzene) to obtain a higher-grade, anhydrous ethanol. In addition to ethanol, there is also an aqueous solution leaving the distillation process, that is called residue.

Molasses obtained from sugar cane are the most important raw material for the purposes of bioethanol production. In recent years, however, there have been rising prices and restrictions on the availability of molasses, which have strongly influenced the production of bioethanol (Quintero et al., 2008). Figure 3 shows the flow chart for bioethanol, energy and sugar production from sugar cane.

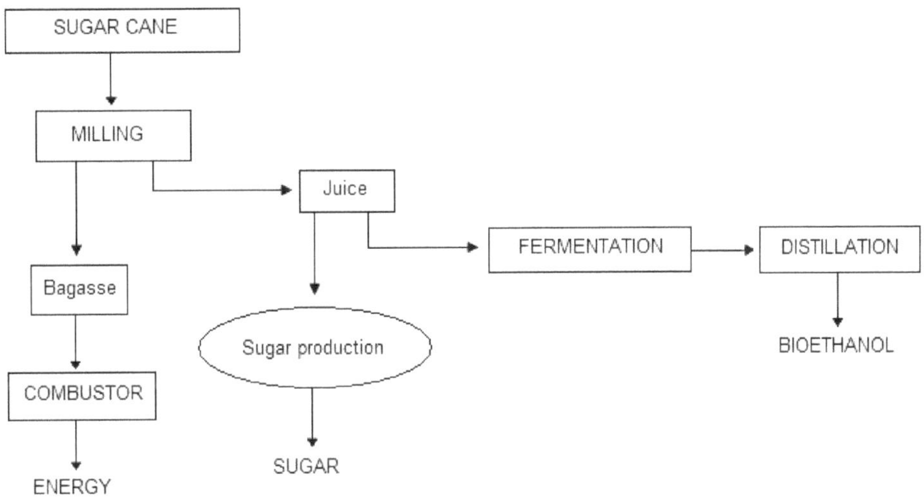

Figure 3.Flow chart for bioethanol, energy and sugar production from sugar cane

3.3.2. Sugar beet (beta vulgaris)

Like sugar cane, sugar beet can also be used to obtain bioethanol by fermenting and distilling its juice. The beet is first cut into thin slices, then placed in contact with a medium (water or juice extracted from a previous process) and brought up to a temperature of about 70-80°C. In the case of sugar beet, temperature is a fundamental extraction parameter because it must be high enough to rupture the proteins in the cell walls containing the sugars, which has the effect of allowing the sugars to spread through the medium. Once this process has been completed, the sugar beet pulp is dried and sold as animal feed or to the pharmaceutical industry for use in the production of citric acid and its esters. The beet juice proceeds instead through the stages that convert it into bioethanol. At plants where sugar and bioethanol are both produced together, the juice can either be used directly or it can be concentrated in evaporators and stored for several months. Both the fresh and the concentrated sugar juice can be used in production processes involving cold crystallization and fermentation. The fermentation process relies on the use of yeasts (preferably Saccharomyces cerevisiae) or bacteria such as Zymomonasmobilis (Linde et al., 1998), which is only used in the case of a discontinuous fermentation. The great interest focusing on the bacteria is due to their capacity to convert the glucose into ethanol more efficiently than yeasts succeed in doing. Figure 4 shows the flow chart for the production of bioethanol and byproducts from sugar beet.

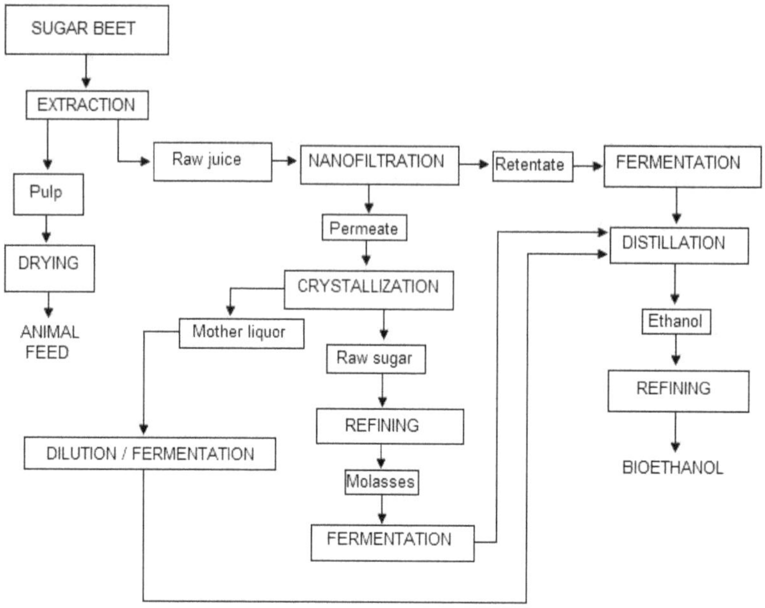

Figure 4. Flow chart for the production of bioethanol and byproducts from sugar beet

3.4. Comparing the various raw materials

The choice of the raw materials to use to produce ethanol depends largely on local climatic conditions. North America and Europe, for instance, have based their ethanol production on materials containing starch, because of their particular farming and ecological conditions, which make it unfeasible to sugar cane adequately, although this plant offers a higher ethanol yield. In these areas, the most often grown energy crops are cereals. Using these raw materials poses some energetic sustainability limits (Patzek et al., 2005; Pimentel, 2003). The yield per ton of raw material is higher for sugar beet molasses than for cereals, so although growing sugar beet is less productive in quantitative terms than growing cereals, the annual ethanol yield from beet is higher than from cereals. The importance of analyzing the geographical position of crops helps us to see that growing the same type of cereal in tropical regions would produce a distinctly lower yield than could be achieved from the same plant grown in more suitable areas (Espinal et al., 2005). The lignocellulose materials represent the future as concerns raw materials for ethanol, because of their excellent energy value, great availability, low cost and high bioethanol yield.

These materials cannot be used to produce food, but they provide important secondary products such as methanol, syngas, hydrogen and electricity. The choice of which lignocellulose material also depends on the nature of the waste products in a given country (Kim & Dale, 2004). Cereals that are discarded during the distribution process can be destined to ethanol production, together with farming waste and sugar cane bagasse. The drawback of these raw materials consists in the complexity of the phenomena involved in converting the biomass into ethanol. Various studies have been conducted on the process of bioethanol production starting from various raw materials, including lignocellulose materials, cereals (McAloon et al., 2000; Cardona et al., 2005), and sugar cane (Quintero et al., 2008).

3.5. Converting syngas into ethanol

Bioethanol can also be obtained by means of chemical processes (Sánchez & Cardona, 2008; Demirbas, 2005), which may or may not demand the presence of microorganisms in the fermentation stage. Gasification of a biomass to obtain syngas ($CO + H^2$), followed by the catalytic conversion of the syngas, has the potential for producing ethanol in large quantities. The catalysts most often used and studies are those based on rhodium (Rh) (Holy & Carey, 1985; Yu-Hua et al., 1987; Gronchi et al.; 1994).

The geometrical structure of the active site seems to be:

$$\left(Rh_x^{\,0}Rh_y^{\,+}\right)-O-M^{n+} \tag{9}$$

where part of the Rh occurs as Rh + and the promoter ion (M n+) is in close contact with these Rh species. The carbon monoxide is then hydrogenated to form an absorbed species -CHx- that is then inserted in the absorbed CO. Hydrogenation of these absorbed species leads to the formation of ethanol (Subramani & Gangwal, 2008).

Another mechanism considered valid for ethanol formation involves the use of acetate (acetaldehyde formation followed by reduction) and is known, in the cases of Rh-based catalysts, to be promoted by manganese (Luo et al., 2001).

In this case, ethanol is formed by direct hydrogenation of tilt-absorbed CO molecules, followed by CH2 insertion on the surface of the CH2-O species to form an absorbed intermediate species. Ethanol is produced by hydrogenation of the intermediate species of CH2-O. Acetaldehyde is formed by the insertion of CO on the surface of the CH3-Rh species, followed by hydrogenation. The catalyst's performance can be improved by modifying its composition and preparing the ideal conditions for the reaction (Subramani & Gangwal, 2008). Manganese (Lin et al., 1995), Samarium and Vanadium (Luo et al., 2001) can also be used as promoter ions in processes involving Rh.

4. ENVIRONMENTAL ISSUES

The greenhouse gases (GHGs) are gases occurring in the Earth's atmosphere that absorb in the infrared field (carbon dioxide, ozone, methane, nitrogen oxides, carbon monoxide and so on). This feature enables them to trap the heat of the sun reflected back from the Earth's surface.

The GHG that occurs in the largest quantities is carbon dioxide, and that is why it attracts so much attention. In fact, the carbon cycle is a delicate balance between carbon accumulation, release and recycling that enables vegetable and animal species to survive. Problems linked to CO2 began to emerge at the start of the industrial era: the ever-increasing use of fossil fuels as a source of energy meant that the carbon dioxide trapped for centuries in the fossils was being put back into the atmosphere, with no correspondingly reinforced recycling mechanism, which relies on chlorophyllic photosynthesis).

In addition to reducing carbon dioxide emissions, bioethanol can be seen as a no-impact fuel because the amount of CO2 released into the atmosphere is compensated by the amount of CO2 converted into oxygen by the plants grown to produce the bioethanol (Ferrel & Glassner, 1997).

4.1. Carbon sequestering

In the analysis of the environmental impact of bioethanol (and other biofuels too), some of the key factors concern the impact of the increasing quantities of dedicated crops on soil carbon levels and subsequent photosynthesis: these changes will also influence the atmospheric concentrations of GHG such as CO_2 and CH_4.

The main problem concerns the fact that, when a system in equilibrium experiences persistent changes, it can take decades before a new equilibrium with a constant carbon level is reached. Taking the current situation in Europe as concerns wheat and sugar beet crops, there is an estimated depletion of approximately 0.84 t of C or 3.1 t CO2 equivalent ha-1 years-1 from the ground. If no crops were grown on the soil, this depletion would be even greater, i.e. 6.5 t of C each year for sugar beet and 4.9 t of C for wheat. Apart from the effects on ground carbon levels, there are also signs of other adverse effects indirectly linked to crops grown for energy purposes, such as the increase in the amount of C in the atmospheric levels of GHG. Irrigation with good-quality water also exacerbates carbon sequestering: the water used for irrigation contains dissolved calcium and carbon dioxide (in the form of $HCO3^-$); Ca and $HCO3^-$ react together, giving rise to the precipitation of $CaCO3$ and the consequent release of $CO2$ into the atmosphere. In the typical dry conditions of the USA, further reactions take place and irrigation is responsible for the transfer of $CO2$ from the ground into the atmosphere (Rees et al., 2005). An important type of crop that can be used to reduce soil carbon sequestering is defined as "zero tillage", which means that it can be grown year after year without disturbing the soil. Seed crops (such as wheat) may be zero tillage, but not root crops (such as Panicum virgatum). Zero tillage has variable effects, and in some cases carbon sequestering in the soil may even increase, but this phenomenon can be completely overturned by a one-off application of conventional tillage. If only the carbon in the soil is considered, zero tillage leads in the long term to less global warming than growing conventional crops in damp climates, but in areas with dry climates, there is no certainty of any such beneficial effect (Six et al., 2004). Using straw from cereals can increase the carbon levels in the soil. Such residue is useful in maintaining soil carbon levels (Blair et al., 1998; Blair and Crocker, 2000) because it has a low rate of breakdown, so it is important for the residue to go back into the ground in order to keep the farming system sustainable. Since removing the residue from the ground has other negative effects too, such as an increased soil erosion and a lesser availability of macro- and micronutrients, some have suggested in the United States (Lal, 2005) that it would be advisable to remove only 20-40% of the residue for the purposes of bioethanol production, whereas it was claimed (Sheehan et al., 2004) that if up to 70% of the residue were removed to produce bioethanol, the carbon levels would initially decline and then remain stable for about 90 years. Increasing the land used to grow energy crops would have a substantial impact on the concentrations of carbon-containing gases in the atmosphere. If areas covered with forest were converted into arable land, the carbon sequestering would go from values of around 50-145 t ha^{-1} to approximately 50-200 t ha^{-1}, assuming a 60-year rotation (Reijinders&Huijbregts, 2007).

4.2. Emissions

Blending bioethanol with petrol, even in small amounts, enhances the fuel's octane rating and decreases the proportion of aromatic and cancer-causing compounds, as well as the emissions of NOx, smoke, CO, SOx, and volatile organic compounds (VOC). However, there is also a rise in the release of formaldehyde and acetaldehyde. However, contemporary bioethanol production systems often exhibit an energy ratio, also known as net usable energy, ranging from 2 to 7. This ratio is contingent upon the specific crops and procedures employed. The composition of petrols can impact the release of organic compounds. Petrols that contain aromatic hydrocarbons like benzene, toluene, xylene, and olefins tend to have higher levels of reactive hydrocarbons. On the other hand, petrols formulated with oxygenated compounds, such as those mixed with bioethanol, may have lower amounts of aromatic compounds. The issue with petrols containing large levels of aromatic compounds is their strong tendency to release unburned hydrocarbons, which pose challenges for catalytic converters to oxidize and also serve as precursors to photochemical pollution. All fuels that contain oxygen have the capacity to decrease the release of carbon monoxide (CO) and unburned hydrocarbons. These hydrocarbons are also less reactive in terms of photochemistry compared to the hydrocarbons found in regular gasoline. When ethanol is added to petrol in an internal combustion engine equipped with a three-way catalytic converter, it functions as an oxygenating agent for the exhaust gases. This results in a significant 10% decrease in the emission of CO, as well as a reduction in the overall emissions of aromatic hydrocarbons (Poulopoulos et al., 2001). The utilization of four-stroke engines, equipped with four cylinders and electronic injection, and powered by different ethanol and petrol blends (Al-Hasan, 2003), resulted in a significant decrease of around 46.5%

in CO emissions. The anti-detonation properties of gasoline are crucial and primarily dependent on its chemical composition. Life cycle analysis, using the "well to wheel" approach, demonstrated that the greenhouse gas (GHG) emissions from bioethanol derived from sugar beet are around 40-60% lower than the emissions from petrol derived from fossil fuels (Reijinders&Huijbregts, 2007). Blending bioethanol with diesel oil enhances the combustion of the fuel (Lapuerta et al., 2008) and decreases the size of the particles in the exhaust without causing an increase in their number. The utilization of an E10 blend diminishes the overall hydrocarbon emissions due to the higher heat of vaporization exhibited by ethanol. The addition of moderate amounts of ethanol to diesel oil leads to an increase in CO emissions, however the proportion of ethanol increases results in a decrease in CO emissions (Li et al., 2005). In contrast, the emissions of nitrogen oxides (NOx) drop when a small or moderate amount of ethanol is used, but increase when a larger amount of ethanol is added. The total hydrocarbons (THC) also vary in accordance with the varying quantities of ethanol and varied speeds.

5. CONCLUSIONS

While bioethanol is indeed a viable substitute for fossil fuels and has minimal ecological consequences, its utilization nevertheless presents challenges concerning the consumption of essential raw materials like grains, which are crucial for the food industry. Expanding the agricultural land dedicated to cultivating energy crops for biofuel generation results in direct competition with food crops. Several research have endeavored to evaluate the necessity of agricultural land for cultivating crops used in ethanol production. The bioethanol yield per hectare is determined by the choice of crops. However, we can refer to the average productivity in Europe, which is currently estimated to be approximately 2790 liters per hectare. This estimation is based on the average yield of 7 tons of seeds per hectare, with each ton producing 400 liters of bioethanol. While bioethanol production can be achieved in temperate regions, tropical climes offer superior conditions for achieving high productivity. In Brazil, the cultivation of sugar cane yields an estimated production of around 6200 liters per hectare. This calculation is based on a crop output of 69 tons per hectare and a conversion rate of 90 liters per ton. In India as well, the production of bioethanol derived from sugar cane is significant, reaching roughly 5300 liters per hectare. Should bioethanol derived from sugar cane become a globally traded commodity, South America, India, Southeast Asia, and Africa have the potential to emerge as significant exporters. Current research is primarily investigating alternative options for raw materials, with a specific focus on innovative sources like Miscanthus Giganteus, a non-edible plant that possesses a remarkably high calorific value of approximately 4200 Kcal/kg of dry matter. Additionally, filamentous fungi such as Trichoderma reesei are being studied for their ability to break down the complex lignocellulose molecules. This article provides a concise overview of the primary raw materials utilized in the manufacturing of bioethanol, ranging from conventional to cutting-edge options, as well as the key production methods employed. Additionally, it examines the matters pertaining to emissions and carbon sequestration.

REFERENCES

Al-Hasan, M. (2003). Effect of ethanol unleaded gasoline blends on engine performance and exhaust emission. *Energy Conversion Management,* Vol.44, No.9, pp.1547-1561.

Balat, M.; Balat, H. & Öz, C. (2008). Progress in bioethanol processing. *Progress Energy Combustion Science,* Vol.34, No.5, pp.551-573.

Blair, N.; Chapman, L.; Whitbread, A.M.; Ball-Coelho, B.; Larsen, P. & Tiessen, H. (1998). Soil carbon changes resulting from sugarcane trash management at two locations in Queensland, Austria, and in N-E Brazil. *Australian Journal of Soil Research,* Vol.36, pp.873-881.

Blair, N. & Crocker, G.J. (2000). Crop rotation effects on soil carbon and physical fertility of two Australian soils. Australian Journal of Soil research, Vol.38, No.1, pp.71-84.

Bouton, J.H. (2007). Molecular breeding of switchgrass for use as a biofuel crop. *Current Opinion in Genetics & Development*, Vol.17, No.6, pp.1-6.

Cardona, C.A.; Sánchez, O.J.; Montoya, M.I. & Quintero, J.A. (2005). Analysis of fuel ethanol production processes using lignocellulosic biomass and starch as feedstocks. *Proceeding of the Seventh World Congress of Chemical Engineering*, Glasgow, Scotland, UK, 2005.

Chandel, A.K.; Chan, E.S.; C, Rydravaram, R.; Narasu M,L.; Rao, L.V. & Ravindra, P. (2007). Economics and environmental impact of bioethanol production technologies: an appraisal. *Biotechnology and Molecular Biology Review*, Vol.2, No.1, pp.14-32.

Chum, H.L.; Johnson, D.K.; Balck, S.K. & Overend, R.P. (1990). Pretreatment-catalyst effects and combined severity parameter. *Applied Biochemistry and Biotechnology*, Vol.24-25, No.1, pp.1-14.

Curt, M.D.; Aguado, P.; Sanz, M; Sánchez, G. & Fernández, J. (2006). Clone precocity and the use of Helianthus tuberosus L. stems for bioethanol. *Industrial Crops and Products*, Vol.24, No.3, pp.314-320.

Da Silva, R.; Cataluña, R.; Menezes, E.W.; Samios, D. & Piatnicki, C.M.S. (2005). Effect of additives on the antiknock properties and Reid vapor pressure of gasoline. *Fuel*, 2005; Vol.84, No.7-8, pp.951-959.

DEFRA, *Planting and growing Mischanthus. Policy for Energy Crops Scheme.* London: DEFRA Publications; 31 January 2001, Available from: http://www.defra.gov.uk/

Demirbas, A. (2005). Bioethanol from cellulosic materials: a renewable motor fuel from biomass. *Energy Sources*, Vol.27, pp.327-337.

Dien, B.; Jung, H.J.G.; Vogel, K.P.; Casler, M.D.; Lamb, J.F.S.; Iten, L.; Mitchell, R.B. & Sarath, G. (2006). Chemical composition and response to dilute acid pretreatment and enzymatic saccharification, *Biomass and Bioenergy*, Vol.30, No.10, pp.880-891.

Espinal, C.F.; Martínez, H.J. & Acevedo, X. (2005). The chain of cereals balanced feed, aviculture, and pig farming in Colombia. An overview of its structure and dynamics 1991-2005. *Working report n. 87, 2005.* Observatorio Agrocadenas Colombia, Ministry of Agriculture and Rural Development.

Eurec Agency, *New improvements for lingo-cellulosic ethanol*, 31 January 2011, Available from: www.nile-bioethanol.org

European Union, *Fuel the future with bioethanol. A realistic large scale alternative to fossil fuels,* 31 January 2001, Available from: www.best-europe.org

Ferrel, J. & Glassner, D. (1997). *Bioethanol-the climate-cool fuel.* NREL, produced for the U.S. Department of Energy (DOE) 1997, DOE/GO-10097-515.

Gressel, J. (2008). Transgenics are imperative for biofuel crops. *Plant Science*; Vol.174, No.3, pp.246-263.

Gronchi, P.; Tempesti, E. & Mazzocchia, C. (1994). Metal dispersion dependent selectivities for syngas conversion to ethanol on V2O3 supported rhodium. *Applied Catalyst A: General*, Vol.120, No.1, pp.115-126.

Hahn-Hägerdal, B.; Galbe, M.; Gorwa-Grauslund, M.F.; Lidèn, G. & Zacchi, G. (2006). Bio– ethanol - the fuel of tomorrow from the residues of today. *Trends in Biotechnology*, Vol.24, No.12, pp.549-556.

Hansen, A.C.; Lyne, P.W.L. & Zhang, Q. Ethanol-diesel blends: a step towards a biobased fuel for diesel engines. (2005) *ASAE paper 01-6048.*

Heaton, H.; Voigt, T. & Long, S.P. (2004). A quantitative review comparing the yields of two candidate C4 perennial biomass crops in relation to nitrogen, temperature and water. *Biomass and Bioenergy*, Vol.27, No.1, pp.21-30.

Holy, N.L. & Carey, T.F. (1985). Ethanol and n-propanol from syngas. *Applied Catalysts*, Vol.19, No.2, pp.219-223.

Kim, S. & Dale, B.E. (2004). Global potential bioethanol production from wasted crops and crop residues. *Biomass and Bioenergy*, Vol.26. No.4, pp.361-375.

Lal, R. (2005). World crop residues production and implications of its use as a biofuel. *Environment International*, Vol.31, No.4, pp.575-584.

Lamnet, *Latin America Thematic Network on Bioenergy*, 31 January 2001, Available from: http://www.bioenergy-lamnet.org/

Lapuerta, M; Armas, O & Garcìa-Contreras, R. (2007). Stability of diesel-bioethanol blends for use in diesel engines. *Fuel*, Vol.86, No.10-11, pp.1351-1357.

Lapuerta, M.; Armas, O. & Herreros, J.M. (2008). Emissions from a diesel-bioethanol blend in an automotive diesel engine. *Fuel*, Vol.87, No.1, pp.25-31.

Larsson, S.; Palmqvist, E.; Hahn-Hägerdal, B.; Tengborg, C.; Stenberg, K.; Zacchi, G. & Nilvebrant, N.O. (1999). The generation of fermentation inhibitors during dilute acid hydrolysis of softwood. *Enzyme Microbial Technology*, Vol.24, No.3-4, pp.151-159.

Laser, M.; Schulman, D.; Allen, S.G.; Lichwa, J.; Anta, Jr M.J. & Lynd, L.R. (2002). A comparison of liquid hot water and steam pretreatments of sugar cane bagasse for bioconversion to ethanol. *Bioresource Technology*, Vol.81, No.1, pp.33-44.

Li, D.; Zhen, H.; Xingcai, L.; Wu-gao, Z. & Jian-guang, Y. (2005). Physico-chemical properties of ethanol-diesel blend fuel and its effect on performance and emissions of diesel engines. *Renewable energy*, Vol.30, No.6, pp.967-976.

Lin, P.Z.; Liang, D.B.; Lou, H.Y.; Xu, C.H.; Zhou, H.W.; Huang, S.Y. & Lin, L.W. (1995). Synthesis of C2+-oxygenated compounds directly from syngas. *Applied Catalyst A: General*, Vol.131, No.2, pp.207-214.

Linde, R.; Bekers, M.; Ventina, E.; Vina, I.; Kaminska, H.; Upite, D.; Scherbaka, R. & Danilevich, A. (1998). Ethanol and Fructose from Sugar Beets. Biomass for Energy and Industry, *Proceeding of the C.A.R.M.E.N. International Conference*, pp.464-467, Würzburg, Germany, June 8-11, 1998.

Luo, H.Y.; Zhang, W.; Zhou, H.W.; Huang, S.Y.; Lin, P.Z.; Ding, Y.J. & Lin, L.W. (2001). A study of Rh-Sm-V/SiO2 catalysts for the preparation of C2-oxygenates from syngas. *Applied Catalyst A: General*, Vol.214, No.2, pp.161-166.

Lynd, L.R.; van Zyl, W.H.; McBride, J.E. & Laser, M. (2005). Consolidated bioprocessing of cellulosic biomass: an update. *Current Opinion in Biotechnology*, Vol.16, No.5, pp.577-583.

Marek, N.J. & Evanoff J. The use of ethanol-blended diesel fuel in unmodified compression ignition engines: an interim case study. *Air and Waste Management Association (AWMA) Annual Conference* ,Orlando, Florida, USA, June 2001.

Martinez, D.; Berka, R.M.; Henrissat, B.; Saloheimo, M.; Arvas, M.; Baker, S.E., Chapman, J.; Chertkov, O.; Coutinho, P.M.; Cullen, D.; Danchin, E.G., Grigoriev, I.V.; Harris, P.; Jackson, M.; Kubicek, C.P.; Han, C.S.; Ho, I.; Larrondo, L.F.; de Leon, A.L.; Magnuson, J.K.; Merino, S.; Misra, M.; Nelson, B.; Putnam, N.; Robbertse, B.; Salamov, A.A.; Schmoll, M. Terry A, Thayer N, Westerholm-Parvinen A, Schoch

CL, Yao J, Barbote R, Nelson MA, Detter C, Bruce D, Kuske CR, Xie G.; Richardson, P.; Rokhsar, D.S.; Lucas, S.M.; Rubin, E.M.; Dunn-Coleman, N.; Ward, M. & Brettin, T.S. (2008). Genome sequencing and analysis of the biomass-degrading fungus Trichoderma reesei (syn. Hypocrea jecorina). *Nature Biotechnology*, Vol.26, No.5, pp.553-60.

McAloon, A.; Taylor, F.; Yee, W.; Ibsen, K. & Wooley, R. (2000). Determining the cost of producing ethanol from corn starch and lignocellulosic feedstocks. *Technical report NREL/TP-580-28893*. Golden, Co (USA), National Renewable Energy Laboratory 2000.

Overend, R.P. & Chornet, E. (1987). Fractionation of lignocellulosics by steam-acqueous pretreatments. *Philosophical Transactions for the Royal Society of London. Series A, Mathematical and Physical Sciences*,Vol.321, No.1561, pp.523-536.

Palmieri, G.; Giardina, P. & Sannia, G. (1997). A novel white laccase from Pleurotus ostreatus. *The journal of biological chemistry*, Vol.272, No.50, pp.31301-31307.

Palmqvist, E.; Hahn-Hägerdal, B.; Galbe, M. & Zacchi, G. (1996). The effect of water-soluble inhibitors from steam-pretreated willow on enzymatic hydrolysis and ethanol fermentation. *Enzyme and Microbial Technology*, Vol.19, No.6, pp.470-476.

Palmqvist, E. & Hahn-Hägerdal, B. (2000). Fermentation of lignocellulosic hydrolysates. II: inhibitors and mechanisms of inhibition. *Bioresource Technology*, Vol.74, No.1, pp.25- 33.

Patzek, T.W.; Anti, S.M.; Campos, R. Ha, K.W.; Lee, J.; Li, B.; Padnick, J. & Yee, S.A. (2005). Ethanol from corn: Clean renewable fuel for the future, or drain on our resources and pockets? *Environment, Development, and Sustainability*, Vol.7, No.3, pp.319-336.

Prasad, S.; Singh, A. & Joshi, H.C. (2007). Ethanol as an alternative fuel from agricultural, industrial and urban residues. Resources. *Conservation and Recycling*, Vol.50, No.1, pp.1-39.

Petersson, A.; Thomsen, M.H.; Hauggaard-Nielsen H. & Thomsen, A.B. (2007). Potential bioethanol and biogas production using lignocellulosic biomass from winter rye, oilseed rape and faba bean. *Biomass and Bioenergy*, Vol.31, No.11-12, pp.812-819.

Pimentel, D. (2003). Ethanol fuels: energy balance, economics, and environmental impacts are negative. *Natural Resources Research* , Vol.12, No.2, pp.127-134.

Poulopoulos, S.G.; Samaras, D.P. & Philippopoulos, C.J. (2001). Regulated and unregulated from an internal combustion engine operating on ethanol-containing fuels.*Atmosphere Environment*, Vol.35, No.26, pp.4399-4406.

Quintero, J.A.; Montoya, M.I.; Sánchez, O.J.; Giraldo, O.H. & Cardona, C.A. (2008). Fuel ethanol production from sugarcane and corn: comparative analysis for a Colombian case. *Energy*, Vol.33, No.3, pp.385-399.

Rees, R.M.; Bingham, I.J.; Baddeley, J.A. & Watson, C.A. (2005). The role of plants and land management in sequestering carbon temperate arable and grassland systems. *Geoderma*, Vol.128, No.1-2, pp.130-154.

Reijinders, L. & Huijbregts, M.A.J. (2007). Life cycle greenhouse gas emissions, fossil fuel demand and solar energy conversion efficiency in European bioethanol production for automotive purposes. *Journal of Cleaner Production*, Vol. 15, No.18, pp.1806-1812.

Robertson, G.H.; Wong, D.W.S.; Lee, C.C.; Wagschal, K.; Smith, M.R. & Orts, W.J. (2006). Native or raw starch digestion: a key step in energy efficient biorefining of grain. Journal of Agricultural and Food Chemistry, Vol.54, No.2, pp.353-365.

Six, J.; Ogle, S.M.; Breidt, F.J.; Conant, R.T.; Mosiers, A.R. & Paustian, K. (2004). The potential to mitigate global warming with no-tillage management is only realized when practiced in the long term. Global Change Biology, Vol.10, pp.155-160.

Rooney, L.W.; Blumenthal, J.; Bean, B. & Mullet, J.E. (2007). Designing sorghum as a dedicated bioenergy feedstock. Biofuels, Bioproducts and Biorefining, Vol.1, No.2, pp.147-157.

Rosenberg, A.; Kaul, H.P.; Senn, T. & Aufhammer, W. (2001). Improving the energy balance of bioethanol production from winter cereals: the effect of crop production intensity. Applied Energy, Vol.68, No.1, pp.51-67.

Sánchez, Ó. & Cardona, C.A. (2008). Trends in biotechnological production of fuel ethanol from different feedstocks. Bioresource Technology, Vol.99, No.13, pp.5270-5295.

Sassner, P.; Galbe, M. & Zacchi, G. (2008a). Techno-economic evaluation for bioethanol production from three different lignocellulosic materials. Biomass and Bioenergy, Vol.32, No.5, pp.422-430.

Sassner, P.; Mårtensson, C.G.; Galbe, M. & Zacchi, G. (2008b). Steam pretreatment of H2SO4- impregnated Salix for the production of bioethanol. Bioresource Technology, Vol.99, No.1, pp.137-145.

Sheehan, J.; Aden, A.; Paustian, K.; Killian, K.; Walsh, M. & Nelson, R. (2004). Energy and environmental aspects of using corn stover for fuel ethanol. Journal of Industrial Ecology, Vol.7, No.3-4, pp.117-146.

Shevchenko, S.M.; Chang, K., Robinson, J. & Saddler, J.N. Optimization of monosaccharide recovery by post-hydrolysis of the water-soluble hemicellulose component after steam explosion of softwood chips. Bioresource Technology, Vol.72, No.3, pp.207-211.

Silverstein, R.A.; Chen, Y.; Sharma-Shivappa, R.R.; Boyette, M.D. & Osborne, J. (2007). A Comparison of chemical pretreatment methods for improving saccharification of cotton stalks. Bioresource Technology, Vol.98, No.16, pp.3000-3011.

Soccol, C.; Marin, B.; Raimbault, M. & Lebeault, J.M. (1994). Breeding and growth of Rhizopus in raw cassava by solid state fermentation. Applied Microbiology and Biotechnology, Vol.41, No.3, pp.330-336.

Subramani, V. & Gangwal, S.K. (2008). A Review of Recent Literature to Search for an Efficient Catalytic Process for the Conversion of Syngas to Ethanol. Energy & Fuels, Vol.22, No.2, pp.814-839.

Sun, Y. & Cheng, J. (2002). Hydrolysis of lignocellulosic materials for ethanol production: a review. Bioresource Technology, Vol.83, No.1, pp.1-11.

Varvel, G.E.; Vogel, K.P.; Mitchell, R.B. & Kimble, J.M. (2007) Comparison of corn and switchgrass on marginal soils for bioenergy. Biomass and Bioenergy , Vol.32, No.1, pp.18-21.

Verma, G.; Nigam, P.; Singh, D. & Chaudhary, K. (2000). Bioconversion of starch to ethanol ion a single-step by co-culture of amylolytic yeasts and Saccharomyces cerevisiae 21. Bioresource Technology, Vol.72, No.3, pp.261-266.

Wyman, C.E. (1994). Ethanol from lignocellulosic biomass: technology, economics, and opportunities. Bioresource Technology, Vol.50, No.1, pp.3-16.

Yu-Hua, D.; De-An, C. & Khi-Rui, T. (1987). Promoter action of rare earth oxides in rhodium/silica catalysts for the conversion of syngas to ethanol. Applied Catalyst A, Vol.35, No.1, pp.77-92.

CHAPTER-27

THE IMPACT OF A VARIETY OF FRIED DISHES ON CERTAIN CHARACTERISTICS OF UTILIZED PLANT OILS AS PRIMARY RESOURCES FOR THE CREATION OF AN ALTERNATIVE TO DIESEL FUEL

1. INTRODUCTION: AN OVERVIEW

The utilization of post-frying plant oils, which are waste products from the operation of establishments that serve fried goods, has become an increasingly difficult problem of technological, ecological, and economic character over the years. The strategies for tackling this problem have been extensively researched by Alcantara (2000), Buczek and Chwiałkowski (2005), Dzieniszewski (2007), and Leung and Guo (2006). The idea of using post-frying plant oils as components for producing an alternative for diesel fuel appears to be promising. However, it is necessary to thoroughly examine the characteristics of these oils in order to determine the most efficient methods for their application. The question pertains to the impact of a variety of fried items on the quality of the oil used for frying, and its practicality, with a focus on the differences in the utilization of specific batches of oil derived from frying different food products, which appears to be the most important worry. The most prevalent technique for cooking food at culinary establishments is deep frying. In this frying method, processed food is fully immersed in a frying medium and comes into touch with oil or fat, covering most of its outside surface. The primary function of the frying medium is to maintain the processed food in the correct position relative to the heat source and to transfer an appropriate quantity of thermal energy to the fried product [Drozdowski, 2007; Ledóchowska& Hazuka, 2006]. The combination of frying oil, a medium used for frying, and food products that undergo culinary preparation creates a distinct system where there is a partial penetration of these two compounds and a two-way transfer of energy and weight. During the frying process, the product undergoes dehydration and loses a substantial amount of water. Additionally, depending on its composition, certain chemicals such as food colors, taste and flavor compounds are also lost and partially transferred to the frying fat, which consists of lipids. The frying fat used in fried food is typically substituted for a certain quantity, the exact proportion of which can vary considerably and potentially reach up to 40% [Ledóchowska and Hazuka, 2006]. The water present in processed products and released during submersion frying has a varied and multidirectional impact on the changes that occur in the oil. One of these changes is the partial increase in the acid number (AN) of the oil, which is caused by fat hydrolysis. Furthermore, the transportation of heat that is emitted along with the released water vapors helps to lower the temperature of fried food and partially hinders the oxidation process of fat by displacing oxygen within it [Ledóchowska and Hazuka, 2006]. The presence of dissolved oxygen and water vapor in frying fat are important elements in thermooxidative reactions, which have not been completely understood. These modifications lead to the production of various compounds with intricate and undetermined structures. These are substances that come before and lead to further changes. The resulting products can generally be grouped into two categories: compounds that can easily turn into gas (such as hydrocarbons, fatty acids, and carboxylic compounds) and compounds that do not easily turn into gas (such as monomers, dimers, polymers, aldehydes, ketones, and fatty acids with altered melting points) [Drozdowski 2007, Paul and Mittal 1996, Blumenthal 1991, Choe and Min 2006, Clark and Serbia 1991, Hoffman 2004, Ledóchowska and Hazuka 2006]. Gastronomical fryers are typically large vessels that contain a significant amount of oil. The oil is deposited near the bottom of the fryer, while the surface layer of the fryer plays a crucial role in determining the qualities of the food being cooked. The lower area of a fryer, often located near the heat source, has a lower concentration of oxygen and water vapor. This environment promotes the production of free radicals or the polymerization of unsaturated fatty acids present in the frying oil. The predominant outcome of these alterations is the generation of many, intricately organized, non-polar thermal polymers. The macroscopic outcome of these processes includes an increase in viscosity and darkening, as well as a rise in the melting point of the frying medium. Consequently, there is a change in its state of aggregation. The principal constituents of the dark brown deposits visible on the walls of a fryer are the products of these conversions. These deposits might cause various issues connected to the use of the oil [Hoffman 2004]. The direction and intensity of frying fat changes are influenced by various elements that occur during the frying of food products. In the field of literature, Ledóchowska and Hazuka (2006) identify several kinds of factors. The essential factors for conducting the process, such as its duration, temperature, periodicity, and the level of unsaturation of fatty acids in the triglycerides of the fat, are specified. Several parameters, including oxygen availability, the amount and type of chemicals produced from food (such as vitamins and antioxidants), and the presence of water, have a major impact on the properties of frying medium during the frying process [Ledóchowska and Hazuk'a 2006].

2. ASSESSMENT OF USABILITY OF POST-FRYING EDIBLE OILS AS A RAW MATERIAL FOR PRODUCTION OF DIESEL FUEL SUBSTITUTE

2.1. Materials and methods

2.1.1. Preparation of samples for Investigation

This study aimed to compare the effects of different fried dishes, specifically potato chips and breadcrumbs coated fish fingers, on the physicochemical properties and quality of post-frying plant oils. These oils are intended to be used as raw materials for the production of fatty acids methyl esters, which can be used as a substitute for diesel fuel. The primary objective of the study was to assess the impact of various fried food items on the quality of the resulting cooking oils (rapeseed, sunflower, and soybean), specifically in relation to their potential use as a raw material for the generation of biofuel for engines. The applicability of post-frying waste oils as raw materials for the manufacture of fatty acids methyl esters was investigated under controlled laboratory settings. The research utilized three prevalent edible oils, namely rapeseed, sunflower, and soybean, as the primary materials. A sample for laboratory tests was taken from each of the raw oils. The item was assigned a value of "0" and utilized as a reference sample. The remaining quantity of each oil was divided into three equal portions and placed into individual containers. Batch number 1 was prepared using a cyclic heating process, involving five repetitions, without the use of fried product. A specific cycle in this batch involves heating the entire quantity of oil to a temperature of around 180º C, and then keeping it at that temperature for 10 minutes. Subsequently, the oil was allowed to reach ambient temperature before extracting a sample for laboratory analysis. The sample was labeled as "heating I - without fried product". After a 24-hour period, all of the aforementioned operations were repeated, resulting in a sample labeled as "heating II - without fried product". The entire procedure of heating, cooling, and sampling was repeated, resulting in samples labeled as "heating without fried product" with the relevant labels of III, IV, and V, depending on the number of cycles. The preparation of oil from batch number 2 differed from the previous presentation in only one aspect. After being heated to a temperature of 180º C, potato chips made from raw potatoes obtained from the store and chopped to match the size and form of commercially available frozen potato chips were fried in each of the three oils under investigation. Following the frying and separation of the chips, the oil was allowed to cool to the ambient temperature. Subsequently, samples were collected for research purposes. The chips frying procedure was labeled as "heating I". By repeating the entire process, we were able to collect samples labeled III, IV, and V, corresponding to the number of cycles. The third portion of oil (batch no. 3) was subjected to the same heating process as batch no. 2. However, in this instance, commercially bought fish fingers covered with breadcrumbs were used as the fried product. Following the frying and separation of fish fingers covered in breadcrumbs, the oil was allowed to cool to room temperature before samples were collected for study purposes. The fish fingers coated in breadcrumbs were being fried as part of the heating process. By repeating the entire process, we were able to collect samples labeled III, IV, and V, corresponding to the number of cycles.

2.2. Laboratory test

The oil samples, collected according to the selected approach, were analyzed in the laboratory. The studies included determining the peroxide number (PN), acid number (AN), and fatty acid content. The peroxide number (PN) was determined according to ISO 3960 by titrating the iodine produced from potassium iodide by the peroxides in the sample, and then calculating it based on their weight unit. The analyses yielded results reported as millimoles of oxygen per unit weight of the sample. The acid number (AN) was determined according to the [PN-ISO 660] standard by titrating and evaluating the acidity of a sample. The result was stated as the number of milliliters of a 0.1M solution of sodium hydroxide required per unit weight of the analyzed oil. The fatty acid content was determined using a gas chromatography method as described by Krełowska – Kułas in 1993. A sample of fat was treated with a solution of sodium hydroxide in methanol under dry conditions using alkaline hydrolysis. As a consequence of this reaction, the fatty acids present in the studied oil underwent a transformation into a combination of sodium soaps. These soaps were then treated to an esterification reaction with an anhydrous solution of hydrogen chloride in methanol, resulting in the formation of a mixture of fatty acid methyl esters. The methyl esters that were obtained were separated using a chromatographic column, and then their contribution to the total amount of fatty acids was calculated [Krełowska – Kułas, 1993]. The process of chromatographic separation was carried out using a gas chromatograph equipped with a nitrogen carrier gas, a packed column (2.5 meters long) containing a stationary phase of PEGA - polyethylene glycol adipate on a GAZ-ChROM-Q carrier, and a flame ionization detector.

2.3. Engine tests

Samples of soybean oil which remained after laboratory samples had been taken from each of three batches, differentiated by type of initial preparation (frying potato chips, frying breadcrumbs coated fish fingers and heating without fried product), were separately subjected to esterification with methanol. Fatty acids methyl esters were obtained by method analogous to the one used in investigation of fatty acids composition by means of gas chromatography. Fuel obtained this way was used in engine tests including main engine work parameters. Four mixtures were prepared, each containing 90% diesel fuel and 10% addition of fatty acids methyl esters obtained in research and marked as:

- $\mathbf{M_1}$- esters obtained from purchased fresh soybean oil,
- $\mathbf{m_2}$- esters obtained from soybean oil subjected to five-time cyclic heating, without addition of fried product,
- $\mathbf{m_3}$- esters obtained from soybean oil, previously used for five-time cyclic frying of potato chips,
- $\mathbf{M_4}$- esters obtained from soybean oil, previously used for five-time cyclic frying of breadcrumbs coated fish fingers.

Results of internal combustion engine running on diesel fuel (DF) were used as reference for determination of work parameters of engine powered with fuel blends. Above mentioned fuel mixtures, were used for powering 2CA90 diesel engine installed on dynamometric stand for purpose of conducting measurements of its energetic work parameters. Test bed comprised of following devices:

- Internal combustion diesel engine 2CA90;
- Dynamometric stand composed of eddy-current brake AMX210 and control-measurement system AMX201, AMX 211;
- Fuel consumption measuring system;
- System measuring engine parameters: exhaust gasses temperature - tsp, engine oil temperature - tol, oil pressure -pol;
- System measuring state of environment: temperature of environment - tot, atmospheric pressure - pa, and air humidity - φ.

Measurements for each of investigated fuels were conducted and obtained results of energetic parameters were elaborated. Data yielded by measurements was used to draw external characteristics of the engine for rotational speed ranging from minimal to nominal. Carried out research included kinematic and dynamic parameters of the engine: torque - Mo, rotational speed - n, time in which set amount of investigated fuel was used - τ. Amount of fuel used for purpose of this characteristic was 50 g. Methodology of measurements and methods of measurements and results reduction of power and torque, were in conformity with norms: PN-88/S-02005, BN-79/1374-03.

3. RESULT OF INVESTIGATION

Raw, purchased plant oils characterised with typical properties, fulfilling requirements of recommended in Poland norm [PN – A - 86908] with regard to peroxide number (PN) and acid number (AN) (fig. 2 and 3).

Heating edible oils in conditions corresponding to frying potato chips, breadcrumbs coated fish fingers and heating without a product lead to significant changes of investigated oils properties. It caused mainly distinct changes of acid number (AN) and peroxide number (PN).

Differences in properties of oils subjected to cyclic heating without fried product and in which potato chips or breadcrumbs coated fish fingers were fried may result from course of temperature changes for various investigated batches (Fig. 1).

The highest temperature for each of investigated oils and in each of five heating cycles was observed in case of samples heated without fried products, in which temperature remained at 180º C. Changes of temperature of oil heated in the process of frying potato chips or breadcrumbs coated fish fingers had dynamic course, reaching the lowest value in approximately beginning of fifth minute. However, value of this minimum was depended on weight of fried product but main factor was fried product to frying medium weight ratio (fig. 1).

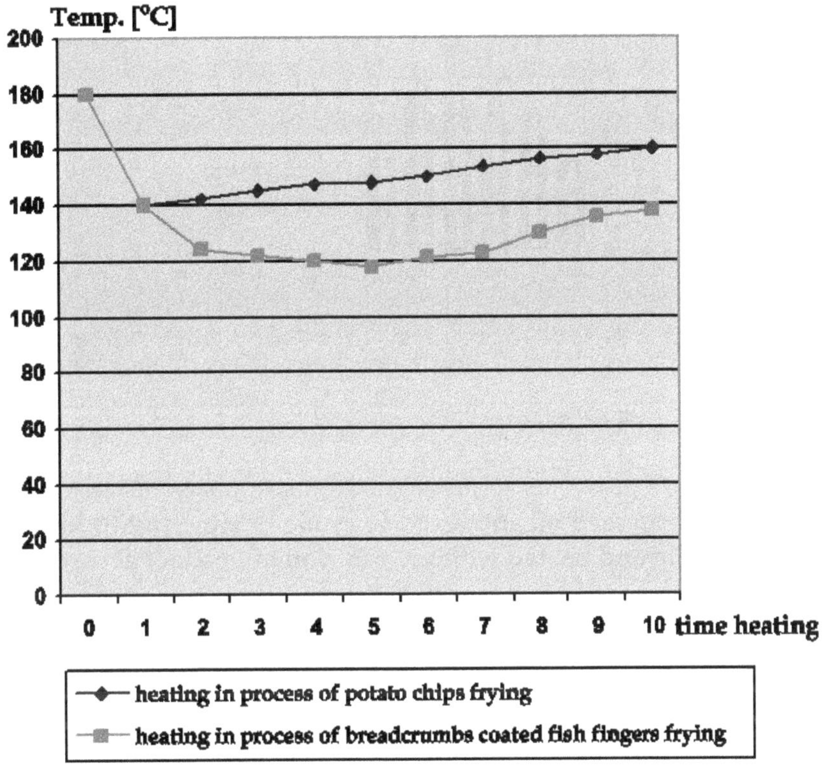

Figure 1.Course of rapeseed oil temperature changes in relation to time of potato chips and breadcrumbs coated fish fingers frying (presented data based on authors own research [Szmigielski et al. 2009])

Conducted research show that heating plant oils caused noticeable increase of peroxide number (PN) value, when compared to samples not subjected to thermal processing. It must be noted that diverse course and intensity of these changes were observed in case of samples heated without product, samples heated in process of potato chips and breadcrumbs coated fish fingers frying (Fig. 2).

Typical course of peroxide number changes in relation to number of frying cycle was presented in Fig. 2. In case of each of five heating cycles, highest value of peroxide number in rapeseed and soybean oils was observed in samples heated without a product [Szmigielski et al. 2008]. It was characteristic, that in these samples peroxide number value increased fast until third or fourth cycle, after which decrease of its value was noted (Fig. 2). Most probable cause of such course of peroxide number changes, in relation to heating cycles, is formation of oxidation products, which partially evaporate from the environment of reaction in form of volatile products. An exception to the rule were analyses conducted for samples of soybean oil (firs and second cycle of heating), in which temporarily highest value was observed in samples heated in process of breadcrumbs coated fish fingers frying [Szmigielski et al. 2011]. Most probably it results from influence of fat present in fried product on a final result of determination.

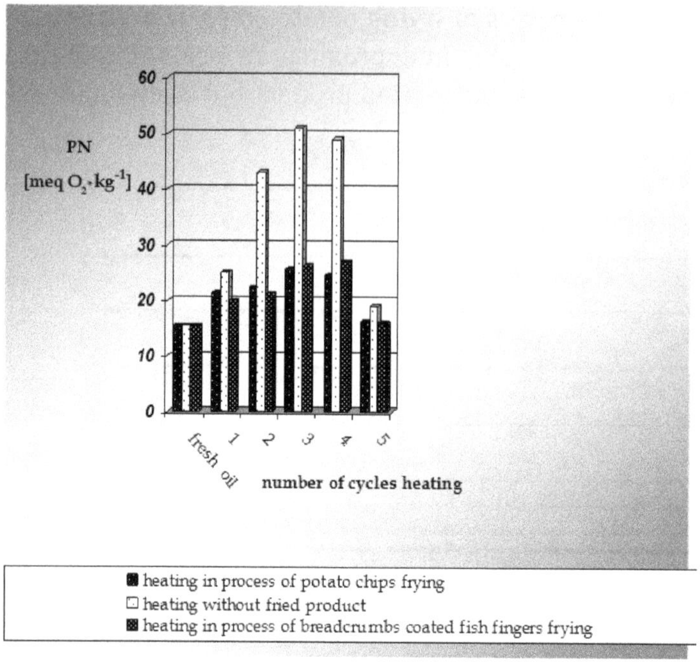

Figure 2.Peroxide number of rapeseed oil subjected to cyclic heating [mMO kg-1]/ data for oil heated in process of potato chips frying and heated without addition of product according to Szmigielski et al. 2008/

Samples of oil heated in the process of potato chips frying, characterised with lower values of peroxide number, for each of five heating cycles, when compared to samples heated without the fried product. Stabilizing effect of potato chips, caused by sorption of oxidation products on their surface or partial absorption of frying fat, is most commonly mentioned probable cause of such course of PN changes in these samples [Maniak et al. 2009, Szmigielski et al. 2008, 2009, 2011]. It should be noted (Fig. 1) that PN level in samples of rapeseed and soybean oil heated in process of potato chips frying [Szmigielski et al. 2008], had similar course, stabilizing respectively at approx. 2 mMo/100g and approx. 1,5 mMo/100g [Szmigielski et al. 2011] (with exemption of null samples and first cycle of soybean oil heating). Results of investigation of soybean and rapeseed oil samples heated without fried product differed significantly - reaching almost two times higher value of peroxide number (PN) than respective samples heated in process of potato chips frying. As opposed to this research, heating sunflower oil in process of potato chips frying caused only slight decrease of its peroxide number (PN) when compared to samples heated without fried product [Maniak et al. 2009].

Typical course of acid number (AN) changes of heated oil samples in relation to number of frying cycles was presented in Fig. 3. Acid number of heated oil samples was higher than in raw oil, however, heating in process of potato chips frying caused stabilization of acid number value (AN) at similar level (0,02 mgKOH/g) regardless of number of oil heating cycles, while heating without the product caused systematic increase of AN. Very similar course of acid number changes of investigated post-frying oils was also observed in analogous research on rapeseed oil [Szmigielski et al. 2008] and sunflower oil samples [Maniak et al. 2009]. It is believed, that the most probable cause of observed changes of acid number of these samples is sorption of oxidation products on surface of, subjected to culinary processing, potato chips or partial absorption of oil surrounding the product into its deeper, more distant from surface of investigated raw product layers.

Acid number (AN) of plant oils (rapeseed and soybean) heated in the process of frying breadcrumbs coated fish fingers was increasing systematically. It should be noted that AN for first two cycles of heating remained at level similar or lower than AN determined in respective samples heated in the process of potato chips frying. However, starting from the third heating cycle AN exceeded this value and was systematically increasing with each of heating cycles, reaching values lower than in respective samples of soybean oil heated without fried product (fig. 3). It is believed that two opposing processes were the most probable cause of above described course of changes of acid number (AN) in samples of oils heated in process of breadcrumbs coated fish fingers frying. Increase of AN value should probably be explained with oxidation of fatty acids and hydrolytic effect of water vapour, released from product as a result of frying, while reduction of its level occurred as an effect of sorption of oxidation products on surface of fried product [Szmigielski et al. 2009; 2011]. Five-time cyclic heating of plant oils caused significant changes in composition of fatty acids, which can be simply characterised as significant decrease of fatty acids content. It concerns mainly unsaturated fatty acids, and significant increase of oxidation products content, what can be easily observed on example of soybean oil (fig. 4-6). Similar course of fatty acids composition changes of investigated post-frying oils was also observed in research of, subjected to cyclic heating, samples of rapeseed oil [Szmigielski et al. 2008;2009] and sunflower oil [Maniak et al. 2009]. Five-time cyclic frying of breadcrumbs coated fish fingers or potato chips caused partial stabilization of fatty acids composition, what can be noted in case of two, dominating in soybean, fatty acids i.e. oleic and linolic. Their content in typical raw soybean oil often exceeds 75% (fig. 3-5), [Staat and Vallet 1994, Tys et al. 2003]. Heating this oil only slightly changed proportion of oleic to linolic acid, for in raw oil, on one particle of oleic acid approx. two particles of linolic acid are found. After process of heating, this rate is approx. 1,5 - from 1,4 for sample heated without fried product to 1,50 for sample heated in the process of frying potato chips, and up to 1,63 when sample of oil heated in the process of frying breadcrumbs coated fish fingers is investigated.

Similar effect, when ratio of unsaturated fatty acids (oleic and linolic) is taken into consideration, was also observed in research of sunflower oil used as frying fat in cyclic frying of potato chips.

Fresh sunflower oil usually contains over 80% of these fatty acids, while, statistically on one particle of oleic acid 2,47 particles of linolic acids are found. After five-time cyclic heating in process of potato chips frying this proportion remains unchanged, while it changes only in case of oil heated without fried product [Maniak et al. 2009].

In fresh rapeseed oil, proportion of linolic acid to oleic acid is 1 : 2,72. Five-time cyclic heating in process of potato chips frying caused significant change of this proportion to 1 : 2,37, while, for example, effect of disturbance of this fatty acids ratio occurring during similar cycle of heating without fried product reached 1:3,77 [Szmigielski et al. 2008]. The same processes of heating caused also slight changes of saturated fatty acids ratio. In fresh soybean oil, on one particle of stearic acid 2,66 particles of palmitic acid are found, while after five cycles of heating this ratio was from 1 : 2,1 in oil heated without product (Fig. 4), 1 : 2,31 in oil subjected to heating in process of potato chips frying (Fig. 6) to 1 : 2,38 in oil subjected to heating in process of breadcrumbs coated fish fingers frying (Fig. 5).

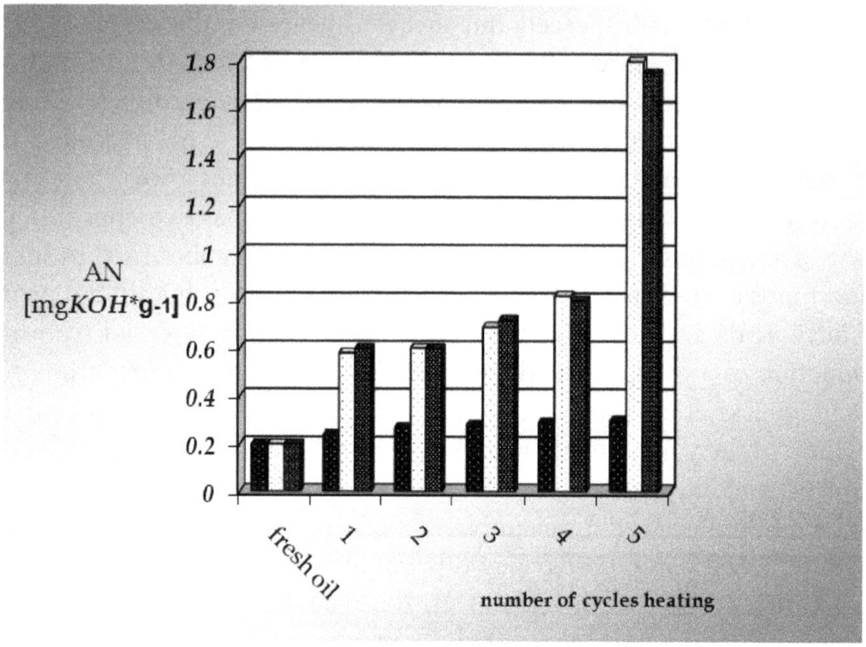

Figure 3.Acid number of rapeseed oil subjected to cyclic heating [mgKOH g-1]/ data for oil heated in process of potato chips frying and heated without addition of product according to Szmigielski et al. 2008/

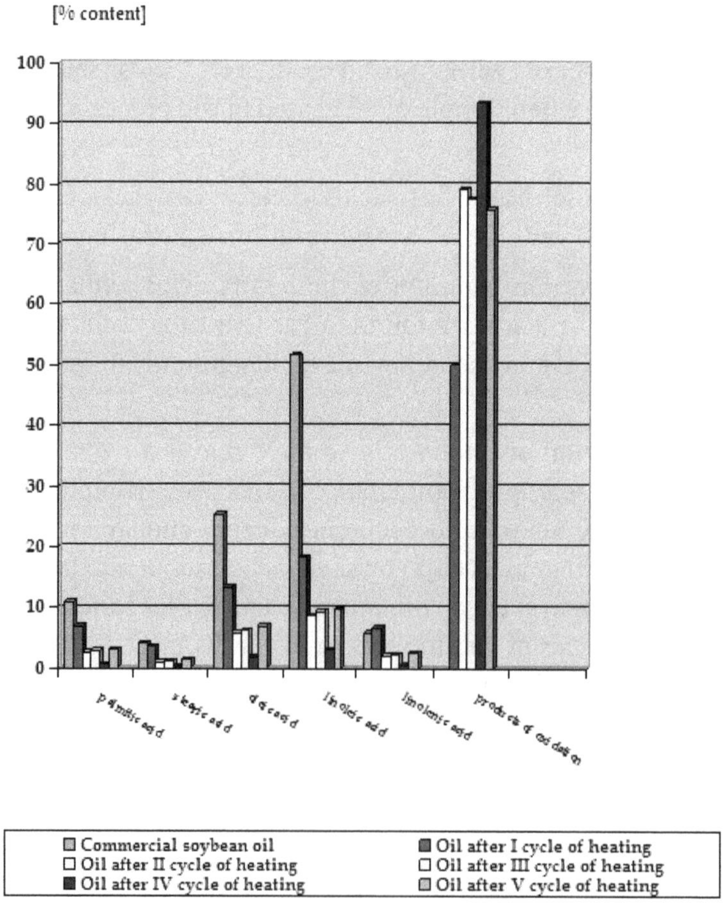

Figure 4.The composition of fatty acids of soybean oil treated five-time cyclic heating, heating without fried product /presented data based on authors own research [Szmigielski et al. 2011]/

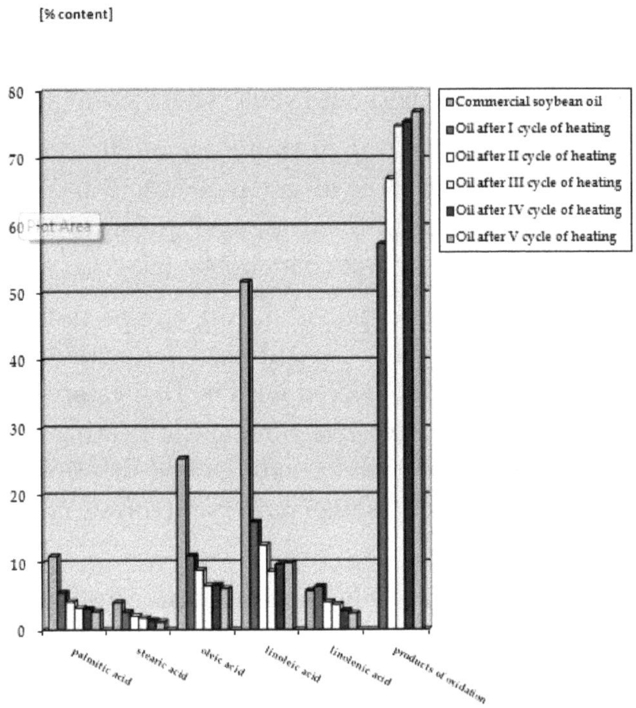

Figure 5.The composition of fatty acids of soybean oil treated five-time cyclic heating, heating in process of breadcrumbs coated fish fingers frying /presented data based on authors own research [Szmigielski et al. 2011]/

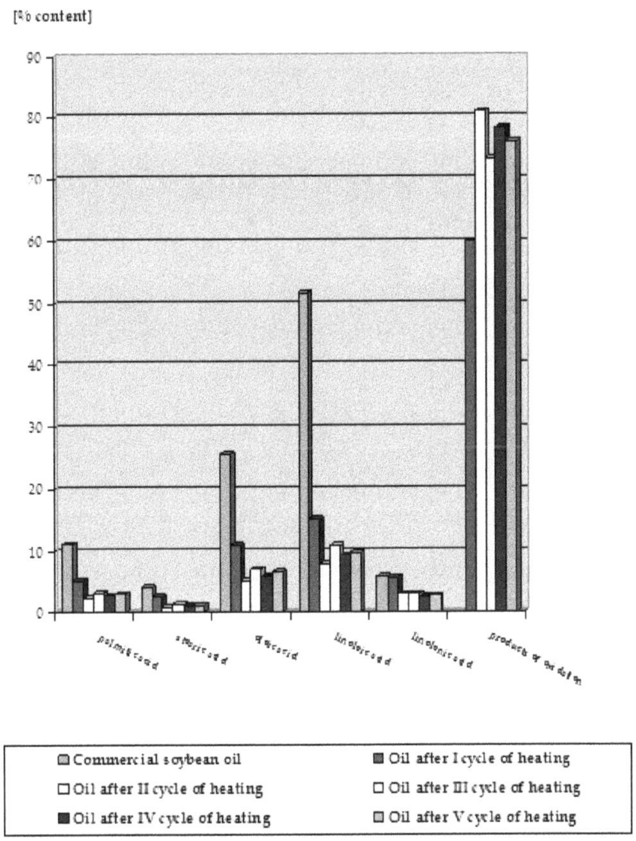

Figure 6.The composition of fatty acids of soybean oil treated five-time cyclic heating heating in process of potato chips frying /presented data based on authors own research [Szmigielski et al. 2011]/

Similar, slight fluctuations of stearic and palmitic acids ratio were noted after five-time cyclic heating of rapeseed oil, and ranged from 1 : 2,97 in fresh oil, to 1 : 3,03 after heating in process of cyclic potato chips frying and 1 : 2,94 after cyclic heating without fried product [Szmigielski et al. 2008].

It should be noted that similar cyclic heating of sunflower oil did not cause change of ratio of two main saturated fatty acids present in investigated oil e.g. stearic acid and palmitic acid. The ratio was 1 : 1,69. Both in fresh sunflower oil and in oil after five-time cyclic heating without fried product or heated in process of potato chips frying, this ratio did not change [Maniak et al. 2009].

Presented in graphs 7-10 research data, obtained during engine tests, in which mixtures of diesel fuel containing 10% of fatty acids methyl esters were utilised, indicate on similar character of changes of investigated parameters of 2CA90 engine powered with methyl esters obtained from purchased raw soybean oil and post-frying oils obtained after five-time cyclic heating without fried product as well as five-time cyclic frying of potato chips or breadcrumbs coated fish fingers. Mixtures containing 10% addition of esters have similar influence on changes of power and torque of investigated engine in relation to its rotational speed (Fig. 7 and 8).

The curves depicting the specific and hourly fuel consumption for the fuel mixtures under investigation, which include a 10% addition of fatty acids methyl esters, exhibit greater values of energetic parameters in comparison to diesel fuel. This holds true for all five rotational speeds that were examined. It is important to observe that they possess similar characteristics and follow a very similar path, indicating that the distinctions between them are insignificant. Szmigielski et al. [2009] obtained similar research findings when they tested rapeseed oil samples under comparable settings.

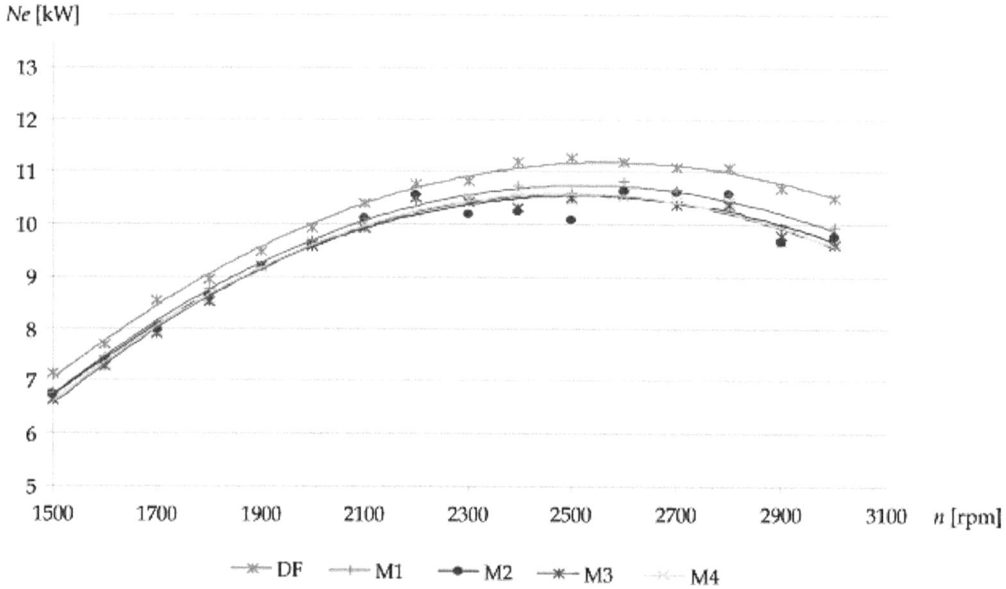

Figure 7. Changes of the course of engine power 2CA90 powered by diesel fuel (ON) and mixtures containing 90% diesel fuel and 10% methyl esters of fatty acids and diesel fuel: M1 - esters obtained from purchased fresh soybean oil, m2 - esters obtained from soybean oil without addition of fried product, m3 - esters obtained from soybean oil frying of potato chips, M4 - esters obtained from soybean oil frying of breadcrumbs coated fish fingers. /presented data based on authors own research /[Szmigielski et al. 2011]/

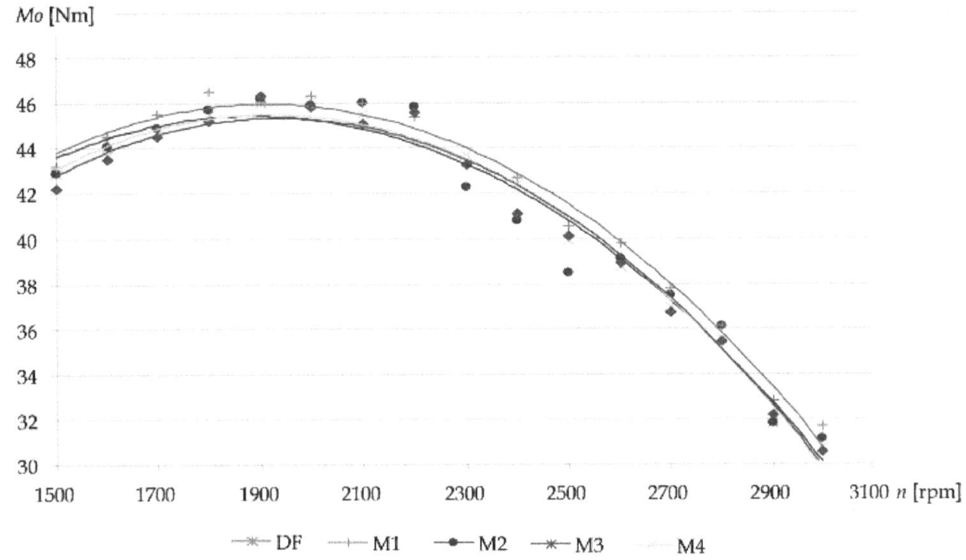

Figure 8.Changes of the course of engine torque 2CA90 powered by diesel fuel (ON) and mixtures containing 90% diesel fuel and 10% methyl esters of fatty acids and diesel fuel: M1 - esters obtained from purchased fresh soybean oil, m2 - esters obtained from soybean oil without addition of fried product, m3 - esters obtained from soybean oil frying of potato chips, M4 - esters obtained from soybean oil frying of breadcrumbs coated fish fingers. /presented data based on authors own research /[Szmigielski et al. 2011]/

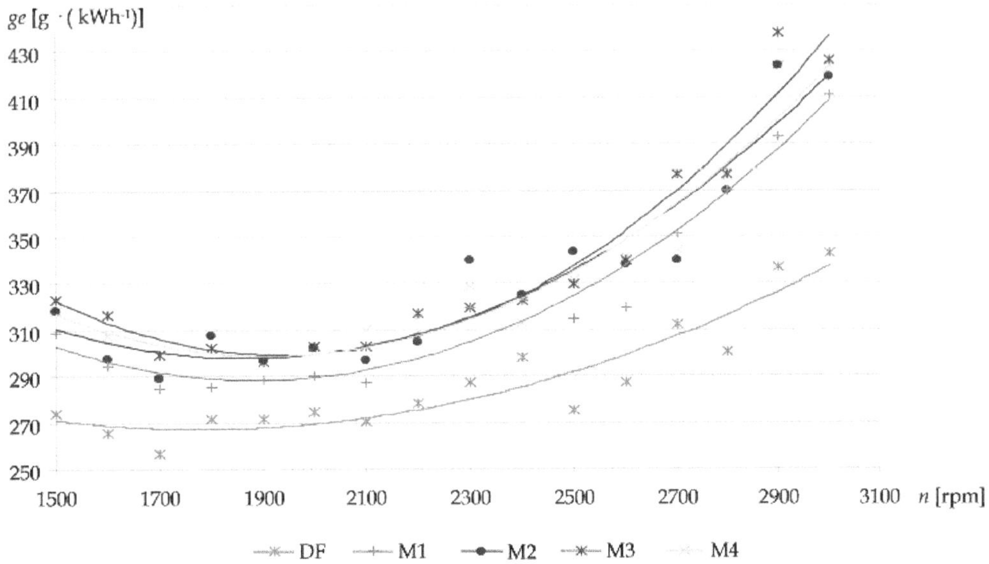

Figure 9.Changes of the course of unitary fuel consumption engine 2CA90 powered by diesel fuel (ON) and mixtures containing 90% diesel fuel and 10% methyl esters of fatty acids and diesel fuel: M1 - esters obtained from purchased fresh soybean oil, m2 - esters obtained from soybean oil without addition of fried product, m3 - esters obtained from soybean oil frying of potato chips, M4 - esters obtained from soybean oil frying of breadcrumbs coated fish fingers. /presented data based on authors own research /[Szmigielski et al. 2011]/

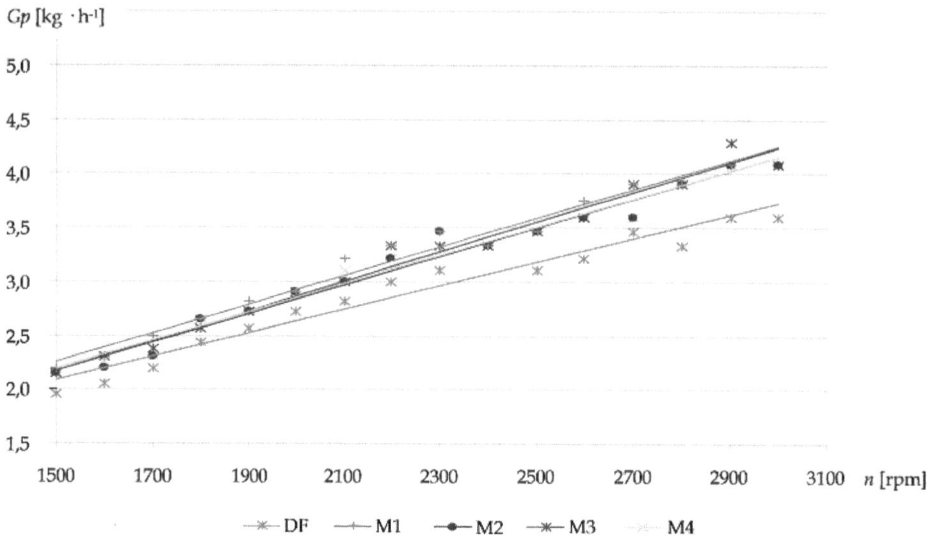

C

Figure 10.Changes of the course of hourly fuel consumption engine 2CA90 powered by diesel fuel (ON) and mixtures containing 90% diesel fuel and 10% methyl esters of fatty acids and diesel fuel: M1 - esters obtained from purchased fresh soybean oil, m2 - esters obtained from soybean oil without addition of fried product, m3 - esters obtained from soybean oil frying of potato chips, M4 - esters obtained from soybean oil frying of breadcrumbs coated fish fingers. /presented data based on authors own research /[Szmigielski et al. 2011]/

4. CONCISE SUMMARY

- The process of cyclic heating of plant oils, particularly during the first three cycles, leads to substantial alterations in the content of their fatty acids. Notable fluctuations in the peroxide number (PN) and acid number (AN) of the oils under investigation were observed. The level of unsaturated fatty acids falls, whereas there is a rise in oxidation products.
- Heating plant oils during the frying process of products such as breadcrumbs coated fish fingers or potato chips affects the stabilization of peroxide products in the oil. This leads to a decrease in the peroxide number (PN) of the oil compared to heating without fried products.
- The acid number (AN) of post-frying oils remained stable when frying potato chips, but increased gradually when frying breadcrumbs covered fish fingers and heating oil without any fried product.
- Frying fish fingers and potato chips with breadcrumbs helps maintain a stable proportion of fatty acids in the oils used for frying. This proportion is similar to the one found in commercially available raw oils.
- The alteration of post-frying plant oils during the chemical conversion to fatty acids methyl esters leads to the standardization of biofuel properties made from different batches of post-frying oils. This process also promotes the use of post-frying oils that have been found to be suitable for biofuel production, similar to fresh vegetable oils.
- Unidentified oxidation products undergo comparable changes during the synthesis of fatty acids methyl esters and do not pose a significant hindrance to the proper functioning of diesel engines fueled with this biofuel.
- The research results have indicated that post-frying plant oils may be effectively used as a raw material for producing biocomponents for diesel fuel.

- Currently, it is imperative to develop effective methods for recovering post-frying fats from tiny gastronomy establishments, as well as the technology for purifying and utilizing them as fuel components for diesel engines.

REFERENCES

Alcantara R., Amores J., Canoira L., Fidalgo E., Franco M.J., & Navarro A. (2000). Catalytic production of biodiesel from soy-bean oil, used frying oil and tallow. Biomass and Bioenergy 18, 515-527, ISSN.0961-9534

Blumenthal, M. M. (1991). A new look at the chemistry and physics of deep-fat frying. Food Technology, 45, 68-94, ISSN 0015-6639.

Buczek B.& Chwiałkowski W. (2005). Wpływ modyfikacji węgla aktywnego na jego zdolność do oczyszczania zużytego oleju smażalniczego. Żywność, Nauka, Technologia, Jakość. 4(45) Supl. 85-99, ISSN 1425-6959.

Choe E. & Min D. B. (2006). Mechanisms and factors for edible oil oxidation. Comprehensive Reviews in Food Science and Food Safety, 5, 169-186, ISSN 1541-4337.

Clark W L & Serbia G. (1991). Safety aspects of frying fats and oils. Food Technology, 45, 84- 94, ISSN 0015-6639.

Dzieniszewski G. (2007). Wybrane problemy zasilania silnika diesla zużytym olejem roślinnym. Inżynieria Rolnicza 9, 49-55, ISSN 1429-7264.

Drozdowski B. (2007). Reakcje chemiczne kwasów tłuszczowych i acylogliceroli. In: Chemia żywności T. 2. 112-140. Praca zbiorowa pod red. Z.E. Sikorskiego WNT ISBN 978-83- 204-375-6 Warszawa.

Hoffman M. (2004) Niebezpieczne przemiany. Przegląd Gastronomiczny. 10/2004, 14-16, ISSN 0033-2119

Krełowska-Kułas M. (1993) Metody ilościowego oznaczania i badania tłuszczów In: Badanie jakości produktów spożywczych 63-90. PWE, ISBN 83-208-0902-9 Warszawa..

Ledóchowska E. & Hazuka Z. (2006) Przemiany termooksydatywne wybranych olejów oliwkowych i oleju rzepakowego zachodzące w czasie ogrzewania i smażenia. Tłuszcze Jadalne 41, 193-196, ISSN 0371-9227.

Leung D.Y.C. & Guo Y. (2006). Transesterification of neat and used frying oil: Optimization for biodiesel production. Fuel Processing Technology 87, 883-890, ISSN 0378-3820.

Maniak B., Szmigielski M., Piekarski W. & Markowska A. (2009). Physicochemical changes of post frying sunflower oil. International Agrophysics 23, 243-248, ISSN 0236-8722.

Paul S. & Mittal G. S. (1996). Dynamics of fat/oil degradation during frying based on optical properties. Journal of Food Engineering, 30, 389-403, ISSN 0260-8774.

Polish Norm PN-A- 86908, Oleje i tłuszcze roślinne oraz zwierzęce – Rafinowane oleje roślinne, PKN 2000.

Polish Norm PN-ISO 660, Oleje i tłuszcze roślinne oraz zwierzęce – Oznaczanie liczby kwasowej i kwasowości, PKN 1998.

Polish Norm PN-ISO 3960, Oleje i tłuszcze roślinne oraz zwierzęce – Oznaczanie liczby nadtlenkowej, PKN 1996.

Polish Norm PN-S-02005, Silniki samochodowe -- Badania stanowiskowe –Wyznaczanie podstawowych parametrów pracy, PKN 1988.

Polish Norm BN-1374-03, Silniki samochodowe -- Badania stanowiskowe – Wykonywanie charakterystyk, PKN 1979.

Staat F. & Vallet E. (1994). Vegetable oil methyl ester as a diesel substite. Chemistry and Industry, 7, 863-865, ISSN 0009-3068.

Szmigielski M., Maniak B. & Piekarski W. (2008). Evaluation of chosen quality parameters of used frying rape oil as fuel biocomponent. International Agrophysics 22, 361-364, ISSN 0236-8722

Szmigielski M., Maniak B., Piekarski W. & Zając G. (2009): Wybrane cechy posmażalniczego oleju rzepakowego jako surowca do produkcji komponentów paliw silnikowych w aspekcie zróżnicowanego asortymentu dań smażonych. In: Biopaliwa i smary ekologiczne. 41-51 pod red. naukową W. Piekarskiego, ISBN 83-89503-75-1, Warszawa.

Szmigielski M., Piekarski W., Maniak B., Zając G., Masłowski A.& Biernacka B. (2011) Influence of fried meals assortment on selected postfrying soya-bean oil features as components to produce diesel biocomponents International Agrophysics (in review), ISSN 0236-8722.

Tys J, Piekarski W, Jackowska I, Kaczor A, Zając G & Starobrat P. (2003). Technologiczne I ekonomiczne uwarunkowania produkcji biopaliw z rzepaku. Acta Agrophysica, Rozprawy i Monografie 99, 1-162, ISSN 1234-4125.

CHAPTER-28

ADVANCEMENTS IN MINIATURIZED ENZYMATIC BIOFUEL CELLS

1. INTRODUCTION: AN OVERVIEW

The annual global energy demands have experienced a substantial growth, rendering the current dependence on fossil fuels unsustainable, given their availability from the environment is limited. Furthermore, the utilization of fossil fuels leads to the emission of pollutants and contributes to the phenomenon of global warming. Fuel cells provide an alternative resolution to this problem. A fuel cell is an electrochemical device that transforms the chemical energy stored in a fuel into electrical energy. Within a fuel cell, an oxidation reaction takes place at the anode, producing electrons that then move to the cathode via the external circuit. Simultaneously, a reduction reaction occurs at the cathode. Conventional fuel cells can generate energy by utilizing hydrogen or methanol (MeOH) as fuel sources, resulting in the production of water and carbon dioxide as byproducts. Nevertheless, the gaseous nature of hydrogen poses safety concerns during storage and transportation. In addition, a significant number of the alternative fuels suitable for fuel cells still depend on petroleum-based resources. Hence, there is a widely acknowledged need for alternative sources of sustainable energy. Many attempts have been made to provide alternative power sources that can operate under normal bodily settings for extended periods without needing to be recharged. In recent times, there has been a growing interest in miniaturized medical implants such as pacemakers, defibrillators, insulin pumps, sensor-transmitter systems for animals and plants, nano-robots for drug delivery, and health monitoring systems. This has resulted in an increase in research and development in micropower sources, particularly biofuel cells (Ramanavicius, 2005; Liu & Dong, 2007; Zhu et al., 2007; Moehlenbrock& Minteer, 2008 and Wang et al., 2009). A biofuel cell is a specific type of fuel cell that converts biological energy into electrical energy through the use of biocatalysts (Palmore & Whitesides, 1994). There are two primary categories of biofuel cells: microbial fuel cells and enzymatic biofuel cells. Microbial fuel cells utilize microorganisms as catalysts to transform chemical energy into electricity, whereas enzymatic biofuel cells employ enzymes to catalyze the redox reaction of the fuels. This chapter will begin by providing an introduction to both types of biofuel cells, including the catalysts utilized, the method of electron transfer, the materials used for electrodes, and the performance of the cells. Next, we will provide a concise overview of the latest advancements in tiny biofuel cells, which present potential applications for implanted devices inside the human body. The chapter also emphasizes the use of miniature enzymatic biofuel cells that are based on carbon-microelectromechanical systems (C-MEMS).

1.1. History

The earliest discovery between biology and electrical energy was demonstrated by Galvani in 1791 showing the frog leg twitching from an electric current (Galvani, 1791). The first fuel cell, which involved electrolysis of water, was discovered by Grove in 1839. An electrical stimulation can induce a biological reaction and vice-versa a biological process can also generate electricity. The first half-cell using microorganism (E.coli) was demonstrated by Potter at University of Durham (Potter, 1910). Further development of half-cell by Cohen from University of Cambridge led to one of the major types of biofuel cells, i.e., microbial fuel cells. Cohen applied a number of microbial half-cells connected in series, which generated over 35 volts (Cohen, 1931). In the late 1950s and early 1960s, the interest in development of biofuel cells received a boost by the USA space program, which led to the application of microbial fuel cells as an advanced technology for waste disposal treatment in space flights. Also, in the late 1960s, a biofuel cell using cell-free enzyme systems was discovered aiming to permanently power medical implants by utilizing specific body fluids as fuel (Yahiro et al., 1964). The concept of using microorganism as a biocatalyst in microbial fuel cells was widely applied since the 1970s (Suzuki, 1976 and Roller et al., 1984) and in the 1980s it was found out that power output could be greatly improved by using electron mediators (Vega & Fernandez, 1987; Habermann & Pommer, 1991 and Allen & Bennetto, 1993). However, the toxicity and instability of mediators limited the cell performance. A breakthrough was made when some microorganisms were found to transfer electrons directly to the electrode which led to the mediator-less microbial fuel cells first used in wastewater treatment and electricity generation (Kim et al., 1999; Chaudhuri & Lovley, 2003). These microorganisms are stable and yield a high Coulombic efficiency which facilitates the direct electron transfer (Scholz & Schroder, 2003). Shewanella putrefaciens (Kim et al., 2002), Geobacteraceaesulferreducens (Bond & Lovley, 2003), Geobactermetallireducens (Min et al., 2005) and Rhodoferaxferrireducens (Chaudhuri & Lovley, 2003) are all bioelectrochemically active microbes and can transfer electrons directly through the membrane. On the other hand, since the first enzymatic biofuel cell was reported in 1964, noticeable developments have been made in the terms of the power density, cell lifetime, operational stability (Bockris& Srinivasan, 1969; Govil & Saran, 1982 and Palmore & Whitesides, 1994). However, the output potential generated from enzymatic biofuel cells was still far beyond the demand of commercial application. Therefore, instead of considering enzymatic biofuel cells as a conventional power source, most of the researches on enzymatic biofuel cells have been aimed toward special applications such as implantable medical devices (Katz & Willner, 2003; Barton et al., 2004 and Heller, 2004). In the past ten years, cell performances on both types of biofuel cells have been improved significantly and we will discuss the detailed development in the following sessions.

1.2. Microbial fuel cells

A microbial fuel cell (MFC) converts chemical energy to electrical energy by the catalytic processes of microorganisms. Microorganisms in the MFC oxidize organic substrates and generate both electrons and protons on the anode. Electrons transfer from the anode to the cathode through an external circuit and simultaneously the protons migrate to the cathode and reduce the oxygen with the electrons available at the cathode surface. Various kinds of microorganisms are reported in association with electrodes in MFC systems. For example, brevibacillus sp. PTH1 is one of the most abundant microorganisms in a MFC system. Pure cultures used for generating current in a MFC include firmicutes, acidobacteria, proteobacteria and yeast strains Saccharomyces cerevisiae and hansenulaanomala (Allen & Bennetto, 1993). These microorganisms interact with fuels through a variety of direct and indirect processes to generate energy. Microbial biofuel cells have major advantage of thorough oxidation of the fuels due to the use of microorganism as catalyst system and they can be typically operated for long lifetimes.

Besides, a MFC has no intermediated processes thus it is a very efficient energy producing process. In addition, as a fuel cell, a MFC does not need charging during operation (Willner et al., 1996; Katz et al., 2003 and Calabrese et al., 2004). However, the bottlenecks of MFC still remain. Power generation of a MFC is affected by many factors including microbe type, fuel biomass type and concentration, ionic strength, pH, temperature, and reactor configuration.

The principle cell performance of MFCs lies in the electron transfer from microbial cells to the anode electrode. The direct electron transfer from the microorganism to electrodes is hindered by overpotential due to transfer resistance. The overpotential lowers the potential of a MFC and significantly affects the cell efficiency. In this case, the practical output potential is less than ideal because the electron transfer efficiency from the substrate to the anode varies from microbe to microbe. Microorganism species do not readily release electrons and hence the redox mediators are needed. A desirable mediator should have a whole range of properties: Firstly, its potential should be different from the microorganism potential to facilitate electron transfer. Secondly, it should have a high diffusion coefficient in the solution. Lastly, it is suitable for repeatable redox cycles in order to remain active in the electrolyte. Widely used Dye mediators such as neutral red (NR), methylene blue (MB), thionine (Th), meldola's blue (MelB) and 2-hydroxy-1,4-naphthoquinone (HNQ) can facilitate electron transfer for microorganism such as Proteus, Enterobacter, Bacillus, Pseudomonas and Escherichia coli. In the electron transfer process, these mediators are reduced by interacting with electron generated within the cell then these mediators in reduced form diffuse out of the cell to the anode surface where they are electrocatalytically oxidised. The oxidised mediator is then capable to repeat this redox cycle.

Better performing electrodes can improve the cell performance of a MFC because different anode materials can result in different activation of a polarization loss, which is attributed to an activation energy that must be overcome by the reactants. Carbon or graphite based materials are widely used as electrodes due to their large surface area, high conductivity, biocompatibility and chemical stability according to Table 1. Also, platinum and gold are popular as electrode system although they are expensive. Compared with carbon based electrode materials, platinum and gold electrodes are superior in the performance of the cells based on the Table 1. Besides, they have a higher catalytic kinetics towards oxygen compared to carbon based materials and hence the MFCs with Pt based cathodes yielded higher power densities than those with carbon based cathodes (Moon et al., 2009).

Electrode modification is another way to improve MFC performance of cells. (Park &Zerkeis, 2003) reported an increase of 100-folds in current compared to the previous results by using (neutral red) NR-woven graphite and Mn^{4+}-graphite anode instead of the woven graphite anode alone. Four times higher current was reported in 2004 using the combination of Mn^{4+}-graphite anode and Fe^{3+}-graphite cathode (Niessen et al., 2004). NR and Mn^{4+} doping ions serve as mediators in their MFC systems and also catalyze the cathodic reactions to facilitate electricity generations. Electrodes modifications including adsorption of AQDS or 1,4-naphthoquinone (NQ) and incorporation with Mn^{2+}, Ni^{2+}, Fe_3O_4 increased the cell performance of MFCs in their long-term operations (Lowy et al., 2006). In addition, the fluorinated polyanilines, poly (2-fluoroaniline) and poly (2, 3, 5, 6-tetrafluoroaniline) outperformed polyaniline were applied for electrode modification (Niessen et al., 2006). These conductive polymers also serve as mediators due to their structural similarities to conventional redox mediators.

Table 1.Summary of microbial biofuel cells

Fuel	Organism	Electrode (cm_2)	Electron transfer	OCV (V)	Current density ($\mu A\ cm^{-2}$)	Reference
Sugar/ferricyanide	Suspended Proteus vulgaris/anaerobic	RVC anode (30.4), platinum cathode (16)	MET	0.52	5.26	Kim et al. (2000)
Glucose/ ferricyanide	Suspended E. coli/ anaerobic	Woven graphite	MET	0.85	5.3	Park &Zeikus (2000)
H2/O2	Desulphovibrio vulgaris,	Carbon felt mat (5.1)	MET	1.17	176	Tsujimura et al. (2001)
Lactate/O2	Suspended Shewanella putrefaciens and E. coli , anaerobic	Graphite felt (56)	DET	0.5	0.02	Kim et al. (2002)
Marine sediment con- stituents/ seawater constituents	Mixed natural bacteria	Drilled graphite discs	DET & MET	0.75	3.2	Tender et al. (2002)
Lactate/O2	Suspended Shewanella putrefaciens/ anaerobic	Mn4+ graphite plate (80) anode, Fe3+ modified graphite plate cathode (50)	MET	0.6	0.94	Park &Zeikus (2002)
Glucose/ ferricyanide	Rhodoferax ferrireducens/ anaerobic	Graphite rod (0.65)	DET	N/A	N/A	Chaudhuri & Lovley (2003)
Glucose /O_2	Suspended E. coli/ anaerobic	NR-woven graphite (80) or Mn4+graphite plate anode (80), woven graphite or Fe3+ graphite plate (80) cathode	MET	N/A	N/A	Park &Zeikus (2003)
Glucose /O2	Mixed culture	Graphite plate electrodes (50)	MET and DET	N/A	36	Rabaey et al. (2003)
Glucose/ ferricyanide	Suspended E. coli/ anaerobic	Woven graphite cloth	DET	0.895	120	Schroder et al. (2003)
Glucose/ ferricyanide	Clostridium butyricum	Woven graphite cloth	MET	0.759	200	Niessen et al. (2004)
Glucose/ ferricyanide	Mixed culture	Graphite plate electrodes (50)	MET & DET	N/A	231	Rabaey et al. (2004)
Glucose/ ferricyanide	Shewanella oneidensis DSP-10	Graphite Felt (610)	DET	N/A	110	Ringeisen et al. (2006)
Glucose /O2	Saccharomyces cevevisiae	Gold (0.51)	MET	N/A	15	Chiao et al. (2006)
Glucose /O2	Mixed bacterial culture	Carbon cloth (7)	MET	N/A	90	Fan et al. (2007)
Glucose/ ferricyanide	Geobacter Sulfurreducens	Pt (7.8)	DET	N/A	688	Richter et al. (2008)

Fuel	Organism	Electrode (cm_2)	Electron transfer	OCV (V)	Current density ($\mu A\ cm^{-2}$)	Reference
Glucose /O2	Shewanella oneidensis MR-1	Pt (1.2)	MET	N/A	302	Siu & Chiao (2008)
Glucose/ ferricyanide	Shewanella oneidensis MR-1	Pt (0.49)	MET	N/A	370	Hou et al. (2009)
Glucose/ ferricyanide	Shewanella oneidensis MR-1	Gold (0.15)	MET	N/A	130	Qian et al. (2009)

Proton exchange membrane (PEM) can also significantly affect a MFC system's internal resistance and concentration polarization loss because the internal resistance of MFC decreases with the increase in the PEM surface area (Oh & Logan, 2006). Nafion (DuPont, Wilmington, Delaware) is the most popular proton exchange membrane material due to its highly selective permeability of protons (Min et al., 2005). Compared with the performance of MFC using a PEM or a salt bridge, the power density using the salt bridge MFC was 2.2 mW/m2 that was an order of magnitude lower than that attained using Nafion. However, side effect is unavoidable with the use of PEM. For example, the concentration of cation species such as Na^+, K^+, NH^{4+}, Ca^{2+}, Mg^{2+} is much higher than that of proton so that transportation of cation species dominates. In this case, Nafion used in the MFCs is not an efficient proton specific membrane but actually a cation specific membrane (Rozendal et al., 2006). Subsequent studies have implied that anion-exchange or bipolar membranes has better properties than cation exchange membranes regarding to cell performance (Zhang et al., 2009).

Two promising applications of MFCs in the future are wastewater treatment and electricity generation (Feng et al., 2008 and Katuri& Scott, 2010). Although some noticeable development has been made in the MFC research, there are still a lot of challenges to be overcome for large-scale applications. The primary challenge is how to improve the cell performance in terms of power density and energy efficiency. In addition, catalytic effect of bioelectrodes needs to be further enhanced to solve the problems caused by enzyme activity loss and other degradation processes. Moreover, the lifetime of the MFC must be significantly improved.

1.3. Enzymatic biofuel cells

Enzymatic biofuel cells (EBFCs) employ redox enzymes, such as glucose oxidase (GOx) and laccase, as catalysts to assist the creation of electrons between substrates and the electrode surface. This process results in the production of an output potential. There exist two distinct methods for electron transfer: direct electron transfer (DET) and mediator electron transfer (MET). In DET-based EBFCs, the substrate undergoes enzymatic oxidation at the anode, resulting in the generation of protons and electrons. These electrons are then immediately transferred from the enzyme molecules to the surface of the anode. At the cathode, the oxygen has a reaction with electrons and protons, resulting in the production of water. Nevertheless, the interaction between an enzyme and an electrode has only been documented for a limited number of enzymes, including cytochrome c, laccase, hydrogenase, and several peroxidases (Schuhmann, 2002 and Freire et al., 2003). Certain enzymes possess a protein shell that is not conductive, resulting in poor electron transport. In order to surmount this obstacle, MET was employed to augment the conveyance of electrons. The selection and mechanism of microbial electrolysis cells (MECs) in electrochemically-biased fuel cells (EBFCs) closely resemble those of microbial fuel cells (MFCs) previously studied. Furthermore, the utilization of MET in EBFCs still faces certain obstacles, including inadequate dispersion of mediators and a lack of uninterrupted provision. Consequently, there has been significant interest in altering bioelectrodes to achieve direct electron transfer (DET) in enzymatic biofuel cells (EBFCs). Power density and longevity are critical criteria that dictate the performance of the cell in EBFCs applications. Notable advancements have been achieved in the past ten years (Katz et al., 2003; Calabrese et al., 2004; Zhu et al., 2007; Moehlenbrock& Minteer, 2008; Wang et al., 2009; Lee et al. 2010 and Saleh et al. 2011). Significantly, these gains have primarily been accomplished through the modification of electrodes with enhanced performance, the enhancement of enzyme immobilization methods, and the optimization of cell architecture. Table 2 displays the latest advancements in enzymatic biofuel cells. The effectiveness of electrodes for EBFCs mostly relies on electron transfer kinetics, mass transport, stability, and reproducibility. The electrode mostly consists of gold, platinum, or carbon (Katz & Willner, 2003). In addition to traditional materials, biocompatible conducting polymers are extensively utilized due to their ability to enhance electron transport and co-immobilize enzymes simultaneously (Schuhmann & Muenchen, 1992; Haccoun et al., 2006; Nagel et al., 2007). To optimize cell efficiency, researchers have utilized mesoporous materials in several experiments due to their large surface areas, which enables the attainment of high power density. In addition, numerous endeavors have been undertaken to employ nanostructures, including nanoparticles, nanofibers, and nanocomposites, as electrode materials in the production of electrodes for EBFCs. The utilization of these nanostructures with a substantial surface area allows for a significant enzyme loading, resulting in an enhanced power density of the cells. Carbon nanotubes have been used to modify electrodes in EBFCs, which is considered a major breakthrough. (Wang et al., 2009; Lee et al., 2010; Tanne et al., 2010; and Saleh et al., 2011.) Various research endeavors have focused on the utilization of a hybrid system consisting of single wall carbon nanotubes. The directed assembly of short single-walled carbon nanotubes perpendicular to electrode surfaces was achieved through the covalent bonding of the nanotubes to the electrode surface. According to reports, the use of single-walled carbon nanotubes (SWNT) allows for a strong electrical connection between the surface assembled glucose oxidase (GOx) and the electrode. The SWNT acts as conductive nanoneedles, effectively linking the enzyme's active site to the transducer surface. Additional research has been conducted to enhance the electrochemical and electrocatalytic properties, as well as the rapid electron transfer kinetics, of carbon nanotubes (CNTs). Enhanced enzyme activity was reported when comparing to analogous enzyme-containing composites that did not utilize single-walled carbon nanotubes (SWNTs). It was suggested that using SWNTs, which have a large specific surface area, could efficiently absorb enzyme molecules and keep them within the polymer matrix. In contrast, other types of enzyme-composites may experience enzyme loss when exposed to water-based solutions. While the current progress in electrode modification shows promise in enhancing cell function, it is crucial to conduct further research and address the issues of biocompatibility and nanotoxicity.

Table 2.Summary of enzymatic biofuel cells.

Fuel	Enzyme	Electrode	Electron transfer	OCV (V)	Current density (μA cm-2)	Reference
Glucose/ O_2	GOx/laccase	Carbon fiber electrodes	MET	0.8	64	Chen et al. (2001)
Glucose/ O_2	GOx/BOx	Carbon fiber electrodes	MET	0.84	432	Mano et al. (2002)
Glucose/ O_2	GDH/BOx	Glassy carbon disc electrodes	MET	0.44	58	Tsujimura et al. (2002)
Glucose/ O_2	GOx/COx	Gold electrodes coated with Cu	MET	0.12	4.3	Katz & Willner (2003)

Fuel	Enzyme	Electrode	Electron transfer	OCV (V)	Current density (μA cm-2)	Reference
Glucose/ O2	GOx/BOx	Carbon fiber electrodes	MET	0.68	50	Kim et al. (2003)
Glucose/ O_2	GOx/BOx	Carbon fiber electrodes	MET	0.8	440	Mano et al. (2003)
Glucose/ O_2	GOx/BOx	Carbon fiber electrodes	MET	0.63	244	Mano & Heller (2003)
Glucose/ O_2	GOx/laccase	Carbon fiber electrodes	MET	1.0	350	Heller (2004)
EtOH to CH_3 CHO to CH_3COOH	ADH, ADH +AldDH, formaldehyde dehydroge- nase + FDH	Carbon coated with poly(methylene)	MET	0.62	1160	Akers et al. (2005)
Glucose/ O_2	PLL-VK3 / PDMS	Pt	MET	0.55	130	Togo et al. (2007)
Ethanol/H_2O_2	QH-ADH/AOx	Pt	DET	0.24	30	Ramanavicius et al. (2008)
Glucose/ O_2	GDH/PDMS	Pt	DET	0.80	11000	Sakai et al. (2009)
Glucose/ O_2	GOx/laccase	Silicon/ SWNTs	DET	N/A	30	Wang et al. (2009)
Glucose/ O_2	GOx/laccase	Au/SWNTs	DET	0.46	960	Lee et al. (2010)
Glucose/ O_2	PQQ-GDH/BOD	Au/MWNTs substrates	DET	0.60	200	Tanne et al. (2010)
Glucose/O_2	GDH/NB	Glass carbon/ SWNTs	DET	0.35	100	Saleh et al. (2011)

Successful immobilization of the enzymes on the electrode surface is considered as another critical factor that affects cell performance. The immobilization of enzyme can be achieved physically or chemically. There are two major types of physical methods, physical absorption and entrapment. The first one is to absorb the enzymes onto conductive particles such as carbon black or graphite powders. For example, hydrogenase and laccase were immobilized by using physical absorption on carbon black particles to construct composite electrodes and the EBFCs could continuously work for 30 days. Another physical immobilization method is based on polymeric matrices entrapment, which usually shows more stabilized enzyme immobilization (Mano et al., 2002; Mano et al., 2003, Heller, 2004 & Soukharev et al., 2004). For example, Soukharev utilized redox polymers to fabricate enzymatic biofuel cells system. The electrodes were built by casting the enzyme-polymer mixed solution onto 7 μm diameter, 2 cm length carbon fibers. It showed that the glucose–oxygen biofuel cell was capable of generating a power density up to 0.35 mW/cm2 at 0.88 V (Soukharev et al., 2004). Compared with the physical immobilization which is unstable during the operation, the chemical immobilization methods with the efficient covalent bonding of enzymes and mediators are more reliable. Katz et al. reported a biofuel cell using co-immobilized enzyme-cofactor-mediator composites on metal electrodes to functionalize the electrode surface with a monolayer then integrate with enzymes via bioaffinity (Willner et al., 1998; Katz et al., 2001 and Katz et al., 2003). Another example is that a redox monolayer was covalently grafted with pyrroloquinoline quinone (PQQ) to Au-electrode. Then GOx-FAD electrode was assembled with PQQ as mediators (Willner et al., 1996). Other widely used materials to functionalize electrode surface have also been reported, such as nitrospiropyran (Blonder et al., 1998), rotaxane (Katz et al., 2004), C-60 (Patolsky et al., 1998) and Au nanoparticles (Xiao et al., 2003).

Rapid development on EBFCs has been achieved in the past decade with the arised demands for reliable power supplies for implantable medical device. It has shown particular advantages over conventional batteries because of the specific biocatalysts and the possibility of miniaturization. However, there are still challenges for further development of long term stability of the enzymatic bioelectrodes and efficient electron transfer between enzymes and electrode surfaces. Recent efforts have been given to protein engineering, reliable immobilization method and novel cell configuration.

2. MINIATURE BIOFUEL CELLS

Miniature power systems using biocatalysts have received increased attention associated with demand for micro-scale power supplies for implantable medical devices. Development of miniature biofuel cells offers a great opportunity to serve as long-term power sources in implantable device where frequent replacement of battery is not practical. The ability of biocatalyst in converting available indigenous fuels into electrical energy makes miniature biofuel cells attractive to enable long-term and self-sustained power system. The success of medical implants is akin with the effective miniaturization of power sources. This can be achieved by miniaturization of different functional components such as electrodes, power supply, and signal processing units. Some of the effective techniques for miniaturization involve fabricating microfluidic systems using photolithography, etching, polymer molding, and metal deposition (Kim et al., 2008). For example, Siu and Chiao (Siu & Chiao, 2008) applied photolithography and polymer molding to fabricate polydimethylsiloxane (PDMS) electrodes. It was also used by Hou et al. (Hou et al., 2009) to fabricate gold electrode arrays for the microbe screening. Besides polymer molding, etching can also be used to transfer micro-patterns onto device-building substrate. Chiao (Chiao et al., 2006) applied wet etching to construct silicon-based chambers containing serpentine channels. Additionally, C-MEMS microfabrication technique for 3D microstrustures, involving the pyrolysis of patterned photoresist has been developed which can be used as microelectrodes for miniature biofuel cells (Wang & Madou, 2006). With current microfabrication processes, the miniature biofuel cells offer unique advantages such as large surface area to volume ratio, short distance between the electrode, fast response time and low Reynolds number. In the following section, we will discuss the developments of both miniature MFCs and EBFCs.The experimental demonstration of miniature biofuel cells, along with the discussion of the key challenges and opportunities for realizing the practical potential of miniaturized biofuel cells for medical implants will be discussed.

2.1. Miniature microbial biofuel cells and its state of the art

One of the early efforts on miniature microbial biofuel cells reported a surface power output of 0.023 mW/m_2 and current density of 150 mA/m^2 based on the 10μm diameter circular anodic electrode (Chiao et al., 2006). The miniature microbial biofuel cells were limited by relatively low volumetric power density and coulombic efficiency due to their high internal resistances. Compared with macro scale microbial biofuel cells using the same microbes (Ringeisen et al., 2006), miniature microbial biofuel cells generated similar volumetric current density but significantly lower volumetric power density, which is insufficient for the anticipated applications. It was pointed out that the internal resistances of miniature microbial biofuel cells were around 40 fold higher than that in the macro scale microbial biofuel cells. The ohmic loss was higher in the micropillar devices with the same catholyte and anolyte; however, they generated higher volumetric power density (32 A/m^3) than the serpentine-channel devices (0.5 A/m^3) (Ringeisen et al., 2006). The high surface area-to-volume ratio and good microbe adaptivity of the micropillar electrodes decreased the anode resistance and resulted in higher volumetric power output. Carbon based anodes are known for high surface area-to-volume ratio and easy adaptation of microorganism and they are widely used in macro scale microbial biofuel cells. The recent investigations using carbon nanotubes (CNT) as electrodes (Qiao et al., 2007 and Timur et al., 2007) provide promising solutions for constructing carbon-based anodes in miniature microbial biofuel cells. The CNT based electrodes showed great improvement in the electricity generation and biocompatibility. Its maximum power density was 42 mW/m2 using E.coli as the microbial catalyst.

In the pursuit to improve the miniature microbial biofuel cell performance, different strategies were employed such as increasing the anode surface area, improving coupling of microorganism to anode surface, developing electrochemically active microbes and decreasing proton diffusion resistance. In summary, the enhancement strategies resulted in enhanced mass transport, improved reaction kinetics, and reduced ohmic resistance. Based on these developments, the ability to generate sufficient current and power from miniature devices was realized, thus breaking the conventional concept that small scale microbial biofuel cells would perform unsatisfactorily due to limited amount of substrates and microorganism. Since the development of the first miniature microbial biofuel cells in 2006, the volumetric power density and coulombic efficiency have been increased over 5 times. Although the output potential from the miniature MFCs is still insufficient for powering conventional equipment, they are promising options for on-chip power sources, especially for medical implants, which only require several millivolts to operate. Given the evidence that volumetric current density of the miniature MFC was achieved to be 2400 mA/m3 and required power from the cell was therefore 960 mW/m3, which is sufficient for existing devices (Wang & Lu, 2008). However, higher current density can result in excessive ohmic heating and electrolysis during the operation. Therefore, study in optimizing current density, overall output voltage and stability of the miniature MFCs as well as electrode design and device configuration for implantation rejection, microbe leakage, and analysis of the composition and distribution of internal resistances is necessary before further implementation in practical applications.

2.2. Miniature enzymatic biofuel cells and its state of the art

The first micro-sized enzymatic biofuel cells reported in 2001 (Chen et al., 2001). A glucose/O2 biofuel cell consisted of two 7 μm diameter, 2 cm long electrocatalyst-coated carbon fibers operating at ambient temperature in a pH 5 aqueous solution. The areas of the anode and the cathode of the cell were about 60 times smaller than those of the smallest reported fuel cell and 180 times smaller than those of the previously reported smallest biofuel cell. The power density of the cell is 64 $\mu W/cm^2$ at 23º C and 137 $\mu W/cm^2$ at 37º C, and its power output is 280 nW at 23º C and 600 nW at 37º C. The results revealed that the miniature enzymatic biofuel cells could generate sufficient power for small power-consuming CMOS circuit. Later, a miniature enzymatic biofuel cell with the same carbon fibers operating in a physiological buffer was reported (Mano et al., 2002). In a week operation the cell generated 0.9 J of electrical energy while passing 1.7°C charge. Based on this result, Mano developed a miniature compartment-less glucose/O2 biofuel cell operating in a living plant. Implantation of the fibers in the grape leads to an operating biofuel cell producing 2.4 μW at 0.52 V, which is adequate for operation of low-voltage CMOS/SIMOX integrated circuits. The performance of the miniature enzymatic biofuel cell was upgraded to 0.78 V operating at 37º C in ph 5 buffer later on (Mano et al., 2003). In 2004, a miniature single-compartment glucose/O2 biofuel cell made with the novel cathode operated optimally at 0.88 V, the highest operating voltage for a compartmentless miniature fuel cell (Soukharev et al., 2004). The enzyme was formed by "wiring" laccase to carbon through an electron conducting redox hydrogel, its redox functions tethered through long and flexible spacers to its cross-linked and hydrated polymer, which led to the apparently increased electron diffusion coefficient. The latest report on miniature glucose/O2 biofuel cells demonstrated a new kind of carbon fiber microelectrodes modified with single-wall carbon nanotubes (CNTs) (Li et al., 2008). The power density of this assembled miniature compartment-less glucose/O2 BFC reaches 58l Wcm-1 at 0.40 V. When the cell operated continuously with an external loading of 1 M resistance, it lost 25% of its initial power in the first 24 h and the power output dropped by 50% after a 48 h continuous work. Although from the practical application point of view, the performance and the stability of the current emzymatic biofuel cells remain to be improved, the miniature feature and the compartment-less property as well as the tissue-implantable biocapability of enzymatic biofuel cell essentially enable the future studies on in vivo evaluation of the cell performance and stability in real implantable systems.

In an effort to miniaturize the EBFCs, we have developed a versatile technique based on C-MEMS process for the miniaturization of electrodes. Our research focuses on the fabrication of 3D microelectrodes for miniature enzymatic biofuel cells. First, the functionalization methods for EBFCs enzyme immobilization were studied. Then we apply finite element approach to simulate the miniature EBFCs to attain the design rule such as electrode aspect ratio, configuration as well as orientation of the chip. Building an EBFC based on the design rule we obtained is on-going.

3. C-MEMS 3D ARCHITECTURE ELECTRODES

The surface area of biofuel cells determines its amperage, meaning that cell power is directly proportional to the electrode surface area. A conventional 2D power system is typically a parallel arrangement of a planar cathode and an anode separated by a solid or liquid electrolyte. More recently, carbon-microelectromechanical (C-MEMS) fabrication technology has offered the flexibility to fabricate complex carbon-based EBFCs, with 3D dense microelectrode arrays. C-MEMS, describes a manufacturing technique in which carbon microstructures are fabricated by baking UV sensitive polymers at high temperatures in an inert environment. It has been demonstrated that 3D high-aspect-ratio carbon structures can be made from carbonizing (pyrolysis) patterned NANOTM SU-8 negative photoresist (Wang et al., 2005). The four steps involved in converting organic polymer to pyrolytic carbon is shown as schematic in Fig. 1. Positive photoresist (AZ4620, AZ1518) as well as negative photoresist (SU-8) can be converted to carbon by pyrolysis depending on the application. Electrodes based on 3D microstructures are expected to offer higher surface area and significant advantages in comparison to thin-film devices for powering MEMS and miniaturized electronic devices.

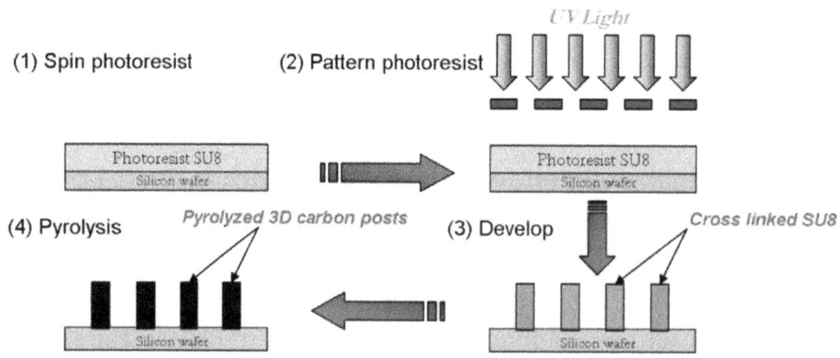

Figure 1.Schematic showing the typical C-MEMS process.

3.1. C-MEMS microelectrodes

Fig. 2. shows the various carbon architectures possible using C-MEMS technique. The versatility in the technique gives us the opportunity to integrate nanofeatures such as suspended carbon nanowires, carbon nanofibers with microelectrodes (Wang et al., 2005; Malladi et al., 2006).

Figure 2.SEM images of different carbon architectures possible by C-MEMS technique.

Besides, in order to increase the surface area, we reported a modified C-MEMS process using a block copolymer F127 as porogen capable of producing porous carbon microelectrodes (Penmatsa et, al., 2010). These results indicated that porous carbon thin film electrodes derived from 10% F127 mixed in SU-8 had an Aeff 185% compared to the conventional photoresist derived carbon electrode. This fabrication approach can be employed to produce reproducible high aspect ratio carbon microelectrodes with different shapes for various electrochemical devices.

Although the 3D structures compared to 2D planar electrodes or thin films have advantages, such as an increase in the surface area and power density for same foot print area, there are yet certain important issues which need to be addressed in order to use these structures effectively. Anandan and Godino have studied the mass transport phenomenon in micro and nano-electrodes by finite element analysis approach. They suggest that in order to accommodate the specific analyte species in terms of reaction kinetics and mass transport, it is necessary to optimize the geometry of nanopillars (their diameter, spacing and height), to reap the true benefit of using micro-nanostructured electrodes for enhancing the performance of biosensors. They reveal that the glucose immediately react with the top portions of the nanopillars due to higher reaction rate of enzymes and hence the bottom portion of the pillars lack the diffusion of glucose, which may not be favorable to improve the performance of EBFC. Jeffrey suggests that in contrast to the 2D electrodes, in which uniform current density is naturally obtained over the surface of the cathodes and anodes, the current density in the 3D microelectrode array suffer from a non-uniform primary current distribution. These non-uniform currents result in utilization of the electrode materials and are thus associated with lower cell efficiencies, reduced electrodes stability due to non-uniform stresses, and non-uniform heat dissipation. Therefore, it is essential to optimize the geometries of electrodes to homogenize the current density distribution around microelectrodes surfaces.

In the later section, we introduced finite element method based simulative approach to understand the effect of 3D design rule and spatial distributions of the microelectrodes in the arrays with respect to the mass transport of glucose, enzymatic reaction rate and open circuit output potential.

4. SURFACE FUNCTIONALIZATION OF C-MEMS

There are several factors regulating the lifetime of biofuel cells, which has always been a concern for their practical application. In most cases, the stability of enzyme determines the lifetime of biofuel cells. Immobilization of enzyme through covalent bonding on solid surface has attracted great attention for applications in catalytic processes. Therefore, in our research, covalent attachment of enzyme on supports was studied to promote rigidification of enzyme structure of the immobilized enzyme.

We have studied three different types of covalent surface functionalization for enzyme immobilization in EBFCs – (1) Direct amination; (2) Diazonium grafting; (3) Diamine grafting. In all these methods, the functional groups are realized on the pyrolyzed carbon surface.

4.1. Direct amination on pyrolyzed carbon

Direct amination was conducted by functionalizing its surface with ultraviolet (UV) irradiation under ammonia gas (Yang et, al., 2009). Quantified amino groups on the carbon surface were estimated by X-ray photoelectron spectroscopy in Fig. 3. It is found out that the amino groups exist at surface by amination processes due to the high density of carbon. Ammonia gas forms as C-NH2 on carbon substrate because C-H bonds react easily with ammonia gas and undergo photochemical reaction on exposure to UV irradiation although steric limitations will limit the amine group coverage on the surface. The results showed that the amino groups were successfully formed on pyrolyzed carbon surface by direct amination.

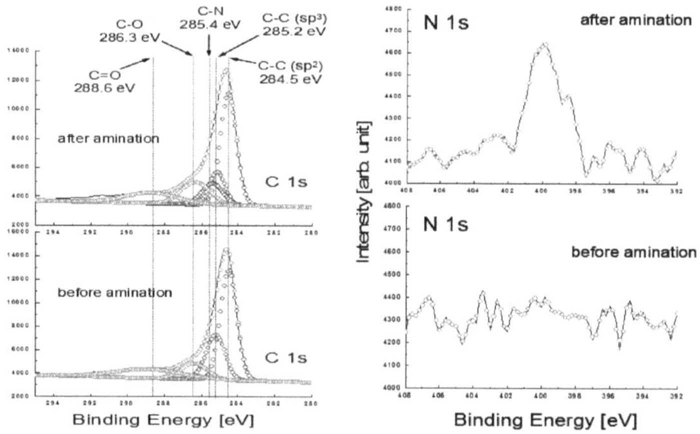

Figure 3.X-ray photoelectron spectra of pyrolyzed carbon surface. (a) High-resolution scan of the C 1s peak and (b) high-resolution scan of the N 1s peak are compared before and after direct amination.

4.2. Surface functionalization of carbon surface by diazonium grafting

Diazonium grafting is a promising alternative to conventional electrode functionalization method. In this approach the electrochemical reduction of diazonium forms an aryl centered radical. The resulting aryl radical can then form a covalent bond with conducting and semiconducting surfaces. The CV results shown in Fig. 4. indicate the first cycle of electro-reduction process from NO2 to NH2 at different diazonium concentrations. The reduction of NO2 to NH2 occurs on the first negative-going sweep in a range of potential from -1.0 V to -1.1 V forming a clear irreversible anodic peak. From the results, it was confirmed that the amino groups are successfully grafted on the carbon surface for further immobilization.

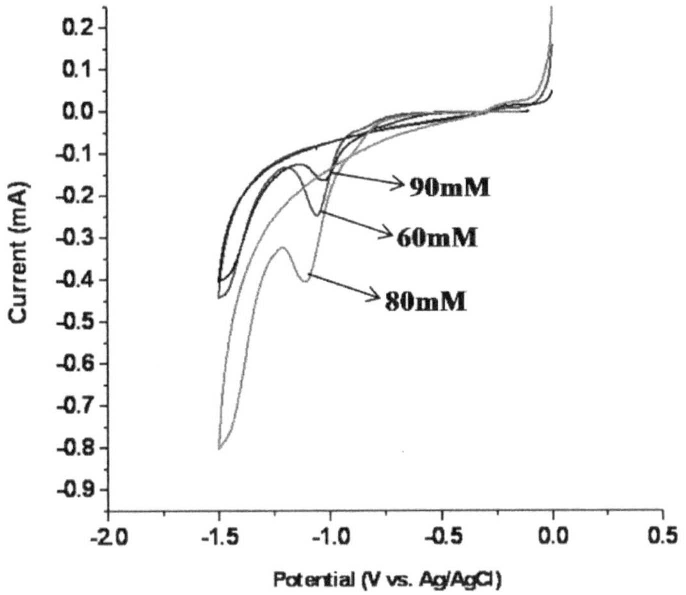

Figure 4.CV curves showing the electroreduction from NO2 to NH2 at different concentrations of diazonium. The irreversible peak indicates the electroreduction process.

4.3. Functionalization of carbon surface by diamine grafting

Ethylene diamine grafting on the carbon surface was conducted using the electrochemical reduction. As the potential sweeps to negative direction during the first cycle, the oxidation of amino group in ethylene diamine is reduced on pyrolyzed carbon surface at potentials between -0.8 V and -1.4 V, leading to a clear irreversible anodic peak. It also shows that any peaks by oxidation and reduction do not exist after first reduction sweep even though we notice that the anodic current of electrode is decreased. This result indicates that the ethylene diamine was grafted on the surface of pyrolyzed carbon by applying potential and pyrolyzed surface is successfully functionalized by the electrochemical method.

We have introduced three types of functionalization on carbon surface in this session. In the future work, we will immobilize different biomolecules based on these functionalization methods for EBFCs device. Work on building a prototype EBFC consisting of glucose oxidase immobilized anode and a laccase immobilized cathode using C-MEMS based interdigitated electrode arrays is underway.

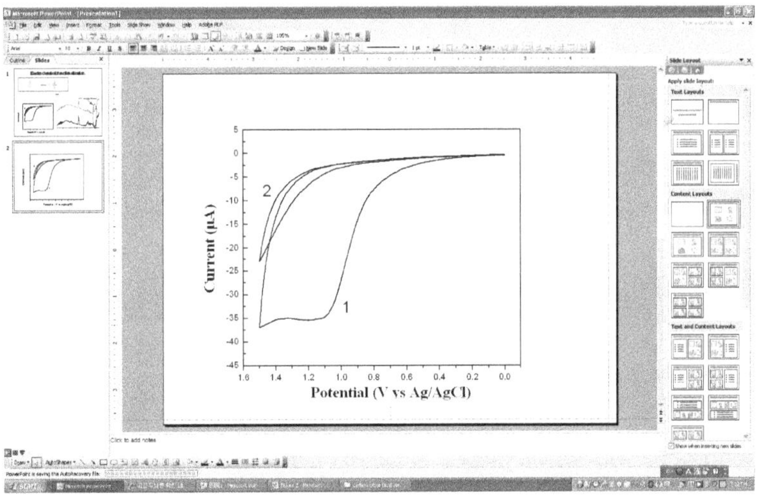

Figure 5.Cyclic voltammograms showing the first and second cycle confirming the surface functionalization completed in the first irreversible cycle.

5. SIMULATION OF C-MEMS BASED EBFCS

5.1. Finite element approach for optimization of electrodes design

For our simulation approach, we used commercially available COMSOL 3.5 software multiphysics software, which solves partial differential equations (PDEs) by finite element technique. In the model we assume that 3D carbon microelectrode arrays were uniformly immobilized with glucose oxidase and laccase on anode and cathode respectively with out the use of any mediators. The proposed implantable membraneless EBFC is assumed to be placed inside a blood artery of the human body thus utilizes the glucose extracted from blood as a fuel. In principle, glucose oxidase reacts with glucose and produces gluconolactone and hydrogen peroxide. This hydrogen peroxide oxidizes on the anode to generate electron and hydrogen ions. The hydrogen ions travel from electrolyte to cathode, while electrons flow through an external load and generate electricity. On cathode, dissolved oxygen is reduced via laccase enzyme and by combining with electrons and hydrogen ions forms water.

We applied Michaelis-Menten theory in our 2D model to analyze phenomenon between enzyme kinetics on the electrode surface and glucose diffusion and thus optimize the electrode microarray design rule according to the enzyme reaction rate In order to determine the output potential in developing biofuel cell, we also incorporated Nernst equation. The numerical simulations have been performed with various electrodes heights and well widths (distance between any two electrodes) to obtain the relation between design rule and EBFCs performance. Various 2D models are investigated for same foot print length (600 μm) of SiO2, with fixed electrode diameter of 30 μm and fixed enzyme layer thickness of 10 μm. The height of electrodes is chosen as 60 μm, 120 μm and 240 μm for different well widths (WW-distance between any two electrodes) of 10μm, 20 μm, 40 μm, 60 μm, 80 μm, 100 μm, 120 μm, 140 μm, 160 μm, 180 μm and 200 μm.

The quantification of reaction rates of enzymes on anode and cathode is showed in Fig. 6.

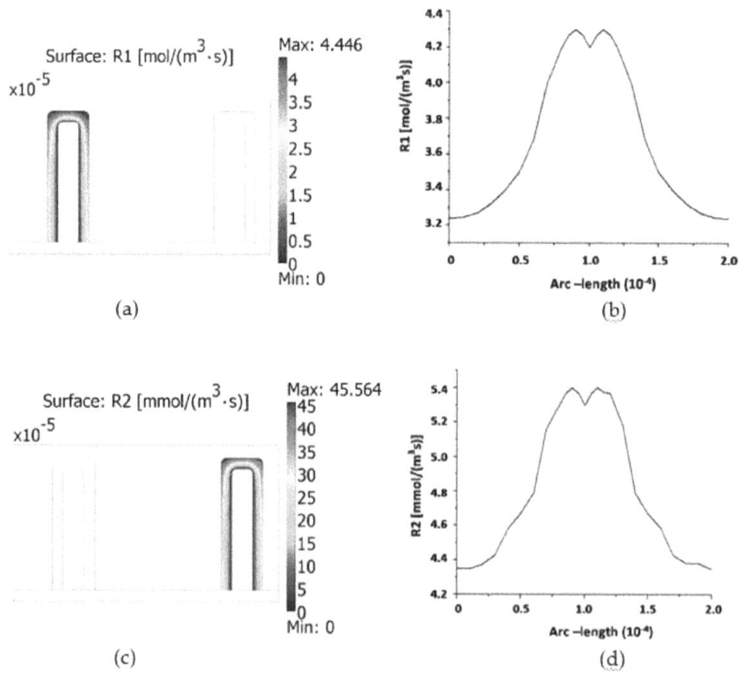

Figure 6.a) Subdomain plot of anode reaction rate (R1); (b) reaction rates from the whole surface of anode. (c) Subdomain plot of cathode reaction rate (R2); (d) reaction rates from the whole surface of cathode.

From the results, we observe that the reaction rate decreased from the top to the bottom along the surface of both electrodes due to the lack of diffusion of the substrate as we go towards the bottom; also the outer surfaces of the electrodes have the larger reaction rate in the enzyme layer. The reaction rate along the surface of both electrodes is plotted in Fig. 6. The reaction rate is increased from the bottom to the top along the electrode surface and reached the maximum at edge of the top due to the edge effect. The maximum reaction rates of GOx enzymes vs. different well widths is shown in Fig. 7. for three different heights of electrodes: 60 μm, 120 μm and 240 μm, with 10 μm, 20 μm, 40 μm, 60 μm, 80 μm, 100 μm and 120 μm well widths, respectively. In the case of 60 μm height of electrodes, the maximum reaction rate is obtained when the well width is about 30 μm. For the height of 120 μm and 240 μm, reaction rate reached the highest at the well width of 60 μm and 120 μm respectively. From all these three sets of models both in anode and cathode, we can conclude that the reaction rates of one pair of electrodes reach the maximum when the well width is half as the height of electrodes.

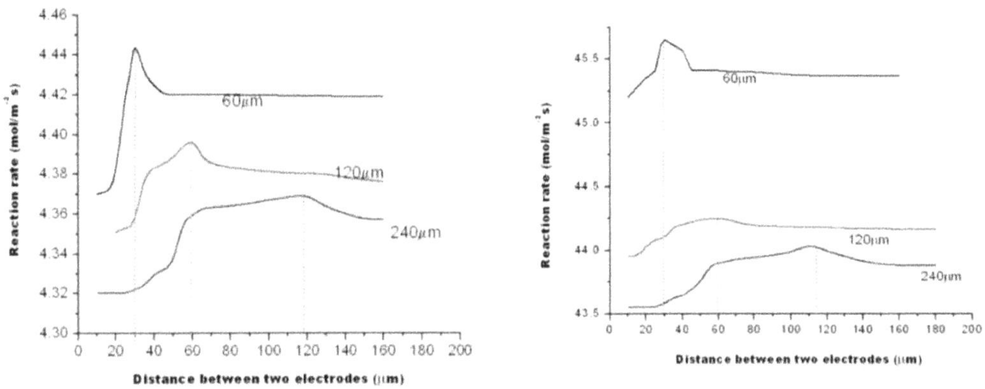

Figure 7.a) Anode reaction rate curves vs. well width at different ratio of electrode dimensions; (b) Cathode reaction rate curves vs. well width at different ratio of electrode dimensions.

The open circuit output potential also has been simulated for the same heights and well widths of electrodes by applying the Nernst equation. The current collectors are assumed at the bottom of the electrodes and hence these potentials are calculated from the bottom. Fig. 8. shows the open circuit output potential vs. well width of electrodes at different height of electrodes. From the results of simulation, we could find out an empirical relationship between electrodes height and well width to achieve optimized output potential is when height of electrodes is twice than that of well width which is in agreement to the results we obtain for the diffusion of the substrate.

5.2. Finite element approach for optimization of orientation of microelectrodes chip for enzymatic biofuel cells

Until now, majority of the research was focused on in-vitro experiments by mimicking physiological conditions. The additional complex problems may arise when a BFC chip is placed inside a blood artery. The first is with implantation process itself, which involves a surgery for the insertion of a BFC, and other necessary electronics components. The second is the stability of this chip inside an artery and how/where this chip can be fixed such that it can survive against the blood flow. Third problem is the clotting of the blood. The goal is to put this EBFC chip in such a way that it does not obstruct the flow of blood and lead to substantial pressure drop inside an artery. The fixation of this chip with the blood artery also should not harm the blood vessel walls (Parikh et al., 2010).

In order to improve mass transport around microelectrodes by optimizing the positioning of an EBFC chip, we have adopted the finite element analysis approach to look into the stability of an EBFC inside a blood artery. On the initial stage, we have analyzed only two orientations: horizontal position (HP) and vertical position (VP). The stability of the chip in these positions, diffusion and convectional fluxes around microelectrodes has been finely investigated. We have proposed a novel chip design, with holes in between all electrodes on the substrate, which can drastically improve the diffusion in between microelectrodes.

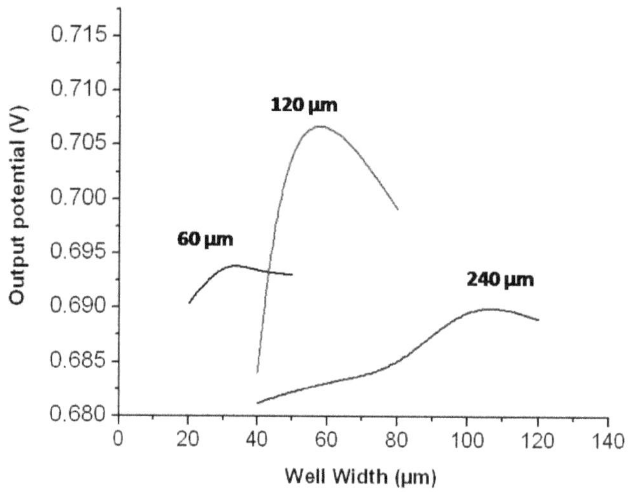

Figure 8.Output potential vs. well width for different ratio of electrode dimensions.

The diffusion between the microelectrodes has shown in Fig. 9, where Fig. 9a and b shows the simulation profiles for diffusive flux along with the streamlines around microelectrodes in HP and VP, respectively. In HP, it is observed that the diffusive flux is less near the central electrodes and increases when going towards outer electrodes. However, the diffusive flux is almost same on top of all electrodes in VP. It is observed that in both the positions, the diffusive flux is following laminar pattern. The diffusive flux from bottom of an electrode to top of an electrode is investigated in HP and VP as shown in Fig. 9c and d, respectively. The flux is not uniform from the central to outer electrodes. The electrodes located at the circumference of a chip are having more flux compared to those located in the centre of the chip. The variation of the diffusive flux distribution around inner to outer electrodes is high in HP. The flux is not constant at every instance, but it is oscillating as shown in inset figures. The diffusive flux profiles in these figures are considered at the time, when the flux reaches its maximum value. This is also evident from Fig. 9e and f, the flux is higher exactly at the top of electrodes while lesser in the vicinity between any two electrodes. In comparison of HP and VP, the diffusive flux is 8 orders larger in case of VP than in HP.

Total flux is the combination of a diffusive flux and a convective flux. Fig. 10 depicts the total flux data for (a) HP and (b) VP of a chip. In HP, flux is negligible up to almost 275 µm height of electrodes and then increasing at the top. Total flux is highest at the top of outer most electrodes and then reducing to the central electrodes. In case of VP, the flux is almost uniform on top of all electrodes, with negligible value in between electrodes up to 200 µm height and then gradually increasing to about 2000–3500 mmol m-2 s-1 at the top of all electrodes.

Based on the results, the new design with the holes in between all microelectrodes has been inspected precisely and compared with the prototype design. The diffusive flux (Fig. 11a, c, e) and convective flux (Fig. 11b, d, f) profiles for the new design are compared with diffusive flux and convective flux profiles of the prototype model, respectively. The streamlines present the lines of motion of glucose at a particular instance.

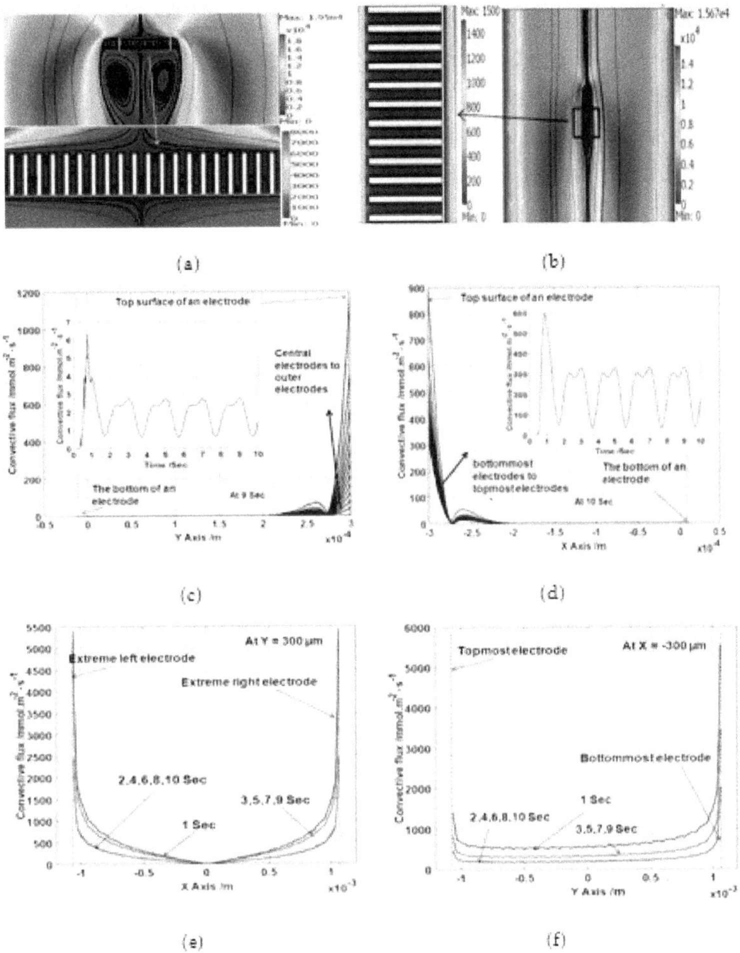

Figure 9.Surface plot with streamlines for convective flux of glucose around microelectrodes for a) HP, b) VP, convective flux in between all 24 electrodes from bottom of electrodes to up to 300 μm height is shown for c) HP and d) VP, convective flux at top of all the electrodes from leftmost to right most electrodes for 0 – 10 secs in e) HP and f) VP.

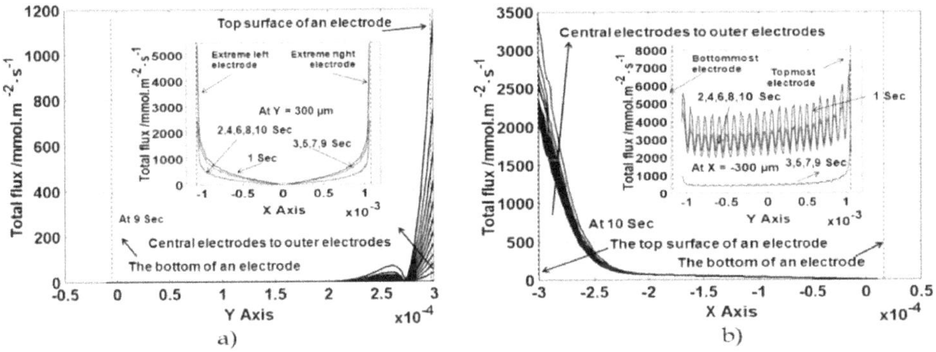

Figure 10.Total fluxes in between micro-electrodes for a) HP and b) VP. Insets provide the total flux on top of all electrodes.

Based on the information provided in Figure 11, it can be deduced that the overall flux (which includes both diffusive and convective flux) has been enhanced for all microelectrodes in terms of both their values and uniformity in the chip with the holes. The increased movement of particles around microelectrodes is crucial for improving the performance of an EBFC. This proposed design could potentially have the potential to be beneficial in preventing blood clotting. The composition of human blood is mostly comprised of red blood cells and white blood cells. The dimensions of various cells, including red blood cells (6 μm), lymphocytes (7–8 μm), neutrophils (10–12 μm), eosinophils (10–12 μm), basophils (12–15 μm), and monocytes (14–17 μm), are predominantly less than 20 μm, which is the size of the apertures on the chip. These cells are able to traverse the gaps between microelectrodes without obstructing the pathway. The size of these holes can be increased based on the specific need. The enhanced convection between microelectrodes may also exert sufficient force to prevent the production of bubbles. Nevertheless, the biomechanical process and hemodynamic process exhibit greater complexity compared to convection and diffusion, particularly at the micro-scale level. Cell proliferation and coagulation phenomena are interconnected with various factors, including biocompatibility, arterial flexion, platelet aggregation, and protein constituents. Further comprehensive study with biologists is necessary to acquire more substantial and valuable information and to achieve a higher level of applicability for the EBFCs.

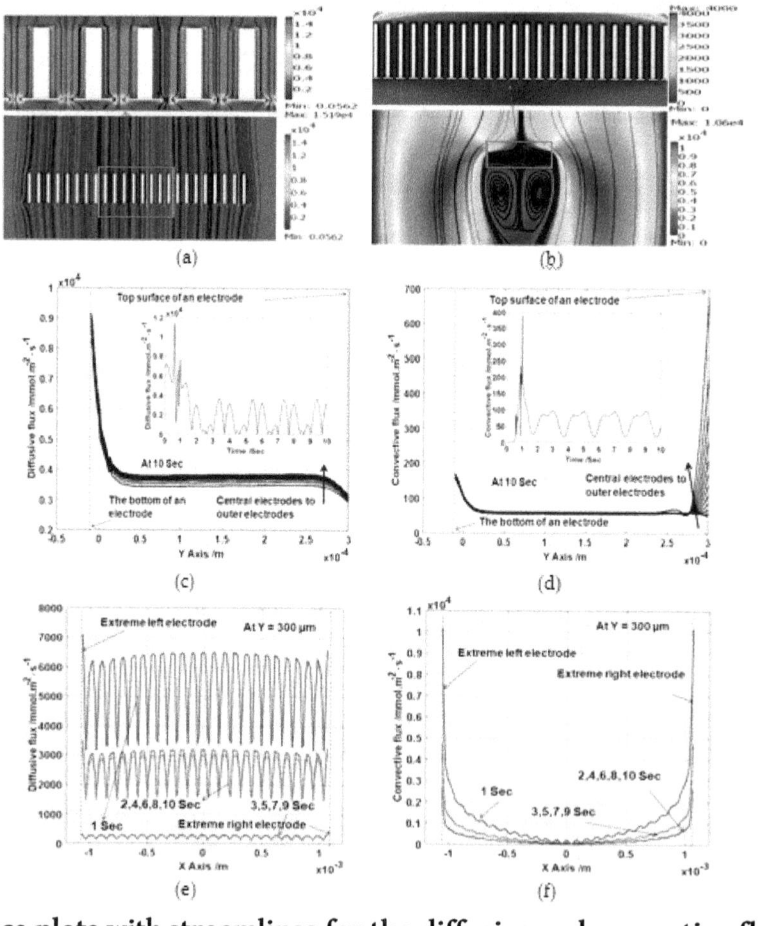

Figure 11 shows surface plots with streamlines for the diffusive and convective flux of glucose around microelectrodes. Panel (a) displays the diffusive flux, while panel (b) shows the convective flux. Panel (c) illustrates the diffusive flux and panel (d) shows the convective flux between all 24 electrodes, from the bottom of the electrodes up to a height of 300 μm. Panel (e) represents the diffusive flux, and panel (f) shows the convective flux at the top of all the electrodes, ranging from the leftmost to the rightmost electrodes, for a duration of 0 to 10 seconds.

6. CONCISE SUMMARY

This chapter has presented the two primary types of biofuel cells: microbial fuel cells and enzymatic biofuel cells. Substantial progress has been made in the past ten years in the advancement of both biofuel cells. In response to the need for dependable power sources for implantable medical devices, significant endeavors have been undertaken to develop miniature biofuel cells. Previous experimental findings demonstrated that enzymatic micro biofuel cells have the capability to provide an ample amount of energy suitable for slower and less power-intensive CMOS circuits. Furthermore, we have demonstrated through simulation findings that the theoretical power generated by C-MEMS enzymatic biofuel cells is sufficient to meet the energy requirements of current implantable medical devices. Nevertheless, there exist certain obstacles that need to be overcome in order to achieve further progress in tiny biofuel cells. The primary concerns encompass enduring stability and inadequate electricity generation. The successful advancement of biofuel cell technology necessitates the collaborative endeavors of various fields: biology for comprehending biomolecules, chemistry for acquiring knowledge on electron transfer mechanisms, material science for creating innovative materials with strong biocompatibility, and chemical engineering for designing and implementing the system.

REFERENCES

Aelterman, P.; Versichele, M.; Marzorati, M.; Boon, N. & Verstraete, W. (2010). Loading rate and external resistance control the electricity generation of microbial fuel cells with different three-dimensional anodes, Bioresource Technology 99 (18), pp. 8895–8902.

Akers, N.L.; Moore, C.M. & Minteer, S.D. (2005). Development of alcohol/O2 biofuel cells using salt-extracted tetrabutylammonium bromide/Nafion membranes to immobilize dehydrogenase enzymes, Electrochim. Acta 50 (12), pp. 2521–2525.

Allen, R.M. & Bennetto, H.P. (1993). Microbial Fuel Cells: Electricity Production from Carbohydrates. Appl. Biochem. Biotechnol., 39/40, pp. 27-40.

Barton, S.C.; Gallaway, J. & Atanassov, P. (2004). Enzymatic biofuel cells for implantable and microscale devices, Chem Rev 104, pp. 4867–4886.

Blonder, R.; Willner, I. & Bueckmann, AF. (1998). Reconstitution of apo-glucose oxidase on nitrospiropyran and FAD mixed monolayers on gold electrodes: photostimulation of bioelectrocatalytic features of the biocatalyst. J Am Chem Soc 120, pp. 9335– 41.

Bockris, JOM & Srinivasan, S. (1969). Fuel cells: their electrochemistry. NewYork; McGraw-Hill.

Bond, DR.& Lovley (2003). Electricity production by Geobacter sulferreducens attached to electrodes. Appl. Environ. Microbiol. 69, pp. 1548-1555.

Calabrese Barton, S.; Gallaway, J. & Atanassov, P. (2004). Enzymatic Biofuel Cells for Implantable and Microscale Devices.Chem. Rev. 104 (10), pp. 4867–4886.

Chaudhuri, S.K. & D.R. Lovley. (2003). Electricity generation by direct oxidation of glucose in mediatorless microbial fuel cells ,Nat. Biotechnol. 21, (10), pp. 1229–1232.

Chen, T.; Barton, S.C.; Binyamin, G.; Gao, Z.Q.; Zhang, Y.C.; Kim, H.H. & Heller, A. (2001). J. Am. Chem. Soc. 123 (35), pp. 8630–8631.

Chiao, M., Lam, K.B. & Lin, L.W. (2006). Micromachined microbial and photosynthetic fuel cells, Journal of Micromechanics and Microengineering 16 (12), pp. 2547–2553.

Cohen, B. J. (1931). Bacteriol. 21, pp. 18.

Crittenden, S.R.; Sund, C.J. & Sumner, J.J. (2006). Mediating electron transfer from bacteria to a gold electrode via a self-assembled monolayer, Langmuir 22 (23), pp. 9473– 9476.

Debabov, V.G. (2008). Electricity from microorganisms, Microbiology 77 (2), pp. 123–131.

Feng, Y.; Wang, X.; Logan, B.E. & Lee, H. (2008). Appl. Microbiol. Technol. 78, pp. 873–880. Fan, Y.Z., Hu, H.Q. & Liu, H. (2007). Enhanced coulombic efficiency and power density of air-cathode microbial fuel cells with an improved cell configuration, Journal of Power Sources 171 (2), pp. 348–354.

Freire, R.S.; Pessoa, C.A.; Mello, L.D.; & Kubota, L.T. (2003). Direct electron transfer: an approach for electrochemical biosensors with higher selectivity and sensitivity, J Braz Chem Soc 14, pp. 230–243.

Galvani, L. (1791). De bononiensi scientiarum et artium instituto atque academia Comentarrii 7, pp. 363–418.

Ghangrekar, M.M. & Shinde, V.B. (2007). Performance of membrane-less microbial fuel cell treating wastewater and effect of electrode distance and area on electricity production, Bioresource Technology 98 (15), pp. 2879–2885.

Gorge, G; Kirstein, M & Erbel, R. (2001). Microgenerators for Energy Autarkic Pacemakers and Defibrillators: Fact or Fiction. Herz 26, pp. 64–68.

Govil, G & Saran, A. (1982). Biochemical fuel cells. J Indian Chem. Soc. (1982), 59: p. 1226– 8. Grove, W.(1839). Philos. Mag. Ser. 3 (14), pp. 127.

Heller, A. (2004). Miniature biofuel cells. Phys. Chem. Chem. Phys. 6, pp. 209–216. Katz, E.; Shipway, A.; & Willner, I. (2003). In: Handbook of Fuel Cells—Fundamentals, Technology and Applications Chichester, W. Vielstich, H.A. Gasteiger and A. Lamm, England, John Wiley & Sons Ltd., pp. 355–381.

Habermann, W. & Pommer, E.H. (1991). Biological Fuel Cells with Sulphide Storage Capacity. Appl. Microbiol. Biotechnol., 35, pp. 128-133.

Haccoun, J.; Piro, B.; Noel, V. & Pham, M.C. (2006). Bioelectrochemistry 68, pp. 218–226. Heller, A. (2006). Potentially implantable miniature batteries. Anal. Bioanal. Chem. 385, pp. 469–473.

Hou H.J., Li L., Cho Y., de Figueiredo P. & Han, A. (2009). Microfabricated microbial fuel cell arrays reveal electrochemically active microbes, Plos One 4 (8), pp. e6570.

Ieropoulos, I. Winfield, J. & Greenman, J. (2010). Effects of flow-rate, inoculum and time on the internal resistance of microbial fuel cells, Bioresource Technology 101 (10), pp.3520–3525.

Katuri, K.P. & Scott, K. (2010). Biotechnol. Bioeng. 107, pp. 52–58. Katz, E.; Buckmann, A.F. & Willner, I. (2001). J. Am. Chem. Soc. 123 (43), pp. 10752–10753.

Katz, E.; Shipway, A.N. & Willner, I. (2003). Biochemical fuel cells. In: Handbook of Fuel Cells —Fundamentals, Technology and Applications, W. Vielstich, H.A. Gasteiger and A. Lamm, Fundamentals and Survey of Systems vol. 1, John Wiley & Sons, Ltd., Hoboken, NJ, pp. 355–381.

Katz, E. & Willner, I. (2003). Biofuel cells based on monolayer-functionalized biocatalytic electrodes. In: K.E. Geckeler, Editor, Advanced macromolecular and supramolecular materials and processes, Kluwer Academic/Plenum Publishers, New York, pp. 175–196.

Katz, E.; Sheeney-Haj-Ichia, L. & Willner, I. (2004). Electrical contacting of glucose oxidase in a redox-active rotaxane configuration. Angew Chem Int Ed 43, pp. 3292–300.

Kim, B.H.; Kim, H.J.; Hyun, M.S. & Park, D.H. (1999). Direct Electrode Reaction of Fe (III)-reducing Bacterium, Shewanella putrefaciens. J. Microbiol. Biotechnol. 9, pp.127-131.

Kim, H.J. H.S. Park, M.S. Hyun, I.S. Chang, M. Kim & B.H. Kim. (2002). A mediator-less microbial fuel cell using a metal reducing bacterium, Shewanella putrefaciens Enzyme Microb. Technol. 30 (2), pp. 145–152.

Kim, N. Y. Choi, S. Jung & S. Kim. (2000). Development of microbial fuel cells using Proteus vulgaris, Korean Chem. Soc. 21 (1), pp. 44–48.

Kim, H.H.; Mano, N.; Zhang, X.C. & Heller, A. (2003). J. Electrochem. Soc.150 (2), pp. 209–213.

Kim, P.; Kwon, K.W.; Park, M.C.; Lee, S.H.; Kim S.M. & K.Y. Suh. (2008). Soft lithography for microfluidics: a review, Biochip Journal 2 (1), pp. 1–11.

Lee, J.; Shin, H. Y.; Kang, S.W.; Park, C. & Kim, S.W. (2010). Use of bioelectrode containing DNA-wrapped single-walled carbon nanotubes for enzyme-based biofuel cell, Journal of Power Sources 195, pp. 750–755.

Li, X.; Zhou, H.; Yu, P.; Su, L.; Ohsaka, T. & Mao, L. (2008). A Miniature glucose/O2 biofuel cell with single-walled carbon nanotubes-modified carbon fiber microelectrodes as the substrate, Electrochemistry Communications. 10 (6), pp. 851-854

Liu, Y. & Dong, S.J. (2007). A biofuel cell harvesting energy from glucose-air and fruit juiceair. Biosens. Bioelectron. 23, pp. 593-597.

Logan, B.E. (2008). Microbial Fuel Cells, John Wiley & Sons, Inc.

Lowy, D.; Tender, L.; Zeikus, J.; Park, D. & Lovley, D. (2006). Harvesting energy from the marine sediment-water interface II - Kinetic activity of anode materials. Biosensors & Bioelectronics 21, pp. 2058-2063

Lovley, D.R. (2008). The microbe electric: conversion of organic matter to electricity, Current Opinion in Biotechnology 19, pp. 564–571.

Malladi, K.; Wang, C. & Madou, M. (2006). Microfabrication of Suspended C-MEMS Structures by EB Writer and Pyrolysis. Carbon. 44(13), pp. 2602-07.

Mano, N.; Mao, F. & Heller, A., (2002). J. Am. Chem. Soc. 124 (44), pp. 12962–12963.

Mano, N.; Mao, F & Heller, A. (2003). Characteristics of a Miniature Compartment-less Glucose/O2 Biofuel Cell and Its Operation in a Living Plant. J. Am. Chem. Soc. 125, pp. 6588–6594.

Mano, N. & Heller, A. (2003). J. Electrochem. Soc. 150 (8), pp. A1136–A1138.

Min, B; Cheng, S & Logan, BE. (2005) Electricity generation using membraneand salt bridge microbial fuel cells. Water Res, 39, pp. 1675–1686.

Moehlenbrock, M.J. & Minteer, S.D. (2008). Extended lifetime biofuel cells. Chem. Soc. Rev., 37, pp. 1188-1196.

Moon, H.; Komlos, J. & Jaffe, P. (2009) Biogenic U(IV) oxidation by dissolved oxygen and nitrate in sediment after prolonged U(VI)/Fe(III)/SO4 2- reduction. J Contam Hydrol 105,pp. 18–27.

Nagel, B.; Warsinke, A. & Katterle, M. (2007). Langmuir 23, pp. 6807–6811.

Niessen, J.; Schroder, U. & Scholz, F. (2004). Exploiting complex carbohydrates for microbial electricity generation - A bacterial fuel cell operating on starch, Electrochem. Commun. 6 (9), pp. 955–958.

Niessen, J.; Harnisch, F.; Rosenbaum, M.; Schroder, U. & Scholz, F. (2006). Heat treated soil as convenient and versatile source of bacterial communities for microbial electricity generation. Electrochem Commun 8, pp. 869–73.

Oh, S.E.; Min, B. & Logan, B.E. (2004). Cathode performance as a factor in electricity generation in microbial fuel cells. Environ. Sci. Technol. 38, pp. 4900–4904.

Palmore, G & Whitesides, G.M. (1994). Microbial and enzymatic biofuel cells. In: Enzymatic Conversion of Biomass for Fuels Production, E. Himmel, Editor,vol. 566, American Chemical Society, pp. 271–290.

Parikh, Y.; Yang, J. H. & Wang, C. (2010). Optimizing the mass transport phenomenon around micro-electrodes of an enzymatic biofuel cell inside a blood artery via finite element analysis method, J Power Sources. 195 (15), pp. 4685-4694.

Park, D.H. & Zeikus, J.G. (2000). Electricity Generation in Microbial Fuel Cells Using Neutral Red as an Electronophore, Appl. Environ. Microbiol. 66 (4), pp. 1292–1297.

Park, D.H. & Zeikus, J.G. (2002). Impact of electrode composition on electricity generation in a single-compartment fuel cell using Shewanella putrefaciens, Appl. Microbiol. Biotechnol. 59 (1), pp. 58–61.

Park, D.H. & Zeikus, J.G. (2003). Improved fuel cell and electrode designs for producing electricity from microbial degradation, Biotechnol. Bioeng. 81 (3), pp. 348–355.

Patolsky, F.; Tao, G.; Katz, E. & Willner, I. (1998). C60-Mediated bioelectrocatalyzed oxidation of glucose with glucose oxidase. J Electroanal Chem 454, pp. 9–13.

Penmatsa, V.; Yang, J. H.; Yu, Y. & Wang, C. (2010). Fabrication of porous carbon micropillars using a block copolymer as porogen, Carbon. 48 (14), pp. 4109–4115.

Potter, M.C. (1910). Proceedings of the Royal Society B, vol. 84, p. 260.

Qian, F.; Baum, M.; Gu, Q. & Morse, D.E. (2009). A 1.5 μL microbial fuel cell for on-chip bioelectricity generation, Lab on a Chip 9 (21), pp. 3076–3081.

Qiao, Y.; Li, C.M.; Bao, S.J & Bao, Q.L. (2007). Carbon nanotube/polyaniline composite as anode material for microbial fuel cells, Journal of Power Sources 170 (1), pp.79–84.

Rabaey, C.; K. Lissens.; G. Siciliano S.D. & Verstraete, W. (2003). A microbial fuel cell capable of converting glucose to electricity at high rate and efficiency, Biotechnol. Lett. 25, pp. 1531–1535.

Rabaey, K.; Boon, N.; Siciliano, S.D.; Verhaege, M. & Verstraete, W. (2004). Biofuel Cells Select for Microbial Consortia That Self-Mediate Electron Transfer, Appl. Environ. Microbiol. 70 (9), pp. 5373–5382.

Ramanavicius, A.; Kausaite, A. & Ramanaviciene, A. (2005) Biofuel cell based on direct bioelectrocatalysis. Biosens. Bioelectron., 20, pp. 1962-1967.

Ramanavicius, A.; Kausaite, A. & Ramanaviciene, A. (2008). Enzymatic biofuel cell based on anode and cathode powered by ethanol. Biosens. Bioelectron., 24, pp.761-766.

Richter, H.; McCarthy, K.; Nevin, K.P.; Johnson, J.P.; Rotello, V.M. & Lovley, D.R. (2008). Electricity generation by Geobacter sulfurreducens attached to gold electrodes, Langmuir 24 (8), pp. 4376–4379.

Ringeisen, B.R.; Henderson, E.; Wu, P.K.; Pietron, J.; Ray, R.; Little, B.; Biffinger, J.C. & Jones-Meehan, J.M. (2006). High power density from a miniature microbial fuel cell using Shewanella oneidensis DSP10, Environmental Science & Technology 40 (8), pp. 2629–2634.

Roller, S.B.; Bennetto, H.P.; Delancy, G.M.; Mason, J.R.; Stirling, J.L. & Thurston, C.F. (1984). Electron-transfer coupling in microbial fuel cells: 1. comparison of redox-mediator reduction rates and respiratory rates of bacteria. J. Chem. Technol. Biotechnol. 34B,pp. 3–12.

Roundy, S. (2005). On the Effectiveness of Vibration-based Energy Harvesting.Intell. Mater. Syst. Struct. 16, pp. 809–823.

Rozendal, R. A.; Hamelers, H. V. M. & Buisman, C. J. N. (2006) Effects of membrane cation transport on pH and microbial fuel cell performance. Environ. Sci. Technol. 40(17), pp. 5206–5211.

Saleh, F. S.; Mao, L. & Ohsaka, T. (2011).Development of a dehydrogenase-based glucose anode using a molecular assembly composed of nile blue and functionalized SWCNTs and its applications to a glucose sensor and glucose/O2 biofuel cell, Sensors and Actuators B 152, pp. 130-135.

Sakai, H.; Nakagawa, T.; Tokita, Y.; Hatazawa, T.; Ikeda, T.; Tsujimura, S. & Kano, K. (2009). A high-power glucose/oxygen biofuel cell operating under quiescent conditions, Energy Environ.Sci. 2, pp. 133-138.

Schroder, U.; Niessen J. & Scholz, F. (2003). A Generation of Microbial Fuel Cells with Current Outputs Boosted by More Than One Order of Magnitude, Angew. Chem. Int. Ed. 42 (25), pp. 2880–2883.

Schuhmann, W. (2002). Amperometric enzyme biosensors based on optimized electrontransfer pathways and non-manual immobilization procedures, Rev Mol Biotechnol 82, pp. 425–441.

Schuhmann, W. & Muenchen, T.U. (1992). DECHEMA Monograph. 126 (1992), pp. 237–253. Siu, C.P.B. & Chiao, M. (2008). A microfabricated PDMS microbial fuel cell, Journal of Microelectromechanical Systems 17 (6), pp. 1329–1341.

Soukharev, V.; Mano, N. & Heller, A. (2004). J. Am. Chem. Soc. 126 (27), pp. 8368–8369 Suzuki, S.(1976). Hosp. Hyg. Gesundheitswesen Desinfect., pp. 159.

Tanne, C.; Göbel, G. & Lisdat F. (2010). Development of a (PQQ)-GDH-anode based on MWCNT-modified gold and its application in a glucose/O2-biofuel cell, Biosensors and Bioelectronics 26, pp. 530-535.

Tender, L.M.; Reimers, C.E.; Stecher, H.A.; Holmes, D.E.; Bond, D.R.; Lowy, D.A.; Pilobello, K.; Fertig, S.J. & Lovley, D.R. (2002). Harnessing microbially generated power on the seafloor, Nat. Biotechnol. 20 (8), pp. 821–825.

Timur, S.; Anik, U.; Odaci, D. & Gorton, L. (2007). Development of a microbial biosensor based on carbon nanotube (CNT) modified electrodes, Electrochemistry Communications 9 (7), pp. 1810–1815.

Togo, M.; Takamura, A.; Asai, T.; Kaji, H. & Nishizawa, M. (2007). An enzyme-based microfluidic biofuel cell using vitamin K-3-mediated glucose oxidation. Electrochim. Acta, 52, pp. 4669-4674.

Tsujimura, S., Kano, K., Ikeda, T., 2002. Electrochemistry 70 (12), pp. 940–942.

Tsujimura, S.; M. Fujita, H. Tatsumi, K. Kano & T. Ikeda. (2001). Bioelectrocatalysis-based dihydrogen/dioxygen fuel cell operating at physiological pH, Phys. Chem. Chem. Phys. 3 (7), pp. 1331–1335.

Vega, C.A. & Fernandez, I. (1987). Mediating Effect of Ferric Chelate Compounds in Microbial Fuel Cells with Lactobacillus Plantarum, Streptococcus lactis and Erwinia dissolvens. Bioelectrochem. Bioenerg., 17, pp. 217-222.

Wang, C.; Jia, G.; Taherabadi, L. H. & Madou, M. J. (2005). "A Novel Method for the Fabrication of High Aspect Ratio C-MEMS Structures," J. of Microelectromechal Sys.14, pp. 348

Wang, C. & Madou, M. J. (2006). Sensors: C-MEMS Based Microbattery Arrays for Miniature Sensors. 2006 NSF Design, Service, and Manufacturing Grantees and Research Conference.

Wang, S.C.; Yang, F.; Silva, M.; Zarow, A.; Wang, Y.B. & Iqbal, Z. (2009). Membrane-less andmediator-free enzymatic biofuel cell using carbon nanotube/porous silicon electrodes. Electrochem. Commun., 11, pp. 34-37.

Watkins, C; Shen, B & Venkatasubramanian, R. (2005) Low-grade-heat energy harvesting using superlattice thermoelectrics for applications in implantable medical devices and sensors. Proceedings of the 24th International Conference on Thermoelectronics (ICT 2005) Clemson, USA, pp. 265–267.

Willner, I.; HelegShabtai, V; Blonder, R; Katz, E. & Tao, G.L. (1996). Electrical Wiring of Glucose Oxidase by Reconstitution of FAD-Modified Monolayers Assembled onto Au-Electrodes. J. Am. Chem. Soc. 118 (42), pp. 10321–10322.

Willner, I.; Arad, G. & Katz, E. (1998). A biofuel cell based on pyrroloquinoline quinone and microperoxidase-11 monolayer-functionalized electrodes. Bioelectrochem. Bioenerget. 44 (2), pp. 209–214.

Yahiro, A.T.; Lee, S.M. & Kimble, D.O. (1964) Bioelectrochemistry: I. Enzyme utilizing biofuel cell studies. Biochim. Biophys. Acta. 88 (1964), pp. 375–383.

Yang, J.; Penmatsa, V.; Tajima, S.; Kawarada, H. & Wang, C. (2009). Direct amination on 3-dimensional pyrolyzed carbon micropattern surface for DNA detection, Materials Letters. 12, pp. 2680-2683.

Zhang, X.; Cheng, S.; Wang, X.; Huang, X. & Logan, B.E. (2009). Environ. Sci. Technol. 43, pp. 8456–8461.

Zhu, M.W.; Gautam, A.; Nazor, J.; Momeu, C.; Prodanovic R. & Schwaneberg, U. (2007). Directed evolution of glucose oxidase from Aspergillus Niger for ferrocenemethanol mediated electron transfer. Biotechnol. J., 2, pp. 241-248.

CHAPTER-29

PROCESSING LIGNOCELLULOSIC BIOMASS USING FEEDSTOCK IMPREGNATION FOR BIOREFINING. SWIFT AND CONSECUTIVE APPLICATION OF STEAM TREATMENT

1. INTRODUCTION: AN OVERVIEW

The initial iteration of biofuels, either from starch (ethanol) or triacyglycerol (biodiesel), relies on costly homogeneous raw materials (such as sugar cane, corn, wheat, and edible oils) in conjunction with well-established and affordable industrial methods that have been in use for many years. First generation biofuels have received negative publicity due to their high water and energy consumption, particularly the substantial amount of energy required for producing the fertilizers used in agriculture. Additionally, these biofuels have been subject to criticism regarding the competition between fuel production and food production. The growing utilization of biofuels necessitates finding alternate sources of biomass that reduce water and energy usage and do not compete with food resources. The utilization of lignocellulosic biomass, derived from either forestry or agriculture, presents significant possibilities. Cellulose is a component of the lignocellulosic matrix, which is composed of several macromolecules such as lignin, hemicelluloses, and extractives. Cellulose is the most abundant carbohydrate found on Earth. The cellulose macromolecule consists of glucose units connected by β-1,4-glycosidic linkages (also known as acetal bonds), which create lengthy chains. These chains combine to form fibrils and, ultimately, fibers. Cellulose has water resistance due to the orientation of its polar hydroxyl groups, which makes contact with polar solvents challenging. Natural macromolecules typically exist in two forms in nature: crystalline and amorphous. A typical fibril consists of alternating crystalline and amorphous zones. Although the crystalline form is resistant to disassembly by hydrolyzing agents, the amorphous phase, characterized by a certain level of disorder, allows for relatively simple entry and action of hydrolyzing agents, including enzymes and ionic species. The additional macromolecules present in the lignocellulosic matrix are also noteworthy. Lignin is a large molecule made up of phenylpropane units that are mostly connected by ether bonds, although the carbon-carbon interactions between the units are also important. Lignin possesses a low oxygen concentration, resulting in a high energetic value. Hemicelluloses are complex carbohydrates, similar to cellulose, that form macromolecules. Through hydrolysis, the C6 part of these carbohydrates can be efficiently transformed into ethanol by fermenting it with traditional yeast strains (Gírio, 2009). Research has demonstrated that it is possible to ferment all the carbohydrate components of hemicelluloses, including both C6 and C5 sugars, using nonconventional microbes (Agbogbo& Coward-Kelly, 2008; Casey et al., 2009; Chu & Lee, 2007). Furthermore, it is widely recognized that the hydrolysis of hemicelluloses results in the release of organic acids, which have the potential to impede the fermentation process of the carbohydrate.

While cellulosics offer an alternative for ethanol production, it is crucial to take into account the cost of the feedstock. Lignocellulosics used for sugar production and fermentation can be classified into three groups that are interconnected: homogeneous, quasi-homogeneous, and non-homogeneous. These classifications are based on factors such as biomass cost, quality, and transformability. The first category consists of structural and furniture wood as well as chips used for pulp manufacture, which necessitates the use of a single species or a combination of similar species. This homogeneous biomass also possesses a uniformly consistent chemical makeup, making it suitable for high-quality products that have well-established markets. The cost of homogeneous biomass in the NorthEast of America is high, exceeding $US 100 per tonne (on an anhydrous base equivalent) as of 2011. This pricing range information is provided by CRB Innovations. In addition to its cost, biomass also possesses a significant market value, particularly in the structural wood sector and the pulp and paper business. Quasi homogenous biomass typically consists of a blend of different species and, to some degree, various tissues. This category encompasses the remaining lignocellulosic biomass that is generated as a byproduct of forest or agricultural activities. The cost range for this biomass in 2011, in the NorthEast of America, when delivered to the conversion facility, is between $US 60 and 80 per tonne on a dry basis. The variation in cost is primarily due to transportation expenses within a maximum radius of 100 km from the harvesting activities. These price estimates have been provided by CRB Innovations. In contrast to the homogeneous feedstock, which has a broad and varied market yet is highly competitive, the quasi-homogeneous feedstock is less desirable due to its lower homogeneity and higher ash content. Hence, this biomass holds the potential to serve as the primary raw material for the emerging cellulosic ethanol industry due to its limited competition, non-interference with food supply, and convenient proximity to major urban centers worldwide. The final classification of biomass is non-uniform. It has worse quality compared to the previous categories and typically has a price close to 0 USD (in certain situations, there may even be an additional fee for service). The affordability is undoubtedly appealing, but the conversion process will necessitate the utilization of biomass as a comprehensive blend, transforming it into a more uniform intermediate form. This chapter will primarily focus on residual lignocellulosic material derived from existing forest and agricultural sectors, which is considered quasi-homogeneous biomass. Additionally, plantation biomass will also be discussed. Plantation biomass, often known as "energy crops," refers to crops that are planted specifically for the purpose of sequestering carbon and converting it into carbon-based macromolecules. These crops are advantageously cultivated on marginal soils and can be utilized to produce bioenergy. Several North American cultures, including willows, poplars, miscanthus, switchgrass, panic, reed canary grass, and others, have garnered attention in the past decade. Depending on the specific market being targeted, energy crops can be focused on achieving high yields of cellulose (if the primary goal is to cater to the ethanol market) or high yields of lignin with lower ash content (if the focus is on the combustion market). The primary considerations of a cellulosic ethanol plant are the expense and makeup of biomass. An important technological challenge is finding efficient methods to convert carbohydrates into monomeric sugars at a cheap cost and with high yields. The pulp and paper industry has been extracting cellulose from plants for many years. They have created large-scale facilities that efficiently convert a significant amount of lignocellulosic biomass into pulp and paper. Nevertheless, the current methods for extracting cellulose require significant amounts of water, which places a strain on water resources. In addition, the pulp and paper industry does not utilize hemicellulose and lignin for any purpose other than for combined heat and power (CHP) generation. In recent decades, research worldwide has been primarily focused on developing technologies that efficiently extract and utilize the carbon found in biomass. The aim is to establish biomass refineries that can produce a variety of valuable goods. It is necessary to carefully examine which biomass to utilize in order to obtain valuable multiproducts, and which biomass to utilize for heat and power generation. The primary technological obstacle in the manufacturing of cellulosic ethanol is the depolymerization of cellulose to get optimal yields of glucose. Cellulose, especially its crystalline parts, is a dense macromolecule that needs certain enzymes or chemicals to break down the β-1,4-glycosidic linkages through hydrolysis. The availability of enzymes and chemical hydrolytic agents depends on the three-dimensional structure of cellulose. Hence, it is imperative to have a thorough understanding of the composition of the initial feedstock and the intricate arrangement of its isolated cellulosic fraction before proceeding with the manufacture of cellulosic ethanol. This knowledge is essential for tailoring the hydrolytic processes to suit the specific ultrastructure. This chapter specifically examines three issues that are intricately connected to the manufacture of second generation ethanol. The first section will focus on reviewing the composition of various substrates in terms of their cellulosic, hemicellulosic, and lignin concentrations. This data is crucial for modifying the subsequent stages of a biorefinery's process. The second section of this chapter will focus on reviewing the steam treatments used in the Feedstock Impregnation Rapid and Sequential Steam Treatment (FIRSST) process. This process was developed over a period of twenty-five years by our team, which includes experts in academia, engineering, and technology through our spin-off company, CRB. The third section of this chapter will provide an overview of the chemical treatments used for cellulose hydrolysis. These treatments will be compared with the CRB decrystallization and depolymerization process, which was created by our team at the Université de Sherbrooke.

2. BIOMASS

Lignocellulosic biomass is an easily accessible raw material that can be obtained annually from forestry and agricultural activities. Forest leftovers consist of non-utilizable parts of tree trunks, branches, and tree tops. The composition of these residues is comparable to that of typical wood chips, as illustrated in Table 1. Agricultural residues refer to the inedible portion of a plant that remains in the field after harvesting. Typically, these residues consist of straw and stalk. It also includes the discarded portions of cultivated plants that result from industrial procedures. An example of biomass, among others, is maize cobs. Agricultural residues are the most likely source of feedstock for producing ethanol from lignocellulosic materials. This is because they are readily available, abundant, and conveniently located near the existing grain-to-ethanol platform. Energy crops, which are non-conventional plantation crops, should also be regarded as potential feedstock for biorefineries. Only a few of these crops have been produced on an industrial scale in North America. However, there has been a growing quantity of information published in recent years regarding their chemical composition. The pricing for this biomass has been assessed at $100-120 per metric ton on a dry basis, with data provided by CRB Innovations. However, the price tends to decrease due to a decrease in fertilizer usage and the utilization of marginal areas instead of high-value agricultural land. Energy crops, particularly perennial grasses, are appealing for ethanol production due to their abundant cellulose and hemicellulose content, as well as their beneficial environmental impact, subject to certain limitations.

Lignocellulosic biomass consists of cellulose, hemicelluloses, lignin, extractives, and ashes. The specific amounts of each fraction are provided below for a wide variety of lignocellulosic materials, such as agricultural residues, energy crops, and forest residues. The forest residues are further categorized into leafy hardwoods and coniferous families (Table 1). Cellulose is the main component of lignocellulosic plants, making up 30-50% of their composition by weight. This substance is a polymer consisting of D-glucose. In contrast to cellulose, hemicelluloses are a diverse polymer primarily made up of pentoses (β-D-xylose, α-L-arabinose), hexoses (β-D-mannose, β-D-glucose, α-D-galactose), and/or uronic acids (α-D-glucuronic, α-D-4-O-methylgalacturonic, α-D-galacturonic acids). Xylans and glucomannans are the predominant chemicals among them. Hemicelluloses constitute 15-35% of the plant's weight. Lignin is an amorphous polymer composed of phenylpropane units that is relatively hydrophobic. The macromolecule is mainly found between the fiber cells, serving as a binding substance that provides the wood with its stiffness and ability to withstand impact. Lignin is consistently linked to hemicelluloses by carbon-carbon and ether connections (Xu et al., 2008) and can be categorized into two primary classes as outlined by Gibbs and Thimann (1958). (1) Guaiacyl is the main component of lignins found in softwoods (gymnosperms). (2) Guaiacyl-syringyl is the combination of lignins found in hardwoods (angiosperms) and lignins found in non-woody or herbaceous crops such as grasses (angiosperms). Extractives consist of resins, lipids and fatty acids, phenolics, phytosterols, terpenes, salts, and minerals. This fraction is excluded from the manufacturing of ethanol, and for evident reasons, the ashes are also not utilized. The latter refers to the residual residue after complete combustion. The composition of this substance includes elements such as silicon, aluminum, calcium, magnesium, potassium, and sodium. Within a single biological species, the amount of each fraction can vary due to factors such as soil composition, water supply, and weather patterns. Additionally, the quantification of sugar present in the holocellulose (which is the sum of hemicelluloses and cellulose) becomes challenging during the plant's growth. The conversion of lignocellulosic materials, specifically the sugars found in holocellulose, into ethanol is achieved through fermentation. Each plant has distinct composition (Table 1); nonetheless, as previously explained, they all have the same principal components. All the biomasses exhibit similar features (Table 2) in terms of quantity, with the following order: The order of abundance of polysaccharides in general is glucan, xylan, and mannan-galactan-arabinan. However, coniferous forest leftovers have a significant amount of mannan, causing a change in the relative proportions of glucan and xylan. Table 2 also displays the mean value of the sugars containing six carbon atoms (C6), which can undergo fermentation by various types of yeast, such as S.cerevisiae, resulting in the production of ethanol. Agricultural residues, energy crops, and leafy forest leftovers have average weights of 41.5%, 46.6%, and 48.2% respectively. Additionally, the residues from coniferous forests have the potential to yield a greater amount of ethanol due to their high concentration of C6 sugars (56%), which can be attributed to a higher proportion of mannose compared to other species (about 10% more). In 2008, North America accounted for 46% of global biofuel production, according to the International Energy Agency (IEA, 2009). Additionally, research and development (R&D) activities for second generation biofuels have predominantly focused on ethanol production. Based on the findings shown in Table 2, it is feasible to calculate the amount of ethanol produced directly from C6 sugars using S. cerevisiae. We did not include the production of ethanol from C5 sugar in our study because it necessitates the employment of certain microbes. C5 sugars, despite their difficulty in undergoing fermentation to produce ethanol, can be transformed into ethyl levulinate, a valuable product that is considered a component of the extended P-fuel pool. This conversion process involves consecutive steps of dehydration, reduction, and ethanolysis. Table 3 presents a comparison of the potential ethanol production from forest residues, agricultural residues, and unutilized forest biomass in Quebec, Canada, and North America. This is contrasted with the current production of ethanol from energy crops (first-generation biofuel) and the consumption of gasoline. Only a quarter (25%) of the forest and agricultural leftovers have been considered in the estimation of ethanol production, as the remaining biomass is already allocated for other purposes. Our study provides a result for the unexploited forest zone that can be utilized without harming biodiversity, as stated by the Ministère des Resources Naturelles et de la Faune (MNRF) in Quebec in 2009.

Table 1.Chemical composition of various lignocellulosic materials

	Cellulose (wt%)	Hemicellulose (wt%)	Lignin (wt%)	Extractives (wt%)	Ashes (wt%)
Agricol residues					
Rice strawa	41.2	19.5	21.9	-	-
Wheat strawa	39.7	36.5	17.3	-	-
Rye strawb	37.9	36.9	17.6	-	-
Triticalstrawv	34	31.7	17	12.4	
Flax strawc	53.8	17.1	23.3	-	3.6
Sun flower stalka	37.6	29.3	10.3	-	-
Sorghum stalka	41.5	24.4	15.6	-	-
Cotton stalka	58.5	14.4	21.5	-	-
Corn stoverd	38	26	19	-	6
Barley strawe	42	28	-	-	11
Vine shootsa	41.1	26	20.4	-	-
Olive pruningsa	35.7	25.8	19.7	-	-
Energy crops					
Switchgrass	37	28	16.4	15	3.7
Miscanthus	40	18	25	-	-
Big bluesternf,g,h	37	28	18		6
Little bluesternh	35	31	-	-	7
Prairie cordgrassi	41	33	-	-	6
Indian grassf,h	39	29	-	-	8
Intermediate wheatgrassg	35	29	-		6
Reed canarygrassj,k	24	36	-	-	8
Smooth bromegrassj,k	32	36	-	-	6
Tymothybl,m	28	30	-	-	6
Tall fescuen	25	25	14	-	11
Sundan grassj,k	33	27	-	-	8
Jatropha stemh	37.1	30.6	22.3	-	-
Jatropha seed cakeo	13.5	26.8	12.4	-	-
Cannabis sativav	43	26	14.5	16	3.2
Salix viminalisv	30.2	33.5	29.2	8.8	-
Forest residues					
Leafy					
Soft maple	41	35	24	-	-
Red oakp	35.5	18.8	29	-	-
European oakq	38	29	25	4.4	0.3
White oakr	44	24	24	5.4	1
Chesnut oakr	41	30	22	6.6	0.4
Post oakr	38	30	26	5.8	0.5
White birchr	45	33	18	5	0.3
Yellow birch	40	39	21		

	Cellulose (wt%)	Hemicellulose (wt%)	Lignin (wt%)	Extractives (wt%)	Ashes (wt%)
Quaking aspenr	49	29	19	6	0.4
White elm	49	27	24	-	-
Beech	42	36	22	-	-
Basswood	-	-	-	-	-
Poplarwood	41.4	23.7	24.5	-	-
Eucalyptusa	52.8	27.7	20	-	-
		Coniferous			
Pinus pinastera	55.9	13.7	26.2	-	-
Pinus radiate chipsc	53	15.8	23.7	-	-
Cedars	43.5	20.3	-	-	-
Eastern Hemlock	42	26	33	-	-
Eastern white cedar	44	25	31	-	-
White spruce	44	29	27	-	-
Jack pine	41	30	29	-	-
Tamarack	43	28	29	-	-
Sprucet	54.1	21.4	24.4	-	-
Loblolly pineu	43.6	21.2	26.8	3.2	0.4
Balsam-fir	44	27	29	-	-

a(Rodríguez et al., 2010); b(Sun et al., 2000); c(Schafer & Bray, 1929); d(Lee et al., 2007); e(Mani et al., 2008); f(Lee & Owens, 2008); g(Owens et al., 2006); h(Jefferson et al., 2004); i(Boe& Lee, 2006); j(Jung et al., 1997); k(Jurgens, 1997), l(Alvo et al., 1996), m(Claessens et al., 2004); n(Department of energy, 2006); o(Liang etal., 2010); p(Mazlan et al., 1999); q(Bednar & Fengel, 1974); r(Pettersen, 1984); s(Yamashita et al., 2010); t(Yildiz et al., 2006); u(FrederickJr et al., 2008); vMesured in our laboratory.

In the case of forest residues, we can assume that for 1 m3 of roundwood exploited, 0.6 m3 of residual biomass is left behind (Smeets & Faaij, 2007). In the province of Quebec, forest residues have been estimated to 6.9 millions of tons per year (Goyette & Boucher, 2009). Thus, the production of ethanol from glucose fermentation can be determinated assuming that the average of this sugar in such materials is about 52.4% (calculated from the Table 2) and that the maximum yield is equal to 0.51g of ethanol per g of glucose. Thus, 584 millions liters of ethanol could be produced in Quebec. To put such a value in perspective, consumption of refined petroleum in Quebec reached 9 billion liters in 2007 (Ministères de RessourcesNaturelles et de la Faune, MNRF, Quebec, 2009). The production of ethanol from forest residues is sufficient to reach the objective fixed by the government (5 vol% in gasoline in 2012) since it represents 6.5 vol%. The North American consumption of gasoline for transport was estimated in 2008 at 518 739 millions liters with respectively 479 243 millions liters for the United States of America and 39 496 millions liters for Canada (IEA energy statistic, 2010). More ethanol can be produced by using agricultural residues, energy crops and unexploited forest zone. In Quebec, the latter represents 14,100,000 m3 or 5.6 millions tons assuming an average density of 400kg/m3. Thus, 1 638 millions liters of ethanol can be produced per year. Quebec will be able to replace 24.7 vol% of its gasoline by ethanol just by using exploited forest zone and forest residues, thus using only residual biomass. The comparison between the gasoline consumption and the possible production of ethanol from residues shows undeniably the importance of such source of raw material. Furthermore, the production of ethanol coming from the latter could rise above the production of ethanol coming from the first generation of biofuel. As an example, in North America, the nameplate production of ethanol from grain represents 53 949 millions of liters per year while the exploitation of residues could give more than 60 000 millions of liters per year.

Table 2.laboratoryDetails of carbohydrates and lignin amounts present in various lignocellulosic materials

	Glucan (wt%)	Xylan (wt%)	Mannan (wt%)	Galactan (wt%)	Arabinan (wt%)	Lignin (wt%)
Agricol residues						
Corn stovera	39	14.8	0.3	0.8	3.2	13.1
Rice strawa	41	14.8	1.8	0.4	4.5	9.9
Rice hullsa	36.1	14	3	0.1	2.6	19.4
Wheat strawa	36.6	19.2	0.8	2.4	2.4	14.5
Triticale strawj	43.5	17.7	-	-	2.3	17
Sugar cane	41.3	21.8	0.3	0.5	1.8	-
C6 Average	39.5		1.2	0.8		
Energy crops						
switchgrass	35.2	21.7	0.2	0.9	2.8	27.4
Miscanthusb	44	21	-	-	-	-
Hempj	51	14.3	1.5	0.7	1.3	14.5
Sweet sorghumc	44.6	25.3	-	-	-	18
Bagasse fiber	38.1	13	8	-	2	20
C6 Average	42.6		3.2	0.8		
Forest residues						
Leafy						
Populus tristisa	40	13	8	-	2	20
Oak	45.2	20.3	4.2	-	-	-
Red Mapplei	46.0	19.0	2.4	0.6	0.5	24
Aspend	45.9	16.7	1.2	0	0	23
Salixe	41.4	15.0	3.2	2.3	1.2	26.4
Yellow poplarf	42.1	15.1	2.4	1	0.5	23.3
Eucalyptusf	48.1	10.4	1.3	0.7	0.3	26.9
C6 Average	44.1		3.2	0.9		
Coniferous						
Spruceg	43.2	5.7	11.5	2.7	1.4	28.3
Lodgepole pineh	42.5	5.5	11.6	2.1	1.6	27.9
Ponderosa pineh	41.7	6.3	10.8	4.7	1.8	26.9
Douglas-firi	44	2.8	11.0	4.7	2.7	32
Loblolly pinei	45	6.8	11.0	2.3	1.7	28
Red pinei	42	9.3	7.4	1.8	2.4	29
C6 Average	43.1		10.6	3		

a(Lee, 1997); b(Sørensen et al., 2008); c(Ballesteros et al., 2004); d(Wang et al. 2008); e(Sassner et al., 2008); f(Zhu & Pan, 2010); g(Zhu et al., 2009); h(Youngblood et al., 2009); i(Pettersen, 1984); jMesured in our our laboratory

Table 3.Comparison between the actual possible production of ethanol from C6 sugars contained in the lignocellulosic biomass, the operational ethanol production from energy crop and the gasoline consumption

	Possible Production of ethanol (Millions of liters per year)from			Production of ethanol operational from energy crop (Millions of liters per year)	Gasoline consumption (Millions of liters per year)
	Agricultural residues (25%)	Forest residues (25%)	Unexploited Forest		
Quebec	-	584	1638	155(a)	9000(d)
Canada	5097(e)	3353	-	1821(a)	39496(f)
North America	41483(e)	20322(c)	-	53949(b)	518739(f)

(Canadian Renewalable Fuels Association, 2010), (b) (Renewable Fuels Association, 2010) (c) Estimated from the production of roundwood; (d) (Natural Ressources Canada, 2007) (e) Estimated from FAOSTAT (FAOSTAT, 2010) thanks to the coeffient of residues proposed by D. Bellerini (Bellerini, 2006); (f) IEA energy statistic (IEA, 2011)

In this estimation of the possible production of ethanol, marginal lands have not been taken into account. The potential of surplus land for the cultivation of energy crops like willows, poplars, miscanthus, switchgrass, panic, reed canary grass (second generation of biofuel) considerably depends on the regions. Numerous constraints exist for the implementation of new energy crops, making the estimation of biofuel production very approximative. The first one is food competition. In fact, in major countries, populations are growing, consequently increasing the food demand therefore reducing surplus land, which overall limits the additionnal production of energy crops. Among the others constraints we can mention, the water shortages, the implementation of indigeneous species (could be a probleme for biodiversity), the type of plant, improvement of agricultural system (allowing the cultivation of other land), etc. Depending on the scenario, disparate results are obtained; bioenergy could reach between 39EJ to 204 EJ in 2050, furthermore, Smeets et al. 2007 show that the biggest energy apport will be done by dedicated energy crops with 20-174 EJ of biomass against 6-11 EJ of agricultural residues and 6 EJ of forestry residues.

More than replacing gasoline coming from fossil ressources, the employement of biofuels in well defined conditions can contribute to reduce the Greenhouse Gas (GHG) emissions. The GHG balance varies significantly following the choice of biomass, the technology employed throughout the full « fuel cycle » from biomass production to final fuel consumption, the caracteristic of the land and climate, the crop management, etc. Thus, the choice of biomass is essential. As for ethanol production, the potential of lignocellulosic biomass to reduce GHG is comprised between 60 and 120 % and it is comparable to the high diminution of GHG observed with sugar cane (90%). In comparison, production of ethanol from wheat grain brings a lesser gain of 20 to 50% (IEA, 2004). The reduction, especially for lignocellulosic biomass, is due to the compostion of plant itself, to fertilizer loading and to the efficiency of vehicles. The high reduction of carbon dioxide emissions in the case of lignocellulosic biomass (cellulose to ethanol) essentially comes from the use of the other part of the plant (mainly lignin) as a source of energy for the process. However the previous estimations do not take into account the modifications affecting the lands. In fact, the GHG balance of the second-generation biofuels are closely related to the land use change (LUC) and the indirect use change (ILUC) which could in certain case conduct to a negative GHG balance. When a prior land-use like forest is replaced by culture for biofuel production, a direct land-use change occurs which can change the carbon stock of that land. This aspect has been widely studied and factors of changing balance can be found in the literature. However, the changes on GHG are induced by ILUC (takes place when land use change implies the displacement of the previous activity on another land). Thus, the replacement of sparsely vegetated and certain grass land by energy crops could generate a positive effect on GHG and in the mean time participate to the stockage of carbon in soil. Contrarily, the estimation made by Farrel and O'hare (2008) on the ILUC GHG emission shows that if the actual crops of soybean are used for the production of ethanol, the result could lead to the expansion of soybean for food into forests and will conduct to more emission than the use of fossil ressources (6 times more). These cultivations can also have several positive or negative impacts on soil, water and biodiversity. ILUC are normally less important if residues are used as feedstock since there is no need for additional land to be cultivated.

The second-biofuel generation, and in particular in the case of lignocellulosic ethanol production, should start with a sustainable development of agricultural and forestry residues which are at that time very interesting in terms of productivity (as it was shown on the Table 3) and environment. Even if some species contains much more C6 sugars like coniferous and notably loblolly Pine, the use of a wide range of biomass genotypes is advised and needed. All the species presented show an interest for ethanol production. In fact, the use of just the best species would be catastrophic for biodiversity. In general, lignocellulosic biomass is a promising source of fuel as it is shown on the figure 1. As an example, in about ten years the production of biofuel via the lignocellulosic biomass in the US could almost reach the same production as from the other sources of biofuel.

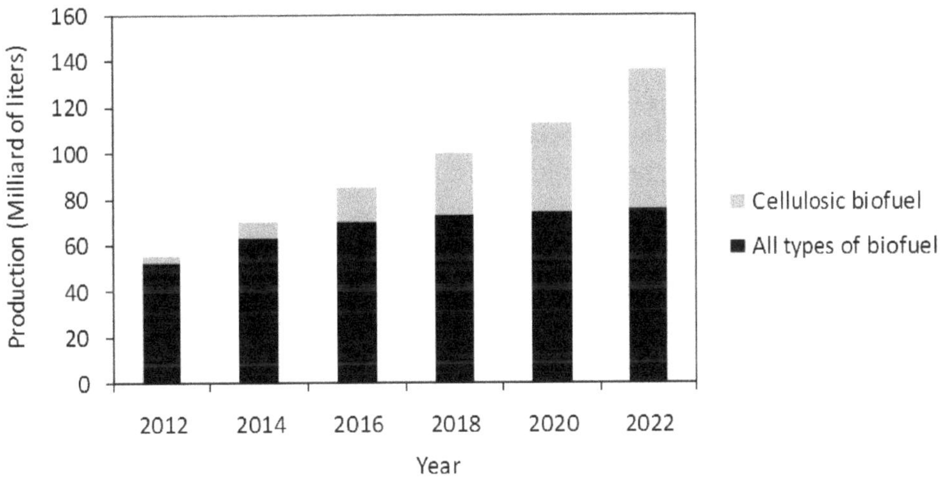

Figure 1.Biofuel Mandate from lignocellulosic materials in the United States Renewable Fuels Standard

3. FRACTIONATION

3.1. Basics informations about steam treatments

The first technological challenge restraining the commercialisation of the cellulosic biomass industry is the fractionation of the lignocellulosic biomass to isolate the cellulose macromolecule. Similar processes have been applied by the pulp and paper industry for decades but there is a new objective now, the complete utilisation of the carbon-based structures of the biomass as well as a reduction of the water consumption, hence, the concept of biorefinery. Although the traditional pulping processes are being remodelled to fit with the biorefinery approach, other processes are also investigated among which are the organosolv process and steam treatments. There is a significant variety of different steam treatments which all rely on the same concept: biomass is first saturated with a solvent (usually water) with or without the utilisation of a catalyst (acid or basic depending on the targeted macromolecule). The mixture is then "cooked" by addtion of steam in a pressure-resistant vessel for a certain period after which a valve is open, and the vapor phase exits the vessel through a nozzle entraining the solids. The exiting vapor reaches very high velocities as a function of the geometry of the nozzle thus reaching a sonic velocity. A "explosion" takes place while induced by the sonic field. The water saturating the biomass in its pores rapidly expands to vapour causing cell changes which vary from simple fibrilar disaggregation to fragmentation. During cooking, water, at high pressure and temperature, has a high dissociation constant leading to the occurence of a larger quantity of hydronium ions directly formed in the saturated pores of the biomass. Hemicelluloses, which are highly ramified and relatively easy to hydrolyse (in comparison to cellulose), will be affected by the increasing concentration of ions in the solution. The reaction of water and biomass is of course a major concern when considering steam treatments and it was found that in the absence of a catalyst, the two most important factors that were related to fractionation of the biomass and the hydrolysis of hemicelluloses were temperature and cooking period. The relationship between both parameters has been related to a mathematical equation called the "severity factor". This equation, reported first by Overend and Chornet (1987) has been from this point forward a significant contribution for the homogeneization of the steam processes.

$$R_0 = \int_0^{t[\min]} \exp\left(\frac{T[^\circ C] - 100}{14.75}\right) * dt \quad (1)$$

$$S_o = \log R_o \quad (2)$$

Where So is the severity fractor, T is the temperature (expressed in degree celcius) and the overall equation relates on the integral of the temperature curve between the start and the end of the cooking period, including the preliminary heating leading to operating temperature. Severity factors were also related to the relative hydrolysis of the hemicelluloses macromolecule and (Overend and Chornet., 1987) showed that at a severity factor of 4, hemicelluloses were completely hydrolysed and there was starting to be an impact on the cellulosic fiber. This concept, can serve as a guide for other substrates although it was shown to vary from one feedstock to another, mostly because of the varying nature and amounts of hemicelluloses found in the biomass (Lavoie et al., 2010a, 2010b, 2010c). Impact of the calculated severity factor has also been reported for other feedstocks as residual cotton and recycled paper (Shen et al., 2008), aspen wood (Li et al., 2005), douglas fir (Wu et al., 1999), from rice husk and straws (Gerardi et al., 1999), from yellow poplar, from peanut hulls and from sugar cane (Glasser et al., 1998). In most of the cases reported previously, the ideal severity factor for the isolation of cellulose and hydrolysis of hemicelluloses was found to be between a severity factor of 3 and 4. A non-catalytic steam process of biomass usually leads to a brown lignocellulosic fibre as depicted in Figure 2 below:

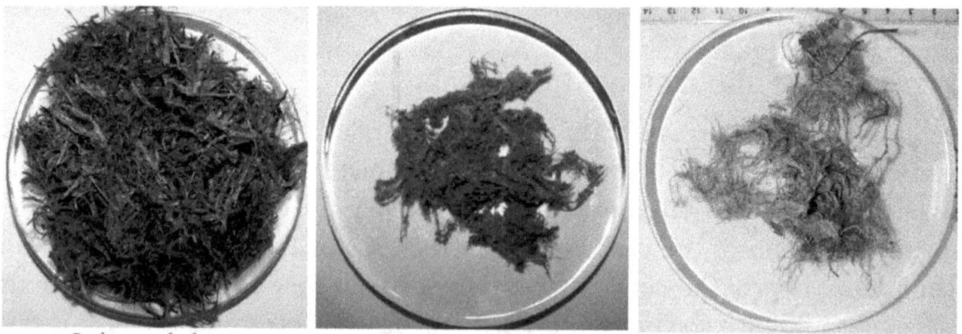

Figure 2.Lignocellulosic matrix obtained after a non-catalytic steam treatment of different biomasses with a severity factor between 3 and 4

Although hydrolysis of hemicelluloses can be performed only using the natural dissociative potential of water, many researches have investigated the effects of including a catalyst on the overall outcome of the process. Both acid and basic catalyst has been considered and each will have the tendency to target one type of macromolecule more than the other. Whilst acids will have a more pronounced effect on cellulose, bases will have a more significant effect on lignin. Among the acids that were used for catalysis of steam explosion reaction, sulphuric acid is one of the most common but also one of the less expensive at an industrial level. Utilisation of the latter has been reported repetitively in literature (Lawford et al., 2003; Emmel et al., 2003; Ballesteros et al., 2001). Directly comparable to sulphuric acid, sulphur dioxide was also widely used as a catalyst for steam explosion. The latter will interact with biomass and react with water to produce in turn sulphuric acid. The main difference might be that utilisation of SO2 would allow a more homogeneous distribution of the acid catalyst in the biomass since the diffusion of the gas should be higher than the sulphuric acid molecule. Such a treatment has been effectively applied on lodgepole pine (Ewanick et al., 2007), poplar (Lu et al., 2009), aspen (De Bari et al., 2007) and eucalyptus (Ramos et al., 1999). The acid catalyst will have a direct effect on the hydrolytic potential of the mixture increasing the natural hydrolytic potential of water considerabily. As for the previously mentioned severity factor, researchers have tried to translate this phenomenon into an equation. Abatzoglou et al. (1992) were able to introduce the concentration of acid into the calculation of the severity factor as depicted below:

$$R_{0H} = \exp\left(\frac{X - X_{ref}}{\lambda X_{ref}}\right)\exp\left\{\frac{T - T_{ref}}{\omega'}\right\}t_R \quad (3)$$

Where X and Xref is the acid loading (g of acid/g of dry biomass) and reference (acid loading, g of acid/g of dry biomass) respectively, λ is a parameter expressing the acid catalyst role in conversion of the system, ω' is parameter expressing the temperature role in conversion of the catalysed reaction system and tR is the reaction time. A couple of years later, Montane and co-workers (Montane et al., 1998) developed a new version of the equation which included slight modifications over the equation proposed by Abatzoglou et al. The equation is depicted below:

$$R_0 = \exp\left(\frac{1}{\omega_0}\left(1 - \frac{T_{ref}}{T}\right)\right)\frac{t^{\gamma}}{\gamma} \quad (4)$$

In this equation the ω0 parameter express the energetic of the process respect to a reference reaction temperature, T and t remains the temperature and time whilst γ defines the shape of the distribution of activation energies. The research also showed that it was possible, for a specific species, to estimate the whole conversion of the process using a single equation:

$$(1 - f) = \exp\left(-1.06x10^{10}\exp\left(\frac{-9733}{T}\right)C^{0.674}\frac{t^{0.608}}{0.608}\right) \quad (5)$$

Where f is the conversion parameter and C is the catalyst concentration. Equation 5 has been developed by Montane et al. using birch as substrate for the steam explosion process. Utilisation of an acid catalyst, for similar temperatures and times, should allow a more complete hydrolysis of the hemicelluloses but should also attack the cellulose molecule which is overall sensitive to the occurence of protons. It has been mentioned and it is still widely studied that the interactions between the hydronium and the cellose macromolecule may lead to a more efficient hydrolysis to glucose when used as a pretreatment for an enzymatic treatment (Dererie et al., 2011; Khunrong et al., 2011; Zhang et al., 2009). In most of the previously mentioned situations, utilisation of the catalyst leaded to increased value for conversion following the enzymatic hydrolysis, although in some specific cases, even if the conversion to glucose was increased, the fermentation was strongly inhibited by the production of furfural-derived compounds. Dehydration of xylose to furfural is depicted in Figure 3.

Dehydration of carbohydrates is strongly induced by acid catalysts at temperature higher than 150º C with a classical inorganic catalyst although lower operating conditions were reported for the utilisation of ionic liquids (Tao et al., 2011). Five-carbon carbohydrates will dehydrate to furfural whilst dehydration of C6 sugars will lead to 5-hydroxymethylfurfural (5-HMF) (Zhang et al., 2010). The latter will usually be less concentrated in a steam explosion process since it will require an isomerisation of the aldohexose sugars to a ketohexose form. Furthermore, under acid catalyst, 5-HMF has been reported to undego spontaneous hydrolysis to levulinic acid and formic acid which are both fermentation inhibitors. The minimal concentration at which furfural starts to inhibit fermentation has been reported to be at 2-3 g/L (Palmqvist et al., 1999) whilst as for 5-HMF, it has been reported that the concentration that causes 50% inhibition of fermentation was of 8 g/L.

Base-catalysed steam explosion may also be a potential pathway since the occurence of hydroxide ions, as in the case of kraft pulping, would lead to the hydrolysis of the hemicelluloses as well as the lignin whilst allowing the isolation of cellulose. Utilisation of NaOH as a catalyst for steam explosion has been reported in literature (Zhuang et al., 1997; Li et al., 2005), although less frequently in comparison to the acid-catalysed reaction. Another process called ammonium fiber explosion is also slightly comparable to a base-catalysed steam explosion since it will allow defibration of the feedstock in a first time, then the interaction of ammonia with water can be directly related to an hydroxide ion catalyst although part of the hydrolysis process could be related to the ammonia itself although it is highly soluble in water and it interacts in an classical acido-basic reaction to produce ammonium hydroxide. This concept was efficiently tested on rice straws (Vlasenko et al., 1997) and the process itself has been patented by Dale et al. (2008).

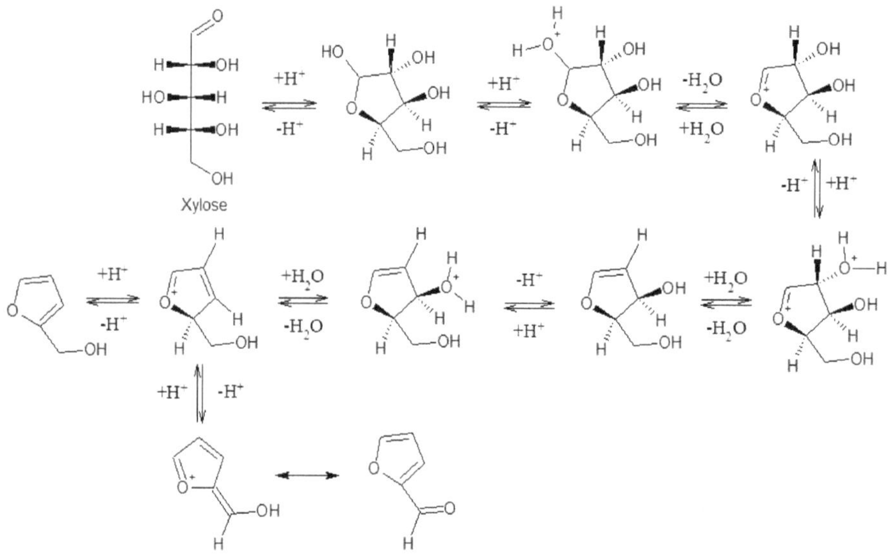

Figure 3.Example of dehydration of a carbohydrate to furfural under acid catalyst (in this case xylose)

3.2. The Feedstock Impregnation Rapid and Sequential Steam Treatment (FIRSST)

Since a non-catalytic and an acid steam treatment allowed targeting the carbohydrate-based macromolecules from the biomass and the based-catalysed reaction allowed partial depolymerisation and solubilisation of lignin, our group has developed the two step FIRSST (Feedstock Impregnation Rapid and Sequential Steam Treatment) process. The biomass is first reduced in size to a range 3 to 6 cm long. It then follows the process flow diagram depicted below (Figure 4).

Biomass is first extracted with water, solvent and/or a mixture of both to extract the secondary metabolites. Two reasons justify the preliminary extraction, first some of the compounds could have a bioactive potential thus leading to applications in cosmetics and pharmaceuticals. Secondly, the extractives could act as inhibitors for fermentation and depending what is the targeted application for the broth obtained after the first steamexplosion, it might be beneficial to remove such compounds. After extraction, biomass is rinced with a minimal amount of water to remove traces of the residual solvent or to ensure maximal removal of extracts. Typically, at the bench scale level, a 5/1 massic ratio of water/ biomass is used at this point. Biomass is then impregnated with water to ensure maximal penetration of the aqueous medium in the biomass' pores. Impregnation could be performed with or without pressure, either positive or negative.

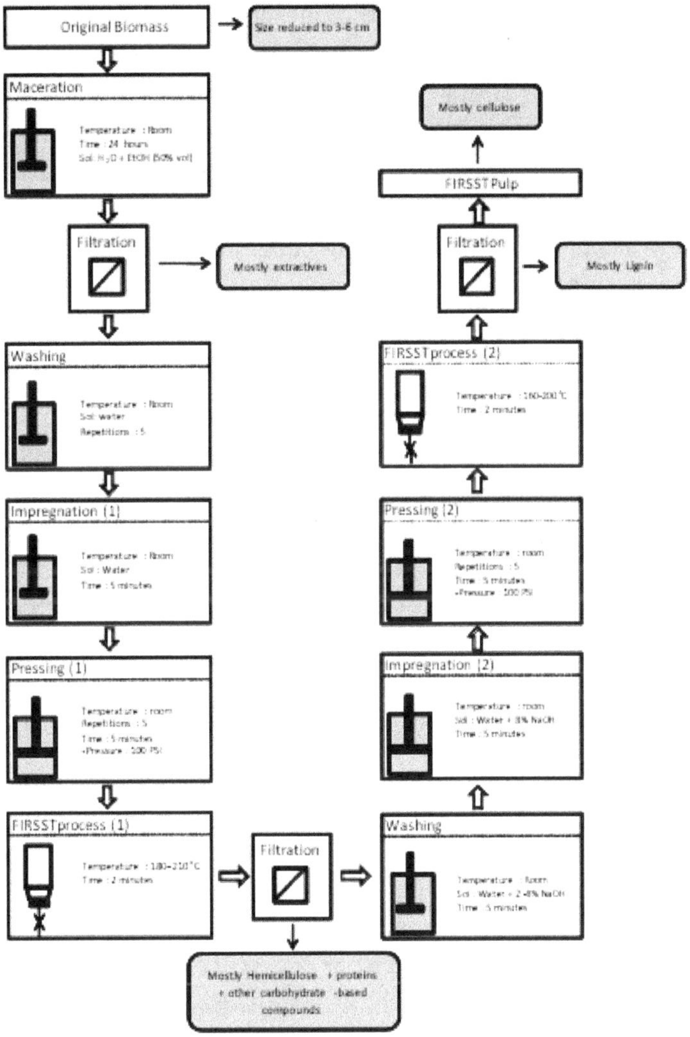

Figure 4.Process flow diagram for the two step FIRSST process allowing isolation of the cellulose fibre as well as lignin, hemicelluloses and extractives with minimal purification required.

Saturation with water is one of the key elements for performing an efficient steam process and has to be monitored carefully. Impregnation could be performed by letting the biomass soak in water for a time period (typically up to 24h) allowing the water molecules to fill the small pores via capillarity. Whilst both positive and negative pressure might be used, utilisation of a positive pressure to ensure water penetration is by far the most efficiently scalable approach. After impregnation, excess water has to be removed, a pressure of 100 psi is sufficient both for the pressurized impregnation process as well as the following excess water removal. Once excess water is removed, the biomass is transferred in the FIRSST reactor where it is cooked for 2-4 minutes whilst monitoring the severity factor of the whole process. In a two step FIRSST process, one must ensure that the severity factor of the first process is not excessively high or the following delignification process, although efficient, will lead to excessive conversion and a lignin-carbohydrates broth after the second process.

Once the first FIRSST process is completed, the biomass is once more impregnated but this time with an aqueous diluted NaOH solution (typically 1-10%). Impregnation as well as removal of excessive solution was performed at 100 psi and room temperature. The residual lignocellulosic matrix is then cooked at temperature comparable to the first process although typically in a 10º C inferior temperature range. Once the second cooking period is completed, biomass is rinced with a 10/1 ratio water/fibre to ensure removal of the remaining sodium ions. Such process has been tested on different feedstock including energy crops (Lavoie et al., 2010a), residual forest biomass (Lavoie et al., 2010b) and different agricultural residues (Lavoie et al., 2011). When using based-catalysed steam process, the catalyst itself, as in the case of acid hydrolysis, becomes an important factor of the reaction. Increasing the concentration of the basic catalyst showed to have a direct impact on lignin removal and using the same conditions (time and temperature), it was shown that an increasing alkali concentration in the mixture allowed the production of a whiter fibre. The texture of the fibresare strongly affected by the severity of the steam processes, example of the different textures of fibres produced according to the two-step FIRSST process used on triticale straws are depicted below (Figure 5).

At this point, the two-step FIRSST process has allowed the isolation, in high yields, of all the fractions of lignocellulosic biomass whilst producing pulp with good mechanical properties. Example of mechanical properties obtained from FIRSST pulps are depicted in the Table 4 for Salix vimimalis (Lavoie et al., 2010a), for a mixture of softwood (Lavoie et al., 2010b) and for Cannabis sativa (Lavoie et al. unpublished results). Opportunity to produce quality pulp is yet another advantage of this technique which would, from the same feedstock, allow the production of many derived products including ethanol and cellulose.

For the first step of the steam explosion and as mentionned earlier, no catalyst is used since the hemicelluloses and/or protein found in the lignocellulosic matrix were shown to be directly affected by water at the operating conditions. The first step of the FIRSST process is usually performed between temperatures of 180-230º C depending on the nature of the biomass used as a feedstock. As an example, residual forest biomass was shown to require more severe conditions in comparison to residual agricultural biomass as triticale or hemp. Cooking period is usually ranging from 2-4 minutes, but it most of the cases investigated by our team, a 2 minute cooking period was shown sufficient. The uncatalyzed steam explosion process can be related to the severity factor that is calculated from the cooking temperature and time, needless to underline the fact that similar severity could be obtained by increasing the cooking period whilst decreasing the temperature. The 2 minutes cooking time usually excludes the heating period where biomass is heated to the operating conditions, nevertheless, this heating period is taken into consideration when calculating the severity factor. The heating period for the FIRSST process varies from 10-30 seconds and the temperature of biomass is monitored with thermocouples strategically located in the steam explosion reactor and recorded on an acquisition system allowing control and downstream calculations with regards to the conditions of operation.

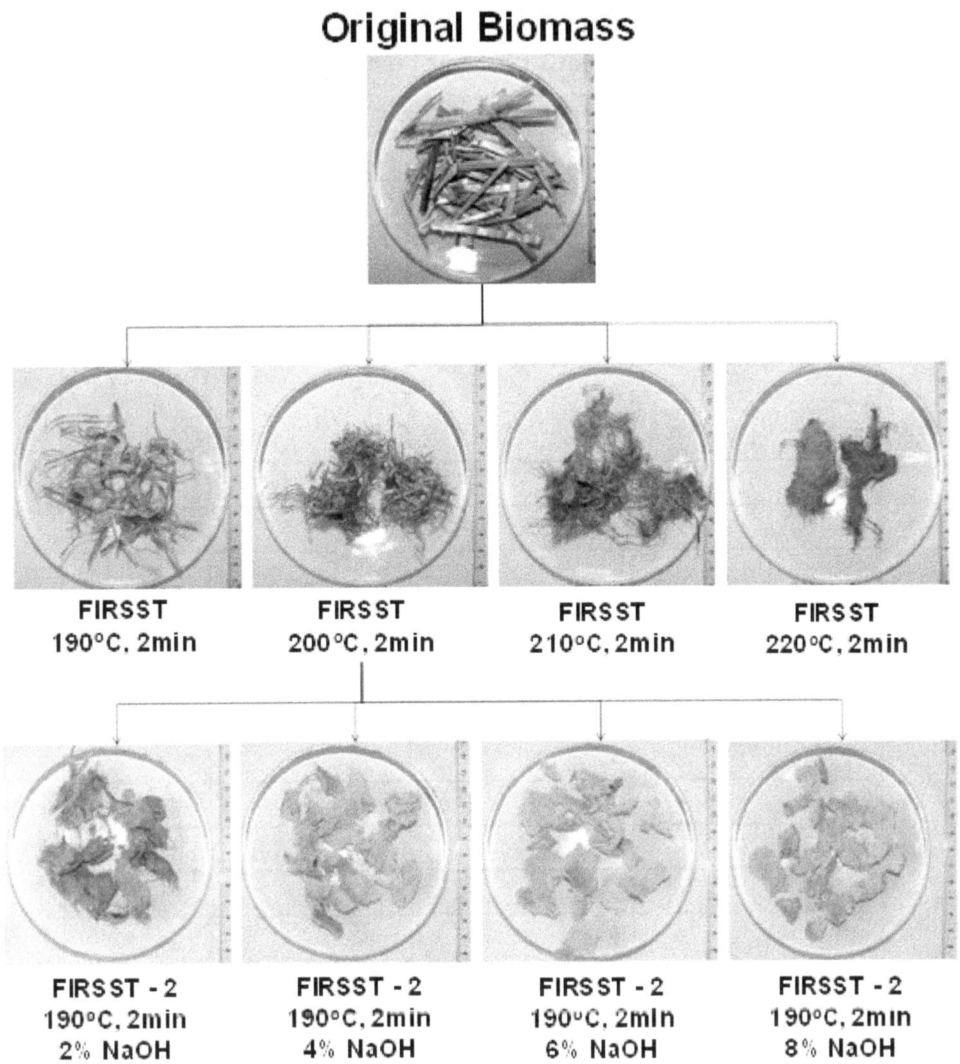

Figure 5.Lignocellulosic and cellulosic fibresobtaineds with the two step FIRSST process applied to triticale straws

Another important factor that has to be taken into consideration when applying the 2-steps FIRSST process concerns the production of high quality pulp which may be tricky because of the severity of the reaction. Therefore, even if the second steam process is often based catalyzed, the severity of the treatment will lead to an indirect attack on cellulose which will reduce the fibre length if not suitably controlled. If ethanol production is intented, then the quality of the fibres produced after the FIRSST process is of lesser interest and both FIRSST treatments can be relatively severe. When dealing with quality cellulose fibres production, one must think about reducing the strength of the first FIRSST treatment in order to cope with the severity of the second treatment. So far the approach that has been developed in order to cope with such problem was to evaluate the composition of the lignocellulosic matrix at different severities for a first uncatalyzed FIRSST process. Optimal severity for the first process would lead to a drastic decrease of the hemicelluloses content whilst alterating minimally the long cellulose fibres (which are determined by quantification of the hemi, holo and α-cellulose). During the first FIRSST process, the holocellulose concentration will progressively decrease whilst α-cellulose will remain constant. We identify the maximal severity at the point where α-cellulose starts to significantly decrease. An example of such is depicted in Table 5.

Table 4.From Lavoie et al. 2010a 2From Lavoie et al. 2010b Mechanical properties of FIRSST for different types of feedstock.

Standard test	Willow[1]	Softwood[2]	Hemp
ATPPC C.1	454	664	407
ATPPC C.12	0.16	5.58	n.d.
ATPPC C.5U + 14 + 28 + 48 + 100 + 200 - 200	n.d.	47.8 20.0 11.8 9.5 4.0 7.0	n.d.
ATPPC B.4P	0.41	2.08	n.d.
ATPPC D.3	60.1	59.6	59.2
ATPPC E.1	33.3	24.8	29.4
ATPPC E.5 L* a* b*	71.45 ± 0.01 2.18 ± 0.01 12.42 ± 0.01	67.26 4.41 18.31	68.90 2.74 13.79
ATPPC E.2	99.5	98.5	99.0
ATPPC D.4	1.79	2.24	2.90
ATPPC D.9	3.14	8.04	11.8
ATPPC D.8	1.56	3.05	<0.34
ATPPC D.34 Lenght of rupture	4.24	5.03	2.29
ATPPC D.34 TEA	19.6	37.8	8.8
1From Lavoie et al. 2010a			
2From Lavoie et al. 2010b			

Table 5.Mechanical properties of FIRSST for different types of feedstock.

Conditions	Severity	Conversion	Lignin	Holocellulose	α-cellulose*
-	Eqn [1,2]	%wt	%wt ± SD	%wt ± SD	%wt ± SD
210º C, 2 min.	3.64	20.03	28 ± 2	51 ± 2	36.2 ± 0.7
220º C, 2 min.	3.95	22.00	29.8 ± 0.5	47 ± 5	37 ± 1
220º C, 4 min.	4.20	23.02	32 ± 3	44 ± 6	34.6 ± 0.4
230º C, 2 min.	4.27	24.94	31 ± 8	43 ± 3	33 ± 1
Percentages are expressed in terms of bone dry biomass. Deviations from 100% closure are due to ash content. * Comprised in the holocellulose					

For the specific case depicted in Table 5, we can assume the severity factor varying from 3.64 to 3.95 only had a minor impact on the alpha cellulose and therefore we can assume that the treatment targeted the hemicelluloses principally. The quantity of removed hemicelluloses can be assimilated to the conversion also presented in the same table which shows an increasing mass content in the lignin broth produced from the FIRSST process. In a situation where quality pulp would be intended, although the best conditions for removing all the hemicelluloses would be at a severity factor of approximatively 4.00, it would be more strategic to use a lower severity first process (210º C, 2 minutes as an example) to ensure that the second steam treatment will not affect to much the pulp quality downstream.

The second process is also tricky since ideally and economically, lignin should be recuperated at >80%. Increasing the severity of the treatment will lead to very low lignin content but will also affect the cellulose fibres. To a certain extent, concentration of lignin inferior to 1% can be obtained using steam treatment. The downside will be that the higher severity will also affect the cellulose fibres and overall, the conversion will be higher and the quantity of pulp recovered downstream will also be lower with regards to the original quantity of biomass used for the process. An example of this is depicted in Table 6 below:

Table 6.Pulp yield and composition (lignin and holocellulose) following delignification of the filtered solids (76.98 wt% of the original biomass, dry basis) obtained from the first steam treatment performed at 220º C during 4 minutes from softwood biomass(From Lavoie et al. 2010b)

Conditions	Yield*	Lignin	Holocellulose
-	%wt	%wt ± SD	%wt ± SD
190º C, 2 min.	48.52	24.8 ± 0.2	75.3 ± 0.2
200º C, 2 min.	41.35	14.5 ± 0.1	85.6 ± 0.1
210º C, 2 min.	37.60	6.9 ± 0.1	93.2 ± 0.1
220º C, 2 min.	37.23	3.2 ± 0.4	96.9 ± 0.4
*Yield is expressed in terms of %wt of dry biomass and can therefore be considered as the rate of pulping for the overall process.			

In light of the results depicted in Table 5, it is clear that increasing severity will also affect the overall conversion which will lead, at higher severity to a lower production of pulp. Other less severe technique could be used for removing the residual lignin and it should be considered to accentuate the pulping yields and to reduce to chemical alterations that the process may have induce to the cellulose fibres. On the other hand, if hydrolysis of glucose is intended, it was previously reported that severe steam explosion process could have a beneficial effect on enzyme hydrolysis and therefore, reaching higher severity may be beneficial to the whole process even if part of the cellulose is solubilized in the lignin broth.

Typical energy consumption for each of the steam process is approximatively 7% of the net biomass calorific value which represents approximatively 1.4 GJ per tonne of biomass process. A two step process could easily lead to a 15 % which now justifies investigations towards a one step process. Using triticale, our group compared the two- and one- step steam process. Since lignin has to be partially hydrolysed and removed, the one step steam process will require utilisation of a base-catalyst.

Althought the process was show efficient for the isolation of cellulose from triticale straws (unpublished results), the downside of such process is that it will generate a broth containing both hemicelluloses and lignin. Although lignin can be precipitated by reducing significantly the pH and mildly heating, it will nevertheless produce an ion-rich solution that will be overall hard to ferment without prior purification. Depending on the targeted downstream processes for hemicelluloses and lignine, the one-step or the two-step process may be more suitable and the choice of one or another should be influenced by the economic of the added-vapue products that will be generated from this biorefinery.

4. HYDROLYSIS

The cellulose macromolecule is composed of glucose units that are linked via a **β** 1-4 acetal bond. Despite the fact that the glucose units composing it are highly polar and completely soluble in water, cellulose itself is rather hydrophobic. The reason explaining such a phenomenon is because the polar functional groups of the cellulose macromolecules in the fibrils are oriented facing one another in such a fashion that render penetration by a polar solvent as water fairly hard to achieve. Furthermore, cellulose is composed of highly structured crystalline form as well as a less structured and fairly more ramified amorphous phase. The amorphous phase is more accessible to hydrolysing agents as acids or enzymes and usually bares the sites where original hydrolysis is going to take place. Although it is fairly uncommon for solvent to slide between the cellulose macromolecules, some solvent have been shown to be able to accomplish such task. The latter are usually highly polar and often have an ionic part.

The key to chemical hydrolysis of cellulose is to break the hydrogen bonding between the cellulose macromolecules in the fibrils. To do so, a compound or a mixture of compound that can efficiently rupture the hydrogen bonding is used. Furthermore, the compound has to be made in a specific shape capable of penetration in the tight hydrogen bonded ultrastructure. The compounds that have shown such potential all have in common an "arrow head" structure the ionic part of the molecule being the tip of the arrow. Once the ion is inserted between the cellulosic fibril layers, much effort has to be made in order to avoid reformation of the previous hydrogen bonding. Therfore, at this point, the utilisation of a larger quantity of solvent is of the essence. A classical example of this theory is the ASTM D1106-56 method which allows the quantification of lignin in a lignocellulosic matrix. The process relies on the acidification of the fibers with a strong sulphuric acid mixture. The mixture is kept at low temperature, allowing the ionic sulphuric acid to progress through the cellulosic matrix after which water is added to the mixture prior to hydrolysis. In this specific case, once the cellulose molecules have been spaced one from the other, water is added in large quantity to avoid reformation of hydrogen bonding between the cellulose macromolecules, thus isolating them one from the other. The solution is still acidic enough to allow hydrolysis and since the process requires energy to activate the bond, heating the solution close to ebullition of water allows the hydrolysis of cellulose to glucose. This process is highly effective and allows the conversion of cellulose up to a 100% into glucose. The downside of this process is directly related to the low cellulose concentration in the solution and the high residual ionic composition which makes this classic ASTM method fairly hard to scale up into an industrial fully functional process.

Sulphuric acid, although probably the more inexpensive of the potential reactants for cellulose hydrolysis is not the only one that has been reported so far. Formic acid, the smallest organic acid, has also been reported to be efficient for the swelling and overall hydrolysis of water. A few researchers have reported the utilisation of formic acid in relation with cotton (Sun et al. 2008). Cotton is a good example of a cellulose rich medium and it was reported that formic acid was shown effective for the hydrolysis of these fibres. Neverheless, formic acid was never efficient at a 100% concentration and for a suitable hydrolysis, it had to be diluted somewhat with water. This organic acid, as sulphuric acid, absolutely requires water as a medium for propagation of the protons. In a pure formic acid medium, the auto-hydrolysis process will generate far less free hydronium ions which will overall reduce the catalytic potential on the mixture. On the other hand, adding too much water will lead to dilution of the acid which will not be able to performed efficiently as a swelling agent in cellulose, since, as mentioned earlier, water is not strong enough to break the actual bonds between the cellulose macromolecules. Other reports have mentioned that they could increase efficiency of a formic acid hydrolysis by adding a specific amount of sulphuric acid in the mixture. In this specific case, sulphuric acid is not used as a swelling agent (small concentration has been reported to be efficient) but only as a proton provider, accentuating the actual contribution made by formic acid. And since sulphuric acid is significantly more acidic than formic acid, there will be available protons in the mixture, allowing hydrolysis of cellulose without however justifying the addition of water to the mixture. Another catalyst that has been reported often for the hydrolysis of cellulose using formic acid as a solvent is chlorhydric acid (Sun et al., 2010). Although it was shown effective for cellulose hydrolysis, such a mixture also has major drawbacks, first of which is the high concentration of formic acid which is a fermentation inhibitor. Production of glucose according to this process would therefore require a very extensive purification process to ensure minimal occurrence of formic acid in the mixture. Another problem that may arise from a formic-acid process would be the removal of sulphuric acid. Althought it may not be too much of a problem at low concentration, removing formic acid would concentrate the sulphuric acid in the mixture leading to a strong acidic catalyst at the end of the distillation process. Even in small amount, the inorganic acid would more likely interact with glucose at the boiling point of formic acid and induce dehydration of the glucose molecule leading yet to another major problem via the occurrence of another major inhibitor, 5-hydroxymethylfurfural (5-HMF).

Another compound that was shown to be effective to separate the cellulose macromolecules is cupryethylene diamine, an organometallic compound currently used to evaluate the polymerisation degree of the cellulose fibres. This compound also demonstrates the "arrow head" structure and the major interaction between this molecule and the cellulose fibrils is made via the amine groups (see Figure 6). The target of cupryethylene diamine is to solubilise cellulose in order to evaluate the polymerisation level of the fibres, therefore, it does not allow hydrolysis of cellulose. Nonetheless, insertion of an acid catalyst may also lead to an efficient hydrolysis although the acid would most probably interact with the strongly basic amine compounds.

In recent literature, other compounds have also been reported to interact favourably with cellulose and most of these compounds are related to the ionic liquids group. Such liquids have gained attention around the 90's and are currently involved in many scientific studies and specially in the field of high efficiency electrochemical cells. There also have been some reports that ionic liquids could have an impact on cellulose, allowing dissolution of the latter. A recent patent by Braun et al. (2010) reported that they were able to produce cellulose beads by dissolving a cellulose medium in a solution containing an imidazolium compound (see Figure 7).

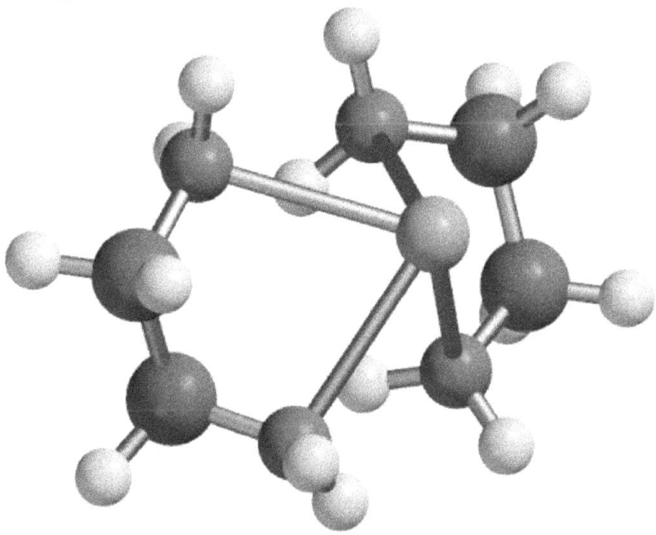

Figure 6.Cupryethylene diamine molecule in aqueous solution (copper is depicted in orange, nitrogen in blue and carbon in black), the model was elaborated with Spartan'10 software

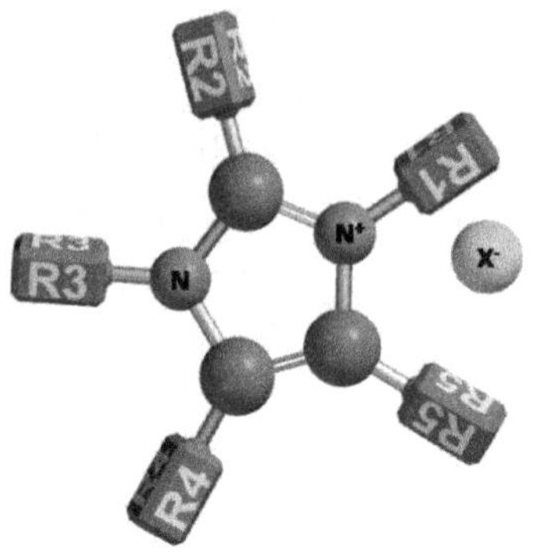

Figure 7.General structure of an imidazolium compound, in which R1 and R3 are each an identical C2-20 org. radical, R2, R4 and R5 are each an H atom, X- is an anion. The model was elaborated with Spartan'10 software. A few reports also mentioned the utilisation of 1-alkyl-3-methylimidazolium chloride for the suitable dispersion of the cellulose macromolecule in solution (Ignatyev et al. 2010 and Vitz et al. 2010).

Our group has developed a process "the CRB-UdeS process" by which cellulose is hydrolysed to glucose (Lavoie et al., 2010a, 2010b; Chornet et al., 2008) being scaled to a demonstration level by CRB. The production of glucose out of lignocellulosics is overall a tricky process and one should keep in mind that before attacking the acetal bonding of the cellulose, it is primordial that the targeted bonds are exposed to the catalyst. The process has been divided in three steps: swelling, hydrolysis and purification. Swelling of cellulose implies the insertion of a suitable molecule, that breaks the strong hydrogen bonding among cellulose macromolecules in the fibrils. A stable gel is formed. Our group has investigated the different parameters which favour an efficient swelling and the most important factors to be taken into account are temperature and time of swelling. The concentration of the swelling agent is of crucial importance if cellulose macromoilecules in the fibrils are to be separated one from the others. Figure 8 below depicts the influence of the concentration of sulphuric acid as well as the acid/cellulose ratio in relationship with total hydrolysis of cellulose.

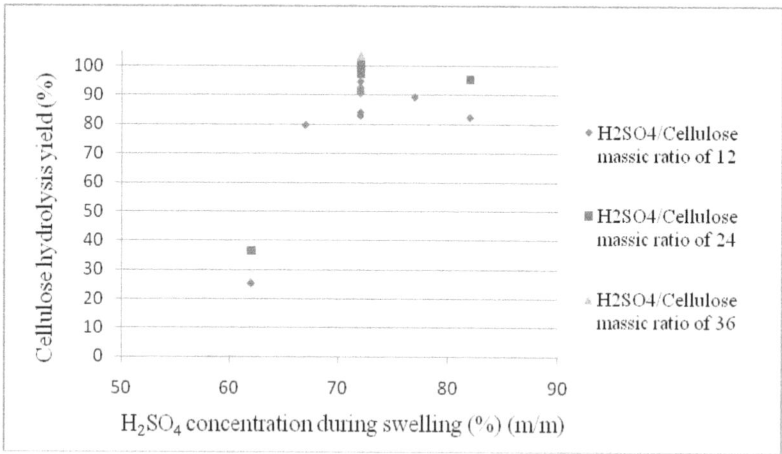

Figure 8.Acid quantity and concentration during swelling vs tritical cellulose hydrolysis yield (for 10min hydrolysis at 121º C in autoclave with H2SO4 concentration of 208-267 g/L and with H+/OH- molar ratio of 2.3 during NaOH 18-23 % addition)

Results shows that the there is an optimal concentration of acid leading to an optimized swelling which facilitates the hydrolysis of cellulose with a yield close to 100% at around 72-74 wt% acid in solution. A 72 %wt of sulphuric acid correspond to a 2.0 molar ratio of water/sulphuric acid which is stoichiometrically sufficient to completely ionisesulphuric acid which leads back to the arrowhead structure. The swelling period is also fairly important for such process, getting optimal after half an hour. Results (depicted in

Figure 9 below) shows that increasing the time of swelling does not lead to significant increases in terms of rate of hydrolysis downstream. The temperature at which the cellulose is swelled can be varied from 20-40º C after which degradation of the cellulose is observed (Belanger, 2005).

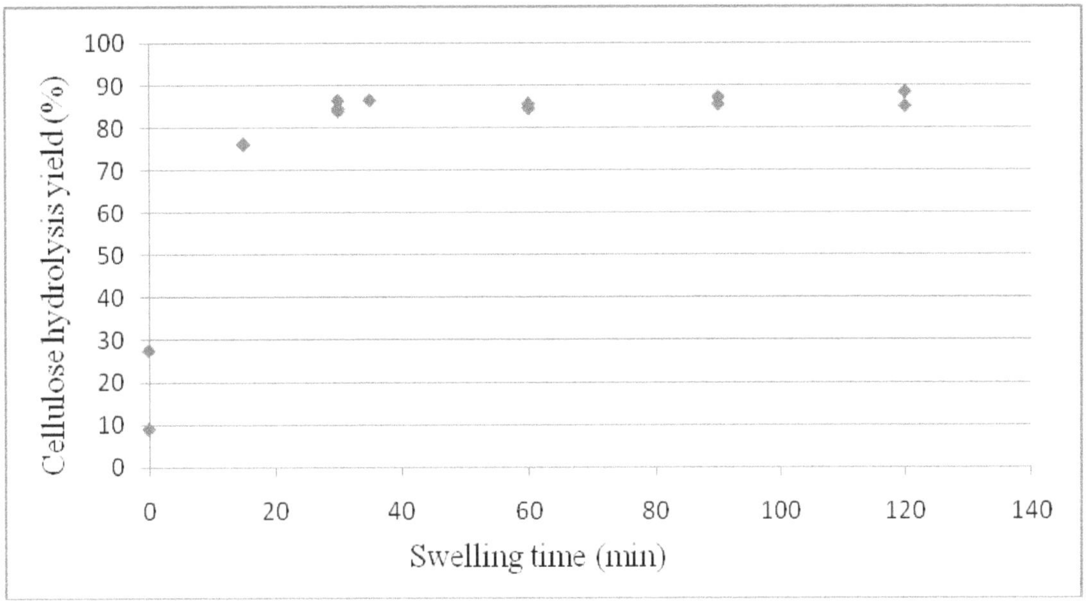

Figure 9.Time of swelling vs tritical cellulose hydrolysis yield (for 72% H2SO4 during swelling at 30º C, 10 min hydrolysis at 121º C in autoclave, H2SO4/Cellulose massic ratio of 36, H+/OH- molar ratio of 2.5 during NaOH 20 % addition)

Once the hydrogen bonding between the cellulose macromolecules is replaced by hydrogen bonding to the aqueous solvent, the next step leading to the production of glucose is hydrolysis. The latter is performed under a decreased free acid concentration. The effect of the pH on the hydrolysis process is depicted in Figure 10 below.

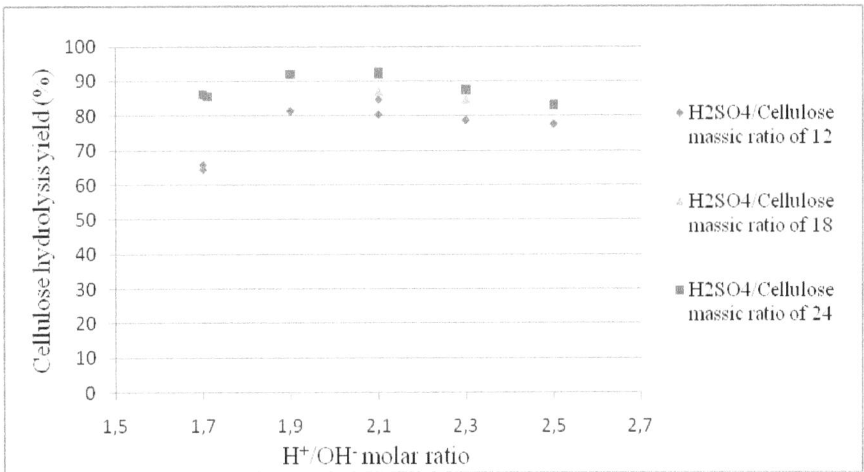

Figure 10.H+/OH- molar ratio vs tritical cellulose hydrolysis yield (for 72% H2SO4 during the 2h at 30º C swelling, with 10 min hydrolysis at 121º C in autoclave, with NaOH 40% addition)

The previously mentioned results show that a minimal amount of hydronium ions are required in order to perform hydrolysis to an optimal value. The concentration of available protons is of course an important factor but temperature and time are also parameters that have to be assessed in order to control optimially the production of cellulosic glucose. Hydrolysis of the acetal bonding requires activation. Figure 11 below shows the effect of temperature on the hydrolysis process.

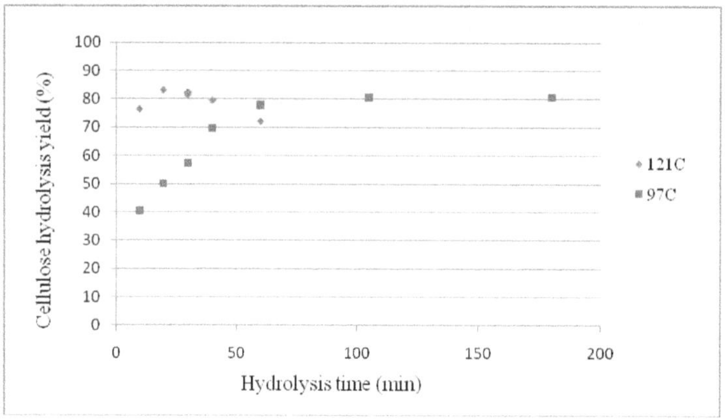

Figure 11.Hydrolysis time and temperature vs tritical cellulose hydrolysis yield (for 72% H2SO4 during the 2h at 30º C swelling, with H2SO4/Cellulose massic ratio of 12, with H+/OH- molar ratio of 2.0 and partial neutralization. All the results at 121º C include 20 min more above 97º C due to the preheat and the cool down of temperature in the autoclave.

Figure 11 demonstrates that a maximum hydrolysis yield is attained at both temperatures. However, the time needed to achieve this hydrolysis differs, which aligns exactly with the kinetics of the hydrolysis reaction. Regarding steam explosion, the hydrolysis of cellulose might also be represented in terms of a severity factor. However, there is currently no reference of any mathematical link in this regard. Undoubtedly, the duration of hydrolysis is a crucial aspect to take into account while contemplating the hydrolysis of cellulose. After cellulose has experienced significant swelling, it becomes fully vulnerable to acid attacks on the acetal bonds of the glucose units. The results illustrated in Figure 12 below unequivocally demonstrate that hydrolysis is rapid and effective at a temperature of 121º C.

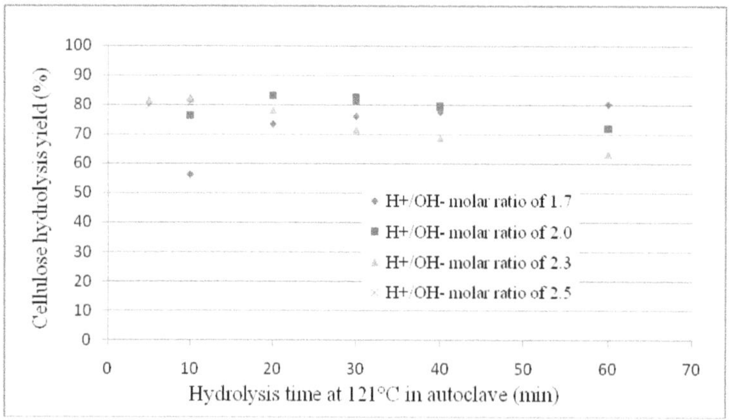

Figure 12.Hydrolysis time and H+/OH- molar ratio vs tritical cellulose hydrolysis yield (for 72% H2SO4 during the 2h at 30º C swelling, with 121º C hydrolysis in autoclave, with H2SO4/Cellulose mass ratio of 12, and partial neutralization with NaOH 32.8%)

5. CONCISE SUMMARY

To produce cellulosic ethanol through the fractionation process, it is necessary to have knowledge about the amount of cellulose present in various lignocellulosic substrates. The presence of cellulose in a biomass can be used to determine the potential ethanol production through the process of glucose fermentation. To convert biomass into glucose, it is necessary to carefully control and optimize two crucial processes: biomass separation to obtain cellulose, and subsequent hydrolysis of the cellulose. Aqueous/ Steam treatments provide a highly effective alternative for the isolation of cellulose. They rely on the use of water, with or without a catalyst, to soak the biomass and, by adding steam, increase the temperature.

This leads to the hydrolysis of the bonds between the different components of the biomass, releasing hemicelluloses and lignin. The effect of steam operations on biomass can be determined by a severity factor, which enables the estimation of the amount of hemicelluloses and, to some extent, cellulose that undergo hydrolysis throughout the process. The use of a base catalyst enables the hydrolysis of lignin and the generation of a relatively pure cellulosic fiber. The two-step FIRSST procedure initially breaks down the hemicelluloses through hydrolysis, while the lignin is subsequently separated from the cellulose in a treatment that is catalyzed by a base. This method enables the separation of biomass into different components, resulting in the production of a superior pulp and smaller fibers known as fines. These fines can be broken down through the infiltration of a potent ion solution, such as concentrated sulphuric acid, which causes the crystalline cellulose to expand. The phenomenon of "swelling" was shown to be more effective when utilizing ionic chemicals that exhibit a "arrowhead" structure. This pattern is characteristic of several chemicals, including sulphuric acid. Hydrolyzing cellulose involves the acid-induced cleavage of the acetal linkages. The latter process can be carried out by a hydronium ion, and the ability to access the acetal bonding of cellulose is directly linked to the effectiveness of the preceding swelling. Experimental evidence demonstrated that once the specific bonds are revealed, the expansion occurs within a few minutes, and this process may be carried out at various temperatures close to the boiling point of water. It serves as a substitute for enzyme hydrolysis.

REFERENCES

Abatzoglou, N.; Chornet, E.; Belkacemi, K. & Overend, R. P. (1992). Phenomenological kinetics of complex systems: the development of a generalized severity parameter and its application to lignocellulosics fractionation. Chemical Engineering Science, Vol. 47, No. 5, (April 1992), pp. 1109-22, ISSN 0009-2509

Agbogbo, F.K. & Coward-Kelly, G. (2008). Cellulosic ethanol production using the naturally occurring xylose-fermenting yeast, Pichia stipitis. Biotechnology Letters, Vol. 30, No. 9,(May 2006), pp.1515–1524, 0141-5492

Alvo, P.; Savoice, P.; Tremblay, D.; Emond, J.P. & Turcotte., G. (1996). A system approach for evaluation of ethanol production based on forages. Bioresource Technology, Vol. 56, No. 1, (April 1996), pp. 61–68, ISSN 0960-8524 Canadian Renewalable Fuels Association CRFA. (2010). In : Industry information, December 2010, http://www.greenfuels.org/en/industry-information/plants.aspx

Ballesteros, I.; Oliva, J. M.; Saez, F. & Ballesteros, M. (2001) Ethanol production from lignocellulosic by products of olive oil extraction. Applied Biochemistry and Biotechnology, Vol. 91-93, No. 1-9, (2001), 91-93, pp. 237-252, ISSN 0273-2289

Ballesteros, M.; Oliva, J.M.; Negro, M.J.; Manzanares, P. & Ballesteros, I. (2004). Ethanol from lignocellulosic materials by a simultaneous saccharification and fermentation process (SFS) with Kluyveromyces marxianus CECT 10875. Process Biochemistry, Vol. 39, No. 12, (October 2004), pp. 1843–1848, ISSN 1359-5113

Bednar, H. & Fengel, D. (1974) Physikalische, chemische und strukturelle Eigenschaften von rezentem und subfossilem Eichenholz. Holz-Roh Werskt, Vol. 32, No. 3, pp. 99-107

Belanger, k. (2005). Controlled depolymerisation and decrystallization of cellulose-rich substrates into glucose. Research master thesis in chemical engineering. Sherbrooke, Canada, August 2005

Bellerini, D. (2006). Les Biocarburants, états des lieux, perspectives et enjeux du développement, Technip, IFP publications, ISBN 9782710808695, Paris, France

Berberi, V.; Turcotte, F.; Lantagne, G.; Chornet, M. & Lavoie, J.-M. (2011). High Yields of Sugars via the Non-Enzymatic Hydrolysis of Cellulose. Energy and sustainability III, Alicante, Espagne, April 2011

Boe, A. & D.K. Lee. (2006). Genetic variation for biomass production in prairie cordgrass and switchgrass. Crop Science, Vol. 47, No. 3, pp. 929-934, ISSN 0011-183x

Boulanger, F.; Bourke, P. & Dunsky. P. (2009). Imaginons le Quebec sans pétrole, In : les rendez-vous de l'energie, February 2011, Available from : <http://www.rdvenergie.qc.ca/wpcontent/uploads/2010/09/Cahier_de_reference_RDV/ Energie1.pdf>

Braun, M.; Guentherberg, N.; Lutz, M.; Magin, A.; Siemer, M.; Swaminathan, V.N.; Linner, B.; Ruslim, F. & Fernadez Ramierz, G.A. (2010). Process for producing cellulose beads from solutions of cellulose in ionic liquid. US Patent Application 20100331222, December 30, 2010

Buranov, A.U. & Mazza, G. Lignin in straw of herbaceous crops. Industrial Crops and Products, Vol. 28, No. 3, (November 2008), pp. 237-259, ISSN 0926-6690

Casey, E.; Sedlak, M.; Ho, N.W.Y. & Mosier N.S. (2010). Effect of acetic acid and pH on the cofermentation of glucose and xylose to ethanol by a geneticallyengineered strain of Saccharomyces cerevisiae. FEMS Yeast Res, Vol 10, (April 2010), pp. 385–393, ISSN 1574-6968

Chornet, E.; Chornet, M. & Lavoie, J.-M. (2008) Conversion of cellulosic biomass to sugar. US Provisional Patent Application #80685-2 filed Oct 8, 2008

Chu, B. & Lee, C.H., (2007). Genetic improvement of Saccharomyces cerevisiae for xylose fermentation. Biotechnology Advances Vol. 25, No. 5, (September 2007), pp. 425–41, ISSN 1873-1899

Claessens, A.; Michaul, R.; Belanger, G. & Mather., D.E. (2004). Characteristics of timothy genotypes divergently selected for fiber traits. Crop Science, Vol. 44, No.1, pp. 81–88, ISSN 0011-183X.

Dale, B.E. & Bals, B. (2008). Separation of proteins from grasses integrated with ammonia fiber explosion (AFEX) pretreatment and cellulose hydrolysis. PCT Int. Appl. (2008), WO 2008020901 A2 20080221

De Bari, I.; Nanna, F. & Braccio, G. (2007). SO2-Catalyzed Steam Fractionation of Aspen Chips for Bioethanol Production: Optimization of the Catalyst Impregnation. Industrial & Engineering Chemistry Research, Vol. 46, No. 23, (October 2007) , pp. 7711-7720, ISSN 0888-5885

Department of Energy (US), DOE. (2006). Biomass feedstock composition and property database, In: Biomass Program. Available from: http://www.eere.energy.gov/biomass/progs/search1.cgi

Dererie, D. Y.; Trobro, S.; Momeni, M. H.; Hansson, H.; Blomqvist, J.; Passoth, V.; Schnuerer, A.; Sandgren, M. & Stahlberg, J. (2011). Improved bio-energy yields via sequential ethanol fermentation and biogas digestion of steam exploded oat straw. Bioresource Technology, Vol. 102, No. 6, (March 2011), pp. 4449-4455, ISSN 0960-8524

Emmel, A.; Mathias, A.L.; Wypych, F. & Ramos, L. P. (2003). Fractionation of Eucalyptus grandis chips by dilute acid-catalysed steam explosion. Bioresource Technology, Vol. 86, No. 2 (January 2003), pp. 105-115, ISSN 0960-8524

Ewanick, S.M.; Bura, R. & Saddler, J.N. (2007). Acid-catalyzed steam pretreatment of lodgepole pine and subsequent enzymatic hydrolysis and fermentation to ethanol. Biotechnology and Bioengineering, Vol. 98, No. 4, (November 2007), pp. 737-746, ISSN 1097-0290

Food and Agriculture Organization of the United Nations FAOSTAT. (2010). In : Production, December 2010, http://faostat.fao.org/site/339/default.aspx

Farrell, A. E. & O'Hare, M. (2008). Greenhouse gas (GHG) emissions from indirect land use change (LUC). Memorandum for the California Air Resources Board. Available from: http://www.arb.ca.gov/fuels/lcfs/011608ucb_luc.pdf

FrederickJr, W.J.; Lien, S.J.; Courchene, C.E.; DeMartini, N.A.; Ragauskas A.J. & Iisa K. (2008). Co production of ethanol and cellulose fiber from Southern Pine: A technical and economic assessment. Biomass and Bioenergy, Vol. 32, No. 12, (december 2008), pp.1293–1302, ISSN 0961-9534

Fitzpatrick, S.W. (2010) Production of formic acid from hydrolysis of carbohydratecontaining compounds such as paper pulp and wood feed stocks. PCT Int. Appl. (2010), WO 2010104722 A1 20100916

Gerardi, V.; Minelli, F. & Viggiano, D. (1998). Steam treated rice industry residues as an alternative feedstock for the wood based particleboard industry in Italy. Biomass and Bioenergy, Vo. 14, No. 3, (March 1998), pp. 295-299, ISSN 0961-9534

Gibbs, R.D. (1958). The Physiology of Forest Trees, Thimann, pp. 269, John Wiley and Sons, New York, USA

Gírio, F.M.; Fonseca, C.; Carvalheiro, F.; Duarte, L.C.; Marques, S. & Bogel-Łukasik, R. (2009). Hemicelluloses for fuel ethanol: A review. Bioresource Technology, Vol. 101, No. 13, (July 2010), pp. 4775-4800, ISSN 0960-8524

Glasser, W. G. & Wright, R. S. (1998). Steam-assisted biomass fractionation. II. Fractionation behavior of various biomass resources. Biomass and Bioenergy, Vol. 14, No. 3, (March 1998), pp. 219-235, ISSN 0961-9534.

Goyette, J. & Boucher, S. (2008). Communication from the engineering firm roche, Etablissement d'une chaufferie central à la biomasse forestière: lignes directrices dans un context Québécois. (November 2008)

Ibrahim, M., & Glasser, W.G. (1999). Steam-assisted biomass fractionation. Part III: a quantitative evaluation of the ``clean fractionation'' concept . Bioresource Technology, Vol. 70, No. 2, (October 1999), pp. 181-192, ISSN 0960-8524

Ignatyev, I.A.; Mertens, P. G. N.; Van Doorslaer, C.; Binnemans, K. & de Vos, D.E. (2010). Cellulose conversion into alkylglycosides in the ionic liquid 1-butyl-3- methylimidazolium chloride. Green Chemistry, Vol. 12, No. 10, pp. 1790-1795, ISSN 1463-9270

International Energy Agency IEA. (2009). Medium Term Oil Market Report, OECD/IEA, Paris. International Energy Agency IEA. (2004). Biofuel for transport an international perspective. Chirat, 9264015124, France.

International Energy Agency IEA. (2011). In : IEA energy statistic, January 2011, Available from: http://www.iea.org/stats/index.asp

Jefferson, P.G.; McCaughey, W.P.; May, K., Woosaree, J. & McFarlane, L. (2004). Potential utilization of native prairie grasses from western Canada as ethanol feedstock. Canadian Journal of Plant Science, Vol. 84, (May 2004), pp. 1067–1075, ISSN 0008-4220

Jung, H.G.; D.R. Mertens & Payne, A.J. (1997). Correlation of acid detergent lignin and klason lignin with digestibility of forage dry matter and neutral detergent fiber. Journal of Dairy Science, Vol. 80, No. 8, (August 1997), pp. 1622–1628, ISSN 1525-3198

Jurgens, M.H. (1997). Animal Feeding and Nutrition, 8th ed. Kendall/Hunt Publishing Company. Dubuque, Iowa.

Khunrong, T.; Punsuvon, V.; Vaithanomsat, P. & Pomchaitaward, C. (2011). Production of ethanol from pulp obtained by steam explosion pretreatment of oil palm trunk. Energy Sources, Part A: Recovery, Utilization, and Environmental Effects, Vol. 33, No. 3, (January 2006) pp. 221-228, ISSN 1556-7230

Lavoie, J.-M.; Capek, E.; Gauvin, H. & Chornet, E. (2010a). Production of pulp from Salix viminalis energy crops using the FIRSST process. Bioresource Technology, Vol. 101, No. 13, (July 2010), pp. 4940-4946, ISSN 0960-8524

Lavoie, J.-M.; Capek, E.; Gauvin, H. & Chornet, E., (2010b). Production of quality pulp from mixed softwood chips as one of the added value product using the FIRSST process in a general biorefinery concept. Industrial & Engineering Chemistry, Vol. 49, pp. 2503-2509, ISSN 1520-5045

Lavoie, J.-M., Capek, E., Chornet, E., (2010c). Evaluation of the co-product pulp from Salix viminalis energy crops. Biomass & bioenergy, Vol. 34, No. 9, (September 2010), pp. 1342-1347, ISSN 0961-9534

Lawford, H. G. & Rousseau, J. D. (2003). Cellulosic fuel ethanol. Alternative fermentation process designs with wild-type and recombinant Zymomonas mobilis. Applied Biochemistry and Biotechnology, Vol. 106, No. 1-3, (March 2003), pp. 105-108, pp. 457-469, ISSN 1559-0291.

Lee, J. (1997). Biological conversion of lignocellulosic biomass to ethanol. Journal of Biotechnology, Vol. 56, No.1, (July 1997), pp. 1–24, ISSN 0168-1656

Lee, D.K. & V.N. Owens. (2005). Biomass production potential of native warm-season grass monocultures and mixtures. Agronomy abstracts. ASA Lee, D.K., Owens, V.N., Boe, A., Jeranyama, P. (2007). Composition of Herbaceous Biomass Feedstocks, February 2011, Available from: <http://ncsungrant1.sdstate.org/uploads/publications/SGINC1-07.pdf>

Li, J.; Henriksson, G. & Gellerstedt, G. (2005). Carbohydrate reactions during hightemperature steam treatment of aspen wood. Applied Biochemistry and Biotechnology, Vol. 125, No. 3, (June 2005), pp. 175-188, ISSN 1559-0291

Liang, Y.; Siddaramu, T.; Yesuf, J. & Sarkany, N. (2010). Fermentable sugar release from Jatropha seed cakes following lime pretreatment and enzymatic hydrolysis. Bioresource Technology, Vol. 101, No. 16, (April 2010), pp.6417–6424, ISSN 0960-8524

Lu, Y.; Warner, R.; Sedlak, M.; Ho, N. & Mosier, N. S. (2009). Comparison of glucose/xylose cofermentation of poplar hydrolysates processed by different pretreatment technologies. Biotechnology Progress, Vol. 25, No. 2, (March 2009), pp. 349-356, ISSN 1520-6033

Mani, S.; Tabil, L.G. & Sokhansanj. S. (2006). Effects of compressive force, particle size and moisture content on mechanical properties of biomass pellets from grasses. Biomass and Bioenergy, Vol. 30, No. 7, (July 2006), pp. 648–654, ISSN 0961-9534

Mazlan I. & W.G. Glasser. (1999) Steam-assisted biomass fractionation. Part III: a quantitative evaluation of the ``clean fractionation" concept. Bioresource Technology, Vol. 70 , No. 2, (November 1999), pp. 181-182, ISSN 0960-8524

Ministères de Ressources Naturelles et de la Faune MRNF. (2009). Chiffres clés du Quebec Forestier, December 2010, Available from : www.mrnf.gouv.qc.ca

Montané, D.; Overend R.P. & Chornet, E., (1998). Kinetic Models for Non-Homogeneous Complex systems with a Time-Dependent Rate Constant. Canadian Journal of Chemical Engineering, Vol. 76, No. 1, (February 1998), pp. 58-68, ISSN 1939-019X

Natural Ressources Canada. (2007). Canadian Petroleum Product Market, December 2010, Available from : http://www.nrcan-rncan.gc.ca/eneene/sources/petpet/reprap/2007-11/supoffeng. php

Overend, R. P. & Chornet, E. (1987). Phil. Trans. R. S. Lond. A321, 523–536 Owens, V.N.; Lee, D.K. & Boe, A. (2006). Manure and harvest timing effects on biomass and seed yield of three perennial grasses in South Dakota. The World Congress on Industrial Biotechnology and Bioprocessing. Biotechnology Industry Organization.

Palmqvist, E. & Grage, H. (1999). Main and interaction effects of acetic acid, furfural, and phydroxybenzoic acid on growth and ethanol productivity of yeasts. Biotechnology and Bioenergy, Vo. 63, No. 1 (April 1999) pp. 46–55, 1097-0290

Pettersen, R.C. (1984). In: 'The Chemistry of Solid Wood', R. Rowell, pp. (57-127), American Chemical Society, ISBN 9780841207967, Washington D.C.

Pienkos, P. T. & Zhang, M., (2009). Role of pretreatment and conditioning processes on toxicity of lignocellulosic biomass hydrolysates. Cellulose, Vol. 16, No. 4, (August 2009), pp. 743-762, ISSN 1572-882X

Ramos, L. P.; Mathias, A.L.; Silva, F.T.; Cotrim, A.R.; Ferraz, A.L. & Chen, C.-L, (1999). Characterization of Residual Lignin after SO2-Catalyzed Steam Explosion and Enzymatic Hydrolysis of Eucalyptus viminalis Wood Chips. Agricultural and Food Chemistry, Vol. 47, No. 6, (May 1999), pp. 2295-2302, ISSN 1520-5118

Renewable Fuels Association. 2010. Ethanol facilities capacity by state, December 2010, Available from: http://www.neo.ne.gov/statshtml/121.htm

Rodríguez, A.; Sánchez, R.; Requejo, A. & Ferrer, A. (2010). Feasibility of rice straw as a raw material for the production of soda cellulose pulp. Journal of Cleaner Production, Vol. 18, No. 10-11, (March 2010), pp. 1084-1091, ISSN 0959-6526

Sassner, P.; Martensson, C.G.; Galbe, M. & Zacchi, G. (2008). Steam pretreatment of H2SO4-impregnated Salix for the production of bioethanol. Bioresource Technology, Vo. 99, No. 1, (January 2008), pp. 137–145, ISSN 0960-8524

Schafer E.R. & Bray M.W. (1929). Pulping Flax Straw. VI. Properties of Flax straw cellulose and its value in the cellulose industries. Industrial Engineering Chemistry, Vol. 21, No. 3, (March 1929), pp. 278–280, ISSN 0888-5885

Shen, J., Agblevor, F. A. (2008). Ethanol production from cotton gin waste by simultaneous saccharification and fermentation. Beltwide Cotton Conferences (2008), pp. 569- 572, Nashville, USA

Sørensen, A.; Teller, P.J.; Hilstrøm, T. & Ahring, B.K. (2008). Hydrolysis of Miscanthus for bioethanol production using dilute acid presoaking combined with wet explosion pre-treatment and enzymatic treatment. Bioresource Technology, Vol. 99, No. 14, (September 2008), pp. 6602–6607, ISSN 0960-8524

Sun, R.C.; Fang, J.M. & Tomkinson, J. (2000). Delignification of rye straw using hydrogen peroxide. Industrial Crops and Products, Vol. 12, No.2, pp. 71–83, ISSN 0926-6690

Sun, Y.; Lin, L.; Pang, C.; Deng, H.; Peng, H.; Li, J.; He, B. & Liu, S. (2007). Hydrolysis of Cotton Fiber Cellulose in Formic Acid. Energy & Fuels, Vol. 21, No. 4, (June 2007), pp. 2386-2389, ISSN 0887-0624

Sun, Y.; Zhang, B.; Lin, L. & Liu, S. (2010). Clean conversion to fermentable glucose from wheat straw. Journal of Biobased Materials and Bioenergy, Vo. 4, No. 1, (March 2010), pp. 27-34, ISSN 1556-6579

Smeets, E. & Faaij, A. (2007). Bioenergy Potentials from Forestry in 2050. Climatic Change, Vol. 81, (November 2006), pp. 353-390, ISSN 0165-0009

Tao, F.; Song, H. & Chou, L. (2011). Efficient process for the conversion of xylose to furfural with acidic ionic liquid. Canadian Journal of Chemistry, Vol. 89, No. 1, (January 2011), pp. 83-87, ISSN 1480-3291

Vitz, J.; Yevlampieva, N.P.; Rjumtsev, E. & Schubert, U.S. (2010). Cellulose molecular properties in 1-alkyl-3-methylimidazolium-based ionic liquid mixtures with pyridine. Carbohydrate Polymers, Vol. 82, No. 4, (July 2010), pp. 1046-1053, ISSN 0144-8617

Vlasenko, E. Y.; Ding, H.; Labavitch, J. M. & Shoemaker, S. P. (1997). Enzymic hydrolysis of pretreated rice straw. Bioresource Technology, Vol. 59, No. 2-3, pp. 109-119, ISSN 0960-8524

Wang, G.S.; Pan, X.J.; Zhu, J.Y. & Gleisner, R.L. (2008). Sulfite pretreatment for biorefining biomass. US Patent Application. Filed by Wisconsin Alumni Research Foundation (WARF) 61/047024. Provisional application 4/22/2008. Utility Conversion, 20090298149, April 2009

Wu, M.M.; Chang, K.; Gregg, D. J.; Boussaid, A.; Beatson, R. P. & Saddler, J. N. (1999). Optimization of steam explosion to enhance hemicellulose recovery and enzymatic hydrolysis of cellulose in softwoods. Applied Biochemistry and Biotechnology, Vo. 77- 79, No. 1-3, (March 1999), pp. 47-54, ISSN 1559-0291

Xu, F.; Sun, J.; Sun, R.; Fowler, P. & Baird, M.S. (2006). Comparative study of organosolv lignins from wheat straw. Industrial Crops and Products, Vol. 23, No. 2, (March 2006), pp. 180–193, ISSN 0926-6690

Xu, F.; Sun, R.C.; Zhai, M.Z.; Sun, J.X.; Jiang, J.X. & Zhao, G.J. (2008). Comparative study of three lignin fractions isolated from mild ball-milled Tamarix austromogoliac and Caragana sepium. Journal of Applied Polymers Science, Vol. 108, No. 2, (April 2008), pp. 1158–1168, ISSN 0021-8995

Yamashita, Y.; Sasaki, C. & Nakamura, Y. (2010). Effective enzyme saccharification and ethanol production from Japanese cedar using various pretreatment methods. Journal of Bioscience and Bioengineering, Vol. 110, No. 1, (July 2010), pp. 79–86, ISSN 1389-1723

Yildiz, S. & Gezerb, E.D. (2006). Mechanical and chemical behavior of spruce wood modified by heat. Building and Environment, Vol. 41, No. 12, (December 2006), pp. 1762–1766, ISSN 0360-1323

Youngblood, A.; Zhu, J.Y. & Scott, C.T. (2009). Ethanol Production from Woody Biomass: Silvicultural Opportunities for Suppressed Western Conifers. 2009 National Silviculture Workshop, June 2009, Boise, Idaho

Zhang, J., Chu, D.; Huang, J.; Yu, Z.; Dai, G. & Bao, J., (2009). Simultaneous saccharification and ethanol fermentation at high corn stover solids loading in a helical stirring bioreactor. Biotechnology and Bioengineering, Vol. 105, No. 4, (March 2010), pp. 718- 728, ISSN 1097-0290

Zhang, W. & Woelk, K. (2010). NMR Investigations into hydrothermal biomass-to-fuel (BTF) conversion using glucose as model substrates for cellulosic biomass. 45th Midwest Regional Meeting of the American Chemical Society, Wichita, KS, United States, October 27-30, MWRM-417

Zhu, J.Y.; Pan, X.J.; Wang, G.S. & Gleisner, R., (2009). Sulfite pretreatment for robust enzymatic saccharification of spruce and red pine. Bioresource Technology, Vol. 100, No. 8, (April 2009), pp. 2411–2418, ISSN 0960-8524

Zhu, J.Y. & Pan X.J. (2010). Woody biomass pretreatment for cellulosic ethanol production: Technology and energy consumption evaluation. Bioresource Technology, Vol. 101, No. 13, (July 2010), pp. 4992–5002, ISSN 0960-8524

Zhuang, Q. & Vidal, P.F. (1997). Hemicelluloses solubilization from Populus tremuloides via steam explosion and characterization of the water-soluble fraction. II. Alkalicatalytic process. Cellulose Chemistry and Technology (1997), Vol. 31, No. 1-2, pp. 37-49, ISSN 0576-9787

CHAPTER-30

PRODUCTION OF BIOMETHANOL FROM FORAGE GRASSES, TREES AND CROP RESIDUES

1. INTRODUCTION: AN OVERVIEW

In 2007, around 12 billion tons of fossil fuels (measured in oil equivalent) were burned globally, as reported by OECD in 2010. These fuels have a significant impact on the generation of acid rain, photochemical smog, and the rise in atmospheric carbon dioxide (CO2) levels. Scientists caution that the Earth's temperature is expected to increase by at least 1°C, and perhaps up to 4°C, if the concentration of CO2 in the atmosphere doubles from pre-industrial levels by the end of the 21st century (Brown et al., 2000). Another worldwide issue is the probable exhaustion of fossil fuels within a few decades, notwithstanding the discovery of fresh oil reserves. In order to tackle these concerns, it is imperative that we pinpoint alternative sources of fuel. The stabilization of the earth's climate relies on the reduction of carbon emissions through the transition from fossil fuels to the utilization of solar energy, either directly or indirectly. Biofuel utilization is the most advantageous among the options mentioned for several reasons. Firstly, biomass, which is used to produce biofuel, is derived from solar energy, making it a sustainable energy resource. Secondly, biofuel helps reduce atmospheric CO2 levels through the process of photosynthesis and carbon sequestration. Although combustion of biofuel does release CO2, it does not contribute to the overall increase in global CO2 levels. Thirdly, bioethanol and biomethanol, which are liquid fuels derived from biofuel, offer viable alternatives to petroleum fuels for various engines and machines. Additionally, biofuel can be managed to eliminate the emission of soot and SOx. Lastly, in terms of storage, biofuel ranks second only to petroleum and is much easier to store compared to batteries, natural gas, and hydrogen. The current utilization of biomass has been significantly restricted, mainly involving the combustion of wood and the conversion of sugarcane in Brazil or maize in the USA into bioethanol. The essential raw materials for bioethanol synthesis by fermentation are derived from crop plants that have a high sugar or starch content. Due to the escalating worldwide population and the consequent growth in demand for food, it is imperative that we refrain from utilizing these crops haphazardly for biofuel production, since they serve as crucial sources of human sustenance. Efforts are being made to ferment lignocellulosic materials, such as poplar wood (Populus spp.) (Wyman et al. 2009), switchgrass (Panicum virgatum) (Keshwani and Cheng 2009), Miscanthus (Miscanthus spp.) (Sørensen et al. 2008), rice straw (Oryza sativa) (Binod 2010), and old oil palm trunks (Elaeisguineensis) (Kosugi et al. 2010). These attempts involve improving the pre-treatment of the materials, as well as the use of yeast and enzymes. However, in order to establish this technology successfully, it will be necessary to achieve low cost and high ethanol yield. A novel technique using partial oxidation has been devised for gasification and the manufacture of biomethanol from carbohydrate sources (Sakai 2001). This technology allows for the utilization of any biomass source as a feedstock for the synthesis of biomethanol. We provide information on the projected gas composition and methanol production rate achieved through the utilization of this innovative method for generating biofuels from a variety of biomass sources, including wood, pasture, and agricultural residues. Data acquired from the operational testing of the plant is also supplied.

2. GASIFICATION TECHNOLOGY AND THE TEST PLANTS

The concept and technology of gasification systems that produce soot and tar are not novel. We have developed gasification technology that utilizes partial oxidation and a novel high calorie gasification approach. Our focus is on achieving optimal gasification at temperatures ranging from 900 to 1,000°C, while avoiding the generation of soot and tar. These technologies provide a high-quality blend of biogases that can be used to generate liquid biofuels using a Zn/Cu-based catalyst or to generate power using a generator through thermo-chemical reaction. The initial prototype, designated as "Norin Green No. 1" (with "Norin" denoting the Ministry of Agriculture, Forestry and Fisheries in Japanese; subsequently rebranded as "Norin Biomass No. 1"), was the first test plant. The first plant was finished on April 18, 2002, and the second plant, called "Norin Biomass No. 3," was built in March 2004, including advanced high-calorie gasification technology.

2.1. Gasification technology of partial oxidation

Figure 1 shows the concept of our new method of gasification by partial oxidation. This production of biomethanol from carbohydrate (Sakai 2001) has been given the term "C1 chemical transformation technology". In this process, the biomass feedstock must be dried and crushed into powder (ca. 1mm in diameter). When the crushed materials are gasified at 900-1000°C with gasifying agent (steam and oxygen), all carbohydrates are transformed to hydrogen (H2), carbon monoxide (CO), carbon dioxide (CO2) and vapor (H2O). The mixture of gases is readily utilized for generating electricity. The mixture of gases is transformed by thermo-chemical reaction to biomethanol under pressure (40-80 atm) with Cu/Zn-based catalyst, too. That is,

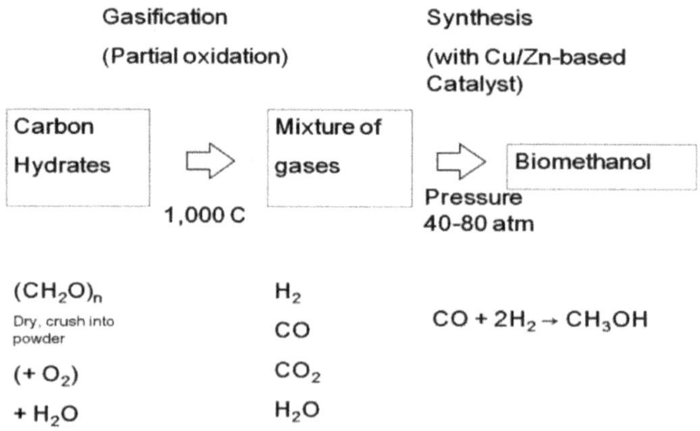

Figure 1.Principle of methanol synthesis by gasification method (the C1 chemical transformation technology)

$$CO + 2H_2 \rightleftarrows CH_3OH + Q \text{ (Radiation of heat)}$$

$$CO_2 + 3H_2 \rightleftarrows CH_3OH + H_2O + Q \text{ (Radiation of heat)}$$

All the ash contained in the materials is collected in the process (Fig. 2). This process enables any source of biomass to be used as a raw material for biomethanol production.

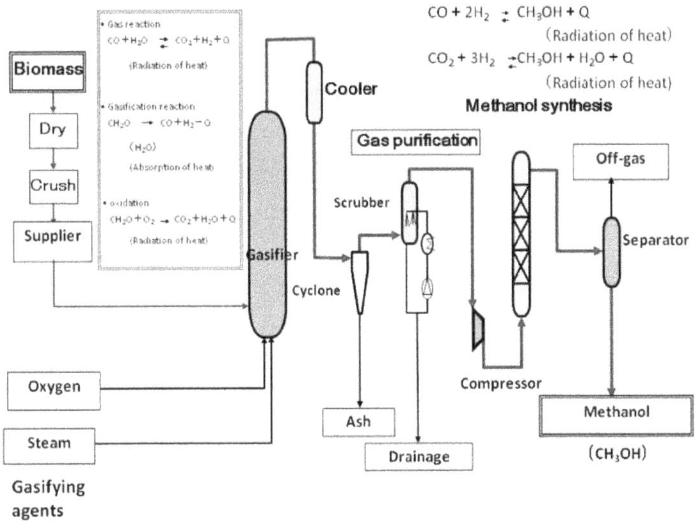

Figure 2.Gasification and biomethanol synthesis system (Nakagawa et al. 2007)

2.1.1. Materials and methods

A total of twenty materials underwent testing: 1) Sawdust refers to the wood of Japanese cedar (Cryptomeria japonica) that has been separated from the bark by using a 2 mm screen sieve. 2) rice bran (Oryza sativa: cv. Koshihikari); 3) rice straw (cv. Yumehitachi: only the inflorescences are harvested in September and the plants were left in the field until cutting in December); 4) rice husks (cv. Koshihikari: rice was threshed in October and kept in plastic bags following typical post-harvest practices); 5) sorghum heads (Sorghum bicolor: var. Chugoku Kou 34 (medium maturing hybrid line between a male sterile grain sorghum line and sudangrass; with mature seeds at ripened stage); 6) leaf and stem of sorghum (ibid.; at ripened stage, cut to a length of 30 cm and dried in a dryerfor 7 days at 70°C); 7) total plant of sorghum (ibid.); 8) sorghum (cv. Kazetachi; extremely late maturing dwarf type; before flowering); 9) sorghum (cv. Ultra sorgo; late maturing tall type; heading stage); 10) sorghum (cv. Green A; medium maturing hybrid between sudangrass and grain sorghum; heading stage); 11) sorghum (cv. Big Sugar: late maturing tall sweet sorghum: milk-ripe stage); 12) guineagrass (Panicum maximum cv. Natsukaze; heading stage); 13) rye (Secale cereale cv. Haru-ichiban; heading stage); 14) Japanese lawngrass (Zoysia japonica cv. Asamoe; before flowering); 15) Erianthus sp. Line NS-1; heading stage; 16) bark of Japanese cedar; 17) chipped Japanese larch (Larix leptolepis); 18) bamboo (Phyllostachys pubescens); 19) salix (Salix sachalinensis and S. pet-susu); 20) cut waste wood: sawn wood and demolition waste (raw material for particle board).

Characteristics important for gasification were evaluated for the above materials: 1) Water content and ash were measured following drying at 107 ± 10°C for 1 hour; then followed by combustion at 825 ± 10°C for 1 hour; 2) Percent carbon (C), hydrogen (H), oxygen (O), nitrogen (N), total sulfur (T-S), and total chloride (T-Cl): C and H weights were estimated by CO2 and H2O weight after combustion at 1,000 ± 10°C by adding oxygen. The estimate of O was calculated by the equation, O = 100 – (C + H + T-S + T-Cl); estimates of N were determined by the amount of ammonia produced by oxidation with sulfuric acid to generate ammonium sulfate. Following distillation, total sulfur was estimated by SO2 following combustion at 1,350°C with oxygen. Total chloride was estimated by the water soluble remains following combustion with reagent and absorption of the gas; 3) The higher heating values were measured by the rise in temperature in water from all the heat generated through combustion. The lower heating value was estimated by the calculation (the higher heating value – (9×h+w)×5.9) [h: hydrogen content (%); w: water content (%)]; 4) Chemical composition (molecular) of the biomass was calculated based on molecular weight of the elements; 5) Size distribution of the various biomass types was measured (diameter, density of materials [g/ml]); 6) Gas yield and generated heat gas were estimated by the process calculation on the basis of chemical composition and the heating value. Heat yield or cold gas efficiency was calculated by (total heating value of synthesized gases)/(total heating value of supplied biomass); 7) The weight and calories generated as methanol, given a production gasifier capacity of 100 tons dry biomass/day, were estimated by the process calculation. These data were obtained in different years.

2.1.2. Results and discussion

Water and ash content for some materials evaluated are shown in Fig. 3. The materials were prepared in various ways. Water contents ranged from 3.4% (wood waste) to 13.1% (bark). Water content of sorghum was low (4.6%) because this material was dried in a mechanical drier. The other materials were not mechanically dried and the water content averaged ca. 10%. Although individual elements are not reported, the ash content of wood materials, such as sawdust, bark, chip, and bamboo was very low, 0.3% for sawdust, 1.8% for bark, and 2.2% for wood waste. Although the ash content of rice straw and husks was very high (22.6% and 14.6%), probably due to the high Si content of rice plants, the ash content of rice bran was much lower (8.1%). The ash content of sorghum plantwas 5.8%.

The percent by weight of some elements in the raw materials are shown in Fig. 4 and Table 1. Carbon content was high in wood materials and averaged 48.3% for wood waste and 51.8% for bark. Rice bran carbon content was 48.3% and sorghum carbon content was ca. 45%. Carbon content of rice straw and husks were lower at 36.9 and 40.0%, respectively. Four sorghum cultivars with different plant types exhibited a narrow range of carbon content (45.5 - 46.1%). Carbon content of the sorghum heads (with seeds), is higher than leaf and stem of sorghum (with lignin) by 2.3%. Rye, Japanese lawngrass and Erianthus exhibited slightly higher carbon content and guineagrass was at the lower end of the range.

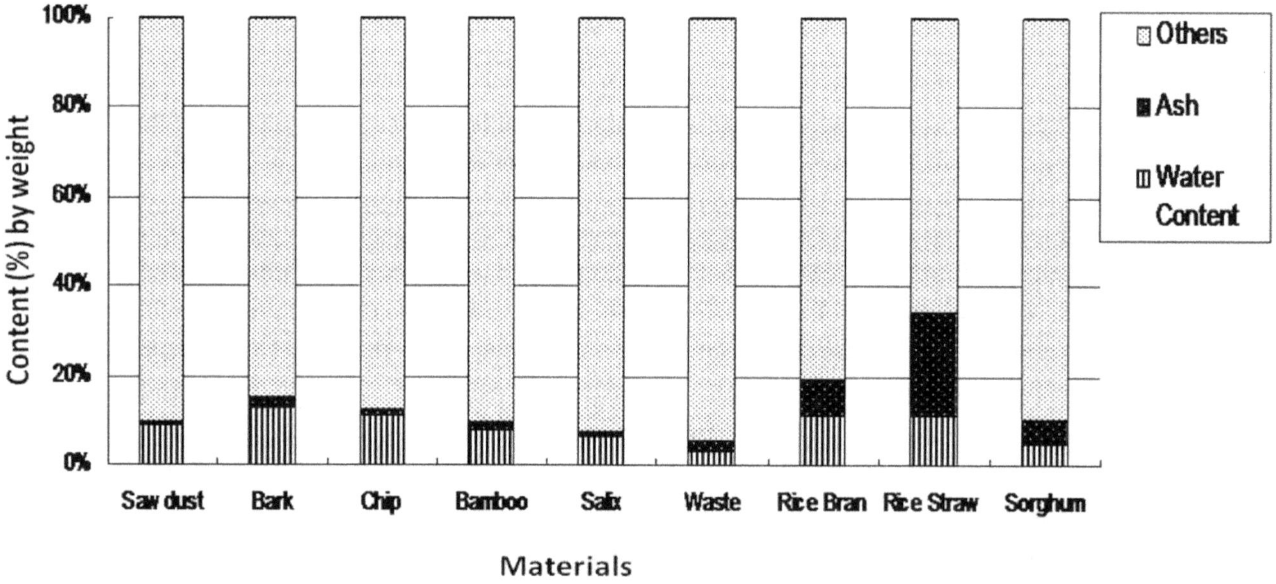

Figure 3.The numbers of materials are same as those in Materials and Methods.Saw dust (1); Bark (16); Chip (17); Bamboo (18); Salix (19); Waste (20); Rice Bran (2); Rice straw (3); sorghum (7).Content of water and ash in materials (Nakagawa et al. 2007).

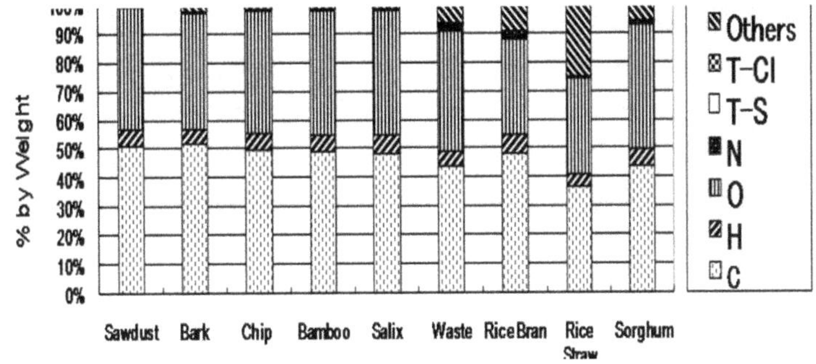

Figure 4.The numbers of materials are same as those in Materials and Methods.C: carbon; H: hydrogen; O: oxygen; N: nitrogen; T-S: total sulfur; T-Cl: total chloride; Saw dust (1); Bark (16); Chip (17); Bamboo (18); Salix (19); Waste (20); Rice Bran (2); Rice straw (3); sorghum (7) Content of some elements in materials without water (% by weight) (Nakagawa et al. 2007).

Hydrogen content ranged from 4.7 to 7.0% for rice straw and rice bran, respectively. Although rice bran had the highest hydrogen content, the others were only marginally different and the range of wood materials was narrow (from 5.6 to 5.9% for bark and salix, respectively). Oxygen content ranged between 32.5% and 43.9% for rice straw and salix, respectively with wood materials and sorghum in the higher range. Nitrogen content was between 0.12% (sawdust) and 2.44% (rice bran), with wood materials exhibiting low values except for wood waste (1.92%). Nitrogen contents of sorghum cultivars ranged from 0.80 to 1.30 % and sorghum heads exhibited 1.68%. The sulfur content was very low in all of the materials and ranged between 0.02% (sawdust) and 0.30% (Japanese lawngrass). Chlorine content ranged from 0.01% (sawdust) to 1.31% (rye). These data demonstrates that these materials are much cleaner than coal and other fossil fuels and, we expect chemical properties of harvested tropical grasses to be similar to the grasses used in this report.

Table 1.C: carbon, H: hydrogen, O: oxygen, N: nitrogen, T-Cl: total chloride, T-S: Total sulfurContent of some elements in dry matter (% by weight). (The numbers of materials are same as those in Materials and Methods)

Biomass Materials	C	H	O	N	T-Cl	T-S	Ash
Sawdust (1)	51.1	5.9	42.5	0.12	0.01	0.02	0.3
Rice bran (2)	48.3	7.0	33.0	2.44	0.05	0.21	8.1
Rice straw (3)	36.9	4.7	32.5	0.30	0.08	0.06	22.6
Rice husk (4)	40.0	5.2	37.3	0.76	0.41	0.22	14.6
Sorghum Head(5)	46.7	6.1	40.7	1.68	0.08	0.14	4.3
Leaf and Stem of Sorghum (6)	44.4	5.8	42.9	0.45	0.23	0.15	5.8
Sorghum 'Kazetachi' (8)	45.7	5.8	39.5	1.30	0.78	0.08	6.2
Sorghum 'Ultra sorgo' (9)	45.5	5.7	41.6	0.80	0.79	0.03	5.4
Sorghum 'Green A' (10)	46.1	5.7	40.6	1.20	0.57	0.04	5.5
Sorghum 'Big sugar' (11)	45.9	5.7	41.2	1.00	0.50	0.05	5.4
Guineagrass (12)	42.8	5.4	37.9	1.50	0.89	0.11	10.4
Rye 'Haruichiban' (13)	45.7	5.8	39.2	1.40	1.21	0.07	6.2
Japanese lawngrass 'Asamoe' (14)	46.4	6.1	37.9	2.15	0.43	0.30	6.4
Erianthus sp. Line 'NS-1' (15)	47.1	6.1	42.3	0.80	0	0.22	3.5
C: carbon, H: hydrogen, O: oxygen, N: nitrogen, T-Cl: total chloride, T-S: Total sulfur							

The higher and lower heating values of materials are shown in Fig. 5 and Table 2. Among the materials tested, the higher heating values of wood materials were high and ranged between 4,570 kcal/kg (sawdust: 19.13 MJ/kg) and 4,320 kcal/kg (bark: 18.08 MJ/kg). Rice bran was also high (4,520 kcal/kg: 18.92 MJ/kg), although rice straw and husks were at the low end, 3,080 kcal/kg (12.89 MJ/kg) and 3,390 kcal/kg (14.19 MJ/kg), respectively. The higher heating value of total sorghum plant of Chugoku Kou 34 was intermediate among the materials evaluated and 3,940 kcal/kg. Sorghum cultivars exhibited mostly similar higher heating value of 17.4 MJ/kg.

Molecular ratios of C, H and O in various materials are shown in Table 3. Most of the materials had similar ratios for CnH2Om (n between 1.28 and 1.54, and m between 0.87 and 0.93) except for rice bran which contains considerable quantities of lipid resulting in an n = 1.15 and m = 0.59. This ratio is important since it will affect the condition of gasification when oxygen and steam are added as gasifying agents.

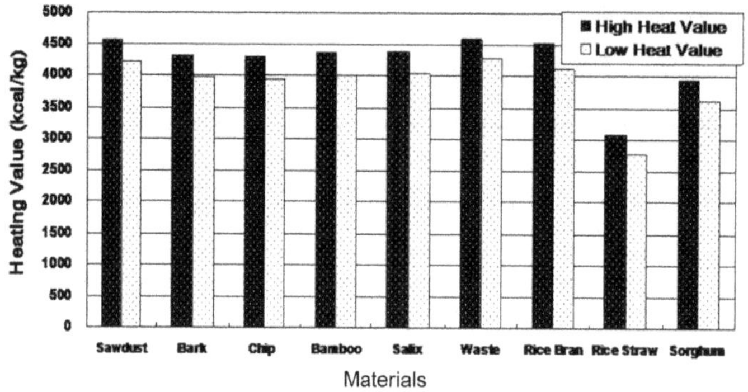

Figure 5.The numbers of materials are same as those in Materials and Methods. Sawdust (1) ; Bark (16); Chip (17); Bamboo (18); Salix (19); Waste (20); Rice Bran (2); Rice straw (3); sorghum (7)Higher and lower heating value of materials (Nakagawa et al. 2007).

Table 2.Higher and lower heating value of Materials. (The numbers of materials are same as those in Materials and Methods)

Biomass Materials	HHV (MJ/kg)	LHV (MJ/kg)
Sawdust (1)	19.13	17.66
Bark of Japanese Cedar (16)	18.08	16.65
Waste Wood (20)	19.08	17.91
Rice Bran (2)	18.92	17.25
Rice Straw (3)	12.89	11.64
Rice Husk (4)	14.19	12.89
Sorghum Head(5)	17.41	15.99
Leaf and Stem of Sorghum (6)	16.49	15.15
Sorghum ‘Kazetachi’ (8)	17.04	15.56
Sorghum ‘Ultra sorgo’ (9)	17.50	16.12
Sorghum ‘Green A’ (10)	17.41	16.03
Sorghum ‘Big sugar’ (11)	17.45	16.11
Guineagrass (12)	15.82	14.48
Rye ‘Haruichiban’ (13)	17.58	15.57
Japanese Lawngrass cv. ‘Asamoe’ (14)	18.59	17.17
Erianthus sp. Line ‘NS-1’ (15)	18.56	17.16

Table 3.Molecular ratios of C, H, O (CnH_2Om) in the materials (The numbers of materials are same as those in Materials and Methods)

Material	C (n)	H	O (m)
Sawdust (1)	1.44	2	0.90
Bark (16)	1.54	2	0.90
Chips (17)	1.39	2	0.88
Bamboo (18)	1.42	2	0.93
Salix (19)	1.38	2	0.93
Waste (20)	1.42	2	0.90
Rice Bran (2)	1.15	2	0.59
Rice Straw (3)	1.31	2	0.87
Sorghum (7)	1.28	2	0.93

The gasification by partial oxidation process yields gas mixtures from diverse materials, with the estimated volume percent for each gas displayed in Figure 6 and Table 4. The hydrogen (H2) and carbon monoxide (CO) are the primary components in the gas mixture that are crucial for methanol synthesis. Despite the small variation in values, the H2 percentage and CO percentage in wood materials are relatively high. Specifically, the H2 percentage ranges from 46.8% for bark, 47.9% for wood waste, 47.3% for salix, and 47.7% for sawdust. Similarly, the CO percentage ranges from 18.3% for bark, 18.2% for wood waste, 17.9% for salix, and 18.7% for sawdust. The hydrogen (H2) content of both rice straw and husks was same, measuring 44.7%. The carbon monoxide (CO) content, on the other hand, differed somewhat, with straw containing 17.1% and husks containing 17.3%. The H2 and CO values of sorghum were found to be in the middle range compared to the other materials examined. Additionally, the H2 values for sorghum varieties ranged from 46.3% to 47.0%, while the CO values ranged from 17.7% to 17.8%.Japanese lawngrass and Erianthus had a slightly higher carbon monoxide (CO) percentage compared to sorghum and were similar to that of the wood components.

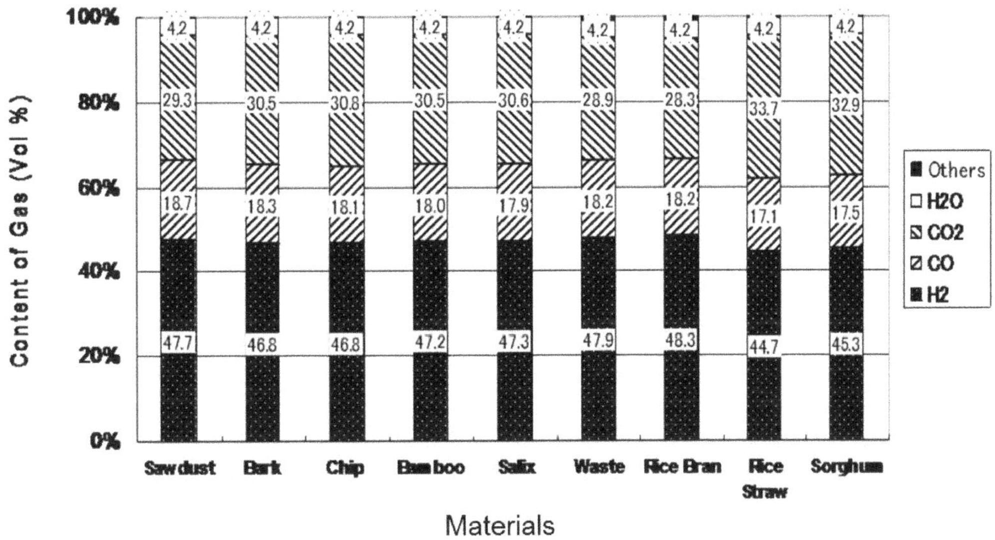

Figure 6.The numbers of materials are same as those in Materials and MethodsSawdust (1) ; Bark (16); Chip (17); Bamboo (18); Salix (19); Waste (20); Rice Bran (2); Rice straw (3); sorghum (7)Volume % of H2, CO, CO2, H2O and other gasses produced from some materials (before the entrance to methanol synthetic system) (Nakagawa et al. 2007).

The methanol yield, both in terms of weight and heating value, for each material examined was determined based on the gas mixtures formed during gasification. These estimates are presented in Figure 7 and Table 5. The values are directly linked to the carbon content and heat emission. The wood materials demonstrated a substantial methanol output by weight, ranging from 53.0% (salix) to 55.8% (sawdust) based on the weight of the dry biomass. Consequently, the gasification of 1 ton of dry salix and sawdust will provide 530 kg and 558 kg of methanol, respectively. Rice bran demonstrated a notable methanol yield potential of approximately 55%, while rice straw and rice husks displayed significantly lesser potentials of 35.8% and 39.4%, respectively. Sorghum cultivars with distinct plant morphologies demonstrated comparable potentials, approximately ranging from 47% to 49%. Surprisingly, the sorghum heads (which include starch) have a methanol yield potential of 48.6%, which is higher than that of the sorghum leaf and stem (which contain more lignin) at the ripening stage (44.1%).

Table 4.HV: higher heating value.Composition and higher heating value of product gas derived from biomass gasification for methanol synthesis. (The numbers of materials are same as those in Materials and Methods)

Biomass Materials	H_2 (vol%)	CO (vol%)	CO_2 (vol%)	H_2O (vol%)	HHV (MJ/kg-wet)
Sawdust (1)	47.7	18.7	29.3	4.2	9.54
Rice Bran (2)	48.3	18.2	28.3	4.2	9.67
Rice Straw (3)	44.7	17.1	33.7	4.2	8.25
Rice Husk (4)	44.7	17.3	33.4	4.2	8.29
Sorghum Head(5)	46.1	17.7	31.3	4.2	8.83
Leaf and Stem of Sorghum (6)	45.3	17.5	32.9	4.2	8.50
Sorghum 'Kazetachi' (8)	46.3	17.7	31.1	4.2	8.92
Sorghum 'Ultra sorgo' (9)	47.0	17.7	30.7	4.2	9.13
Sorghum 'Green A' (10)	46.7	17.8	30.8	4.2	9.00
Sorghum 'Big sugar' (11)	46.6	17.7	31.1	4.2	8.96
Guineagrass (12)	46.2	17.7	31.1	4.2	8.92
Rye 'Haruichiban' (13)	46.9	17.8	30.5	4.2	9.13
Japanese Lawngrass 'Asamoe' (14)	47.6	18.0	29.3	4.2	9.00
Erianthus sp. Line 'NS-1' (15)	46.6	18.6	30.3	4.2	8.96
HHV: higher heating value.					

However, the difference in yield between the two is only 4.5%. This indicates that there is minimal disparity in the production of methanol between starch and lignin. This implies that there may be minimal necessity to employ food starch resources, and competitive methanol production can be achieved by using crude lignocellulosic materials. This suggests that substantial amounts of methanol can be generated from discarded residues and leftovers of agriculture and forest industries. Japanese lawngrass and Erianthus had greater methanol production capabilities compared to sorghum cultivars. While the projected methanol output varied among the materials in terms of weight, the anticipated heat yield, measured in heating value, was rather consistent at 54-59% across the different materials. The heat yield of the several materials examined, irrespective of their heating characteristics, was substantial and clearly demonstrates the efficiency of this method.

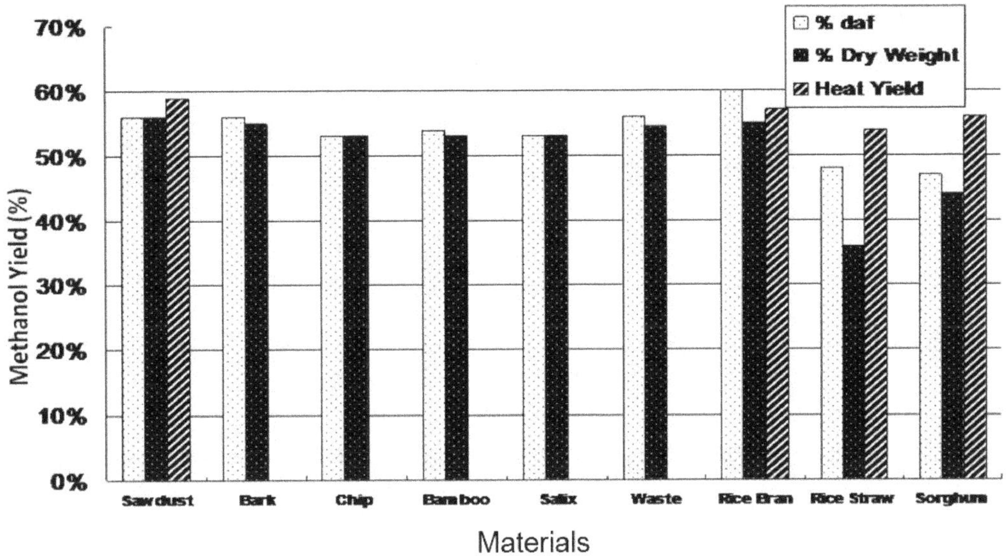

Figure 7.The numbers of materials are same as those in Materials and Methods.Sawdust (1) ; Bark (16); Chip (17); Bamboo (18); Salix (19); Waste (20); Rice Bran (2); Rice straw (3); sorghum (7) daf: percentage of methanol weight to dry biomass weight without ashEstimated methanol yield (weight %) and heat yield of various biomass materials (Nakagawa et al. 2007).

Table 5.Daf: based on dry ash free biomass weight, dry: based on total dry biomass weight with ashEstimated methanol yield (% by weight) (The numbers of materials are same as those in Materials and Methods).

Biomass Materials	methanol yield	
	daf	dry
Sawdust (1)	56.0	55.8
Rice Bran (2)	60.0	54.5
Rice Straw (3)	48.0	35.8
Rice Husk (4)	47.0	39.4
Sorghum Head (5)	51.0	48.6
Leaf and Stem of Sorghum (6)	47.0	44.1
Sorghum 'Kazetachi' (8)	51.1	47.7
Sorghum 'Ultra sorgo' (9)	51.3	48.4
Sorghum 'Green A' (10)	52.0	49.0
Sorghum 'Big sugar' (11)	51.0	48.1
Guineagrass (12)	51.0	45.2
Rye 'Haruichiban' (13)	55.9	52.2
Japanese Lawngrass 'Asamoe' (14)	55.1	51.4
Erianthus sp. Line 'NS-1' (15)	53.4	51.5
daf: based on dry ash free biomass weight, dry: based on total dry biomass weight with ash		

For perfect gasification of any biomass material, it is necessary to convert the materials into ca. 0.1-0.9 mm in diameter powder through micro-crushing. The physical characteristics of the raw materials and the handling procedures needed to prepare these raw materials for biomethanol production are shown in Table 6. As rice bran is very fine, there was no need for any prior preparation. Although the diameter of sawdust is ca. 0.8 mm, we can utilize the powder smaller than ca. 1 mm directly for the gasification. Though the rice straw dried in the field was long in length, it required only micro-crushing without extra-drying in a dryer. Sorghum was harvested at the ripened stage with sickles, cut to a length of 30 cm and dried in a dryer. This procedure made this material very hard to process and both rough-crushing (1.0-3.0 mm) and micro-crushing were needed to prepare sorghum for gasification. Usually, a mechanical harvester is used to cut sorghum plants into lengths of less than 10 cm. This latter harvest method will require much less subsequent preparation than was needed in this study.

Table 6.Size and handling characteristics of various materials (The numbers of materials are same as those in Materials and Methods) (Nakagawa et al., 2000).

Material	Size (mm)		Density	Handling
	Diameter	Length	(g/ml)	Characteristics
Sawdust (1)	0.78	-	0.07	No micro-crushing needed
Rice Bran (2)	0.31	-	0.31	No micro-crushing needed
Rice Straw (3)	3.0-4.0	400	-	Micro-crushing needed
Rice Husk (4)	2.05	-	0.11	Micro-crushing needed
Sorghum head (5) Leaf and Stem of Sorghum (6) Sorghum (7)	15.0 20 (width) 0.5 (thickness) 7.9	250 300 50	- - 0.07	Rough and micro-crushing needed Rough and micro-crushing needed Rough- and micro-crushing needed

2.1.3. "Norin Green No. 1" test plant

The test plant, named "Norin Green No. 1" (Fig. 8), was utilized to obtain data for heat yield and methanol yield through the gasification and biomethanol synthesis system shown in Fig. 2. The test plant comprises a supplier of crushed biomass, a gasifier for gasification, and an apparatus for gas purification and methanol synthesis by the use of a Cu/Zn-based catalyst.The practical methanol yield of crushed waste wood (ca. 1 mm in diameter) produced by pin-type-mill was also measured by operating both "Norin Green No. 1" test plant at Nagasaki Research and Development Center, Mitsubishi Heavy Industries ltd. with a gasifier capacity of 240 kg dry biomass/day and another test plant at Kawagoe Power Station of Chubu Electric Power Co., Inc. with a gasifier capacity of 2 t dry biomass/day (Matsumoto et al. 2008).

Table 7 indicates the heat yield and methanol yield of the two test plants (the test plant gasifier can process 240 kg/day (Norin Green No. 1 test plant) or 2t/day (a test plant constructed by Chubu Electric Power Co., Inc) of dry biomass) by operating these plants, when crushed waste wood is utilized as a raw material, and the estimated capacity of a commercial scale plant (a gasifier capable of processing 100 t/day of dry biomass). The commercial scale plant would be large enough (larger than 100 t/day) to maintain critical temperature (900 to 1,000°C) within the gasifier by adding the raw materials into the gasifier without the addition of supplemental heat. Our data indicate that the estimated heat yield of methanol production by commercial scale plants is 54 - 59 % (Fig. 7). However, the real heat yield of a commercial scale plant after reducing the energy needed for crushing of the biomass (1.0 - 5.0 % of the quantity of heat; biomass feedstock with 2-3 mm in diameter is available), operation of the plant (5 - 10 %), and heat loss from the surface of the gasifier (1-2 %), estimated by simulation using the test plant data will be ca. 40 %.

The cold gas efficiency, that represents a percentage of the total heating value of synthesized gases by gasification divided by the total heating value of supplied biomass of the test plant, varies from 65 to 70% and methanol yield varied from 9 to 13% in the "Norin Green No. 1" test plant. Heat yield and methanol yield of another test plant capable of processing 2 t/day constructed by Chubu Electric Power Co., Inc. with the support of New Energy and Industrial Technology Development Organization (NEDO) Project, has, however, achieved ca. 20% of methanol yield by weight by its operation (Ishii et al. 2005; Matsumoto et al. 2008; Ogi et al. 2008).

Table 7.Ability of test plants and estimated practical plant.

Item	Test Plant		Practical Plant
(Gasifier Size: Dry biomass to be processed)	(240kg/day)	(2t/day)	(100t/day)
Heat Yield (Heating Value %)	60-70%	65%	70-75%
Methanol Yield (by weight)	9-13%	20%	40-50%

2.1.4. Concise Summary

The figures above demonstrate that the primary benefit of this technology is its ability to utilize any type of biomass feedstock for the production of H2 and CO. The gas combination can be directly burned to provide both heat and electricity. An efficient Cu/Zn-based catalyst will be used to create high yields of methanol from the gas combination. The system's drawback is in the dimensions of the processing facility. As the size of the plant increases, its efficiency also increases. The biomethanol production from a gasifier with a capacity of 100 tons per day would be more than double that of a gasifier with a capacity of 2 tons per day, using the same raw materials. While it is technically possible to build a biomethanol plant of this magnitude, obtaining the necessary 100 dry matter tons of biomass on a daily basis for its operation could prove to be quite challenging, except perhaps in the case of huge sugarcane mills and the palm oil sector in Southeast Asian countries. Furthermore, the necessity of obtaining permissions and a license for operating a gasifier of this magnitude in Japan would result in further expenses. Prof. Sakai, from the Nagasaki Institute of Applied Science and one of the authors of this report, has created a new type of plant at the university called "Norin Biomass No. 3" test plant (Fig. 9). This plant was developed using a gasification technology known as "high-calorie gasification reaction", which is explained in the following sentence.

2.2. "Norin Biomass No. 3" test plant

The "Norin Biomass No. 3" Test Plant (Fig. 9, 10), which represents a "Suspension/external heating type" with a boiler (1-2 Dry t/day) was newly developed for improving the disadvantages associated with the "Norin Green No. 1" test plant through the introduction of a new type of gasification called "high-calorie gasification reaction".

Figure 8.“Norin Green No. 1”, a test plant of gasification and biomethanol production, located in Nagasaki Research & Development Center, Mitsubishi Heavy Industries ltd, Nagasaki, Japan.

Figure 9. “Norin Biomass No. 3” test plant located in Sakai Kougyo, ltd., Isahaya City, Nagasaki, Japan.

2.2.1. Essential of the high-calorie gasification reaction

In the high-calorie gasification method, finely crushedbiomass of 1-3 mm in diameter is subjected to the gasification reaction together with steam in an atmosphere of 800-1000°C within the reaction tube. At this time, the reaction tube is heated using high-temperature combustion gas that is separately combusted using additional biomass. The introduced biomass raw materials leave only ash, and all organic content is gasified, resulting in a clean, high-calorie gas (ca. 12MJ/Nm3) composed of H2, CO, CH4, etc. The basic principle of the technology is illustrated in Fig. 10.

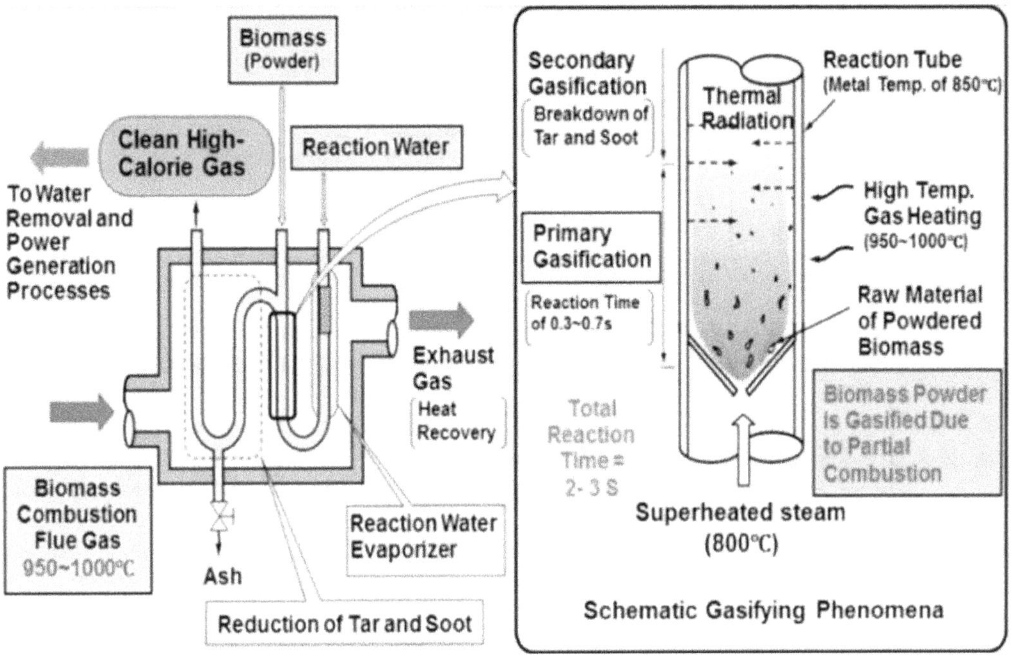

Figure 10.Suspension/external heating type high-calorie gasification.

The gas composition varies with gasification reaction conditions such as reaction temperature, residence time (reaction time), and the [steam]/[biomass carbon] mode ratio, but an example is represented by the following reaction formula.

(Endothermic reaction)

$$C_{1.3}H_2O_{0.9} + 0.4\,H_2O \rightarrow 0.8H_2 + 0.7CO + 0.3CH_4 + 0.3CO_2 + 165.9\,kJ/mol$$

(Biomass) (Steam) (Resulting gas fuel) (Absorbed heat)

In this process, the total biomass material reacts with steam and is converted to an [H2, CO, CH4, CO2] gas mixture.The application of external heat is required due to the fact that the gasification reaction is endothermic. However, the potential heat stored in the gas mixturegenerated in the reaction is greater than that contained in the raw biomass material, such that the cold gas efficiency surpasses 100%. In the formula shown above, a figure of ca. 115% is obtained by solving for cold gas efficiency (Ec). On the other hand, the externally supplied heat used in the reaction, is not considered in the calculation of Ec, and the total gasification efficiency is ca. 85% when this external heat is taken into account.

While the previous biomass gasification technology of "Norin Green No. 1" test plant mentioned above uses the partial oxidationtechnology, thishigh-calorie gasification technology enables the production of a high-calorie gas fuel that was not possible with the conventional method due to the formation of an exhaust gas. The principle is illustrated in Fig. 11.

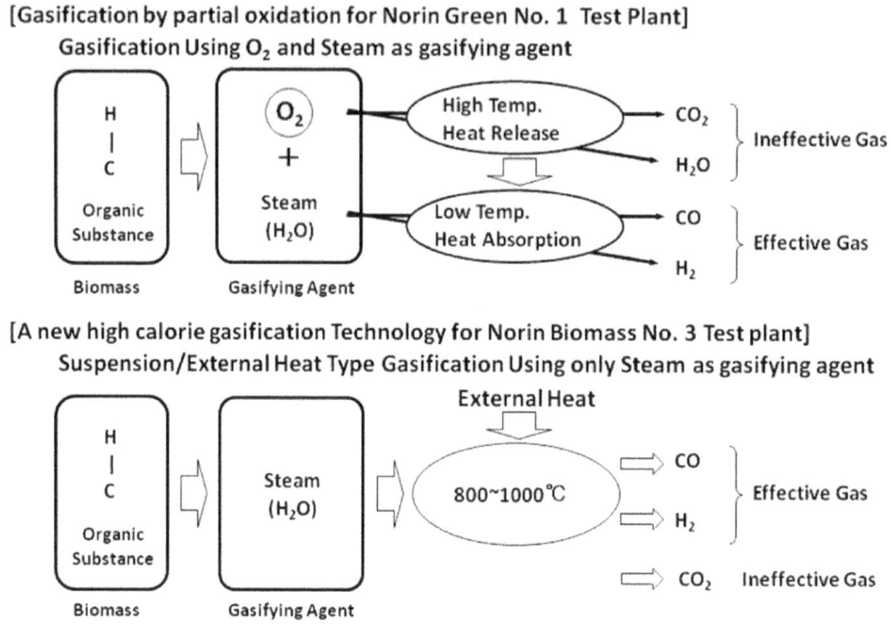

Figure 11.Principle of a new high calorie gasification technology compared with gasification by partial oxidation technology.

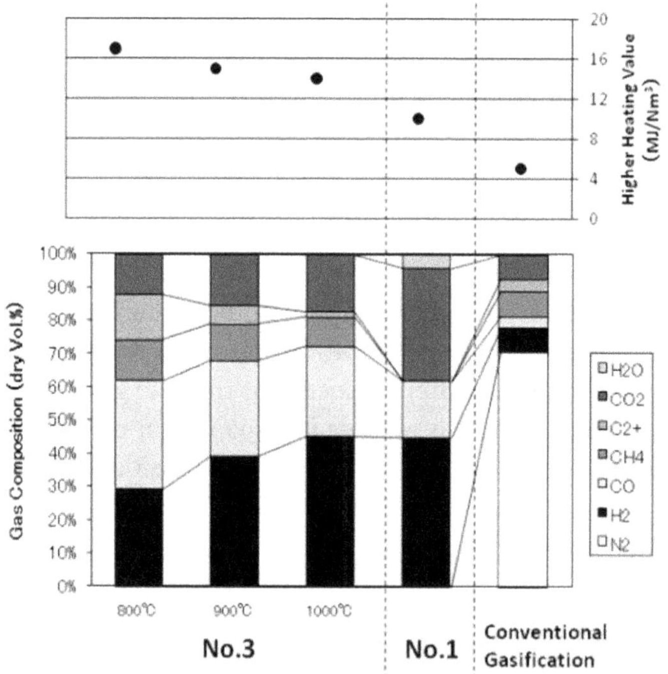

Figure 12.Composition (bar chart) and higher heating value (line chart) of high calorie gasification gas (No. 3), partial oxidation (No. 1) and conventional gas generated by partial oxidation with air as gasifying agent.No.1: "Norin Green No. 1": gasification by partial oxidation using O2 and H2O as gasifying agents. No. 3: "Norin Biomass No. 3": highcalorie gasification using only H2O as a gasifying agent at 800, 900, and 1000°CConventional gasification: gasification using air as a gasifying agent.

2.2.2. Basic high-calorie gasification reaction

Based on the operation's findings, it has been verified that the gas mixture produced has the qualities specified in Figure 12. The figure demonstrates that high-calorie gas with a range of 15-18MJ/Nm3 cannot be achieved through gasification by the "Norin Green No. 1" test plant using partial oxidation (approximately 10 MJ/Nm3 with O2 and H2O as gasifying agents) or conventional gasification using air as a gasifying agent (approximately 5 MJ/Nm3). However, this high-calorie gas can be produced by maintaining a reaction temperature of 800-900°C, an H2O/C mole ratio lower than 5.0, and a reaction time of approximately 2 seconds. Furthermore, this gas combination comprises more than 20% hydrogen (H2). The value exceeds the 10% requirement for ignition and combustion rate, making the gas mixture a high-quality fuel for gas engines and micro gas turbines. In addition, because to the larger compositional ratio of H2 to CO, the threshold combustion temperature is elevated by 90°C compared to methane. Figure 13 illustrates a comparison between the theoretical combustion temperatures of this gas mixture and different fuels, including methane, gasoline, propane, methanol, and ethanol.

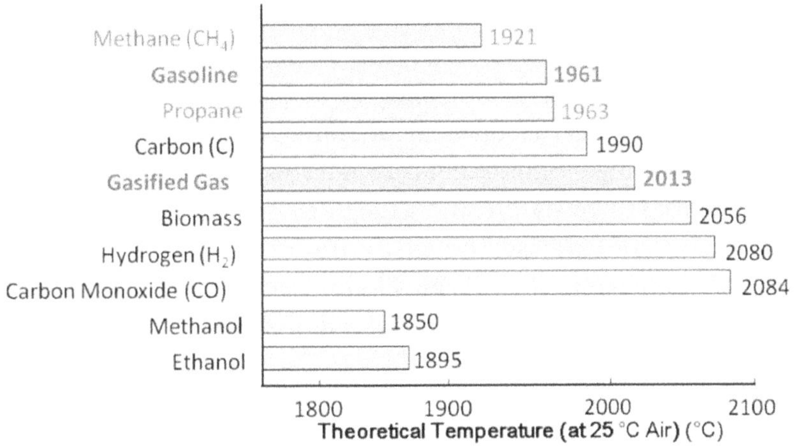

Figure 13.Comparison of theoretical combustion temperature for gasified gas generated by Norin Biomass No. 3 Test Plant and various fuels. Biomass means any form of lignocellulosic materials.

3. CONCISE SUMMARY

The gas mixture produced by the "Norin Green No. 1" test plant and the high-calorie gas generated by the "Norin Biomass No. 3" test plant using high-calorie gasification technology can be stored in a cold gas state and utilized in a manner similar to natural or city gas, with a wide range of applications. The "Norin Biomass No. 3" plant, which successfully converts biomass into a high-calorie gas mixture using a compact system, can be readily utilized as fuel for gas engines and micro gas turbines. Consequently, it is suitable for small-scale power generation and co-generation purposes. Consequently, it is possible to produce power generation that is both highly efficient and on a small scale. The potential uses of the gas mixture produced through high-calorie gas generation via gasification are as follows:

- Co-generation in buildings, hospitals, industrial parks, factories, etc.
- Commercial power (targeted efficiency of 25-35%, with at least 1 million kWh/year)
- Peak cut (reduction of contracted power) and emergency use for large-scale factories
- Gas fuel for industrial parks (e.g. ceramics and porcelain)
- Supplementary fuel for incinerator (dioxin countermeasure for industrial waste processing)
- Fuel for boilers of greenhouse agriculture
- Fuel for food processing industries by the use of residues produced in the process.
- Synthesis of biomethanol for BDF production, for batteries of direct methanol fuel cell (DMFT), and a liquid fuel mixed with gasoline for flexible fuel vehicles (FFV).

Private firms and local government are constructing three "Norin biomass No. 3" plants that will process 4-6 dry tons per day of biomass feedstock. The Government is providing 50% financial support for these projects. This study shows that the gasification of easily accessible biomass materials, using both partial oxidation technology and high calorie gasification technology, can be optimized to produce gas mixtures consisting mainly of H2 and CO. This process can also yield methanol, with theoretical yields ranging from approximately 40% to 60% by dry weight. A pilot facility employing gasification by partial oxidation with a 2-ton per day gasifier can produce around 20% methanol from the biomass feedstock (by weight). This presents an opportunity to employ a diverse selection of crops, such as Erianthus and Miscanthus, that have great productivity but low levels of sugar and starch. Non-edible lignocellulosic byproducts, such as sawdust and agricultural leftovers like straw and rice husks, can also be effectively utilized in numerous sectors. Sawdust, rice bran, sugarcane mill residue (such as bagasse), and rice husks are very appealing and readily available sources of biofuel. The cultivation and use of biomass crops are expected to be appealing as carbon-neutral biomass feedstocks for future biofuel production. The economic benefits of producing biomethanol through the cultivation of grasses and trees on unused land in Japan have the potential to be significant, as highlighted by Nakagawa (2001) and Harada (2001). Promoting biofuel production system based on gasification technologies enables the reduction of CO2 emissions, recycling of abandoned upland and paddy fields, recycling of agricultural waste, and woodland in mountainous locations. This method is highly appealing since biomethanol may be derived from a diverse array of biomass feedstocks.

REFERENCES

Binod, P., Sindhu, R., Singhania, R. R., Vikram, S., Devi, L., Nagalakshimi, S., Kurien, N., Sukumaran, R. K., Pandey, A. (2010) Bioethanol production from rice straw: an overview, Bioresource Technology, 101, 4767-4774.

Brown, L. R. et al. (2000). State of the world 2000, W. W. Norton & Company Ltd., New York, pp. 276.

Harada, Toshirou (2001). Utilization of wooden biomass resources as energy. Farming Japan, 35-2, 34-39.

Ishii, H., Takeno, K., Ichinose, T. (2005) Development of integrated system of biomass gasification using partial oxidizing process and liquid fuel synthesis with catalyst, J. Jpn. Inst. Energy 84: 420-425 (in Japanese with English abstract).

Keshwani, D. R. and Cheng, J. J. (2009) Switchgrass for bioethanol and other value-added applications: a review, Bioresource Technology, 100, 1515-1523.

Kosugi, A., Tanaka, R., Magara, K., Murata, Y., Arai, T., Sulaiman, O., Hashim, R., Abdul, H. Z. A., Azri, Y. M. K., Mohd, Y. M. N., Ibrahim, W. A., Mori, Y. (2010) Ethanol and lactic acid production using sap squeezed from old oil palm trunks felled for replanting, J. Biosci. Bioeng., 110-3, 322-325.

Matsumoto, S. Takeno, K., Ichinose, T., Ishii, H., Nishimura, K. (2008) Analysis of a two ton/day test plant with an integrated biomass gasification and liquid fuel synthesis system, Proceedings of 15th European Biomass Conference and Exhibition from Research to Market Development, Berlin, Germany, 7-11 May, 2008.

Nakagawa, H. (2001). Development and cultivation of forage grasses and crops for clean bio-methanol production to keep global environment. Farming Japan, 35-2, 22-31.

Nakagawa, H., Harada, T., Ichinose, T., Takeno, K., Matsumoto, S., Kobayashi, M., Sakai, M. (2007) Biomethanol production and CO2 emission reduction from forage grasses, trees, and crop residues, JARQ, 41(2), 173-180.

Nakagawa, H.; Sakai, M.; Harada, T.; Kitamura, Y.; Taniwaki, K. & Hashimoto, A (2000). Biomethanol production and CO2 emission reduction from various forms of biomass. Proceedings of the fourth International conference on ecobalance, Ecomaterials Forum, Society for Non-Traditional Technology, pp. 405-408, Tokyo, Japan, October 31- November 2, 2000.

OECD (2010). OECD Factbook 2010, economic, environmental and social statistics. OECD, Paris, France, OECD Factbook 2010: Economic, Environmental and Social Statistics - ISBN 92-64-08356-1 - © OECD 2010.

Ogi, T. and Nakanishi, M. (2008) Biomass-derived liquid fuel via gasification), Journal of JSES 34 (5), 13-20 (in Japanese).

Sakai, M. (2001) Technological innovation of production method of liquefied biomass fuels: thermochemical production method of methanol fuels from any type of biomass. Farming Japan, 35-2, 10-15.

Sørensen, A., Teller, P. J., Hilstrøm, T., Ahring B. K. (2008) Hydrolysis of Miscanthus for bioethanol production using dilute acid presoaking combined with wet explosion pre-treatment and enzymatic treatment, Bioresource Technology, 99, 6602-6607.

Wyman, C., Dale, B. E., Elander, R. T., Holtzapple, M., Landisch, M. R., Lee, Y. Y., Mitchinson, C., Saddleer, J. N. (2009) Comparative sugar recovery and fermentation data following pretreatment of poplar wood by leading technologies, Biotechnol. Prog., 25-2, 333-339.

INDEX

A

Adiabatic 69, 302, 326

Adsorbents 66-67, 75-76

Aeration 198, 200-203

Aerobic 199

Agitation 352, 356

Alkali 215, 228, 284, 325, 337, 351, 401, 403, 415, 420

Alkaline 114, 118, 163, 179, 215, 351, 397, 401, 403, 458

Aluminum 351

Ambient 357, 403, 609

Ammonia 185, 322, 588, 632, 656

Ammonium 656

Amylose 594

Anaerobes 112

Animals 89, 144

Antibiotic 269, 273, 275

Antibodies 485

Aquatic 259

Aqueous 168, 177, 286, 375-376, 379

Aquifers 40

Arabinose 166

Aromatic 22, 593, 600

Atmosphere 23, 25-26, 199, 599-600

Autoclave 665-666

Automotive 56, 245, 518

Azeotropic 299-300, 305, 307, 310, 312, 597

B

Bacillus 595

Bagasse 404

Bamboo 676

Benzene 305

Bimetallic 459

Bioanode 550

Bio-crude 172

Biodiesels 224

Bioenergy 33, 51, 54, 61, 91, 162, 177, 180, 229, 261, 412, 416-417, 419, 444, 497, 519, 524, 532, 537-538, 603, 605, 669-671

Bioethanol 28, 30-32, 37, 48, 51, 91, 112-115, 117, 127, 162, 166, 233, 244-246, 299-301, 309-311, 313, 581-583, 585, 591-592, 594-602, 604-605, 671, 673, 688

Biofuels 1, 7, 10, 22-24, 27-28, 30-32, 35-38, 48, 52-53, 55, 58-59, 61, 83, 91, 99-104, 108, 111, 115-117, 120, 122, 127-129, 134, 138, 155, 161, 185, 257-259, 263, 315, 330, 390, 487-489, 513-514, 517-524, 526-528, 531-534, 536, 581-582, 586, 645, 652, 674

Biomass 19, 23-24, 28-31, 33-34, 38-39, 41, 46, 48-49, 51, 53-55, 61, 101, 103-104, 121-122, 137, 161-169, 173, 175-181, 185, 203-204, 210, 229, 257, 259-266, 268, 281-282, 284, 300, 307, 309-311, 316-318, 327, 330-331, 333, 369-370, 378, 389-390, 393-394, 396-398, 400-401, 404-405,

407-409, 411-419, 421-425, 429-430, 432-435, 437-441, 443-445, 473-475, 477, 479, 482, 487-488, 496-497, 499-502, 509-511, 513, 517-518, 523, 527, 529, 531, 534, 537-538, 557-559, 563, 565-567, 569-570, 573-575, 577-578, 583, 586-589, 591, 598, 602-605, 645-648, 650, 652-658, 661, 666-674, 676, 680-685, 687-689

Biomasses 409

Bio-oils 162, 167, 170, 174, 176, 281-285, 287-289, 291-294, 389

Bioreactor 237

Biosens 642

Biosensors 465-466, 470-471, 554, 643

Biosphere 25

Bleomycin 269

Bottoms 304

Brassica 98

Bridgwater 177, 281-282, 288, 389, 537

Broiler 160

Buffer 271, 541, 551, 629

Butanol 233-235, 237-238, 241-244, 247-254, 259, 276-280

Butyrate 238

C

Candida 124, 363, 592

Card-fish 488

Catalysis 35, 60, 118, 131, 134, 177, 332, 362, 364, 389-390

Cathode 463, 540-542, 549-550, 571, 621-622, 629, 634-635

Cellobiose 166, 590

Cellulase 166, 589, 591

Cellulose 163-165, 167, 182, 192, 195, 394-395, 397, 401, 403-404, 411-414, 416, 424, 443, 474-475, 479-480, 483-484, 486, 489, 492, 495-497, 499-500, 502-503, 506, 508-510, 585-591, 645-647, 654-655, 659, 661-669, 671-672

Ceramic 442

Cerevisiae 166, 591-592, 648, 668

Cetane 225

Charcoal 1, 5, 7-11, 244

Chemical 17-19, 34, 57, 65, 89, 114, 117-120, 124, 126, 129, 131, 162, 174, 214, 216, 253, 259, 267, 278, 299-300, 309, 312, 316, 318, 327, 332, 349, 362, 364, 370, 390, 398, 401, 403, 413-416, 418, 457, 466-468, 470, 479, 481-482, 485, 491-492, 497, 523, 544, 554-555, 574, 587-588, 602, 621, 647, 667, 672, 674

Chimera 488

Chlorella 40

Chloride 452, 675, 677

Chlorine 185

Chloroform 126

Cleavage 382, 494

Climate 25, 27, 55-56, 58, 60, 83, 104, 513, 516, 525, 527

Colloids 197

Components 34, 140, 142, 151, 282, 373, 376, 379-380, 388, 397, 401, 424, 481-483, 485-486, 499, 531-532, 540, 543, 561-562, 588, 595, 667, 679

Compressor 302, 305, 307

Condenser 307

Convention 418

Cooking 35, 113, 129, 215, 357, 362, 365, 654, 658

Copper 226

Corrosion 226, 403

Cyclones 322

Cytometry 238-239, 249, 252, 278

D

Density 1, 4, 6, 87, 149, 201, 224, 245, 258, 282, 284, 291, 369, 373, 376, 393, 400, 408-409, 411, 421-423, 425, 430, 432, 434-437, 440-441, 447, 451, 454-455, 462, 540-543, 546-550, 552, 557, 567, 571-572, 625-626, 628-629, 631

Desorption 69-70, 458

D-glucose 379, 382, 468

Diazonium 632

Diffusion 5, 11, 68, 454, 541, 543-544, 548-549, 629, 635-636

Digesters 63, 183-184, 203-204, 519

Dioxygen 450, 469

Dispersion 69

Diversity 484

Dolomite 320

D-sorbitol 382

E

Ecosystems 50

E-diesel 582

E-hybrid 44

Emissions 23-27, 32, 35, 43-44, 59-60, 91-92, 96-97, 101, 104-105, 111, 115, 247-249, 254, 279, 514, 522, 526-527, 531-532, 538, 558, 565-568, 573, 575, 577, 582, 600-601, 603, 652

Emulsions 105, 175, 177

Enthalpy 303, 305, 474-475

Enzymes 43, 55, 119-120, 123, 128, 163-165, 405, 447-451, 462, 469, 539-541, 546, 549-555, 587-591, 625, 627, 647

Epidermal 484

Ethylene 633

Explosion 397-400, 416, 419, 429, 441, 475, 587-588, 655-656, 658, 668, 671-672

F

Farmers 90, 185, 196, 536

Feeding 149, 320, 393, 422

Feedstock 23, 35, 114-116, 165, 167, 219, 235, 241, 259, 283-284, 315-318, 320, 335, 421-422, 564, 569-570, 646-647, 656, 658, 660, 669, 674, 688

Fibrils 662, 664

Forestry 124, 390, 558

Fossil 21-23, 25-27, 30-31, 37, 51, 91, 111, 113, 127, 137, 161, 213, 224, 226, 257, 259-260, 299, 335, 473, 513-518, 521, 527, 533, 574, 601, 621, 673

Frameworks 75, 79-80

Friction 344, 409, 429

Fructose 30

G

Gasoline 30, 32, 36-37, 41, 44, 115-116, 162, 244-250, 283, 514, 520, 524, 531-533, 601, 650, 652, 687

G-bacteria 239

Genome 263, 268

Geothermal 45

G-lignin 477

Glucose 30, 39, 44, 114, 164-166, 241-243, 389-390, 447-448, 450-451, 455-457, 459-471, 539-541, 548, 550-551, 556, 584, 590-591, 593-595, 629, 631, 634, 641-644, 650, 655, 662-666

Glycerides 34, 215

Glycerol 34, 39, 43, 173, 177, 215, 219, 221, 227

Greenhouse 22, 24-25, 35, 43, 58, 97, 101, 104, 115, 258, 439, 513, 522, 526-527, 573-574

H

Hardwood 287

Helium 69

Hybrids 49, 51

Hydrogen 17-18, 29, 36, 38, 40-44, 55, 57, 59-60, 98, 103, 112, 134, 171, 175, 177, 181, 243, 258, 317-318, 323, 330, 338, 371, 373, 376-377, 381-384, 388, 395, 510, 557-558, 589, 621, 634, 662, 665, 675, 677, 679

Hydrolysis 34, 114, 132, 164-165, 179, 182, 191, 194, 210, 221, 403-404, 414-416, 418-420, 583, 586-591, 593-595, 645, 647, 654-656, 661-668, 671-672

Hydropower 55

Hydroxide 163, 337, 359, 404, 609, 656

I

Incubation 195-196

Inhibition 120

Inhibitors 284, 593-594, 604

J

Japonica 675

Jatropha 351, 360, 523

K

Kinetics 187, 499-500, 502, 509-511, 626

L

Lignins 494, 648

Lipase 118, 129-131, 352

Lipids 39-40, 117, 122-127, 131-132, 195, 261, 263

M

Magnesium 320

Manganese 599

Marine 38

Markets 101, 115, 145, 536

Metabolic 166, 235-236, 239, 243, 262-263, 296

Metabolism 235

Metabolite 236

Microalgae 38-40, 49, 121-123, 129, 259-263, 265, 267, 294, 296

Microfuel 554

Microscale 639

Microscopy 480-482, 484, 487, 489, 492, 494

Microwave 282, 285, 344-345, 359, 362-368, 400-401, 416, 420

Nanotubes 541, 626, 629

Nickel 384

Nitrogen 85, 105, 183, 185-186, 192-193, 195, 198, 206-207, 210, 317-318, 583, 585-586, 677

Nutrients 183-184, 192-193, 196-198, 202, 205, 207

O

Octane 248, 582

Organisms 259, 261-262

Osmium 451-452

Oxidase 448, 450-451, 467-468, 470, 540, 634, 644

Oxidation 7, 9, 42, 187-188, 220-221, 226, 246, 319, 450-452, 456-457, 459, 461, 465-466, 468, 474, 539, 546, 551, 556, 612-613, 618, 633, 674, 686, 688

Oxygen 28, 170, 172, 198, 244-247, 267, 316-318, 320, 326, 369-372, 376, 385-387, 450-451, 453-455, 458, 463, 466-468, 470, 539, 548-549, 608, 675, 677

Ozonolysis 404

P

Pellets 13, 19, 185, 285, 393, 411, 417, 422-423, 425, 428-430, 434-438, 440-442, 444-445

Pentose 591

Photons 265

Plasma 29

Plasmids 270

Plastic 432

Plates 125, 272

Platform 162

Platforms 162

Platinum 451, 455-459, 461, 464-465, 623

Policy 50, 54-55, 57-62, 98, 519-520, 533-534

Polish 619

Política 528, 537

Pollutants 247, 249

Pollution 23, 50, 517, 522, 524

Potassium 337, 350, 359

Potential 19, 22, 25, 28, 33-34, 36, 38, 40, 45-46, 60, 74-75, 93, 96, 99, 104, 106-107, 115, 123, 220, 244, 246-247, 258, 316, 372, 450-451, 453, 455-456, 458-459, 461, 513, 517, 522, 536, 571, 586, 623, 632-633, 635, 646, 648, 652, 655-656, 662, 680

Precursors 262

Promoters 270, 274-275

Proteins 44, 151, 196, 462

Protocols 487

Protons 622, 625

Prototype 43, 636

Pyrolysis 1, 4, 7-9, 29, 41, 166-168, 170, 172-182, 281-283, 285, 291, 369-377, 379, 381-383, 387-391, 477, 491, 499, 510-511, 531, 537-538, 559-560, 562-563, 630

Q

Qualities 19, 88, 523

Quality 86, 97, 122, 167, 172, 174, 180, 184, 188, 201, 209, 219, 221, 226, 228, 283, 367, 411, 416, 422-423, 434-435, 437, 440, 515, 608, 646, 659, 661

R

Radiation 264

Reactions 43, 164, 172, 174, 199, 215, 299, 309, 319, 337, 344, 349-350, 372-373, 375-376, 379, 381-385, 390, 474, 559-561

Reactors 19, 203-204, 267, 326, 349-350, 399

Reflection 481

Regression 397

Relaxation 425

Renewable 23, 31, 35, 37-38, 48, 51, 54, 61, 83, 97, 108, 111, 113, 128, 131, 134, 138, 158, 213, 229, 259, 330, 335, 358, 361, 389, 415, 513, 518, 520-522, 524, 534, 569-570, 578, 604

Resins 119

S

Sediments 226

Sensors 470, 643

Sewage 128, 176

Silage 205

Sinensis 586

Sodium 163, 337, 359, 609

Software 634

Solvents 118-120, 126, 172-174, 176, 179, 234-239, 241-242, 244, 250, 284, 343

Statistics 689

Sugarcane 32, 48

Systems 31, 46, 101, 120-121, 125, 133, 149, 154, 224, 261, 266-267, 275, 294, 313, 369, 447, 519, 546, 551, 554, 558, 577, 583, 621-622, 628

T

Thaliana 480

Titanium 351

Tomography 486

U

Ultrasound 349-351, 355-357, 364

V

Viscosity 30, 87, 113, 170, 173-175, 215, 221, 224, 282-284, 286-291, 294, 386, 543

Volatiles 9

Vulgaris 123

W

Wastewater 39, 125-126

Wildlife 26

Wireless 540

Worldwide 25, 45, 47, 101, 257, 513, 525, 535, 646, 673

X

Xylitol 592

Xylose 39, 166, 584, 589-590, 592, 668

Xylulose 592

Y

Yeasts 592, 597

Z

Zeolites 371

Zymomonas 166

www.ingramcontent.com/pod-product-compliance
Lightning Source LLC
LaVergne TN
LVHW062143200926
848484LV00007B/1848